FORMATION, EVOLUTION, AND SURVIVAL OF MASSIVE STAR CLUSTERS

IAU SYMPOSIUM 316

COVER ILLUSTRATION:

This picture, taken by the Advanced Camera for Surveys (ACS) on board the NASA/ESA Hubble Space Telescope, shows the centre of globular cluster Messier 4. The power of Hubble has resolved the cluster into a multitude of stars. M 4 is relatively close to us, lying 7200 light-years away. It contains several tens of thousand stars and is noteworthy in being home to many white dwarfs.

Credit: ESA/Hubble & NASA

INTERNATIONAL ASTRONOMICAL UNION

UNION ASTRONOMIQUE INTERNATIONALE

International Astronomical Union

IAU

FORMATION, EVOLUTION, AND SURVIVAL OF MASSIVE STAR CLUSTERS

PROCEEDINGS OF THE 316th SYMPOSIUM OF THE INTERNATIONAL ASTRONOMICAL UNION HELD IN HONOLULU, HAWAII, UNITED STATES AUGUST 11–14, 2015

Edited by

CORINNE CHARBONNEL

Department of Astronomy, University of Geneva, Switzerland

and

ANTONELLA NOTA

Space Telescope Science Institute, USA

CAMBRIDGE
UNIVERSITY PRESS

CAMBRIDGE UNIVERSITY PRESS
University Printing House, Cambridge CB2 8BS, United Kingdom
1 Liberty Plaza, Floor 20, New York, NY 10006, USA
10 Stamford Road, Oakleigh, Melbourne 3166, Australia

First published 2017

Printed in the UK by Bell & Bain, Glasgow, UK

Typeset in System LaTeX 2_ε

A catalogue record for this book is available from the British Library Library of Congress Cataloguing in Publication data

This journal issue has been printed on FSC™-certified paper and cover board. FSC is an independent, non-governmental, not-for-profit organization established to promote the responsible management of the world's forests. Please see www.fsc.org for information.

ISBN 9781107138179 hardback
ISSN 1743-9213

Table of Contents

Preface xi

Scientific Organizing Comittee xiii

Participants xiv

Young and old massive star clusters and their stellar populations - Theoretical challenges for the next decade 1
C. Charbonnel

Clusters of high-mass protostars: From extreme clouds to mini-bursts of star formation 9
F. Motte, F. Louvet & Q. N. Lu'o'ng

The Imprints Of Galactic Environment On Cluster Formation and Evolution . . . 17
A. Adamo

Formation of young massive clusters from turbulent molecular clouds 25
M. S. Fujii & S. P. Zwart

Submillimeter View of Gas and Dust in the Forming Super Star Cluster in NGC 5253 31
J. L. Turner

Massive stellar systems: observational challenges and perspectives in the E-ELT era 36
G. Bono, V. F. Braga, I. Ferraro, G. Fiorentino, R. Gilmozzi, G. Iannicola, D. Magurno, N. Matsunaga, M. Monelli & S. Rastello

The super star cluster driven feedback in ESO338-IG04 and Haro 11 44
A. Bik, G. Östlin, V. Menacho, A. Adamo, M. Hayes, J. Melinder & P. Amram

The Assembly History of the Milky Way Nuclear Star Cluster 50
A. Feldmeier-Krause, N. Neumayer, R. Schödel, A. Seth, P. T. de Zeeuw, C. J. Walcher, N. Lützgendorf, M. Kissler-Patig, M. Hilker & H. Kuntschner

Is the massive star cluster Westerlund 2 double? - A high resolution multi-band survey with the Hubble Space Telescope 55
P. Zeidler, A. Nota, E. Sabbi, E. K. Grebel, M. Tosi, A. Z. Bonanos, A. Pasquali, C. Christian & S. E. de Mink

Star-Formation Histories of Resolved Galaxies in the Local Volume 61
A. Aloisi

The SUNBIRD survey: characterizing the super star cluster populations of intensely star-forming galaxies 70
Z. Randriamanakoto & P. Väisänen

Dissecting 30 Doradus: Optical and Near Infrared Star Formation History of the starburst cluster NGC2070 from the Hubble Tarantula Treasury Project . . . 77
M. Cignoni & the HTTP team

Nuclear Star Clusters 84
N. Neumayer

Massive Star Clusters in Dwarf Galaxies 91
S. S. Larsen

Stellar population properties of the most massive globular clusters and ultra-compact dwarf galaxies of the Fornax cluster 99
M. Hilker

Mapping out the origins of compact stellar systems 105
A. J. Romanowsky, J. P. Brodie & t. SAGES

Massive star clusters in high-redshift star-forming galaxies seen at a 100 pc scale thanks to strong gravitational lensing 111
M. Dessauges-Zavadsky, A. Cava, V. Tamburello, D. Schaerer, L. Mayer, J. Richard & P. G. Pérez González

NGC 346: Looking in the Cradle of a Massive Star Cluster 117
D. A. Gouliermis & S. Hony

CHaMP: From Molecular Clouds to Massive Young Clusters 123
P. J. Barnes

Kinematic and thermal structure at the onset of high-mass star formation - ISOSS23053 125
S. Bihr, H. Beuther, H. Linz, S. E. Ragan, J. Tackenberg, R. J. Smith, T. Henning & O. Krause

Fundamental Properties of a Large, Unbiased Sample of Massive, Young, Embedded Star Clusters in the Milky Way 127
Y. Dallilar, P. Barnes, E. Lada & S. Ryder

Drama of HII regions: Clustered and Triggered Star Formation 129
J.-Z. Li, J. Yuan, H.-L. Liu, Y. Wu & Y.-F. Huang

HCN Hyperfine Ratio Analysis of Massive Molecular Clumps 131
W. Schap, P. Barnes, A. Ginsburg & A. Ordonez

Massive cold cloud clusters 133
L. V. Toth, S. Zahorecz, G. Marton, I. Ristorcelli, M. Juvela, L. Montier & J.-P. Bernard

Evidence for the non-destruction of the Most Massive Molecular Clouds even after they have given Birth to Massive Star Clusters 135
J. Zaragoza-Cardiel, J. Beckman, J. Font, A. Camps-Faria, B. Garca-Lorenzo & A. Serrano-Borlaff

Do you know the extinction in your young massive cluster? 137
G. De Marchi, N. Panagia & E. Sabbi

Non-thermal emission in Cyg OB2 139
D. Fenech, J. Morford & R. Prinja

Comparing ALMA, VLT, and HST data for Massive, Young Clusters in Grand-Design Spirals 141
P. Grosbøl & H. Dottori

Exploring the UV excess in star clusters of different mass 143
F. Hernández-Pérez, G. Bruzual & G. Magris C.

Galactic and extragalactic giant molecular clouds and massive star clusters: counting and analysis 145
A. S.Hojaev

Efficient Selection and Classification of Infrared Excess Emission Stars Based on AKARI and 2MASS Data 147
Y.-F. Huang, J.-Z. Li, J.-h. Yuan & H.-L. Liu

Strömgren-Hβ photometry of the young open cluster NGC 6913 149
N. Kaltcheva, V. Golev, J. Beaver, S. Lund & M. Briley

The Spectral Energy Distribution of the Earliest Phases of Massive Star Formation 151
R. Klein, J. Cooper, L. Looney, T. Henning, S. Chakrabarti & S. Shenoy

The Orion Nebula in the Far-Infrared: High-J CO and fine-structure lines mapped by FIFI-LS/SOFIA 153
R. Klein, L. W. Looney, E. Cox, C. Fischer, C. Iserlohe & A. Krabbe

Water Masers Outburst in the Massive Stellar Cluster W49A. 155
B. H. Kramer, K. M. Menten, T. Kamiński, B. Zhang, N. A. Patel & A. Kraus

Is there a cluster in the massive star forming region IRAS 20126+4104? 157
V. A. Montes , P. Hofner, C. Anderson & V. Rosero

Westerlund 1: monolithic formation of a starburst cluster 159
I. Negueruela, J. Simon Clark, B. W. Ritchie & S. P. Goodwin

Mid-infrared [NeII] Imaging of Young Massive Star Clusters Near Galactic Nuclei 161
S. C. C. Yeh, C.-W. Tsai, T. R. Geballe & C. N. Herrera

Identifying and Characterizing Stellar Clusters Toward the Galactic Anti-Center from Pan-STARRS1 Survey. 163
C.-C. Lin, W.-P. Chen, J. Hou, L. Chen & Z. Shao

A starburst region around $l = 347° - 350°$ 165
A. Marco, I. Negueruela, M. Monguió, C. González-Fernández, J. M. Apellániz, R. Dorda & J. S. Clark

Evolved massive stars in W33 and in GMC G23.3−0.3 167
M. Messineo, J. S. Clark, D. F. Figer, K. M. Menten, R.-P. Kudritzki, F. Najarro, R. M. Rich, V. D. Ivanov, E. Valenti, C. Trombley, C.-H. R. Chen, B. Davies & J. W. MacKenty

e-MERLIN Radio Continuum Measurements of OB Star Winds in CYG OB2 . . 169
J. Morford, R. Prinja & D. Fenech

Environmental effects on stellar populations of star clusters and dwarf galaxies . 171
S. Pasetto, M. Cropper, Y. Fujita, C. Chiosi & E. K. Grebel

Formation of the twin galactic starburst regions NGC 6357 and NGC 6334 173
K. Torii, K. Hasegawa, A. Ohama, H. Yamamoto, K. Tachihara & Y. Fukui

N131: A dust bubble was born from the disruption of a gas filament? 175
C.-P. Zhang

The influence of feedback from massive stars on the formation and emergence of massive clusters 177
J. E. Dale

Galactic flows and the formation of stellar clusters........................ 184
R. Smilgys & I. A. Bonnell

Stellar feedback from a massive Super Star Cluster in the Antennae merger.... 190
C. N. Herrera & F. Boulanger

Cluster Assembly in Hierarchically Collapsing Clouds 196
E. Vázquez-Semadeni, A. González-Samaniego, M. Zamora-Avilés & P. Colín

Universal and Environmental Influences on Star Clusters 202
B. C. Whitmore

Formation of the super star cluster RCW 38 triggered by cloud-cloud collision .. 208
Y. Fukui

Inverting the dynamical evolution of globular clusters: clues to their origin..... 214
M. Gieles & P. Alexander

The dynamical importance of binary systems in young massive star clusters.... 222
R. de Grijs, C. Li & A. M. Geller

Initial conditions of formation of starburst clusters: constraints from stellar dynamics.. 228
S. Banerjee

Effects of Stellar-Mass Black Holes on Massive Star Cluster Evolution 234
S. Chatterjee, M. Morscher, C. L. Rodriguez, B. Pattabiraman & F. A. Rasio

Intermediate-mass black holes in globular clusters: observations and simulations - Update.. 240
N. Lützgendorf, M. Kissler-Patig, K. Gebhardt, H. Baumgardt, D. Kruijssen, E. Noyola, N. Neumayer, T. de Zeeuw, A. Feldmeier-Krause, E. van der Helm, I. Pelupessy & S. P. Zwart

Early dynamical evolution of young substructured clusters 246
J. Dorval & C. Boily

Self-shielding clumps in starburst clusters 251
J. Palouš, R. Wünsch, S. Ehlerová & G. Tenorio-Tagle

The UCD Population of the Coma Cluster 253
K. Chiboucas, P. Ferguson, R. B. Tully, D. Carter, S. Phillipps & E. Peng

Binary Information from Open Clusters Using SEDS (BINOCS) Project: The Dynamical Evolution of the Binary Populations in Cluster Environments .. 255
P. Frinchaboy & B. Thompson

Dynamical Evolution of Outer-Halo Globular Clusters....................... 257
A. H. W. Küpper, A. H. Zonoozi, H. Haghi, N. Lützgendorf, S. Mieske, M. Frank, H. Baumgardt & P. Kroupa

GalevNB: from simulations to observations . . . 259
X. Pang, C. Olczak & R. Spurzem

Initial Dynamical Evolution of Star Clusters with Tidal Field . . . 261
S.-M. Park, S. P. Goodwin & S. S. Kim

From a star cluster ensemble to its formation history . . . 263
C. Schulz & M. Hilker

Bayesian Inference of Kinematics and Mass Segregation of Open Cluster . . . 265
Z. Shao, X. Xie, L. Chen, J. Zhong, J. Hou & C-C. Lin

Chemical abundances of multiple stellar populations in massive globular clusters 267
A. F. Marino

The *HST* survey of Magellanic-Cloud clusters and of their stellar populations . . 275
A. P. Milone

Formation of globular clusters with multiple populations . . . 281
T. Decressin

Contribution of globular clusters to halos . . . 287
A. Bragaglia

Bimodal regime in young massive clusters leading to subsequent stellar generations 294
R. Wünsch, J. Palouš, G. Tenorio-Tagle, C. Muñoz-Tuñón & S. Ehlerová

A critical assessment of models for the origin of multiple populations in globular clusters . . . 302
N. Bastian

Multiple populations in globular clusters: constraints from kinematics and dynamics . . . 310
V. Hénault-Brunet

Binary interactions and multiple stellar populations in globular clusters . . . 316
D. Jiang, Z. Han & L. Li

Theoretical evolutionary framework for low-mass stars in massive star clusters. . 320
S. Cassisi

Globular clusters and the evolution of their multiple stellar populations . . . 328
W. Chantereau, C. Charbonnel & G. Meynet

Accretion onto Protoplanetary Discs: Implications for Globular Cluster Evolution 334
T. P. G. Wijnen, O. R. Pols, F. I. Pelupessy & S. P. Zwart

Globular Clusters within Dark Matter Halos : Case Studies of 47 Tuc, NGC 1851 and M 15 . . . 336
J. Lee, J. Shin & S. S. Kim

The ODD Old, Super-Metal-Rich Open Cluster, NGC 6791 . . . 338
A. M. Boesgaard, M. G. Lum & C. P. Deliyannis

The Composition of the Bulge Globular Cluster NGC 6273 . . . 341
C. A. Pilachowski, C. I. Johnson, R. M. Rich, N. Caldwell, M. Mateo, J. I. Bailey, III & J. D. Crane

The birthrates of SNe Ia in globular clusters 343
B. Wang & D. Liu

Evidences of tidal distortion and mass loss from the old open cluster NGC 6791 345
G. Carraro & E. Dalessandro

Ages and chemical compositions of massive clusters in NGC147 and M31 347
M. Sharina & V. Shimansky

Twins born in different environments? Nuclei of two dSphs: isolated galaxy KKS3 and E269-66, a close neighbor of NGC5128 349
M. Sharina, A. Kniazev & I. Karachentsev

Variability Survey of ω Centauri in the Near-IR: Period-Luminosity Relations . . 351
C. Navarrete, M. Catelan, R. C. Ramos, F. Gran, J. Alonso-García & I. Dékány

Searching for Possible Members of Star Moving Groups in the Kepler Field 353
C. Y. Chen. W. P. Chen & A. Luo

Synthetic clusters of massive stars to test stellar evolution models 355
C. Georgy & S. Ekström

The Initial Mass Function of Young Open Clusters in the Galaxy: A Preliminary Result 357
B. Lim, H. Sung, H. Hur & B.-G. Park

Modeling Small Stellar Populations Using Starburst99 359
G. A. Vázquez & C. Leitherer

Are Galactic globular cluster AGB stars rich or poor in Sodium? Sodium abundance of AGB and RGB stars in NGC 2808 361
Y. Wang, F. Primas, C. Charbonnel, M. Van der Swaelmen, W. Chantereau & G. Bono

Author Index 363

Preface

Massive star clusters deserve special attention because they are one of the few bridges that connect the interstellar medium, star formation and evolution, mass and energy feedback to evolution of galaxies, and cosmology. These systems are ideal laboratories to probe very localized processes within interstellar gas clouds as well as large-scale dynamics of interstellar matter in interplay with the formation, evolution, and death of stars. Populous clusters are the key sites of high-mass star formation (i.e., the high-mass end of the initial mass function). They host large numbers of compact and exotic objects that form preferentially in dense environments, and they are fundamental benchmarks to study stellar physics and evolution and to explore dynamics in dense systems. They hold clues to how star formation and evolution as well as chemical and dynamical evolution of stellar systems influence each other and vary from one environment to another, from the present to the early Universe. They are the witnesses of the formation, assembly, and evolution of galaxies and of their substructures across time. They play a role in hierarchical cosmology as well as in the reionization of the intergalactic medium. These very complex systems are a prime example of astronomical objects currently being studied in every wavelength region, from radio to x-rays (and even gamma rays), and are also relevant for gravitational wave detectors.

A complete understanding of these systems and of their stellar populations is a challenging task. It requires the exchange of ideas and the collaboration of astrophysicists with observational, theoretical, and numerical expertise in stellar evolution, interstellar matter magnetohydrodynamics, stellar dynamics, formation and evolution of galaxies, cosmology, multidimensional numerical simulations, N-body simulations, and multi-wavelength high-precision photometry, spectroscopy, and astrometry.

In view of the impressive advances made by the various parties since the beginning of the 21st century, 2015 was the right time and the XXIX IAU General Assembly was the prime location to allow a wide community share necessary knowledge, discuss about remaining hot issues in the field, with the hope to come soon to the characterization of the physical processes that support the interpretation of the recent exquisite observations.

The following questions were addressed during IAU Symposium 316 "Formation, evolution, and survival of massive star clusters":

– Which physical properties determine the mode of star cluster formation? What are the initial and boundary conditions that will lead to massive and not-so-massive cluster formation? What is the relationship with the pre-cluster molecular cloud cores? Can we find pre-cluster molecular cloud cores of massive clusters?

– How important are early disruption processes and does their importance depend on environment?

– How, and what kind of stars form and evolve in very dense environments?

– How well do current models explain the presence of multiple populations in massive clusters, and how multiple generation stars are connected to the properties and evolution of the natal cloud cores?

– How important is the role of binary stars and what is the impact on cluster models? How does the binary mass fraction evolve with time during the life of the clusters? Do binary mergers play a role?

– What are the correlations between the presence of various compact objects in globular clusters and other structural parameters? Is there a correlation between the initial mass of a cluster and the mass of its most massive star?

– Which fraction of stars form in bound clusters and how does this vary with environment, metallicity, and time?

All these aspects are addressed in the papers (reviews, oral contributions, and posters) gathered in this book, with observers, theoreticians, and modellers discussing controversial topics and planning the next steps to be tackled in some of the key open areas. This is particularly timely as we enter a golden age for observations and numerical multi dimensional simulations.

IAU Symposium 316 was supported by Division H (Interstellar Matter and Local Universe; coordinating division), Division G (Stars and stellar physics) and Division J (Galaxies and cosmology), and by the Commissions 25 (Astronomical photometry and polarimetry), 28 (Galaxies), 33 (Structure and dynamics of the Galactic system), 34 (Interstellar matter), 35 (Stellar constitution) and 37 (Star clusters and associations) as well as by the WG Massive stars. We wish to warmly thank them as well as the IAU Executive Committee to make IAU Symposium 316 happens. We thank the GA Organizing Committee, as well as our excellent SOC and all the participants for a lively and successful conference.

Corinne Charbonnel and Antonella Nota, chairs of IAU Symposium 316

Co-Chairs of SOC:

- Corinne Charbonnel (Geneva Observatory)
- Antonella Nota (STScI)

Scientific Organizing Comittee

- Corinne Charbonnel (Switzerland & France)
- Antonella Nota (USA)
- Holger Baumgardt (Australia)
- Peter Cottrell (New Zealand)
- Yasuo Fukui (Japan)
- Richard de Grijs (China)
- Bruce Elmegreen (USA)
- Patrick Hennebelle (France)
- Nicole St Louis (Canada)
- Jan Palous (Czech Republic)
- Margarita Sharina (Russian Federation)
- Elena Terlevich (Mexico)
- Eva Noyola (USA)

Participants

First Name	Last Name	Organization Name	Country	Email
Angela	Adamo	Stockholm University, Astronomy Department, Oskar Klein Centre	Sweden	adamo@astro.su.se
Alessandra	Aloisi	Space Telecope Science Institute	United States	aloisi@stsci.edu
Javier	Alonso-Garcia	Universidad de Antofagasta	Chile	jalonso@astro.puc.cl
Philippe	Andre	CEA Saclay	France	pandre@cea.fr
Sambaran	Banerjee	Argelander-Institut fuer Astronomie, University of Bonn	Germany	sambaran@astro.uni-bonn.de
Peter	Barnes	University of Florida	United States	pjb@astro.ufl.edu
Nate	Bastian	Liverpool John Moores University	United Kingdom	N.J.Bastian@ljmu.ac.uk
Andrea	Bellini	STScI	United States	bellini@stsci.edu
Petro	Bertsyk	National Astronomical Observatories of China, Chinese Academy of Sciences	China	berczik@nao.cas.cn
Steven	Berukoff	National Solar Observatory	United States	sberukoff@nso.edu
Arjan	Bik	Department of Astronomy, Stockholm University	Sweden	arjan.bik@astro.su.se
Angela	Bragaglia	INAF-Osservatorio Astronomico di Bologna	Italy	angela.bragaglia@oabo.inaf.it
Shari	Breen	CSIRO Astronomy and Space Science	Australia	shari.breen@csiro.au
Tomasz	Bulik	University of Warsaw	Poland	tb@astrouw.edu.pl
Giovanni	Carraro	ESO	Santiago	gcarraro@eso.org
William	Chantereau	Department of Astronomy, University of Geneva	Switzerland	william.chantereau@unige.ch

First Name	Last Name	Organization Name	Country	Email
Corinne	Charbonnel	Department of Astronomy, University of Geneva	Switzerland	Corinne.Charbonnel@unige.ch
Sourav	Chatterjee	CIERA, Northwestern University	United States	chatterjee.sourav2010@gmail.com
Kristin	Chiboucas	Gemini Observatory	United States	kchibouc@gemini.edu
Jorgen	Christensen-Dalsgaard	Dept. Physics and Astronomy, Aarhus University	Denmark	jcd@phys.au.dk
Michele	Cignoni	STScI	United States	michele.cignoni@gmail.com
Bruce	Clarke	SETI Institute, NASA Ames Research Center	United States	bruce.d.clarke@nasa.gov
Michael	Connelley	University of Hawaii, Institute for Astronomy	United States	msc@ifa.hawaii.edu
Gary	Da Costa	Research School of Astronomy Astrophysics, ANU	Australia	gary.dacosta@anu.edu.au
Scott	Dahm	WM Keck Observatory		sdahn@keck.hawaii.edu
James	Dale	University Observatory and Excellence Cluster Universe, Munich	Germany	dale.james.e@gmail.com
James	De Buizer	USRA-SOFIA	United States	jdebuizer@sofia.usra.edu
Richard	De Grijs	Kavli Institute for Astronomy and Astrophysics, Peking University	China	grijs@pku.edu.cn
Selma E.	De Mink	University of Amsterdam	Netherlands	S.E.deMink@uva.nl
Licai	Deng	National Astronomical Observatories, CAS	China	licai@nao.cas.cn

First Name	Last Name	Organization Name	Country	Email
Andrea	Dieball	University of Bonn	Germany	adieball@soton.ac.uk
Lee	Dongjoo	Korea Astronomy Space Science Institute	South Korea	marin678@kasi.re.kr
Aaron	Dotter	RSAA ANU	Australia	aaron.dotter@gmail.com
Sylvia	Ekstrom	University of Geneva	Swaziland	sylvia.ekstrom@unige.ch
Michael	Fellhauer	Departamento de Astronomia, Universidad de Concepcion	Chile	mfellhauer@astro-udec.cl
Zhang	Fenghui	Yunnan observatory, Chinese Academy of Sciences	China	zhangfh@ynao.ac.cn
Nicolas	Flagey	CFHT Corporation	United States	flagey@cfht.hawaii.edu
Ori	Fox	University of California, Berkeley	United States	ofox@berkeley.edu
Peter	Frinchaboy	Texas Christian University	United States	p.frinchaboy@tcu.edu
Michiko	Fujii	National Astronomical Observatory of Japan	Japan	michiko.fujii@nao.ac.jp
Gary	Fuller	University of Manchester	United Kingdom	gary.fuller@manchester.ac.uk
Iskren	Georgiev	Max Plank Institute for Astronomy	Germany	georgiev@mpia.de
Cyril	Georgy	Keele University	United Kingdom	c.georgy@keele.ac.uk
Andrea	Ghez	UCLA	United States	ghez@astro.ucla.edu
Mark	Gieles	University of Surrey	United Kingdom	m.gieles@surrey.ac.uk
Stephen	Gillam	Indiana University Purdue University Fort Wayne	United States	gillamsd@ipfw.edu
Oleg	Gnedin	University of Michigan	United States	ognedin@umich.edu
Myriam	Gomez	Fundacion Imagen De Chile	Chile	mgomez@imagendechile.cl
Gabriela	Gonzalez	Louisiana State University	United States	gonzalez@lsu.edu
Dimitrios	Gouliermis	University of Heidelberg	Germany	dgoulier@mpia.de
Preben	Grosboel	European Southern Observatory	Germany	pgrosbol@eso.org
Margaret	Hanson	University of Cincinnati	United States	margaret.hanson@uc.edu
Lisa	Harvey-Smith	CSIRO Australia Telescope National Facility	Australia	lisa.harvey-smith@csiro.au

First Name	Last Name	Organization Name	Country	Email
Vincent	Henault-Brunet	University of Surrey	United Kingdom	v.henault-brunet@surrey.ac.uk
Lynne	Hillenbrand	California Institute of Technology	United States	lah@astro.caltech.edu
D. John	Hillier	University of Pittsburgh	United States	hillier@pitt.edu
Joseph	Hora	Harvard-Smithsonian Center for Astrophysics	United States	jhora@cfa.harvard.edu
Narae	Hwang	Korea Astronomy and Space Science Institute	South Korea	nhwang@kasi.re.kr
Lanwei	Jia	Department of Physics, Guangzhou University	China	ivy.lwjia@gmail.com
Dengkai	Jiang	Yunnan Observatories, Chinese Academy of Sciences	China	dengkai@ynao.ac.cn
Kelsey	Johnson	University of Virginia	United States	kej7a@virginia.edu
Paul	Jones	University of New South Wales	Australia	paulcojones@gmail.com
Casey	Jones	University of Hawaii at Hilo	United States	caseymj@hawaii.edu
Nadejda	Kaltcheva	University of Wisconsin Oshkosh	United States	kaltchev@uwosh.edu
Amanda	Karakas	Australian National University	Australia	Amanda.Karakas@anu.edu.au
Marc	Kassis	Keck Observatory	United States	mkassis@keck.hawaii.edu
Akiko	Kawamura	National Astronomical Observatory of Japan	Japan	akiko.kawamura@nao.ac.jp
Markus	Kissler-Patig	Gemini Observatory	United States	mkissler@gemini.edu
Randolf	Klein	SOFIA USRA (NASA Ames)	United States	rklein@sofia.usra.edu
Patricia	Knezek	National Science Foundation	United States	pknezek@nsf.gov
Busaba	Kramer	Max Planck Institute for Radio Astronomy	Germany	bkramer@mpifr-bonn.mpg.de

First Name	Last Name	Organization Name	Country	Email
Hanindyo	Kuncarayakti	Millennium Institute for Astrophysics (MAS) and Dept. Astronomia, Univ. de Chile	Chile	kuncarayakti@gmail.com
Regis	Lachaume	Instituto de Astronomia, Pontificia Universidad Catolica de Chile	Chile	lachaume@astro.puc.cl
Soeren	Larsen	Department of Astrophysic-sIMAPP, Radboud University	Netherlands	s.larsen@astro.ru.nl
Ryan	Lau	Cornell, JPL	United States	ryanl@astro.cornell.edu
Young-Wook	Lee	Yonsei University	South Korea	ywlee2@yonsei.ac.kr
Joowon	Lee	Kyung Hee University	South Korea	joowon.lee@khu.ac.kr
Jinzeng	Li	National Astronomical Observatories, Chinese Academy of Sciences	China	ljz@nao.cas.cn
Yan	Li	Yunnan Astronomocal Observatory, Chinese Academy of Sciences	China	ly@ynao.ac.cn
Beomdu	Lim	Korea Astronomy and Space Science Institute	South Korea	bdlim1210@kasi.re.kr
Chien-Cheng	Lin	Shanghai Astronomical Observatory	China	cclin@shao.ac.cn
Kang	Liu	Department of Astronomy, Beijing Normal University	China	liukang@bnu.edu.cn
Jessica	Lu	IfA, University of Hawaii	United States	jessica.lu@hawaii.edu
Jan	Lub	Leiden Observatory Sterrewacht Leiden	Netherlands	lub@strw.leidenuniv.nl
Nora	Luetzgendorf	European Space Agency (ESA)	Netherlands	nluetzge@cosmos.esa.int

First Name	Last Name	Organization Name	Country	Email
Paola	Marigo	Department of Physics and Astronomy, University of Padova	Italy	paola.marigo@unipd.it
Anna	Marino	Research School of Astronomy Astrophysics	Australia	amarino@mso.anu.edu.au
Francesco	Mauro	Departamento de Astronomia, Universidad de Concepcion	Chile	skz5k2@gmail.com
Liam	McDaid	Sacramento City College	United States	mcdaidl@scc.losrios.edu
Steve	McMillan	Drexel University	United States	steve@physics.drexel.edu
Maria	Messineo	MPIfR	Germany	messineo@mpifr-bonn.mpg.de
Georges	Meynet	Geneva Observatory	Swaziland	georges.meynet@unige.ch
Andrea	Miglio	University of Birmingham	United Kingdom	a.miglio@bham.ac.uk
Danny	Milisavljevic	Smithsonian Astrophysical Observatory	United States	dmilisav@cfa.harvard.edu
Florentin	Millour	OCA - Laboratoire Lagrange	France	fmillour@oca.eu
Shigeki	Miyaji	Chiba University	Japan	miyaji@astro.s.chiba-u.ac.jp
Michael	Montgomery	The University of Texas	United States	mikemon@astro.as.utexas.edu
Brian	Morsony	University of Maryland	United States	bmorsony@gmail.com
Frederique	Motte	AIM Paris Saclay	France	frederique.motte@cea.fr
Francisco	Najarro de la Parra	CAB CSIC-INTA	Spain	najarro@cab.inta-csic.ed
Paul	Nandra	Max-Planck-Institut for Extraterrestrische Physik	Germany	knandra@mpe.mpg.de
Ignacio	Negueruela	Universidad de Alicante	Spain	ignacio.negueruela@ua.es
Nadine	Neumayer	Max-Planck Institute for Astronomy	Germany	neumayer@mpia.de
Jason	Nordhaus	Rochester Institute of Technology	United States	nordhaus@astro.rit.edu
John J.	Oh	National Institute for Mathematical Sciences	South Korea	chiewoo@gmail.com
Akio	Ohama	Nagoya University	Japan	ohama@a.phys.nagoya-u.ac.jp
Akio	Ohama	Nagoya University	Japan	ohama@a.phys.nagoya-u.ac.jp

First Name	Last Name	Organization Name	Country	Email
Chingis	Omarov	National Center of Space Research an Technology Fessenkov Astrophysical Institut	Kazakhstan	chingis.omarov@gmail.com
Toshikazu	Onishi	Osaka Prefecture University	Japan	ohnishi@p.s.osakafu-u.ac.jp
Livia	Origlia	INAF - Bologna Observatory	Italy	livia.origlia@oabo.inaf.it
Deborah	Padgett	NASAGoddard Space Flight Center	United States	deborah.l.padgett@nasa.gov
Xiaoying	Pang	Shanghai Institute of Technology	China	xypang@bao.ac.cn
So-Myoung	Park	Kyung Hee University	South Korea	smpark12@khu.ac.kr
Iurii	Pavlenko	PE Third Planet	Ukraine	iurii.pavlenko@universemagazine.com
Randy	Phelps	National Science Foundation	United States	rphelps@nsf.gov
Giampaolo	Piotto	University Di Padova		giampaolo.piotto@unipd.it
Thomas H.	Puzia	Institute of Astrophysics, Pontificia Universidad Catolica	Chile	tpuzia@astro.puc.cl
Zara	Randriamanakoto	South African Astronomical Observatory	South Africa	zara@saao.ac.za
John	Rayner	University of Hawaii	United States	john.thornton.rayner@gmail.com
Michel	Rieutord	Inst. Recherches en Astro-physique et Planetologie	France	mrieutord@irap.omp.eu
Alana	Rivera Ingraham	ESA-ESAC	Spain	alana.rivera@sciops.esa.int
Massimo	Robberto	Space Telescope Science Institute	United States	robberto@stsci.edu
Aaron	Romanowsky	San Jose State University	United States	aaron.romanowsky@sjsu.edu
Monica	Rubio	Departamento de Astronomia, Universidad de Chile	Chile	mrubio@das.uchile.cl
Elena	Sabbi	STScI	United States	sabbi@stsci.edu

First Name	Last Name	Organization Name	Country	Email
Gozde	Saral	Harvard-Smithsonian Center for Astrophysics	United States	gsaral@cfa.harvard.edu
Paul	Scowen	Arizona State University	United States	paul.scowen@asu.edu
Zhengyi	Shao	Shanghai Astronomical Observatory, CAS	China	zyshao@shao.ac.cn
Margarita	Sharina	Special Astrophysical Observatory of Russian Academy of Sciences	Russia	sme@sao.ru
Romas	Smilgys	University of St Andrews	United Kingdom	rs202@st-andrews.ac.uk
Verne	Smith	NOAO	United States	vsmith@noao.edu
Nathan	Smith	University of Arizona	United States	nathans@as.arizona.edu
Linda	Smith	Space Telescope Science Institute	United States	lsmith@stsci.edu
Igor	Soszynski	Warsaw University Observatory	Poland	soszynsk@astrouw.edu.pl
Nicole	St-Louis	Universite de Montreal	Canada	stlouis@astro.umontreal.ca
Kazufumi	Torii	Nagoya University	Japan	torii@a.phys.nagoya-u.ac.jp
Jean	Turner	UCLA	United States	turner@astro.ucla.edu
Maureen	Van den Berg	University of Amsterdam	Netherlands	M.C.vandenBerg@uva.nl
Paolo	Ventura	INAF - Observatory of Rome	Italy	paolo.ventura@oa-roma.inaf.it
Paul	Vreeswijk	Weizmann Institute of Science	Israel	paul.vreeswijk@weizmann.ac.il
Debra	Wallace	University of South Carolina Beaufort	United States	dwallace@uscb.edu
Jun-Jie	Wang	National Astronomical Observatory, Chinese Academy of Sciences	China	wangjj@nao.cas.cn
Bo	Wang	Yunnan Observatories, Chinese Academy of Sciences	China	wangbo@ynao.ac.cn

First Name	Last Name	Organization Name	Country	Email
Liu	Weiping	China Institute of Atomic Energy		wpliu@ciae.ac.cn
Roger	Wesson	European Southern Observatory	Chile	rwesson@eso.org
Bradley	Whitmore	Space Telescope Science Institute	United States	whitmore@stsci.edu
Ralph	Wijers	Anton Pannekoek Institute for Astronomy, University of Amsterdam	Netherlands	Ralph.Wijers@uva.nl
Aida	Wofford	Institut d Astrophysique de Paris	France	wofford@iap.fr
Richard	Wunsch	Astronomical Institute, Czech Academy of Sciences	Czech Republic	richard@wunsch.cz
Jin-Long	Xu	National Astronomical Observatories, Chinese Academy of Sciences	China	xujl@bao.ac.cn
Yongqiang	Yao	National Astronomical Observatories, Chinese Academy of Sciences	China	yqyao@nao.cas.cn
Sherry Chia Chen	Yeh	Subaru Telescope	United States	yeh@naoj.org
Jiaerken	Yeshengbieke	Xinjiang Astronomical Observatory, Chinese Academy of sciences	China	jarken@xao.ac.cn
Jincheng	Yu	Pontificia Universidad Católica de Chile	Chile	yujc@astro.puc.cl
Haotong	Zhang	National Astronomical Observatories, Chinese Academy of Science	China	htzhang@bao.ac.cn
Zhumabek	Zhantayev	National Center of Space Research an Technology	Kazakhstan	chingis.omarov@gmail.com
Jianjun	Zhou	Xinjiang Astronomical Observatory, CAS	China	zhoujj@xao.ac.cn
Alice	Zocchi	University of Surrey	United Kingdom	a.zocchi@surrey.ac.uk

Formation, evolution, and survival of massive star clusters
Proceedings IAU Symposium No. 316, 2015
C. Charbonnel & A. Nota, eds.

doi:10.1017/S174392131600702X

Young and old massive star clusters and their stellar populations - Theoretical challenges for the next decade

Corinne Charbonnel[1,2]

[1]Department of Astronomy, University of Geneva
1290 Versoix - Switzerland
email: Corinne.Charbonnel@unige.ch

[2]IRAP CNRS UMR 5277, Université de Toulouse III
31400 Toulouse, France

Abstract. Several models presented in the literature compete to explain the origin of multiple stellar populations in globular clusters (GC), but they all fail to reproduce the large variety of present-day characteristics of these systems. In parallel, independent clues on GC early evolution may be derived from observations of young massive clusters (YMC) in the Local Group. But are these two populations of clusters related? And can we reconcile the informations and data concerning GCs and YMCs? Here we summarize some open questions on the nucleosynthetic origin of multiple stellar populations in GCs, on the actual evolution and characteristics of GC low-mass stars, and on early gas expulsion from massive clusters. We propose theoretical paths to be explored in the near future.

Keywords. globular clusters: general

1. Introduction

Massive clusters of stars (MSC), young and old, are magnificent astronomical objects, and keys to a broad variety of astronomical and cosmological questions. They are fundamental benchmark for stellar evolution theory, key sites for the formation of the most massive stars, and candidate host of gamma ray bursts. They are unique witnesses and clocks of the formation and assembly of galaxies and of their substructures, from the early to the present-day Universe. They play a key role in hierarchical cosmology and potentially also in the reionization of the Universe. However, their formation, evolution, and survival are far from being understood. Comparative studies of MSCs, young and old, in the Milky Way, Local Group, starburst and interacting galaxies are now possible thanks to exquisite observations performed e.g. with HST, VLT, Herschel, and ALMA. Those try to bring gap(s) and make links between young and old MSCs, revealing precious informations and bringing extraordinary surprises. All these scientific issues were at the heart of the scientific rationale we proposed for IAUS316, and they have been extensively addressed during that international conference at the IAU General Assembly in Honolulu. We refer to the reviews and contributions presented in this Volume for the most recent significant results and literature references.

Many of the conversations at IAUS316 were dominated by the debate on the explanation of multiple stellar populations (MSPs) in globular clusters (GCs). In this paper we summarize a couple of the many theoretical challenges that have been raised by the discovery of these MSPs (§2), in particular the nucleosynthetic origin of the abundance anomalies in GCs (§3), the actual evolution and fate of GC low-mass stars (§4), and the mass budget problem under new insight from extragalactic young MSCs (§5).

2. Multiple stellar populations (MSP) in globular clusters - Open questions

One of the major breakthroughs of the past decade in the domain concerns those MSCs we thought we knew the best, namely the globular clusters (GC). It was long taken for granted that individual GCs were the perfect archetype of a single stellar population, and that they had undergone no internal chemical evolution (except for the most massive ones like Omega Cen, which is not even considered as a GC but rather as the remnant of a small galaxy), nor any drastic modification of e.g. their mass along their life. This classical view is at odds with the new paradigm imposed by modern spectroscopic and photometric observations. Indeed we know now that these old systems actually host MSPs. Those exhibit very peculiar chemical properties and leave unexpected imprints in the clusters color-magnitude diagrams that have never been discovered yet in any other stellar population, nor in the field, nor in open clusters (e.g. the reviews by Bragaglia, Marino, and Piotto, this Volume, and references therein).

Extensive literature is devoted to possible explanations to the ubiquitous presence of MSPs in GCs, as well as their unique spectroscopic and photometric complexity. However, none of the proposed scenarii that call for chemical self-enrichment of GCs in their enfancy is able to reconcile all the available data (e.g. Bastian, this Volume). Therefore, we are left with very basic, but crucial questions: Which type of short-lived massive stars did pollute the intracluster material with only hydrogen-burning products during the early GC infancy? How did the multiple populations of low-mass stars that are still alive today form in each individual GC, how did they inherite their chemical peculiarities, and what are the consequences for their evolution and fate? Why are these MSPs not observed in other types of star clusters, like open clusters? Does this reflect very specific star formation mechanisms in the early universe? Or will we also find similar MSPs in young massive star clusters formed in present-day conditions once we will have the suited observational tools? What was the original mass of GCs, and their contribution to galactic haloes? Old GCs consist today of old low-mass stars and little or no gas. However, there must have been a time when GCs formed as gas-rich objects, hosting numerous massive, short-lived stars. What has been the chemical and dynamical impact of these massive stars on their host clusters? How did the feedback with the intra-cluster medium (ICM) impact the formation and evolution of stars on various temporal and spatial scales, as well as the dynamics of the entire clusters? Did they contribute to gas ejection eventually associated to a significant loss of stars without disrupting the cluster by a drastic change of potential well?

All these questions constitute for sure the most important theoretical challenges in the domain for the next (couple of?) decade(s). Here we point out some of the most critical issues.

3. Have we identified the nucleosynthetic origin of the chemical properties of MSPs in GCs?

About two thirds of GC stars exhibit chemical properties that reflect the signatures of hydrogen-burning at high temperature through CNO-cycle and NeNa and MgAl chains (e.g. Prantzos *et al.* 2007). This proportion of so-called 2d population stars (2P†) is relatively constant from cluster to cluster (e.g. Prantzos & Charbonnel 2006, Carretta 2010); it is almost independent of the stellar evolution stage (main sequence and red giant branch), althought it slightly differs from cluster to cluster among AGB stars (e.g. Wang *et al.*, this Volume). This is currenty interpreted as the result of self-enrichment of GCs in their early infancy, with 2P stars having formed out of the H-ashes of massive 1P stars mixed with original proto-cluster gas. The most-commonly invoked polluters are the so-called Fast Rotating Massive Stars (FRMS, $\geqslant$ 20 $M_\odot$; Decressin *et al.* 2007a, Decressin *et al.* 2007b, Krause *et al.* 2013), and massive AGB stars ($\sim$6 - 6.5 $M_\odot$; e.g. Ventura *et al.* 2001, Ventura *et al.* 2013, D'Ercole *et al.* 2010). Variants and combinations of FRMS and AGB have been proposed (e.g. Sills & Glebbeek 2010, Bastian *et al.* 2013), as well as intermediate-mass binaries (10 - 20 $M_\odot$; De Mink *et al.* 2009) and supermassive stars ($\sim 10^4$ $M_\odot$ stars; Denissenkov & Hartwick 2014).

However, all these propositions face difficulties to explain simultaneously all the observed abundance patterns. For example, all the AGB models computed by independent groups do predict AGB yields at odds with the observed O-Na anticorrelation (e.g. Forestini & Charbonnel 1997, Denissenkov & Herwig 2003, Karakas & Lattanzio 2007, Ventura *et al.* 2013). Additionally, state-of-the-art rotating models of AGB stars predict that those stars should also eject helium-burning products, which is sustained by a wide variety of galactic observations (e.g. Maeder & Meynet 2006, Chiappini *et al.* 2008), but which is at odds with the constancy of C+N+O observed among GC low-mass stars (Decressin *et al.* 2009). Attempts to "save" (quoting the authors) the AGB scenario call for adhoc dilution of AGB ejecta with galactic gas of the exact same composition as that of the original proto-cluster that must be re-accreted by all GCs about 50 to 100 Myrs after the expulsion of the original proto-cluster gas and of the supernovae (SNe) ejecta material (D'Ercole *et al.* 2012). Some call for a major revision of the cross-section of a critical reaction rate destroying sodium as an alternative option (Renzini *et al.* 2015).

In the case of FRMS being the polluters, a revision of the cross section of the reaction ^{24}Mg(p,γ) is advocated for in order to reproduce the amplitude of the observed Mg-Al anticorrelation among GC stars (Decressin *et al.* 2007a). In addition, the different mechanisms that may lead to equatorial mass ejections of H-burning ashes when massive stars reach break-up have not yet been fully investigated, so that the actual mass of the H-ashes that can be released by FRMS, especially very early on the main sequence when the Na-rich material is not yet strongly enriched in helium, might well be underestimated. Intermediate-mass binaries being the polluters also suffers some caveats (D'Antona *et al.* 2012), one being their internal temperature being too cool to explain the observed nucleosynthetic patterns (Prantzos *et al.* 2007).

Importantly, all the current polluter models face difficulties to account for the helium differences between 1P and 2P stars that have been derived from analyses of the spreads in the main-sequence colors in ultraviolet and optical filters for a handful of GCs (Bastian *et al.* 2015, and references therein). In this respect, supermassive stars might be promising from the nucleosynthetic point of view. However, this requires the verification of speculations on how the required mass-loss might be caused by the super-Eddington radiation continuum-driven stellar wind or by the diffusive mode of the Jeans instability. Moreover, no self-consistent scenario linking all the chemical and dynamical constraints has been proposed so far for these hypothetic stars.

† Compare with the 1st population, 1P, that shows chemical composition similar to that of halo stars.

Therefore, the exact nature of the first population GCs polluters is far from being definitively ascertained. New stellar models including internal magneto-hydrodynamical processes (rotation, mass loss, magnetic fields, hydrodynamical instabilities) over the whole initial stellar mass range as well as the whole metallicity range covered by GCs are definitively needed to probe the nucleosynthetic origin of GC stellar properties.

We can therefore conclude that GCs will be for long extraordinary exploration territories for stellar physics.

4. How does the peculiar initial composition of GC low-mass stars impact their evolution?

Old GCs contain today only low-mass stars (below ~ 0.85 $M_{\odot}$) and the dark remnants of the upper end of the initial stellar mass function (IMF). Despite the many differences between the scenarios that have been developed to explain the abundance properties of GCs (§ 3), it is well accepted that 2P GC stars have started their life with a higher helium content than their 1P counterparts. This is confirmed by direct spectroscopic measurements of non-local thermodynamic equilibrium He abundances for a subset of blue horizontal branch stars in the GC NGC 2808 (Marino *et al.* 2014). It is therefore mandatory to account for their non-standard helium content on the evolution of low-mass stars to account for all the spectroscopic and photometric observational constraints, as the helium content of stars does definitively impact their evolution, lifetime, and fate.

Predictions for the extent of helium enrichment along the sodium distribution in 2P stars strongly depend on the nature of the invoked 1P polluters. In the AGB scenario, all 2P stars (i.e., spanning a large range of Na abundances) are expected to be born with very similar helium contents (maximum of 0.36 - 0.38 in mass fraction, e.g. Doherty *et al.* 2014, if no dilution with the ISM matter is taken into account, to be compared with ~ 0.248 for 1P stars). In contrast, the FRMS scenario predicts broad and correlated spreads of both helium and sodium in the initial mixture that formed 2P stars, with initial helium mass fraction ranging between 0.248 (as for the 1P) and 0.8 in the most extreme 2P stars (Decressin *et al.* 2007b). This has noticeable consequences on the evolution and fate of GC 2P stars (details in Chantereau *et al.* 2015; see also Chantereau *et al.*, this Volume). Such initial abundances provide e.g. a straightforward explanation to the lack of sodium-rich 2P AGB stars in GCs like NGC 6752 (Charbonnel *et al.* 2013; observations by Campbell *et al.* 2013) due to the impact of the initial helium content on the stellar lifetime and evolution path in the Hertzsprung-Russel diagram. They also account for the finding of different sodium spreads that are observed along the AGB in the Galactic globular clusters of different ages and [Fe/H] values (Charbonnel & Chantereau 2015). Simultaneously, the FRMS scenario in its current form predicts that at a typical GC age of 13.4 Gyr, 95% of the low-mass stars lying two magnitudes below the turn-off should be born with a helium mass fraction Y_{ini} between 0.248 and 0.4, and only 5 % with Y_{ini} higher than 0.4 (Chantereau *et al.* 2016; Chantereau *et al.*, this Volume); no star born with Y_{ini} higher than 0.4 are expected to lie today on the horizontal branch, in agreement with current interpretations of GC horizontal branch morphologies in connection with the so-called second parameter problem (e.g. D'Antona & Caloi 2004, Gratton *et al.* 2010, Dotter *et al.* 2010).

Although the initial helium distribution among 2P stars is not well known yet, all the polluter models presently fail to account for the extremely low helium enrichment deduced from main sequence color spreads in some GCs (e.g. NGC 6752, Milone *et al.* 2013). As of today, the maximum helium spread for a GC was derived in NGC 2808 (Milone *et al.*

12b), and large differences in He spreads have been found between clusters. This clearly poses a problem to all the polluters proposed in the literature, at least within the current frameworks, and it requires large degrees of stochasticity in the polluter yields (Bastian *et al.* 2015).

One has to note however that a direct and reliable comparison between theoretical predictions and observed GC CMDs is still hampered for several reasons. Going from the theoretical to the observational plane requires indeed the use of model atmospheres and temperature-color transformations suited for the proper helium range and associated peculiar composition of 2P stars. This has already been done up to a certain extent (Sbordone *et al.* 2011, Milone *et al.* 2013, Cassisi *et al.* 2013), but not for large helium abundance spread, as such tools are not yet available. In addition, from the stellar evolution point of view we need to go one step further by computing models of 1P and 2P long-lived low-mass stars taking into account the effects of the atomic diffusion, rotation, and mixing, which are expected to have a non negligible impact on stellar evolution and lifetimes, and on the position of stars within the HRD. Work is in progress in these complementary directions.

5. Is massive gas ejection the solution to the mass budget problem for GCs? Insight from extragalactic YMCs

One critical issue to account for MSPs in GCs is that, when one assumes a classical IMF for the potential stellar polluters, the ratio between 2P and 1P stars is predicted to be 10 to 90%; this is at odds with the observed ratio 70% - 30% quoted above (Prantzos & Charbonnel 2006). Variants and combinations of the different potential polluters have been proposed to alleviate the mass budget issue, but this actually strengthens the nucleosynthetic issues that we briefly recalled in § 3. To enhance the 2P to 1P ratio, one can alternatively call for a flat IMF for the polluters (Smith & Norris 1982, D'Antona & Caloi 2004, Prantzos & Charbonnel 2006, Schaerer & Charbonnel 2011), or for a 1P composed of massive stars only giving birth to all the low-mass stars we observe today in GCs (Charbonnel *et al.* 2014). While these options might open extremely interesting avenues, their many implications have not yet been investigated in detail.

As of today, the most commonly invoked solution to the mass budget issue calls for fast and drastic gas expulsion that would have lead to the loss of the vast majority of 1P low-mass stars while 2P stars formed in the center of GCs would have remained bound to the cluster (e.g. Prantzos & Charbonnel 2006, Decressin *et al.* 2010, D'Ercole *et al.* 2010, Schaerer & Charbonnel 2011, Krause *et al.* 2013, Khalaj & Baumgardt 2015).

Models of gas expulsion considering superbubbles kinematics indicate however that stellar winds and supernovae explosions may not be sufficient to accomplish the task, except for the least massive and most extended clusters. Alternatively, coherent onset accretion onto 1P neutron stars and stellar black holes might do it for the whole range of GC compactness. However in that case, gas ejection would happen only after all 1P massive stars have turned into such dark remnants at the end of the SNe phase, i.e., typically after ~ 35 Myr after cluster formation (see Krause *et al.* 2012, Krause *et al.* 2013 for details).

This delay for gas expulsion is in tension with observations of young MCs (YMC) in the Local Group and in dwarf galaxies. Indeed, some of these objects with ages below 10 Myr, current masses between 8 and 50×10^5 $M_\odot$, and half-mass radii between 1.5 and 18 pc (i.e., very similar to the expected properties of GC progenitors), appear to be gas free by an age of ~ 3 Myr, and show no evidence for multiple epochs of star-formation (Bastian *et al.* 2014). In the case GC formation at high redshift happened the same way as YMC formation today, this implies that the dark remnant scenario for fast gas removal occurs too late to explain the disappearance of the gas at such an early stage. For very early gas expulsion to eventually work, one must then invoke either energy ejection by 10^{52} - 10^{53} erg hypernovae, or a very high star formation efficiency (SFE from 50 % in the most loose clusters, and up to more than 90 % in the most compact ones), the efficiency of gas expulsion being extremely dependent on the cluster compactness and on local SFE (see details in Krause *et al.* 2016).

There is no definitive evidence yet that GC formation at high redshift was similar to YMC formation happening in the local and modern universe. However, if YMCs are the actual modern counterparts of GCs then gas and star expulsion is a serious issue for the current GC self-enrichment scenarii that make classical assumptions for the SFE and for the stellar IMF of the first and second stellar populations. Rather, gas could be cleared for very high SFE, but in that case the total stellar mass would not be changed significantly enough to affect the 1P-to-2P ratio as requested by the mass budget.

What can we do next? It clearly appears that the detection of the O-Na anticorrelation in YMCs could be a definitive clue to the "genetic" relationship between YMCs and GCs. On the theoretical side, analytical and semi-analytical modelling of the expansion and instabilities of self-gravitating superbubbles including for the first time relevant prescriptions from stellar models to describe stellar mass loss, winds, luminosity, and energy, have highlighted the very important role of stellar feedback on MSC dynamics. Refined and detailed 3D-hydrodynamical treatment of the subtle interactions between stellar ejecta (mass, energy, nucleosynthesis products), interstellar matter, and secondary star formation are definitively needed now to probe in detail all the aspects of the vast problem of the formation of MSPs in MSCs in the early phases of their evolution.

6. Conclusions

Old GCs contain today only old low-mass stars and the dark remnants of more massive, short-lived stars. Until about a decade ago, most of the studies related to these objects were implicitly considering that this had always been the case, and that the rest of the IMF had very little impact on the evolution of such clusters that managed to stay gravitationally bound for 10 to 13 billion years. The discovery of ubiquitous MSPs has revolutionized the field. It revealed that all GCs have actually undergone drastic chemical and dynamical processes in their early infancy. This has opened far-reaching questions for stellar and galactic physics, and for cosmology, that none of the scenarios proposed so far can solve. As we are trying to get complementary clues from YMCs in nearby and interacting galaxies, we have no definitive evidence that the formation mechanisms of these local objects happen through the same path than GC formation that occurred in the early Universe.

One of the conclusions of the discussions at IAUS 316 is that none of the presently circulating scenario can explain all the photometric and spectroscopic constraints. We may think we are at an impasse, but this is how science works at its best. We just need innovative thoughts and a different approach, out of the box. For sure, finding the answers will require exchange of ideas and close collaboration of astrophysicists with observational, theoretical, and numerical expertise in stellar evolution, interstellar matter magnetohydrodynamics, stellar dynamics, formation and evolution of galaxies, cosmology, multidimensional numerical simulations, N-body simulations, and multi-wavelength high-precision photometry, spectroscopy, and astrometry. Work is in progress and, hopefully, a robust and consensual explanation will be reached in the near future.

Acknowledgements

I thank IAU for supporting the organization of IAUS 316 on Formation, Evolution, and Survival of Massive Star Clusters, as well as the SOC members. My warmest thoughts go to my fantastic co-chair, A.Nota, for the nice organizational and scientific adventures we shared. I acknowledge support from the Swiss National Science Foundation (FNS) for the project 200020-159543 "Multiple stellar populations in massive star clusters - Formation, evolution, dynamics, impact on galactic evolution" (PI CC). I thank the International Space Science Institute (ISSI, Bern, CH) for welcoming the activities of ISSI Team 271 "Massive star clusters across the Hubble Time (2013 - 2016, team leader CC), as well as my colleagues, M.Krause, N.Bastian and R.Diehl, co-authors of K15 paper, and F.Primas, W.Chantereau, and Y.Wang, co-authors of C14.

References

Bastian, N., *et al.*, 2013, *MNRAS*, 436, 2398
Bastian, N., Hollyhead, K, & Cabrera-Ziri, I., 2014, *MNRAS*, 445, 378
Bastian, N., Cabrera-Ziri, I., & Salaris, M., 2015, *MNRAS*, 445, 378
Bastian, N., & Lardo, C., 2015, *MNRAS*, 449, 3333
Bragaglia, A., *et al.* 2014, *ApJ*, 796, 68
Campbell, S. W., *et al.* 2013, *Nature*, 498, 198
Carretta, E.., *et al.* 2009, *A&A* 505, 139
Carretta, E.., *et al.* 2010, *A&A*, 516, A55
Carretta, E. 2010, *A&A*, 557, A128
Cassisi, S., Mucciarelli, A., Pientriferni, A., Salaris, M., & Ferduson, J. 2013, *A&A*, 554, A19
Chantereau, W., Charbonnel, C., & Decressin, T. 2015, *A&A*, 578, A117
Chantereau, W., Charbonnel, C., & Meynet, G. 2016, submitted to *A&A*
Charbonnel, C. & Krause, M., 2015, IAUS 317 "*The General Assembly of the Galactic Halo*", in press
Charbonnel, C. & Lagarde, N. 2010, *A&A*, 522, A10
Charbonnel, C. & Zahn, J. P. 2007, *A&A* (Letter), 467, L15
Charbonnel, C., *et al.* 2013, *A&A* (Letter), 557, L17
Charbonnel, C., Chantereau, W., Krause, M., Primas, F., & Wang, Y. 2014, *A&A* (Letter), 569, L6 (C14)
Charbonnel, C., & Chantereau, W., 2015, *A&A*, in press
Chiappini, C. 2008, *A&A*, 479, L9
D'Antona, F. & Caloi, V. 2004, *ApJ*, 611, 871
D'Antona, F., *et al.* 2012, *A&A*, 443, 3302
Decressin, T., *et al.* 2007a, *A&A*, 464, 1029
Decressin, T., Charbonnel, C., & Meynet, G, 2007b, *A&A*, 475, 859
Decressin, T., Charbonnel, C., *et al.* 2009, *A&A*, 505, 727
Decressin, T., Baumgardt, H., Charbonnel, C., & Kroupa, P., 2010, *A&A*, 516, A73
De Mink, S., *et al.* 2009, *A&A* (Letters), 507, L1
Denissenkov, P. A. & Herwig, F. 2003, *ApJ*, 590, L99
Denissenkov, P. A. & Hartwick, F. D. A. 2014, *MNRAS* (Letter), 437, L21
D'Ercole, *et al.* 2010, *MNRAS*, 407, 854
D'Ercole, *et al.* 2010, *MNRAS*, 423, 1521
Doherty, C. L., *et al.* 2014, *MNRAS*, 437, 195
Dotter, *et al.* 2010, *ApJ*, 708, 698
Forestini, M. & Charbonnel, C. 1997, *A&A*, 123, 241
Gratton, R. G., *et al.* 2010, *A&A*, 517, A81
Karakas, A. & Lattanzio, J. 2007, *PASA* 24, 103
Khalaj, P. & Baumgardt,H. 2015, *MNRAS*, 492, 924
Gratton, R. G., Carretta, E., & Bragaglia, A. 2012, *ARAA*, 20, 50

Krause, M., Charbonnel, C., *et al.* 2012, *A&A* (Letter), 546, L5
Krause, M., Charbonnel, C., *et al.* 2013, *A&A*, 552, A121
Krause, M., Charbonnel, C., Bastian, N., & Diehl, R. 2016, *A&A*, in press, arXiv:1512.04256 (K15)
Kroupa, P., *et al.* 2013, in *Planets, Stars and Stellar Systems* (Springer Netherlands), 115
Lagarde, N., *et al.* 2012, *A&A*, 543, A108
Larsen, S. S., *et al.* 2014, *A&A*, 565, A98
Lind, K., *et al.* 2009, *A&A*, 503, 545
Maeder, A. & Meynet, G. 2006, *A&A*, 448, L37
Marino, A. F., *et al.* 2014, *MNRAS*, 437, 1609
Marino, A. F., *et al.* 2015, *MNRAS*, 450, 815
Martell, S. L., Molinski, J. P., Beers, T. C., & Grebel, E. K. 2011, *A&A*, 354, A136
Milone, A. P., Marino, A. F., Piotto, G. P., *et al.* 2013, *ApJ*, 767, 120
Milone, A. P., Marino, A. F., Piotto, G. P., *et al.* 2015, *ApJ*, 808, 51
Piotto, G., *et al.* 2002, *ApJ*, 670, 39
Prantzos, N. & Charbonnel, C. 2006, *A&A*, 458, 135
Prantzos, N., Charbonnel, C., & Iliadis, C. 2007, *A&A*, 470, 179
Renzini, A., *et al.* 2015, *MNRAS*, 454, 4197
Sbordone, L., Salaris, M., Weiss, A., & Cassisi, S. 2011, *A&A*, 534, A9
Schaerer, D. & Charbonnel, C. 2011, *MNRAS*, 413, 2297
Sills, A. & Glebbeek, E. 2010, *MNRAS* 407, 277
Simmerer, J., *et al.* 2013, *ApJ* (Letters), 764, L7
Smith, G. H. & Norris, J. 1982, *ApJ*, 254, 594
Ventura, P., *et al.* 2001, *ApJ* (Letter), 550, L65
Ventura, P., *et al.* 2013, *MNRAS*, 431, 3642

Formation, evolution, and survival of massive star clusters
Proceedings IAU Symposium No. 316, 2015
C. Charbonnel & A. Nota, eds.

doi:10.1017/S1743921316007006

Clusters of high-mass protostars: From extreme clouds to mini-bursts of star formation

Frédérique Motte[1,2], Fabien Louvet[3,4] and Quang Nguyễn Lương[5,6]

[1]Institut de Planétologie et d'Astrophysique de Grenoble (IPAG),
Univ. Grenoble Alpes - CNRS/INSU, BP 53, 38041 Grenoble Cedex 9, France
email: frederique.motte@ujf-grenoble.fr

[2]Laboratoire AIM Paris-Saclay, CEA/IRFU - CNRS/INSU - Univ. Paris Diderot,
Service d'Astrophysique, Bât. 709, CEA-Saclay, F-91191 Gif-sur-Yvette Cedex, France

[3]Laboratoire franco-chilien d'astronomie (UMI-FCA), CNRS/INSU, France

[4]Departamento de Astronomía, Universidad de Chile,
Camino El Observatorio #1515, Las Condes, Santiago, Chile

[5]National Astronomical Observatory of Japan, Chile Observatory,
2-21-1 Osawa, Mitaka, Tokyo 181-8588, Japan

[6]Canadian Institute for Theoretical Astrophysics, University of Toronto,
60 St. George Street, Toronto, ON M5S 3H8, Canada

Abstract. *Herschel* revealed high-density cloud filaments of several pc^3, which are forming clusters of OB-type stars. Counting *Herschel* protostars gives a direct measure of the mass of stars forming in a period of $\sim 10^5$ yrs, the "instantaneous" star formation activity. Given their activity, these so-called mini-starburst cloud ridges could be seen as "miniature and instant models" of starburst galaxies. Their characteristics could shed light on the origin of massive clusters.

Keywords. stars: formation, ISM: structure, star clusters, starburst

1. Introduction

Despite their low numbers, massive stars (OB-type, $>10\ M_\odot$ on the main sequence) produce much of the luminosity of galaxies and are the main driver of their evolution. Both their large impact on their environment and how their formation is linked to natal clouds remain poorly understood (see reviews by Elmegreen 2011; Krumholz 2014).

High-mass star formation scenarios currently undergo a change of paradigm, in which this process is no longer quasi-static but simultaneously evolves with both cloud and cluster formation. Theoretically, OB stars could either form through (1) a powerful accretion driven by a high degree of turbulence (e.g., McKee & Tan 2002; Hosokawa & Omukai 2009) or (2) colliding flows initiated by competitive accretion or cloud formation (e.g., Bonnell & Bate 2006; Hartmann *et al.* 2012). From an observational point of view, gravitational streamers and shearing motions have been reported from the cloud to the protostellar scales in a few high-mass star-forming regions (on $10-0.1$ pc scales, Schneider *et al.* 2010; Csengeri *et al.* 2011a). We hereafter show that these pioneering studies are now comforted by sensitive *Herschel* images, cloud kinematic studies, and (sub)millimeter interferometric images, favoring the second family of models.

"HOBYS, the *Herschel* imaging survey of OB Young Stellar objects" is a key program, which is exclusively dedicated to high-mass star formation (Motte, Zavago, Bontemps *et al.* 2010; see http://hobys-herschel.cea.fr). It uses the SPIRE and PACS cameras of the

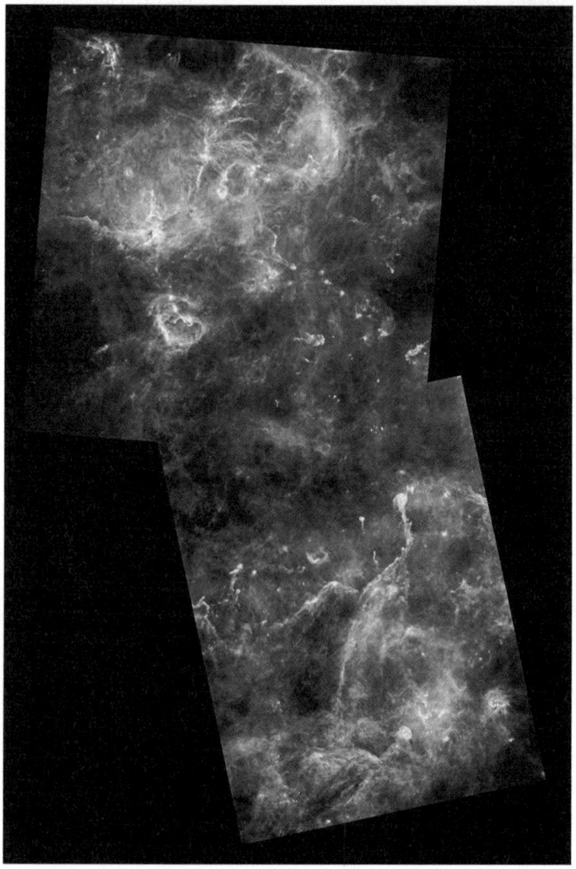

Figure 1. Composite 3-color *Herschel* image of the Cygnus X molecular complex with red $= 250\,\mu$m, green $= 160\,\mu$m, and blue $= 70\,\mu$m (taken from Hennemann *et al.* 2012 and Schneider *et al.* in prep.). The mosaic performed by the HOBYS key program approximately covers a $5^\circ \times 2.5^\circ$ or $120\,\mathrm{pc} \times 60\,\mathrm{pc}$ area. The diffuse blue emission at the center is an H II region powered by the massive Cyg OB2 cluster while earlier stage star-forming sites are seen as red filaments.

Herschel spatial observatory to image essentially all of the regions forming OB-type stars at distances less than 3 kpc from the Sun ($d = 0.7 - 3.2$ kpc). HOBYS images revealed networks of filaments and clusters of burgeoning YSOs (see Fig. 1). Among the HOBYS highlights is the discovery of "mini-starburst ridges", defined as high-density dominating filaments supersonically contracting (see Sect. 2) and efficiently forming clusters of high-mass stars (see Sect. 3). Could these Galactic extreme cloud and star formation events help constrain the physical process at the origin of the formation of extreme extragalactic clusters? To start answering, we will list in Sect. 4 the physical processes acting within ridges and which qualitatively differ from the one advocated to form more classical low-mass star clusters.

2. Ridges & hubs: Hyper-massive clumps forming clusters of high-mass stars

2.1. *Discovery of hyper-massive clumps*

One of the most unexpected results of the *Herschel* mission was in revealing the importance and ubiquity of interstellar filaments (e.g., André *et al.* 2010; Hill *et al.* 2011).

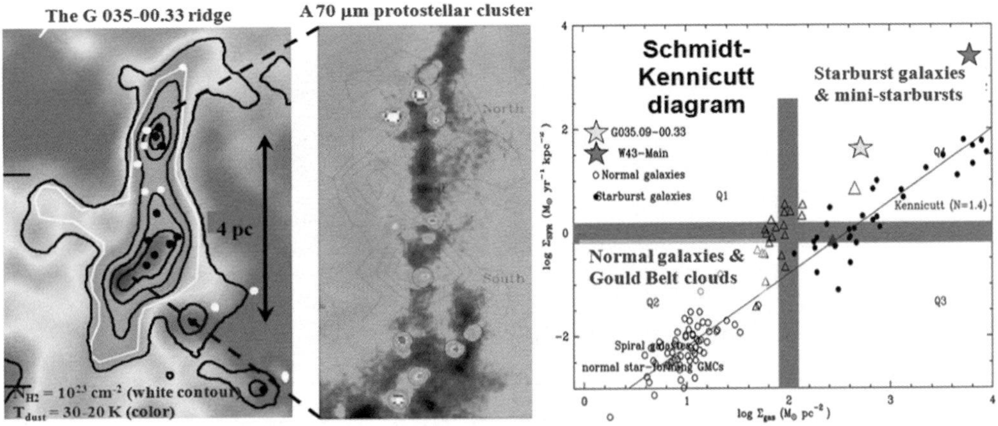

Figure 2. The G035-00.33 ridge and its cluster of high-mass protostars (Left and Center). Ridges are high-density clumps with large star formation activity (Right, see yellow and red star markers). On a Schmidt-Kennicutt diagram, they qualify as mini-starburst regions. Figures extracted from Nguyen Luong *et al.* (2011a) and (in prep.).

They were known from molecular line surveys but could, for the first time, be 1) quantitatively studied thanks to sensitive column density and temperature images and 2) directly compared with the protostellar population (see, e.g., Figs. 2 Left-Center). The HOBYS group has developed a definite expertise in building clean and reliable column density maps, in identifying and analyzing cloud filaments, in properly extracting protostars from their surroundings and characterizing them (e.g., Hill *et al.* 2012; Hennemann *et al.* 2012; Men'shchikov *et al.* 2012; Tigé *et al.* 2016). Within the high-mass molecular complexes imaged by HOBYS, we have discovered high-density dominating elongated clumps, which are the preferential sites to form massive stars and which we quote as ridges (see Figs. 2 Left-Center). The existence of ridges is predicted by dynamical models of cloud formation such as colliding flow simulations (e.g., Heitsch & Hartmann 2008; Federrath *et al.* 2010). The densest ridges should coincide with the precursors of "young massive clusters" that will become bound stellar clusters (e.g., Ginsburg *et al.* 2012).

In more details, Hill *et al.* (2011) studied the Vela C molecular cloud complex and defined the central ~3 pc-long filament with column density above 10^{23} cm^{-2} as a ridge. They showed that the Vela C ridge represents a gravitational well that is dominating and shaping its surroundings, in marked contrast to other more typical filaments (e.g., Arzoumanian *et al.* 2011). In parallel, Hennemann *et al.* (2012) analyzed the structure of the DR21 ridge (see Fig. 3 Left), previously identified by us as hosting a cluster of ~20 high-mass protostars (Motte *et al.* 2007; Bontemps *et al.* 2010). In agreement with the kinematical study and simulations of Schneider *et al.* (2010, see Fig. 3 Right), Hennemann *et al.* (2012) proposed the DR21 ridge is fed by and formed through the merging of several sub-filaments already forming stars. A handful of other ~10^4 $M_\odot$ ridges were recently identified in *Herschel*/HOBYS and Hi-GAL images (Nguyen Luong *et al.* 2013; Hill *et al.* in prep.).

More spherical and less massive, ~10^3 $M_\odot$, clumps were also found to form stellar clusters at the junction of filaments (Schneider *et al.* 2012; Didelon *et al.* 2015). These clumps recall the so-called "hubs" discovered in a few extreme infrared dark clouds (e.g., Peretto *et al.* 2013) and predicted by numerical simulations (e.g., Dale & Bonnell 2011). We propose to generalize the concept of ridges and hubs into that of "hyper-massive

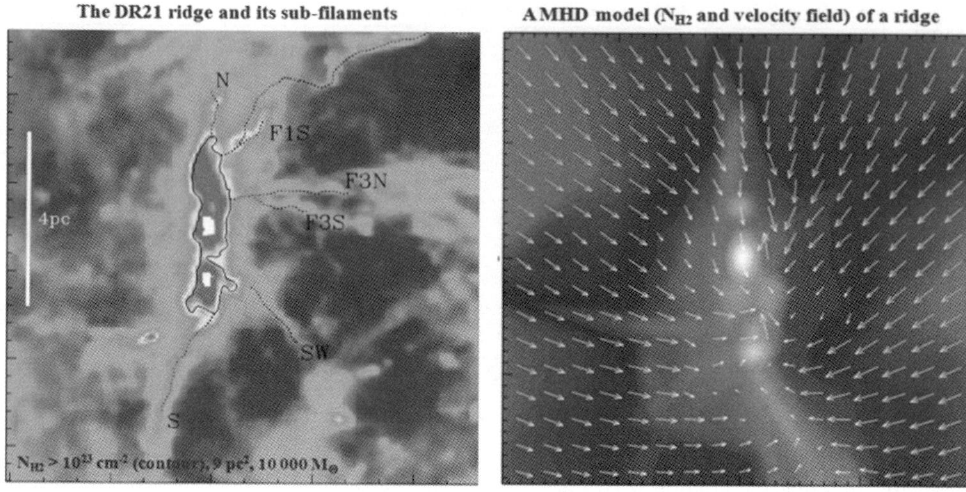

Figure 3. The DR21 ridge and its feeding sub-filaments network (Left), compared to numerical MHD simulations made by P. Hennebelle (Right). Figures are extracted from Hennemann *et al.* (2012) and Schneider *et al.* (2010).

clumps", which could either be spherical or filamentary and either form a couple or a rich cluster of high-mass stars.

2.2. *Dynamical processes at the origin of hyper-massive clouds*

Based on *Herschel* and kinematic studies (Schneider *et al.* 2010; Hill *et al.* 2011; Hennemann *et al.* 2012; Nguyen Luong *et al.* 2011a, 2013), we proposed that most ridges/hubs could have been formed by dynamical scenarios such as converging flows (see Figs. 3 Right). Minier *et al.* (2013) themselves used numerical simulations to show that the ionization of a very nearby H II region/OB cluster could be the main process responsible for shaping some ridges like Vela C (see his Fig. 5). In any case, whatever the origin of the additional pressure, arising from colliding flows or ionization, it supersedes the thermal and micro-turbulence pressure. Those dynamical scenarios for cloud and then subsequent star formation are part of the second family of models shortly described in Sect. 1. They are also consistent with the observational lack of high-mass pre-stellar cores (e.g., Motte *et al.* 2007, 2010; Russeil *et al.* 2010; Tigé *et al.* 2016).

In parallel to these *Herschel* structural studies, we have investigated the kinematics of three ridges identified as such from HOBYS data (Hennemann *et al.* 2012; Nguyen Luong *et al.* 2013). A long-term survey of Cygnus X and an IRAM large program on W43 (W43-HERO, Motte, Schilke *et al.*, see http://www. iram-institute.org/EN/content-page-292-7-158-240-292-0.html) has shown that the DR21, W43-MM1, and W43-MM2 ridges are all undergoing global collapse on the ridge scale with supersonic inward velocities ($1-2$ km s^{-1} over $1-10$ pc^2; Schneider *et al.* 2010; Motte *et al.* in prep.). Such infall motions, sometimes called gravitational focusing, are expected both in colliding flows and ionization compression models (e.g., Vázquez-Semadeni *et al.* 2005; Tremblin *et al.* 2012). We started studying the inner kinematic structure of these ridges, which suggests they are braids/bundles of filaments or gas layers (Louvet *et al.* in prep.; Bontemps *et al.* in prep.). We have also studied the shocks associated with gas shears created by sub-filaments braiding, using SiO imaging and detailed shock modeling (Nguyen Luong *et al.* 2013; Duarte-Cabral *et al.* 2014; Louvet *et al.* 2016). Such extreme kinematics is observed by other groups at the clump scale (e.g. Galván-Madrid *et al.* 2010; Henshaw *et al.* 2014)

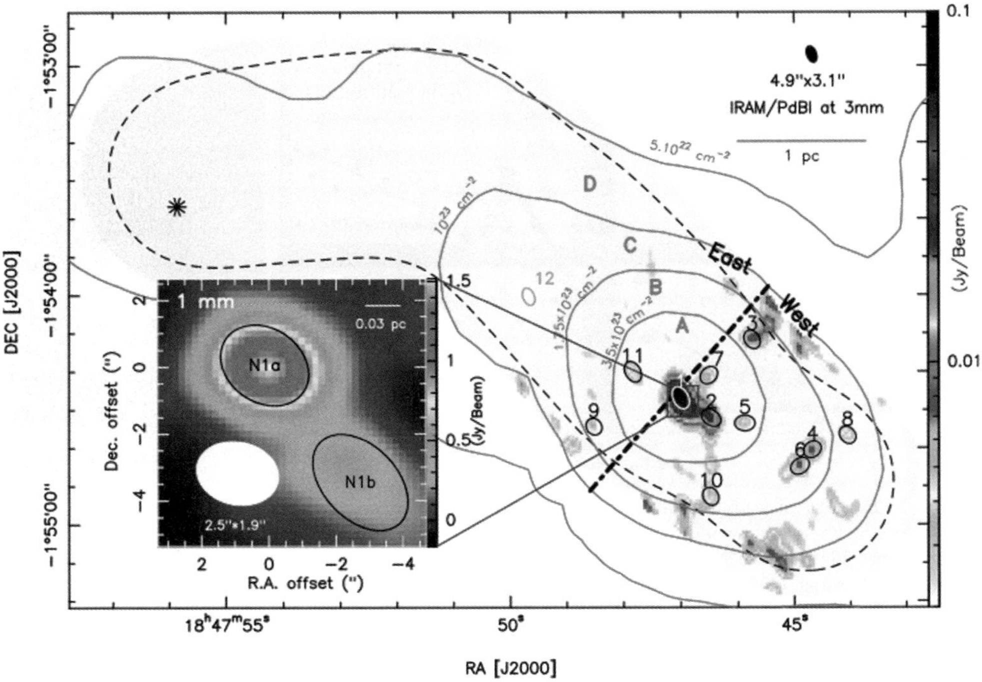

Figure 4. Interferometric mosaic obtained at the IRAM Plateau de Bure toward the W43-MM1 ridge (Louvet *et al.* 2014). A cluster of high-mass protostars, pinpointed by black and white ellipses, forms within the ridge outlined by the 10^{23} cm^{-2} red contour. Clustering properties are evolving through regions A-D, and are different in the eastern and western parts of the ridge.

and is reminiscent to gas flows and shocks observed at the protostellar scale by Csengeri *et al.* (2011a-b).

3. Mini-starburst activity within ridges

The extreme characteristics of ridges and hubs, in terms of density and kinematics (see Sect. 2), could lead to an atypical star formation activity. Star formation efficiency (SFE) is indeed predicted to continuously increase with gas density (e.g., Hennebelle & Chabrier). Analytical theories in fact argue for the star formation rates (SFRs) to be measured as a function of the free-fall time, which itself directly depends on the cloud density (e.g., Krumholz & McKee 2005; Padoan & Nordlund 2011). The clear accumulation of massive protostars we observed along ridges (see, e.g., Fig. 2 Center) traces the present star formation activity (Sect. 3.1), which we will show is intense (Sect. 3.2; Nguyen Luong 2011a; Louvet *et al.* 2014).

3.1. *Various SFR estimates, tracing the history of star formation*

Spitzer/IRAC legacies have measured SFRs in a direct and indirect way by either counting pre-main sequence stars (also called T Tauris) or integrating the diffuse mid-infrared PAH emission attributed to the luminosity impact of recently formed OB stars on the cloud. The first method was successfully applied by the c2d survey of nearby (<500 pc) low-mass star-forming regions (e.g., Heiderman *et al.* 2010). The second method was used when the angular resolution was not sufficient for counting purposes, for example in galaxies (e.g., Wu *et al.* 2005) and more recently for Galactic molecular complexes (Nguyen Luong *et al.* 2011b; see also Eden *et al.* 2012 using *Herschel* 70 μm). If the star

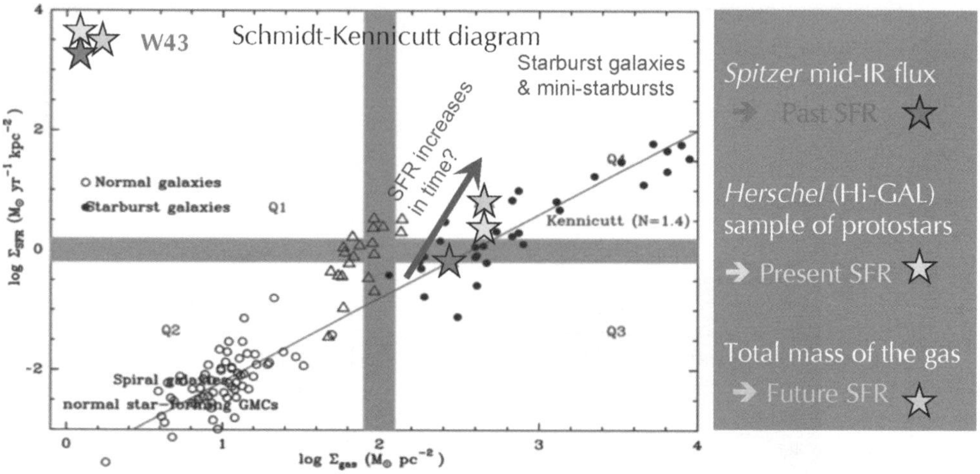

Figure 5. History of star formation in the W43 molecular complex traced by the SFR estimates made from *Spitzer*, *Herschel*, and molecular line surveys. Measures are extracted from Nguyen Luong *et al.* (2011b) and Nguyen Luong *et al.* (in prep.). The Schmidt-Kennicutt relation which is followed by most GMCs and galaxies (examples displayed by filled and open circles) is shown by a black line. The present-day and future SFRs suggest an increase of the SF activity in W43.

formation activity varies with time, these two methods, based on $\sim 10^6$ years-old T Tauri and OB stars, would measure "past integrated" SFRs.

In contrast, counting $\sim 10^5$ years-old protostars allows to evaluate the "present-day instantaneous" star formation activity, meaning the star formation developing for one single free-fall time of the direct progenitors of single stars. We first applied this method to W43-main, a region imaged by ground-based submillimeter radio-telescopes (Motte *et al.* 2003). Members of the HOBYS consortium now use *Herschel* and interferometric images, which more properly identify massive dense cores and their embedded protostars (see Fig. 2 Center and Fig. 4).

If one assumes a constant $1-3\%$ SFE and a $\sim 10^6$ cloud lifetime, the total gas mass of clouds provides a good estimate for their "future integrated" star formation activity. Using SFR values estimated this way in a Schmidt-Kenicutt diagram is meaningless since it will only reflect our assumptions on SFE and cloud lifetime. However, comparing SFRs estimated through the three methods above can give crucial insights into the history of cloud and star formation. As a matter of fact, an increase of the star formation activity has been suggested, for the W43 molecular cloud complex, when using *Spitzer*, *Herschel*, and cloud mass measurements (see Fig. 5, Nguyen Luong *et al.* 2011b, in prep.). This SFR rise could be the consequence of the fact that clouds are currently forming, rather efficiently, within this complex (Motte *et al.* 2014).

3.2. *Galactic mini-starburst events*

On cloud and ridge spatial scales ($1-10$ pc), we measured the present-day instantaneous SFR of the W43-main cloud as well as the G035.39-00.33 and W43-MM1 ridges (Motte *et al.* 2003; Nguyen Luong 2011a; Louvet *et al.* 2014). These cloud structures are the preferential sites for the formation of high-mass star clusters in their host complex (see, e.g., Figs. 2) and W43-MM1 is the most extreme clump of the W43-main cloud. They all have star formation rate densities, $\Sigma_{\rm SFR} \sim 10-100\ M_\odot\,{\rm yr}^{-1}\,{\rm kpc}^{-2}$, on $1-10\ {\rm pc}^2$ areas, worthy of starburst galaxies, usually defined by $\Sigma_{\rm SFR} > 1$ (see Fig. 2 Right). For this reason, W43-main, G035.39-00.33, and W43-MM1 were called mini-starburst regions, i.e. miniature and instantaneous models of starburst galaxies.

These mini-starburst events most probably follow the formation of the ridges, which were proposed to develop through colliding flows (see Sect. 2). Indeed, short mini-bursts of star formation are to be expected after a fast episode of cloud formation (e.g., Vázquez-Semadeni *et al.* 2008) or, equivalently, for a cloud under compressive turbulent forcing (Federrath & Klessen 2012). In this scenario the star formation should gradually settle within ridges. We should thus measure very different SFR levels depending on the evolutionary status of the ridge. To investigate this statement, one must calculate present-day instantaneous SFRs, whose timescale, over which the SFR is integrated, remains much shorter than the ridge formation timescale ($\sim 10^5$ yr vs. $\sim 10^6$ yr). Louvet *et al.* (2014) measured instantaneous SFRs within regions A-D of the W43-MM1 ridge (see Fig. 4) and found a clear correlation of SFR with cloud density or, equivalently, SFR with the evolution of cloud concentration. Moreover, SFR is smaller within the eastern part of the ridge, where shocks associated with more recent cloud formation are the strongest (Louvet *et al.* 2014, 2016). Therefore, the dynamical ridge formation may well be followed by series of intense bursts of star formation.

4. Conclusion

While star cluster properties, among which the initial mass function of stars (IMF, see, e.g., Kroupa 2001) seem universal, our present review suggests that stellar clusters formed in extreme clouds like ridges could have different properties. Indeed, if low-mass stars should accrete their final mass from well-defined pre-stellar cores with a mass distribution mimicking the IMF (Motte *et al.* 1998, Konyves *et al.* 2015), (high-mass) stars within ridges most probably accrete gas from their parental clumps, through sporadic gas flows. Moreover, the star formation activity within ridges may not be constant over 10^6 yrs but present bursts favoring the formation of high-mass stars.

High-resolution studies will be necessary to investigate the core mass function and angular momentum distribution within the protocluster ridges and define if the global properties of the resulting massive star cluster will be classical or not.

References

André, Ph., Men'shchikov, A., Bontemps, *et al.* 2010, *A&A*, 518, L102+
Arzoumanian, D., André, Ph., Didelon, P., *et al.* 2011, *A&A*, 529, L6
Bonnell, I. A. & Bate, M. R. 2006, *MNRAS*, 370, 488
Bontemps, S., Motte, F., Csengeri, T., & Schneider, N. 2010, *A&A*, 524, A18
Csengeri, T., Bontemps, S., Schneider, N., Motte, F., & Dib, S. 2011a, *A&A*, 527, A135 12, 24
Csengeri, T., Bontemps, S., Schneider, N., *et al.* 2011b, *ApJ*, 740, L5
Dale, J. E. & Bonnell, I. 2011, *MNRAS*, 414, 321
Didelon, P., Motte, F., Tremblin, P. *et al.* 2015, *A&A*, in press.
Duarte-Cabral, A., Bontemps, S., Motte, F., *et al.* 2014, *A&A*, 570, A1
Eden, D. J., Moore, T. J. T., Plume, R., & Morgan, L. K. 2012, *MNRAS*, 422, 3178
Elmegreen, B. 2011, in *Star Formation in the Local Universe*, EAS Pub. Series, 51, 45
Federrath, C. & Klessen, R. S. 2012, *ApJ*, 761, 156
Federrath, C., Roman-Duval, J., Klessen, R. S., Schmidt, W., & Mac Low, M.-M. 2010, *A&A*, 512, A81
Galván-Madrid, R., Zhang, Q., Keto, E., *et al.* 2010, *ApJ*, 725, 17
Ginsburg, A., Bressert, E., Bally, J., & Battersby, C. 2012, *ApJ*, 758, L29
Hartmann, L., Ballesteros-Paredes, J., & Heitsch, F. 2012, *MNRAS*, 420, 1457

Heiderman, A., Evans, II, N. J., Allen, L. E., Huard, T., & Heyer, M. 2010, *ApJ*, 723, 1019
Heitsch, F. & Hartmann, L. 2008, *ApJ*, 689, 290
Hennebelle, P. & Chabrier, G. 2011, *ApJ*, 743, L29
Hennemann, M., Motte, F., Schneider, N., *et al.* 2012, *A&A*, 543, L3
Henshaw, J. D., Caselli, P., Fontani, F., Jimez-Serra, I., & Tan, J. C. 2014, *MNRAS*, 440, 2860
Hill, T., Motte, F., Didelon, P., *et al.* 2011, *A&A*, 533, 9
Hill, T., Motte, F., Didelon, P., *et al.* 2012, *A&A*, 542, A114
Hosokawa, T. & Omukai, K. 2009, *ApJ*, 691, 823
Könyves, V., André, Ph., Men'shchikov, A., *et al.* 2015, *A&A*, in press.
Kroupa, P. 2001, *MNRAS*, 322, 231
Krumholz, M. 2014, *Physics Reports*, 539, Iss. 2, 49-134
Krumholz, M. R. & McKee, C. F. 2005, *ApJ*, 630, 250
Louvet, F., Motte, F., Hennebelle, P., *et al.* 2014, *A&A*, 570, A15+
Louvet, F., Motte, F., Gusdorf, A., *et al.* 2016, soon to be submitted to *A&A*
McKee, C. F. & Tan, J. C. 2002, *Nature*, 416, 59
Menshchikov, A., André, P., Didelon, P., *et al.* 2012, *A&A*, 542, A81
Minier, V., Tremblin, P., Hill, T., *et al.* 2013, *A&A*, 550, A50
Motte, F., André, Ph., & Neri, R. 1998, *A&A*, 336, 150
Motte, F., Bontemps, S., Schilke, P., *et al.* 2007, *A&A*, 476, 1243+
Motte, F., Nguyen Luong, Q., Schneider, N., *et al.* 2014, *A&A*, 571, A32
Motte, F., Schilke, P., & Lis, D. C. 2003, *ApJ*, 582, 277
Motte, F., Zavagno, A., Bontemps, S., *et al.* 2010, *A&A*, 518, L77+
Nguyen Luong, Q., Motte, F., Carlhoff, P., *et al.* 2013, *ApJ*, 775, 88
Nguyen Luong, Q., Motte, F., Hennemann, M., *et al.* 2011a, *A&A*, 535, A76
Nguyen Luong, Q., Motte, F., Schuller, F., *et al.* 2011b, *A&A*, 529, A41+
Padoan, P. & Nordlund, A. 2011, *ApJ*, 730, 40
Peretto, N., Fuller, G. A., Duarte-Cabral, A., *et al.* 2013, *A&A*, 555, A112
Russeil, D., Zavagno, A., Motte, F., *et al.* 2010, *A&A*, 515, A55+
Schneider, N., Csengeri, T., Bontemps, S., *et al.* 2010a, *A&A*, 520, A49+
Schneider, N., Csengeri, T., Hennemann, M., *et al.* 2012, *A&A*, 540, L11
Tigé, J., Motte, F., Russeil, *et al.* 2016, soon to be submitted to *A&A*
Tremblin, P., Audit, E., Minier, V., Schmidt, W., & Schneider, N. 2012, *A&A*, 546, A33 100
Vázquez-Semadeni, E., Kim, J., & Ballesteros-Paredes, J. 2005, *ApJ*, 630, L49
Vázquez-Semadeni, E., González, R. F., Ballesteros-Paredes, J., Gazol, A., & Kim, J. 2008, *MNRAS*, 390, 769
Wu, H., Cao, C., Hao, C., *et al.* 2005, *ApJ*, 632, L79

Formation, evolution, and survival of massive star clusters
Proceedings IAU Symposium No. 316, 2015
C. Charbonnel & A. Nota, eds.

doi:10.1017/S1743921315010571

The Imprints Of Galactic Environment On Cluster Formation and Evolution

Angela Adamo

Department of Astronomy, Oskar Klein Centre, Stockholm University, AlbaNova University Centre, SE-106 91 Stockholm, Sweden
email: adamo@astro.su.se

Abstract. Young star clusters (YSCs) appear to be a ubiquitous product of star formation in local galaxies, thus, they can be used to study the star formation process at work in their host galaxies. Moreover, YSCs are intrinsically brighter that single stars, potentially becoming the most important tracers of the recent star formation history in galaxies in the local Universe. In local galaxies, we also witness the presence of a large population of evolved star clusters, commonly called globular clusters (GCs). GCs peak formation history is very close to the redshift ($z \sim 2$) when the cosmic star formation history reached the maximum. Therefore, GCs are usually associated to extreme star formation episodes in high-redshift galaxies. It is yet not clear whether YSCs and GCs share a similar formation process (same physics under different interstellar medium conditions) and evolution process, and whether the former can be used as progenitor analogs of the latter. In this invited contribution, I review general properties of YSC populations in local galaxies. I will summarise some of the current open questions in the field, with particular emphasis to whether or not galactic environments, where YSCs form, leave imprints on the nested populations. The importance of this rapidly developing field can be crucial in understanding GC formation and possibly the galactic environment condition where this ancient population formed.

Keywords. galaxies: star clusters; galaxies: spiral; galaxies: starburst

1. Introduction

YSCs are usually identified as gravitationally bound stellar structures, with radii between 0.5 up to several parsecs and masses between 10^3 and 10^7 $M_\odot$(Portegies Zwart *et al.* 2010). Thanks to the HST spatial resolution and multiband coverage the YSC field has been continuously advancing.

YSCs are easily observed in local galaxies which are able to sustain star formation. This is not always true at the lowest star-forming regimes. For example, in dwarf galaxies, star and cluster formation appears dominated by stochastic processes (Cook *et al.* 2012). The range of star formation rate (SFR) and star formation rate density (Σ_{SFR}) of dwarf galaxies hosting one or few YSCs or none overlaps significantly (e.g. Adamo & Bastian 2015). On the other hand, some local dwarf galaxies are able to form very massive YSCs, in some cases even more massive than usually observed in spiral galaxies like the Milky Way (e.g. Billett *et al.* 2002). This way of forming massive YSCs has been associated to the possibility that dwarf galaxies at high z are the places where low metallicity GCs have formed (Elmegreen *et al.* 2012). However, whether cluster formation in dwarf galaxies is dominated by stochastic events or properties of the galactic environment is still an open question.

YSC formation in spiral galaxies is a continuos process, which reflects the fact that star formation has proceeded with a constant pace for at least several hundreds of Myr. This equilibrium is the result of a regulation process between gas feeding the disk system, the dynamics, and the feedback produced by star formation. Typically spiral arms are the preferential places where star formation takes place in disk galaxies. Similarly to "traffic jams", i.e. regions where rotation speeds slows down, giant molecular clouds (GMCs) entering the spiral arm region can finally grow in mass. Once GMCs move out of the "jam" they will fragment and collapse forming stars and star clusters (e.g. Elmegreen 2011, Dobbs 2014). However, if spiral arms are visually the most dominating features in disk galaxies, dynamically, the presence of bars can also impact star formation often leading to the formation of star-forming circumnuclear rings. In galaxies like the Milky Way, the end of the bar and the central circumnuclear ring are preferential places where massive stars and YSCs form (e.g. Davies *et al.* 2012, Stolte 2013). It is also important to keep in mind that the presence of a large amount of molecular gas is not always a guarantee of constantly efficient star and cluster formation. For example, the central molecular zone of our Milky Way is likely to alternate episodic starburst events to phases of inefficient star formation (e.g. Krumholz & Kruijssen 2015).

Finally, a preferential channel to form numerous and massive YSC populations is during starburst episodes. During a starburst the SFR in the galaxy increases to levels which are impossible to sustain for a Hubble time. In the local Universe, starbursts are usually associated to interaction and merging processes between galaxies, with at least one of the systems involved being gas rich. This is not necessarily true in dwarf galaxies, where star formation appear to be dominated by episodic and, sometimes, extremely efficient star formation processes. This is the case of nearby dwarf starburst galaxies like NGC1705 (Annibali *et al.* 2009); NGC 4449 (Annibali *et al.* 2011), NGC5253 (Calzetti *et al.* 2015b), among many others. Numerous YSC populations, with cluster masses reaching up to a few time 10^7 $M_\odot$ have been easily found in merging systems like NGC7252 and NGC1316 (Bastian *et al.* 2006), and ongoing mergers like the Antennae (Whitmore *et al.* 2010).

At this point, the reader may ask the question whether cluster populations forming in different galaxies share any similarity at all. Another valid question could be related to whether, given a YSC population with certain properties, it is possible to reconstruct the galactic environment where these clusters are forming. As I will summarise in this review, YSC populations forming in the nearby Universe share a large spectrum of statistical properties. There is an underlying universality in the properties that can describe YSC populations in local galaxies. However, several studies, in the last decay, also show that the cluster population of each galaxy carry some imprints that are tightly correlated to the physical conditions of the gas in their host galaxies. Potentially, YSC properties become a key to understand both how star formation proceeds and the undergone formation history of the host galaxy.

2. YSC mass and luminosity function

After almost 3 decay of HST based studies of YSC populations in local galaxies it is fairly accepted that the YSC luminosity function can be described by a power law function, $dN \propto L^{\alpha}$, with $\alpha \sim -2$. This distribution is the result of an underlying power-law distribution of the YSC masses with a similar index. Similar slopes have been observed for the luminosity distribution of increasing stellar aggregate boundaries (Elmegreen *et al.* 2006), confirming that cluster formation is well nested inside the hierarchical star formation process driven by turbulence.

However, this picture is not complete. In the literature, a quick survey of the recovered power-law indexes as a function of the luminosity range above which clusters luminosity are distributed can easily show that the observed α values variate from -1.6 to -2.8 (see Figure 12 by Annibali *et al.* 2011). In particular the steepening of the slopes seams to

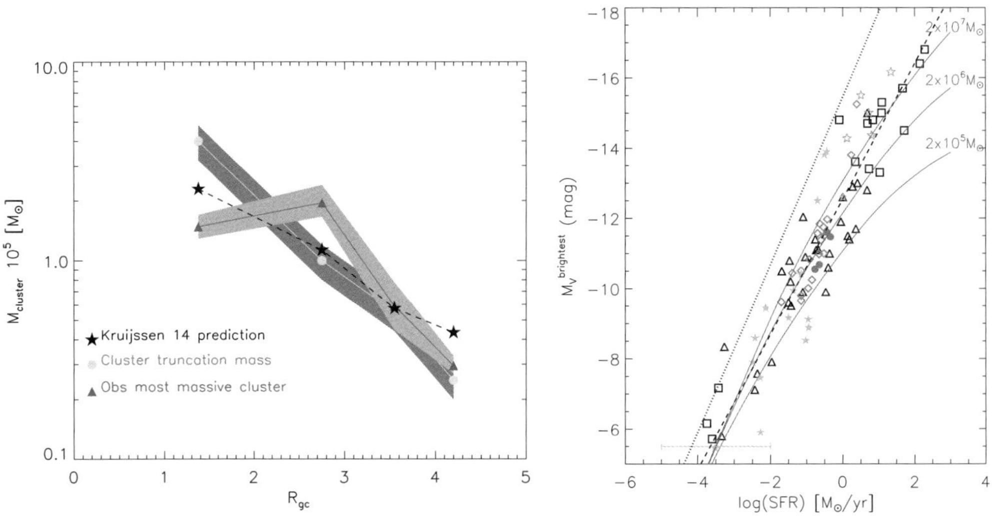

Figure 1. Left: The observed mass of the most massive cluster (red tringles), the recovered $M_\star$ (cyan dots), and the estimated maximum cluster mass (black stars) using models by Kruijssen (2014) found in annuli of increasing galactocentric distances (kpc) in M83. The shadowed regions show the uncertainties. **Right**: The SFR vs. M_V^{brigh} relation. A full description of the compiled data is accessible in Adamo *et al.* (2015). The dashed line is the fit produced by Larsen(2002) to the spiral sample showed with triangle symbols. The green horizontal bar is the range covered in SFR by dwarf galaxies which do not host any bound YSC formed in the last 100 Myr Cook *et al.* 2012. The dotted line shows the region where galaxies would have fallen if 100% of their star formation is happening in YSCs. The observed trend is explained if about 8% of the SFR in galaxy is happening in bound YSCs Bastian 2008. Red solid lines are the regions where galaxies should fall if their mass functions are Schechter functions with increasing $M_\star$ (see values in the insets of the plot). Figures from Adamo *et al.* (2015).

increase at the brightest luminosity bins (Gieles 2010), suggesting that we observe less luminous clusters than expected from a power-law function of index −2.

Another way to address this problem is to look at the underlying YSC mass functions. Does the observed YSC mass function shows a dearth of massive star clusters? Or in other words, is the YSC mass function a power-law with no upper-mass limits or does it have some smooth truncation? The observed steepening of the YSC luminosity function is easily reproducible if the underlying YSC mass function has the shape of a Schechter function, i.e. a power-law function with index −2 and an exponential decline above a certain truncation mass, $M_\star$ (Gieles *et al.* 2006). Using statistical arguments, Larsen(2009) noticed that if the YSC mass function of the Milky Way (as prototype of local spiral galaxies) would be a pure power-law function with no upper limits, then our own Galaxy should have been able to form in the last few Gyr clusters with masses $\sim 10^6$ $M_\odot$. A Schechter mass function with $M_\star \sim 2\times10^5$ $M_\odot$ can easily explain the dearth of such massive YSCs in local spirals. The presence of a truncation is not in contrast with the very massive YSCs found in starburst galaxies. A fit to the YSC mass function of the Antennae shows that a Schechter function with $M_\star \sim 10^6$ $M_\odot$ can reproduce the observed distribution of cluster masses. Therefore, the truncation mass doesn't appear to be a universal value but it is more likely linked to the galactic environment where the cluster population is forming (Larsen 2009).

A similar conclusion has been reached studying the YSC population in the disk of the nearby spiral galaxy M83 (Adamo *et al.* 2015). The YSC population of M83 is better reproduced by a Schechter mass function with $M_\star \sim 2\times10^5$ $M_\odot$. A Schechter function

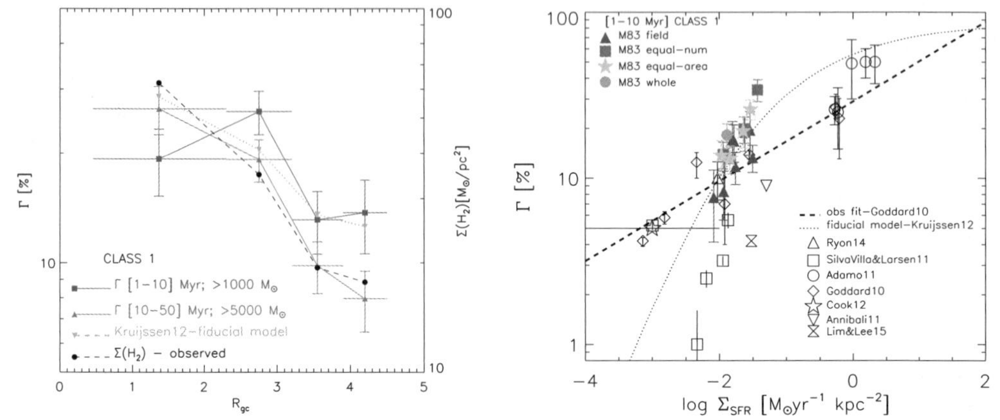

Figure 2. Left: Observed cluster formation efficiency recovered in bins of increasing galactocentric distance. The values have been derived for two different cluster age ranges and SFR indicators (red squares and blue triangles, see inset in the panel). The yellow upside-down triangles show the prediction made with the fiducial model proposed by Kruijssen(2012). The black dots show the observed average gas surface density (unit scales is showed in the right axes) estimated in the same annuli. **Right**: The Γ versus $\Sigma_{\rm SFR}$ relation. The plot includes all the data available in the literature to date (see inset). Figures from Adamo *et al.* (2015).

is also favoured when analysing the mass function of the M83 YSC population as a function of galactocentric distances. The mass of the most massive cluster found in each bin oscillates around the recovered $M_\star$ value, well in agreement with the probabilistic meaning of a truncation mass in the framework of the Schechter function. The change in the $M_\star$ is possibly caused by the change in the ISM physical properties (e.g. gas pressure, density) but further investigation is required.

While our understanding of the driving physical parameter(s) that causes $M_\star$ to change as a function of galactic environment is not well constrained yet, further evidence coming from studies of the mass functions of globular cluster populations reinforce these findings. Jordán *et al.* (2007) analysed the mass function of the GC populations detected in roughly 90 galaxy members of the Virgo cluster. They report that an evolved Schechter function, parametrised to take into account the mass loss of the population, produces a better fit to the observed mass distributions. In particular, the recovered $M_\star$ declines as a function of the total B band luminosity of the galaxy (used as a proxy of the total stellar mass of the galaxy). The decline cannot be explained by the dynamical effects, like friction, and it is probably linked to the properties of the galaxy at the time of the GC formation.

The similarity in the observed properties of YSC and GC population mass functions, although on significantly different mass scales, possibly suggests that the two populations may share the same formation mechanisms under likely different ISM conditions.

3. YSC scaling relations

The YSC scaling relations are important tools to constrain and probe the cluster formation process and how cluster population properties are linked to their galactic environments.

The right panel of Fig.1 shows one of the most well-known relation pointed out in YSC studies, the galactic SFR versus brightest cluster in V-band (SFR vs. $M_V^{\rm brigh}$), in one of the most recent and complete compilation (by Adamo *et al.* 2015). This relation was originally formulated by Larsen (2002). A similar scaling relation has been found if the number of YSCs per galaxy is used instead of the SFR (Whitmore 2000). The ($\log N$) or SFR vs. $M_V^{\rm brigh}$ relation tells us that cluster formation is primarily a stochastic process, and size-of-sample drives the likelihood of forming very massive clusters. Higher is the SFR in the host galaxy and more numerous clusters are formed, therefore, it is more likely to sample the YSC mass function at the high mass bins and form massive (luminous) clusters. In a recent revisitation of this relation, Adamo *et al.* (2015) use monte carlo simulations of YSC populations of increasing $M_\star$ to show that the presence of a truncation in the YSC mass function does not violate the observed trend. As SFR increases so it does the $M_\star$(see red solid lines in the right pane of Fig.1).

Another important information hidden in the the SFR vs. $M_V^{\rm brigh}$ relation is the position that galaxies occupy in the diagram. Bastian (2008) shows using monte carlo simulations of YSC populations that if 100% of star formation happens in bound clusters, that are subsequently destroyed, the observed datapoints should follow the dotted line reproduced in the plot. To explain the current observations, only a smaller fraction of star formation is happening in bound clusters, on average about 8%.

It is possible to directly measure the cluster formation efficiency (CFE or Γ), or in other words, the amount of star formation happening in bound clusters. Γ is the ratio between the total stellar mass found in clusters over a certain age range divided by the total SFR estimated in the same age range. Goddard *et al.* (2010) studying the cluster populations of a sample of nearby galaxies first reported that Γ appears to scale with Σ_{SFR}. The Γ vs. Σ_{SFR} relation has been observed to hold for different type of galaxies (starbursts, Adamo *et al.* 2011; dwarfs, Cook *et al.* 2012; etc.) and even on local scales (e.g. the M83 study, Adamo *et al.* 2015). In the right panel of Fig.2, the Γ vs. Σ_{SFR} relation is shown in one of the most updated compilations available in the literature. The observational fit (dashed line) proposed by Goddard *et al.* (2010) to the initial set is nowadays replaced with the theoretical model (dotted line) proposed by Kruijssen (2012). The scatter around the model is about 3 sigma, but it can easily been accounted for, if one considers that the many tuneable assumptions are made in the model and systematic and random errors in the observational works. The Kruijssen's model predicts that the formation of bound stellar clusters takes place in the highest-density peaks of a hierarchically structured ISM. At these high densities, the gas goes through a large number of free-fall times on a short time-scale and therefore reaches higher star formation efficiency. Therefore, YSCs should form more efficiently at high gas pressures (and hence gas surface densities), because these conditions lead to higher density peaks in the ISM and thus favour bound cluster formation. Indeed, the observed Γ vs. Σ_{SFR} relation indirectly links the CFE to the gas pressure (and surface density) through the observable quantity of the Σ_{SFR}, a well know proxy for gas surface density (Kennicutt & Evans 2012). The left panel of Fig.2 reproduces the decline observed in Γ as a function of galactocentric distances in M83. The decline obtained from two different sets of cluster ages and SFR indicators (see inset in the figure) follows quite closely the drop in gas surface density, $\Sigma(H_2)$, reinforcing the idea that the gas pressure (density) plays an important role in the efficiency of the cluster formation process.

In a recent work, Chandar *et al.* (2015) try to quantify the importance of environmental effects above some "universal" cluster properties as the power-law distribution of the mass function. They derive Γ normalising the observed cluster mass function of 6 galaxies by their respective SFR and plot the recovered residuals as a function of SFR. They suggest that the observed trends are within statistical uncertainties and do not reproduced the expectation of the Kruijssen (2012) model and, more in general, the observed Γ vs. Σ_{SFR} relation. However, this reported disagreement is likely not real (Kruijssen & Bastian

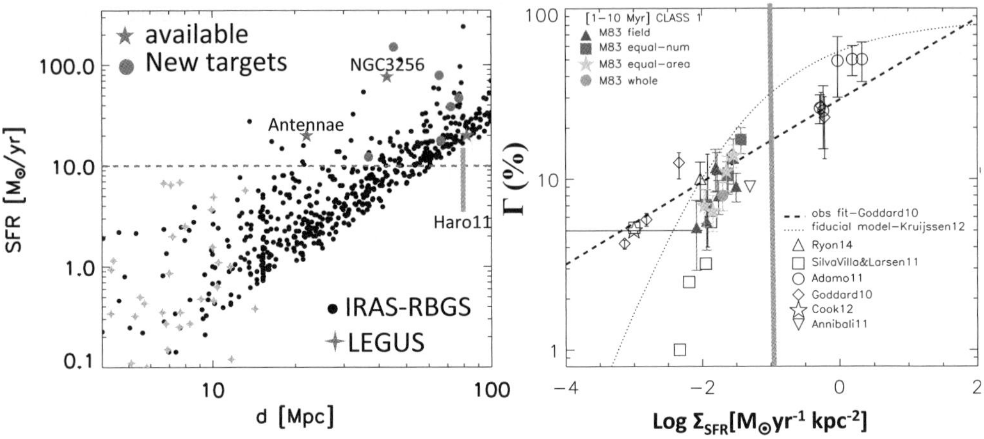

Figure 3. Left: The SFR versus distance space covered by the LEGUS targets (green stars) and Hi-PEEC (magenta stars and dots). The IRAS revised bright galaxy sample is used as a reference (Sanders *et al.* 2003). **Right**: the Γ versus Σ_{SFR} diagram. The blue solid line separates the region that will be probed by LEGUS (left) and by Hi-PEEC (right).

2015). First of all, as discussed in the previous paragraph, there is a well motivated physical link between Γ and the Σ_{SFR}, the latter being an indirect measurement of the gas surface density and thus gas pressure. For this reason, Σ_{SFR} and not the absolute SFR should be used when looking at CFE variations. Secondly, it is important to use cluster catalogues which contains as little as possible contaminations from stellar associations. The effects of using SFR instead of Σ_{SFR}, and catalogues which are contaminated by associations are thoroughly addressed in the recent work by Kruijssen & Bastian (2015).

4. Ongoing surveys and goals for the next few years

The results discussed so far are based on a fairly small sample of galaxies, studied with similar but not homogeneous approaches. To really quantify the role of galactic environment in shaping the YCS formation process we need to access a statistically large sample of galaxies representative of the star-forming system observed in the local Universe. The recently executed Hubble Space Telescope treasury program LEGUS (Legacy ExtraGalactic Uv Survey, GO 13364, PI Calzetti) has ben designed to fully sampled in SFR, specific SFR, morphology, total stellar mass, typical galaxies located within 12 Mpc. The data set of each of the 50 galaxies imaged with the HST contains a homogeneous sampling of 5 UV and optical filters. The access to two UV bands on the left side of the Balmer break and 3 bands on the right part is allowing us to derive cluster properties with tighter constraints on their ages and extinctions values. LEGUS is a laboratory for studying both resolved stellar populations and YSCs (Calzetti *et al.* 2015a). The goal of LEGUS is to revolutionise our knowledge of star formation, which is impossible using a single galaxy, but approachable when a fair sampling of typical local galaxies has been selected. As already described in Calzetti *et al.* (2015a), and soon presented in a forthcoming paper (Adamo *et al.*in prep.), the LEGUS collaboration is constructing high-quality cluster catalogues for most of the targets.

Preliminary results produced by the analysis of the cluster population in the inner and outer pointing of the spiral galaxy NGC628 (Adamo *et al.* in prep.) show that a Schechter function with decreasing $M_\star$(from the inner to the outer region) is statistically a better representation of the observed YSC mass function in the galaxy. This evidence is also supported by the study of the cluster luminosity function in the two regions. We find that the luminosity function shows a steepening as function of increasing wavelengths and a steepening at the highest luminosity bins in each band. The observed trend is in agreement with the expectations by Gieles (2010), where the author assumes that the mass function has a truncation and cluster disruption has not yet been significant.

Another upcoming survey, which has been designed to be complementary to the LEGUS one is the Hubble imaging Probe of Extreme Environments and Clusters (Hi-PEEC, PI Adamo). Hi-PEEC (see left panel of Fig.3) will give us access to star and cluster formation regimes rarely observed in the local Universe but common at high redshift ($z \sim 2-3$), when the bulk of the bulk of the GC populations formed. With Hi-PEEC we will be able to probe the properties of the YSC mass function, putting constraints on the environmental changes of $M_\star$. We will also be able to make a fair comparison with the observed decline in $M_\star$ observed in GC populations by Jordán *et al.* (2007). Results from LEGUS and Hi-PEEC will be fundamental to yield better constraints on the Γ versus Σ_{SFR} relation. The right panel of Fig.3 shows which region of the relation the two surveys will be able to probe.

5. Conclusions

In this short review, I have mainly summarised recent results about cluster formation. Other key aspects related to cluster evolution such as disruption, size distributions, hierarchical properties of YSCs are not covered here, but recent studies can be found in the literature (e.g., Adamo & Bastian 2015, Ryon *et al.* 2015, Grasha *et al.* 2015).

The cluster luminosity and mass functions are powerful tools to understand the role of galactic environment on YSC formations. Recent studies point out that the YCS mass function shows a possible truncation at the high-mass end of the distribution. The truncation seams to change as a function of the galactic environment, but it is not yet understood what galactic physical parameter mainly governs such dependency. Surveys like LEGUS and Hi-PEEC, combined with ancillary data that allow to reconstruct galactic scale properties, like gas surface density, SFR, gas pressure, will produce more stringent constraints on ultimate shape of the YSC mass function.

YSC scaling relations have a key role in linking YSC formation to galaxy evolution, therefore, it is of paramount importance to understand what mechanisms and physical parameters drive these relations.

In a recent numerical work, based on the Illustris cosmological simulations (Vogelsberger *et al.* 2014), Mistani *et al.* (2015) showed that the use of the Γ versus Σ_{SFR} relation maybe can explain why dwarf ellipticals living in over-dense environments have higher GC specific numbers that dwarf irregular living in the field. Dwarf galaxies that have entered a galaxy cluster have experienced a starburst phase before being stripped of their gas reservoirs. During this phase a larger amount of star clusters have been formed. The analog dwarfs living in the field have never experienced such intense starbursts, therefore, their GC specific number is significantly lower.

The change in CFE during starburst phases can also have important impacts in driving not only local but also galactic scale feedback. In a recent work by Bik *et al.* (2015), the authors find that in the starburst galaxy ESO338, galactic scale ionised cones and outflows can directly be linked to the massive YSC that have recently formed in the galaxy. Since YSCs are favourite places to host massive stars, a CFE of about 50 %, like in the case of ESO338, will make a more significant impact on the ISM.

A full description of YSC populations in local galaxies requires us to probe both universal properties and galactic environmental dependencies. Recent and upcoming surveys will play a key role in the advancement of this field. The ultimate goal is to build a complete understanding of cluster formation at any redshift and in any galactic environment.

Acknowledgements

I thank the P. L. Lindahls foundation and the Royal Swedesh Academy (KVA) for the provided generous travel fund. N. Bastian and S. Larsen are thanked for their inputs and comments on this draft. I am very thankful to C. Charbonnel and A. Nota for organising this very interesting symposium.

References

Adamo, A., Östlin, G., & Zackrisson, E. 2011, *MNRAS*, 417, 1904
Adamo, A., Kruijssen, J. M. D., Bastian, *et al.* 2015, *MNRAS*, 452, 246
Adamo, A. & Bastian, N. 2015, "*The lifecycle of clusters in galaxies*", to appear in The Birth of Star Clusters, editor S. W. Stahler, Springer edition
Annibali, F., Tosi, M., Monelli, M., *et al.* 2009, *AJ*, 138, 169
Annibali, F., Tosi, M., Aloisi, A., & van der Marel, R. P. 2011, *AJ*, 142, 129
Bastian, N., Saglia, R. P., Goudfrooij, P., *et al.* 2006, *A&A*, 448, 881
Bastian, N. 2008, *MNRAS*, 390, 759
Bik, A., Östlin, G., Hayes, M., *et al.* 2015, *A&A*, 576, L13
Billett, O. H., Hunter, D. A., & Elmegreen, B. G. 2002, *AJ*, 123, 1454
Calzetti, D., Lee, J. C., Sabbi, E., *et al.* 2015a, *AJ*, 149, 51
Calzetti, D., Johnson, K. E., Adamo, A., *et al.* 2015b, *ApJ*, 811, 75
Chandar, R., Fall, S. M., & Whitmore, B. C. 2015, *ApJ*, 810, 1
Cook, D. O., Seth, A. C., Dale, D. A., *et al.* 2012, *ApJ*, 751, 100
Davies, B., de La Fuente, D., Najarro, F., *et al.* 2012, *MNRAS*, 419, 1860
Dobbs, C. L. 2014, IAU Symposium, 298, 221
Elmegreen, B. G., Elmegreen, D. M., Chandar, R., *et al.* 2006, *ApJ*, 644, 879
Elmegreen, B. G. 2011, EAS Publications Series, 51, 19
Elmegreen, B. G., Malhotra, S., & Rhoads, J. 2012, *ApJ*, 757, 9
Gieles, M., Larsen, S. S., Scheepmaker, R. A., *et al.* 2006, *A&A*, 446, L9
Gieles, M. 2010, Galaxy Wars, 423, 123
Goddard, Q. E., Bastian, N., & Kennicutt, R. C. 2010, *MNRAS*, 405, 857
Grasha, K., Calzetti, D., Adamo, A., *et al.* 2015, arXiv:1511.02233
Jordán, A., McLaughlin, D. E., Côté, P., *et al.* 2007, *ApJS*, 171, 101
Kennicutt, R. C. & Evans, N. J. 2012, *ARA&A*, 50, 531
Kruijssen, J. M. D. 2012, *MNRAS*, 426, 3008
Kruijssen, J. M. D. 2014, Classical and Quantum Gravity, 31, 244006
Kruijssen, J. M. D. & Bastian, N. 2015, arXiv:1511.03286
Krumholz, M. R. & Kruijssen, J. M. D. 2015, *MNRAS*, 453, 739
Larsen, S. S. 2009, *A&A*, 494, 539
Larsen, S. S. 2002, *AJ*, 124, 1393
Mistani, P. A., Sales, L. V., Pillepich, A., *et al.* 2015, arXiv:1509.00030
Portegies Zwart, S. F., McMillan, S. L. W., & Gieles, M. 2010, *ARA&A*, 48, 431
Ryon, J. E., Bastian, N., Adamo, A., *et al.* 2015, *MNRAS*, 452, 525
Sanders, D. B., Mazzarella, J. M., Kim, D.-C., Surace, J. A., & Soifer, B. T. 2003, *AJ*, 126, 1607
Stolte, A. 2013, Massive Stars: From alpha to Omega, 30
Vogelsberger, M., Genel, S., Springel, V., *et al.* 2014, *Nature*, 509, 177
Whitmore, B. C. 2000, arXiv:astro-ph/0012546
Whitmore, B. C., Chandar, R., Schweizer, F., *et al.* 2010, *AJ*, 140, 75

Formation, evolution, and survival of massive star clusters
Proceedings IAU Symposium No. 316, 2015
C. Charbonnel & A. Nota, eds.

doi:10.1017/S1743921316000545

Formation of young massive clusters from turbulent molecular clouds

Michiko S. Fujii[1] **and Simon Portegies Zwart**[2]

[1]Division of Theoretical Astronomy, National Astronomical Observatory of Japan
2-21-1 Osawa, Mitaka, Tokyo, Japan
email: michiko.fujii@nao.ac.jp

[2]Leiden Observatory, Leiden University, Leiden 2300 RA, The Netherlands
email: spz@strw.leidenuniv.nl

Abstract. We simulate the formation and evolution of young star clusters from turbulent molecular clouds using smoothed-particle hydrodynamics and direct N-body methods. We find that the shape of the cluster mass function that originates from an individual molecular cloud is consistent with a Schechter function with power-law slopes of $\beta = -1.73$. The superposition of mass functions turn out to have a power-law slope of < -2. The mass of the most massive cluster formed from a single molecular cloud with mass $M_{\rm g}$ scales with 6.1 $M_{\rm g}^{0.51}$. The molecular clouds that tend to form massive clusters are much denser than those typical found in the Milky Way. The velocity dispersion of such molecular clouds reaches 20km s^{-1} and it is consistent with the relative velocity of the molecular clouds observed near NGC 3603 and Westerlund 2, for which a triggered star formation by cloud-cloud collisions is suggested.

Keywords. star clusters, young massive clusters, open clusters, numerical simulation

1. Introduction

Young massive clusters (YMCs) are more massive and much denser than typical open clusters (Portegies Zwart *et al.* 2010). In the Milky Way YMCs have been found in the last a few decades, but it remains unclear how they formed. In observed star forming regions the stars are spatially distributed in clumps or filaments, rather than in massive spherical distributions. We need to understand the link between such star forming regions and YMCs.

Numerical simulations of star cluster formation have been performed by several groups (e.g., Bonnell *et al.* 2003, 2008; Bate 2012; Federrath & Klessen 2012; Federrath *et al.* 2014; Dale *et al.* 2014). These simulations include stellar growth using sink particles, and the residual gas is eventually removed by stellar feedback. Such numerical simulations are considered realistic, but because of the calculation cost the number of stars formed in such a simulation is limited to $\sim 10^3$.

We adopt a simpler approach in which the hydrodynamical simulations and N-body simulations are separated by assuming instantaneous star formation and gas expulsion at some point in the evolution (see Fujii & Portegies Zwart 2015; Fujii 2015, for the details). This method is motivated by (Moeckel & Bate 2010), and reduces the computational cost such that we can simulate up to $\sim 10^5$ stars. We use this method to perform a series of simulations to study how star clusters form from a variety of turbulent molecular cloud with a mass of 10^4 to $10^6 M_{\odot}$, and a density between 17 and 1700 cm^{-3}.

2. Method

We simulate star cluster formation in three steps:

1. hydrodynamical simulations of the collapse of a turbulent molecular cloud,
2. star formation based on the gas density distribution obtained from the hydrodynamical simulations, and
3. direct N-body simulations of stars based on the structure of the molecular clouds.

The hydrodynamical simulations are performed using the SPH code `Fi` (Hernquist & Katz 1989; Gerritsen & Icke 1997; Pelupessy *et al.* 2004; Pelupessy 2005), using the Astronomical Multipurpose Software Environment (AMUSE) (Portegies Zwart *et al.* 2013; Pelupessy *et al.* 2013). For initial conditions we adopt the isothermal homogeneous gas sphere, as was done in Bonnell *et al.* (2003), with a divergence-free random Gaussian velocity field $\delta \boldsymbol{v}$ with a power spectrum $|\delta v|^2 \propto k^{-4}$ (Ostriker *et al.* 2001).

We adopt the total gas mass of 10^4 to $10^6 M_\odot$ and a mean density of ~ 17 to $\sim 1700\,\mathrm{cm}^{-3}$ (1 to $100 M_\odot \mathrm{pc}^{-3}$) assuming that the mean weight per particle is $2.33 m_\mathrm{H}$. As a consequence, the free-fall time (t_ff) of the gas cloud is 0.83, 2.3, and 8.3 Myr for models with a density of 1, 10, and $100 M_\odot \mathrm{pc}^{-3}$, respectively. We set the gas temperature of 30K and zero total energy (potential plus kinetic energy is zero). We adopt the mass of SPH particles of $1 M_\odot$, which is equal to the mean stellar mass of the following N-body simulations. The gravitational softening length is 0.1 pc. We set the number of particles within a smoothing length as 64. With these settings, the smallest length scale we can resolve is ~ 0.4 pc.

After $0.9 t_\mathrm{ff}$, we stop the hydrodynamical simulations and analyze the density distribution of the collapsed molecular cloud. We replace some SPH particles with stellar particles by adopting the local star formation efficiency (SFE), ϵ_loc, which depends on the local gas density ρ given by $\epsilon_\mathrm{loc} = \alpha_\mathrm{sfe} \sqrt{\rho/(100\, M_\odot \mathrm{pc}^{-3})}$, where α_sfe is a coefficient which controls the star formation efficiency and a free parameter in our simulations, and we adopt $\alpha_\mathrm{sfe} = 0.02$. This SFE is motivated by the recent result that the star formation rate scales with free-fall time (Krumholz *et al.* 2012; Federrath 2013), and they obtained that the SFE per t_ff is 0.015. With these settings, we obtain the global SFE for the entire regions of several per cent and the SFE of dense regions ($>1000 M_\odot \mathrm{cm}^{-3}$) of 20–30 %, which is consistent with observations (Lada & Lada 2003) and also simulations of molecular clouds with star formation (Federrath & Klessen 2013).

After replacing gas particles with stellar particles, we remove all residual gas particles assuming it is expelled instantaneously, and we continue the simulation by means of direct N-body integration with stars only. We stop the simulation after 10 Myr. The N-body simulations were carried out with a sixth-order Hermite scheme (Nitadori & Makino 2008) without any gravitational softening. We include stellar collisions by means of the sticky sphere approximation with stellar radius of zero-age main-sequence following the description of Hurley *et al.* (2000). We also include stellar mass-loss at the end of the main-sequence following Hurley *et al.* (2000) (see Fujii *et al.* 2009; Fujii & Portegies Zwart 2013, for the details). We the a time step criterion of Nitadori & Makino (2008) with an accuracy parameter, $\eta = 0.1 - 0.25$. The energy error was smaller than $\sim 10^{-3}$ for all simulations over the entire duration of the simulations, which is sufficient to warrant an accurate representation of the simulation compared to a real star cluster Portegies Zwart & Boekholt (2014).

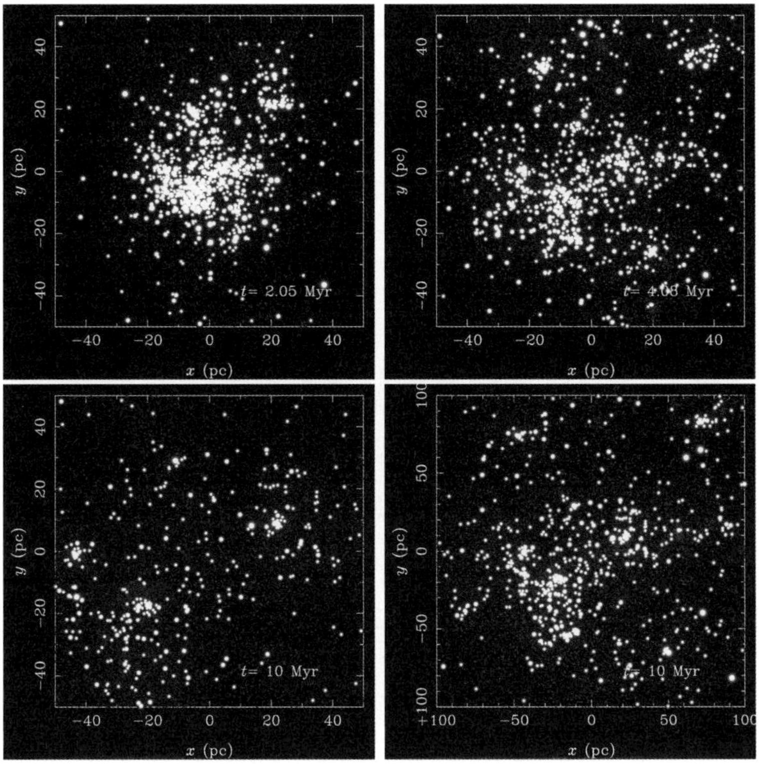

Figure 1. Snapshots of one of our model with a total gas mass of $10^6 M_\odot$ and a mean density of 1700cm^{-3}. This model result in the formation of young massive clusters. The fractal dimension of this model is ~ 1.4 at 2 Myr.

3. Formation of Young Massive Clusters

In Figure 1, we present the stellar distribution that originate from a molecular cloud with a mass of $10^6 M_\odot$ and an initial density of $\sim$1700 cm^{-3}. These form the initial conditions for one of our N-body simulations. The time in this figure indicates the duration time after we stopped the hydrodynamical simulation and when we starting the N-body simulation. These initial condition are clumpy, which reflects the distribution of the gas. We recognize several tens of small stellar conglomerates, quite similar to observed stellar distribution of embedded clusters (Lada & Lada 2003). At later time several of these clumps merge to more massive and larger clusters. Typically one or two clusters survive this hierarchical merger process and evolve to YMCs. This process is more clearly seen in our massive and dense initial conditions of the molecular clouds, those with a mass of 0.4–1$10^6 M_\odot$ and a density of $100 M_\odot$. Such dense and massive molecular clouds typically appear in starburst galaxies, such as M83(Bastian *et al.* 2011) and M51(Chandar *et al.* 2011), but are absent in local disk galaxies such as the Milky Way (Krumholz *et al.* 2012) or M31.

There are only a few YMCs in the Milky Way, and therefore massive and dense molecular clouds must also form in the Milky Way. Recent observation of molecular clouds around YMCs in the Milky Way, such as NGC 3603 and Westerlund 2, suggested that they formed through collisions between giant molecular clouds (GMCs) with a velocity of $\sim$ 20km s^{-1} (Furukawa *et al.* 2009; Ohama *et al.* 2010; Fukui *et al.* 2014). The velocity

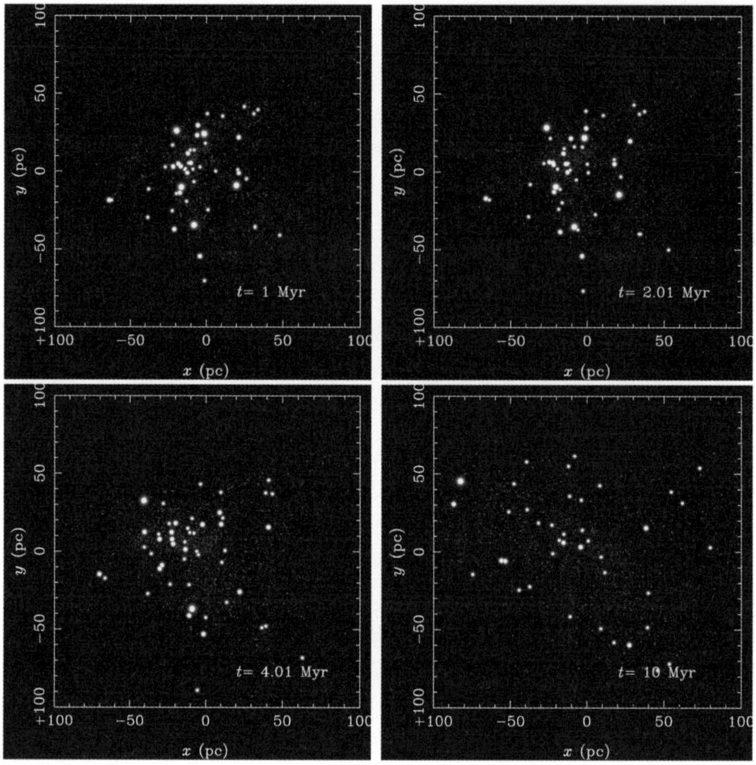

Figure 2. Snapshots of one of our model with a total gas mass of $10^6 M_\odot$ and a mean density of 17cm^{-3}. This model form an association.

dispersion of our initial condition that form YMCs are consistent with these observed values.

Molecular clouds that typically appear in local disk galaxies, which follow Larson's relation Larson (1969), tend to evolve into open clusters and associations, rather than YMCs. In Figure 2, we present the snapshots of a model with a total gas mass of $10^6 M_\odot$ but with a mean density of 17cm^{-3}. This molecular cloud forms fewer stars distributed in a more extended region. The final stellar distribution is less clumpy and does not contain any massive clusters, although some clumps appear very similar to classical open clusters. The distribution of stars from these simulations resembles OB associations, rather than star clusters.

4. Mass Function of Star Clusters

At $t = 2$ and 10 Myr, we interrupt the simulations and search for clumps using HOP, (Eisenstein & Hut 1998) in order to identify clusters. We fit a Schechter function (Schechter 1976):

$$\phi(M) \equiv \frac{dN}{dM} \propto M^{\beta} \exp\left(-\frac{M}{M_{\text{cut}}}\right) \tag{4.1}$$

to the resulting mass distribution. Here M is the mass of clusters and M_{cut} is the cut-off mass. Integration of equation (4.1) results in the cumulative mass function of the form

$$N(>M) \propto M^{\beta+1} \exp\left(-\frac{M}{M_{\text{cut}}}\right). \tag{4.2}$$

For $M_{\rm cut}$ we adopt the mass of the most massive cluster found in the simulations ($M_{\rm c,max}$). The best fit is achieved for a value for $\beta = -1.73$, which is consistent with the mass function for cluster observed in the Carina region (Feigelson *et al.* 2011). For $M_{\rm c,max}$, we find the relation that $M_{\rm c,max} = 6.3 M_{\rm g}^{0.51}$. This is similar to the relation between the mass of the most massive star ($m_{\rm max}$) and the mass of its host star cluster ($M_{\rm c}$); $m_{\rm max} \simeq 0.4 m^{0.67}$ (Pflamm-Altenburg *et al.* 2007).

We superpose the fitted cluster mass function and obtain cluster mass functions for each disk galaxy. For the Milky Way we adopt a power of -1.45 (Planck Collaboration *et al.* 2011) and -0.9 for M31 (Kirk *et al.* 2013). We assume that each individual GMC forms a conglomerate of clusters that follow equation (4.2). The power-law index of the superposed cluster mass function is ~ -2 because of the cut-off in the Schechter mass function, which is consistent with cluster mass functions observed in nearby galaxies (Portegies Zwart *et al.* 2010).

5. Summary

We simulate the formation and evolution of young star clusters using SPH and direct N-body methods. We first perform SPH simulations of turbulent GMCs for one initial free-fall time, and analyze the resulting structure of the cloud. We subsequently replace a density-selected subset of SPH particles with stars by adopting a local star-formation efficiency that depends on the local gas density. The stellar distribution is clumpy and mimics known star forming regions.

We continue to evolve the stars dynamically by means of direct N-body integration and analyze the results at 2 Myr and 10 Myr. The densest and most massive molecular clouds lead to the formation of YMCs via hierarchical merging of smaller clusters. The shape of the cluster mass function that originates from an individual molecular cloud is consistent with a Schechter function with a power-law slope of $\beta = -1.73$ at 2 Myr and $\beta = -1.67$ at 10 Myr, which fits to observed cluster mass function of the Carina region. The superposition of mass functions have a power-law slope of < -2, which is consistent with the observed mass function of star clusters in the Milky Way, M31 and M83. We further find that the mass of the most massive cluster formed in a single molecular cloud with a mass of $M_{\rm g}$ scales with 6.1 $M_{\rm g}^{0.51}$ which also agrees with recent observation in M51.

The molecular clouds which can form massive clusters are much denser than those typical found in the Milky Way. The velocity dispersion of such molecular clouds reaches 20km s^{-1}, which is consistent with the relative velocity of the molecular clouds observed near NGC 3603 and Westerlund 2. We argue that the formation of these clusters was initiated by the collision between molecular clouds.

References

Bastian, N., Adamo, A., Gieles, M., Lamers, H. J. G. L. M., Larsen, S. S., Silva-Villa, E., Smith, L. J., Kotulla, R., Konstantopoulos, I. S., Trancho, G., & Zackrisson, E., 2011, *MNRAS*, 417, L6

Bate, M. R., 2012, *MNRAS*, 419, 3115

Bonnell, I. A., Bate, M. R., & Vine, S. G., 2003, *MNRAS*, 343, 413

Bonnell, I. A., Clark, P., & Bate, M. R., 2008, *MNRAS*, 389, 1556

Chandar, R., Whitmore, B. C., Calzetti, D., Di Nino, D., Kennicutt, R. C., Regan, M., & Schinnerer, E., 2011, *ApJ*, 727, 88

Dale, J. E., Ngoumou, J., Ercolano, B., & Bonnell, I. A., 2014, *MNRAS*, 442, 694

Eisenstein, D. J. & Hut, P., 1998, *ApJ*, 498, 137

Federrath, C., 2013, *MNRAS*, 436, 3167
Federrath, C. & Klessen, R. S., 2012, *ApJ*, 761, 156
—, 2013, *ApJ*, 763, 51
Federrath, C., Schrön, M., Banerjee, R., & Klessen, R. S., 2014, *ApJ*, 790, 128
Feigelson, E. D., Getman, K. V., Townsley, L. K., Broos, P. S., Povich, M. S., Garmire, G. P., King, R. R., Montmerle, T., Preibisch, T., Smith, N., Stassun, K. G., Wang, J., Wolk, S., & Zinnecker, H., 2011, *ApJs*, 194, 9
Fujii, M., Iwasawa, M., Funato, Y., & Makino, J., 2009, *ApJ*, 695, 1421
Fujii, M. S., 2015, *PASJ*, 67, 59
Fujii, M. S. & Portegies Zwart, S., 2013, *MNRAS*, 430, 1018
—, 2015, *MNRAS*, 449, 726
Fukui, Y., Ohama, A., Hanaoka, N., Furukawa, N., Torii, K., Dawson, J. R., Mizuno, N., Hasegawa, K., Fukuda, T., Soga, S., Moribe, N., Kuroda, Y., Hayakawa, T., Kawamura, A., Kuwahara, T., Yamamoto, H., Okuda, T., Onishi, T., Maezawa, H., & Mizuno, A., 2014, *ApJ*, 780, 36
Furukawa, N., Dawson, J. R., Ohama, A., Kawamura, A., Mizuno, N., Onishi, T., & Fukui, Y., 2009, *ApJl*, 696, L115
Gerritsen, J. P. E. & Icke, V., 1997, *A&A*, 325, 972
Hernquist, L. & Katz, N., 1989, *ApJs*, 70, 419
Hurley, J. R., Pols, O. R., & Tout, C. A., 2000, *MNRAS*, 315, 543
Kirk, J. M., Gear, W. K., Fritz, J., Smith, M. W. L., Ford, G., Baes, M., Bendo, G. J., De Looze, I., Eales, S. A., Gentile, G., Gomez, H. L., Gordon, K., O'Halloran, B., Madden, S. C., Duval, J. R., Verstappen, J., Viaene, S., Boselli, A., Cooray, A., Lebouteiller, V., & Spinoglio, L., 2013, ArXiv e-prints
Krumholz, M. R., Dekel, A., & McKee, C. F., 2012, *ApJ*, 745, 69
Lada, C. J. & Lada, E. A., 2003, *AR&A*, 41, 57
Larson, R. B., 1969, *MNRAS*, 145, 271
Moeckel, N. & Bate, M. R., 2010, *MNRAS*, 404, 721
Nitadori, K. & Makino, J., 2008, *New Astronomy*, 13, 498
Ohama, A., Dawson, J. R., Furukawa, N., Kawamura, A., Moribe, N., Yamamoto, H., Okuda, T., Mizuno, N., Onishi, T., Maezawa, H., Minamidani, T., Mizuno, A., & Fukui, Y., 2010, *ApJ*, 709, 975
Ostriker, E. C., Stone, J. M., & Gammie, C. F., 2001, *ApJ*, 546, 980
Pelupessy, F. I., 2005, PhD thesis, Leiden Observatory, Leiden University, P.O. Box 9513, 2300 RA Leiden, The Netherlands
Pelupessy, F. I., van der Werf, P. P., & Icke, V., 2004, *A&A*, 422, 55
Pelupessy, F. I., van Elteren, A., de Vries, N., McMillan, S. L. W., Drost, N., & Portegies Zwart, S. F., 2013, *A&A*, 557, A84
Pflamm-Altenburg, J., Weidner, C., & Kroupa, P., 2007, *ApJ*, 671, 1550
Planck Collaboration, Ade, P. A. R., Aghanim, N., Arnaud, M., Ashdown, M., Aumont, J., Baccigalupi, C., Balbi, A., Banday, A. J., Barreiro, R. B., *et al.*, 2011, *A&A*, 536, A23
Portegies Zwart, S. & Boekholt, T., 2014, *ApJl*, 785, L3
Portegies Zwart, S., McMillan, S. L. W., van Elteren, E., Pelupessy, I., & de Vries, N., 2013, *Computer Physics Communications*, 183, 456
Portegies Zwart, S. F., McMillan, S. L. W., & Gieles, M., 2010, *AR&A*, 48, 431
Schechter, P., 1976, *ApJ*, 203, 297

Formation, evolution, and survival of massive star clusters
Proceedings IAU Symposium No. 316, 2015
C. Charbonnel & A. Nota, eds.

doi:10.1017/S1743921316008735

Submillimeter View of Gas and Dust in the Forming Super Star Cluster in NGC 5253

Jean L. Turner[1]

[1]Department of Physics and Astronomy, UCLA,
Los Angeles, CA 90095-1547 USA
email: turner@astro.ucla.edu

Abstract. A giant molecular cloud has been detected surrounding the supernebula in NGC 5253, revealing details of the formation and feedback process in a very massive star cluster. "Cloud D" was recently mapped in CO J = 3–2 with the Submillimeter Array. The cloud surrounds a currently forming massive cluster of mass $\sim 10^6$ $M_\odot$, and luminosity $\sim 10^9$ $L_\odot$. Cloud D is hot, clearly associated with the cluster, yet kinematically relatively quiescent. The dust mass is $\sim$15,000 $M_\odot$, for a gas-to-dust ratio of $\sim$50, nearly an order of magnitude lower than expected for this low metallicity galaxy. We posit that enrichment by the cluster, leading to a stalled cluster wind, has created the unusual conditions in Cloud D. The absence of current mechanical impact of the young cluster on the cloud, in spite of the presence of thousands of O stars, may permit future generations of stars to form near the massive cluster.

Keywords. galaxies: individual (NGC 5253), galaxies: star clusters, galaxies: starburst, galaxies: dwarf

---

1. Introduction

The process by which the most massive star clusters form is not well understood. The closest massive clusters, such as those in the Galactic Center or R136 in the Large Magellanic Cloud, are relatively evolved and far from their initial molecular conditions. Clusters that are still embedded in their natal clouds can provide more direct information on the interaction of young clusters with their natal clouds, including star formation efficiency and the precise nature of cluster feedback on the clouds. The advantage of submillimeter observations in the study of young clusters is that one can simultaneously trace spectral lines of molecular gas as well as continuum emission from ionized gas and dust, to give a full picture of the gas, dust, and kinematics in young, forming clusters.

The massive cluster in NGC 5253 may be the closest example of a embedded cluster still in the process of formation. Despite having little gas and no disk, this dwarf spheroidal galaxy has formed hundreds of clusters in the recent past (Calzetti *et al.* 1997, 2015, Chandar *et al.* 2005, Harbeck *et al.* 2012, de Grijs *et al.* 2013). This unlikely environment is the location of the "supernebula", a bright and compact radio continuum source associated with a luminous HII region (Beck *et al.* 1996, Turner *et al.* 2000). The HII region core is $\sim$1 pc in radius with an infrared luminosity of $\sim 10^9$ $L_\odot$ at D = 3.8 Mpc, and Lyman continuum rate of $\sim 7 \times 10^{52}$ s^{-1} (7000 O7 star equivalents) (Meier *et al.* 2002). The supernebula is embedded, with $A_v \sim 16$ (Turner *et al.* 2003, Alonso-Herrero *et al.* 2004). Most of the gas in NGC 5253 resides *outside* the galaxy, in a halo of HI filaments (Kobulnicky & Skillman 2008; López-Sánchez *et al.* 2012) and in a molecular streamer of $\sim$200 pc extent detected in CO (Turner *et al.* 1997). The streamer is infalling along the minor axis of the galaxy. No CO(1–0) and only weak CO(2–1) had been detected until recently at the location of the embedded SSC/supernebula (Turner *et al.* 1997; Meier *et al.* 2002; Miura *et al.* 2015).

Submillimeter Array observations of the J = 3–2 transition of CO and dust continuum in NGC 5253 have revealed the giant molecular cloud (GMC) associated with the supernebula and its embedded cluster. "Cloud D" is shown in Figure 1 (Turner *et al.* 2015). Cloud D is unusually bright in CO(3–2), 2.6 times brighter than in CO(2–1). The emission is optically thin and the inferred gas temperature is >200 K, with densities of $\sim 5 \times 10^4$ cm^{-3}. Galactic GMCs, even those actively forming stars, more typically have temperatures of ∼10-15 K on these scales. From its temperature alone, it is clear that Cloud D is closely associated with the young massive cluster.

To determine a star formation efficiency for the cluster we need a gas mass, which is difficult to determine for this strange cloud. From the CO(3–2)/CO(1–0) ratio it is inferred that the emission is optically thin; by contrast, Galactic GMCs are optically thick in CO. Thus we cannot use the standard Galactic conversion factor, X_{CO}, which applies to optically thick and cooler Galactic clouds in the CO(1–0) transition. Cloud D is as yet undetected in CO(1–0). For optically thin emission, instead one counts emitted photons in the line and from that computes the total number of CO molecules from the partition function. However, to get to the total molecular mass requires a relative abundance of CO to H_2, and this abundance depends on the chemistry. The chemistry is uncertain in Cloud D because of the unusually high temperature of the cloud, the high radiation fields, and the low metallicity of NGC 5253, $Z \sim 0.25 Z_{\odot}$ (Kobulnicky *et al.* 1997, López-Sánchez *et al.* 2007). Using the Galactic value of $[CO]/[H_2] = 8.5 \times 10^{-5}$ we infer a total H_2 mass of $M_{H_2} = 5 \times 10^4$ M$_{\odot}$ (Turner *et al.* 2015); however, given the unusual properties of this cloud we have no confidence that the Galactic CO abundance ratio obtains here.

A less slippery number for the gas mass can be determined from subtracting the stellar mass from the dynamical mass of the cloud. The uncertainties are simpler and better constrained than the previous methods; it gives a total gas mass rather than an H_2 mass, which is more appropriate for star formation efficiency. The mass of stars can be obtained from the Lyman continuum rate (Meier *et al.* 2002) and the 3.5 Myr age based on Br γ equivalent widths (Alonso-Herrero *et al.* 2004) and Wolf-Rayet signatures in abundances (Walsh & Roy 1989, Kobulnicky *et al.* 1997, López-Sánchez *et al.* 2007). Calzetti *et al.* (2015) argue that dust emission affects the Br γ EW and that the age of this cluster is closer to 1 Myr, and further divide the stars among two clusters, both within Cloud D. However, this young age would not explain the Wolf-Rayet signatures and local nitrogen enhancement in the region, so we favor an older age. For a 3.5 Myr age, and assuming continuous star formation, the stellar mass based on STARBURST99 models is $M_{stars} = 1.1^{+0.7}_{-0.2} \times 10^6$ M$_{\odot}$. The CO linewidth is $\sigma = 9.2 \pm 0.6$ km s^{-1}; this will give an upper limit to the gravitational mass, which is $M_{vir} = 1.8^{+0.2}_{-0.7} \times 10^6$ M$_{\odot}$. The final gas mass is then $M_{vir} - M_{stars} = 7 \pm 4 \times 10^5$ M$_{\odot}$, giving a star formation efficiency $M_{stars}/(M_{stars} + M_{gas}) \sim 60\%$ (Turner *et al.* 2015). This very high, record efficiency, not so different from the 75±50% inferred from CO(2–1) using X_{CO} by Meier *et al.* (2002), bodes well for the cluster's survival.

2. Cloud D: A Very Dusty Cloud

An alternative path to the gas mass is through dust, but the dust mass of Cloud D is also strange and interesting. About half of the 72 mJy of 870μm continuum emission, 34±14 mJy, is from dust; the rest is free-free emission from the ionized gas. Using the dust opacity of the Large Magellanic Cloud (Galliano *et al.* 2011), and a dust temperature of

Figure 1. Molecular gas in NGC 5253. CO(3-2) emission overlaid on an HST image. The CO image is from the SMA, beam is 4″ x 2 ″. "Cloud D" is the brightest CO source near the center; also detected are another cloud, "F" and the "streamer", which is the extended filament to the east. Turner *et al.* 2015.

46 K (Thronson & Telesco 1986) we obtain $M_{dust} = 1.5 \pm 0.2 \times 10^4$ $M_\odot$. If we scale the gas-to-dust ratio, GTD, of NGC 5253 to match its $Z \sim 0.25 Z_\odot$ metallicity, a common and reasonable assumption, we would estimate that $GTD \sim 500-700$, similar to values in the Magellanic Clouds, and this would give a dust-predicted gas mass of $M_{gas} \sim 10^7$ $M_\odot$. This is more than five times higher than the gas mass could possibly be, based on the dynamical mass of Cloud D. From the inferred gas mass we infer that instead $GTD \sim 50$ in NGC 5253, a factor of three less than the Galactic GTD value (Turner *et al.* 2015).

The reason that NGC 5253 has ten times more dust than expected may be enrichment by the cluster stars, metals that are as yet trapped within the natal cloud. STARBURST99 models for clusters of this luminosity that are more than 3 Myr of age can produce 30,000 $M_\odot$ of metals, twice the dust mass that is seen. So it is not surprising, it is even expected, that enrichment is taking place.

The dust enrichment has potential implications for the evolution of the cluster and its cloud, and the nature of feedback. Cloud D is surprisingly quiescent; its linewidth of $\sim$9 km s^{-1} is not far from the value expected for a cloud of its 25 pc size based on the Galactic size-linewidth relation. Yet its high temperature and precise positional and kinematic coincidence with the massive cluster indicates a close association of the two. Where is the feedback from the 7000 O stars?

Silich *et al.* (2004) and Tenorio-Tagle *et al.* (2007) suggest that enrichment by cluster stars creates a dusty environment that can enhance gas cooling and stall a cluster wind. Based on numerical modeling of the gas dynamics they find that for a cluster of the luminosity exciting the supernebula in NGC 5253, an enrichment of $1.5 Z_\odot$ is sufficient to stall a wind, trapping the enriched products of stellar evolution within the cloud. From the mass of dust it would appear that the enrichment of the gas in Cloud D is roughly twice this amount. The stalled wind model could explain both the quiescence of Cloud D and its dustiness. Given the relative quiescence of this cloud, so similar to Galactic GMCs, it would seem that conditions are ripe for the formation of even more stars at this location, given an influx of gas. This is a very interesting suggestion (Silich *et al.* 2004; Tenorio-Tagle *et al.* 2007) in light of the fact that multiple generations are observed to be present in globular clusters (see also R. Wünsch this volume).

3. Conclusions

Submillimeter Array maps of CO(3–2) and 870μm dust continuum emission from Cloud D associated with the young, embedded massive cluster exciting the supernebula in NGC 5253 allow the determination of dynamical mass, gas mass, and dust mass for this cloud, and allow us to define the interaction of the cluster and its surroundings.

1) The molecular gas in Cloud D is hot; the CO(3–2)/CO(2–1) line ratio suggests that the temperature is unusually warm, > 200 K. The cloud is clearly heated by the cluster.

2) The H_2 mass obtained from the line intensity of the optically thin CO emission is low, particularly compared to the dust mass. This is probably due to the $[CO]/[H_2]$ abundance ratio, which is highly uncertain in this unusually warm and highly irradiated cloud. Instead we obtain a gas mass by subtracting the stellar mass, based on the Lyman continuum rate and STARBURST99 models and a 3.5 Myr age, from the dynamical mass based on the CO size and linewidth of Cloud D. The gas mass of 7×10^5 $(D/3.8\,\mathrm{Mpc})^2$ gives a record high star formation efficiency of 60%.

3) The dust mass is high for the metallicity of the galaxy: gas-to-dust radio is 50:1, sub-solar, much lower than the 500-700 value expected based on the low metallicity of NGC 5253. We posit that the dust mass is almost entirely produced by the stars of the cluster; STARBURST99 models predict twice this yield in metals for a cluster of this luminosity and 3.5 Myr age, so this model is entirely plausible.

4) The numerical models of Silich *et al.* (2004) predict that a cluster wind will catastrophically cool and stall if the gas is enriched to 1.5 $Z_\odot$ for the supernebula cluster in NGC 5253. This condition appears to be met. This could explain why, even in the presence of 7000 O stars and a high cloud temperature, the width of the CO line, 9.2 km/s, roughly that expected for a quiescent cloud of its size based on the Galactic size-linewidth relation. Cluster feedback of the mechanical variety appears to be inhibited in this cluster, which could allow further accretion and future generations of stars to form in this cluster.

Acknowledgements

JLT would like to acknowledge S. Michelle Consiglio for helpful discussions, and the support of NSF Grant AST-1515570.

References

Alonso-Herrero, A., Takagi, T., Baker, A. J., Rieke, G. H., Rieke, M. J., Imanishi, M., & Scoville, N. Z. 2004, *ApJ*, 612, 222

Beck, S. C., Turner, J. L., Ho, P. T. P.., Lacy, J. H., & Kelly, D. M. 1996, *ApJ*, 457, 610

Calzetti, D., Meurer, G. R., Bohlin, R. C., Garnett, D. R., Kinney, A. L., Leitherer, C., & Storchi-Bergmann, T. 1997, *AJ*, 114, 1834

Calzetti, D., Johnson, K. E., Adamo, A., *et al.* 2015, arXiv:1508.04476

Chandar, R. *et al.* Leitherer, C., Tremonti, C. A., Calzetti, D., Aloisi, A., Meurer, G. R., , de Mello, D. 2005, *ApJ*, 628, 210

de Grijs, R., Anders, P., Zackrisson, E., & Östlin, G. , 2013 *MNRAS*, 431, 2917

Galliano, F., Hony, S., Bernard, J.-P., *et al.* 2011, *A&A*, 536, A88
Harbeck, D., Gallagher, J., & Crnojević, D. 2012, *MNRAS*, 422, 629
Kobulnicky, H. A. & Skillman, E. D. 2008, *AJ*, 135, 527
Kobulnicky, H. A., Skillman, E. D., Roy, J.-R., Walsh, J. R., & Rosa, M. R. 1997, *ApJ*, 477, 679
López-Sánchez, Á. R., Esteban, C., García-Rojas, J., Peimbert, M., & Rodríguez, M. 2007, *ApJ*, 656, 168
López-Sánchez, Á. R., Koribalski, B. S., van Eymeren, J., Esteban, C., Kirby, E., Jerjen, H., & Lonsdale, N. 2012, *MNRAS*, 419, 1051
Meier, D. S., Turner, J. L., & Beck, S. C. 2002, *AJ*, 124, 877
Miura, R. E., Espada, D., Sugai, H., Nakanishi, K., & Hirota, A. 2015, *PASJ*, 67, LL1
Silich, S., Tenorio-Tagle, G., & Rodríguez-González, A. 2004, *ApJ*, 610, 226
Tenorio-Tagle, G., Wünch, R., Silich, S., & Palous, J. 2007, *ApJ*, 658, 1196
Thronson, H. A., Jr. & Telesco, C. M. 1986, *ApJ*, 311, 98
Turner, J. L., Beck, S. C., & Ho, P. T. P. 2000, *ApJ*, 532, L109
Turner, J. L., Beck, S. C., & Hurt, R. L. 1997, *ApJL*, 474, L11
Turner, J. L., Beck, S. C., Crosthwaite, L..P., Larkin, J. E., McLean, I. S., & Meier, D. S. 2003, *Nature*, 423, 621
Turner, J. L., Beck, S. C., Consiglio, S. M., Ho, P. T. P., Kovács, A., Meier, D. S., & Zhao, J.-H. 2015, *Nature*, 519, 331
Walsh, J. R. & Roy, J.-R. 1989, *MNRAS*, 239, 297

Discussion

BOESGAARD: How did you get the metallicity?

TURNER: The overall metallicity of NGC 5253, $Z \sim 0.25 Z_{\odot}$, is from observations of the surrounding HII regions. Optical emission lines cannot be detected from the supernebula directly, which is embedded in Cloud D. The extinction is about 16 magnitudes in the visual. The gas-to-dust ratio is directly computed from the gas mass of Cloud D and observed dust mass.

Formation, evolution, and survival of massive star clusters
Proceedings IAU Symposium No. 316, 2015
C. Charbonnel & A. Nota, eds.

doi:10.1017/S1743921315010856

Massive stellar systems: observational challenges and perspectives in the E-ELT era

G. Bono[1,2], V. F. Braga[1], I. Ferraro[2], G. Fiorentino[3], R. Gilmozzi[4], G. Iannicola[2], D. Magurno[1], N. Matsunaga[5], M. Monelli[6,7] and S. Rastello[1]

[1]Universitá di Roma "Tor Vergata", Via della Ricerca Scientifica 1, 00133 Roma, Italy
email: `giuseppe.bono@roma2.infn.it`

[2]INAF-Osservatorio Astronomico di Roma, Via Frascati 33, 00040 Monteporzio Catone, Italy
[3]INAF-Osservatorio Astronomico di Bologna, via Ranzani 1, 40127, Bologna, Italy
[4]European Southern Observatory, K. Schwarzschild-Str. 2, 85748 Garching-Munchen, Germany
[5]Dept. of Astronomy, University of Tokyo, 7-3-1 Hongo, Bunkyo-ku, Tokyo 113-0033, Japan
[6]Instituto de Astrofísica de Canarias, La Laguna, Tenerife, Spain
[7]Departamento de Astrofísica, Universidad de La Laguna, Tenerife, Spain

Abstract. We introduce the empirical framework concerning optical and near-infrared (NIR) photometry of crowded stellar fields. In particular, we address the impact that linear detectors and analytical PSF played in improving the accuracy and the precision of multi-band color-magnitude diagrams (CMDs). We focus our attention on recent findings based on deep NIR images collected with Adaptive Optics (AO) systems at the 8-10m class telescopes and discuss pros and cons of the different approaches. We also discuss the estimate of the absolute age of globular clusters using a well defined knee along the lower main sequence. We mention the role which the current AO-assisted instruments will have in addressing longstanding astrophysical problems of the Galactic center. Finally, we outline the role of first generation of E-ELT instruments upon photometry and spectroscopy of crowded stellar fields.

Keywords. Galaxy: nucleus, globular clusters: general, techniques: photometric, techniques: spectroscopic, stars: distances, stars: evolution, stars: oscillations

1. Introduction

Photometry in crowded stellar fields such as globular clusters (GCs), Galactic Bulge (GB) and the Nuclear Bulge (NB) has been for decades the "battering ram" to constrain their stellar content and to investigate several open astrophysical problems. In this context, accurate absolute stellar ages play a crucial role, since they provide firm constraints on the early formation and evolution of the Galactic spheroid, and on the interaction among the oldest Galactic components (e.g. Vandenberg *et al.* 1996; Athanassoula 2005). The Galactic bulge has been addressed in many papers (Kunder *et al.* 2015; Valenti *et al.* 2015), in the following we focus our attention on GCs and on NB.

Since the pioneering CMDs by Arp *et al.* (1953) and Sandage (1953) GCs have been the subject of extensive photometric surveys. The advent of CCD cameras during the 80s opened a new path (Schild & Kent 1981; Kristian & Blouke 1982): despite their initially small field of view, the new bidimensional sensors, with their high-sensitivity and linear response, allowed for precise photometric measurements of a significantly larger number of individual stars in GCs. However, the new opportunities brought new challenges for the data reduction, and this pushed the development of new algorithms for precise stellar photometry (e.g. DAOPHOT, Stetson (1987); ROMAFOT, Buonanno (1989); DoPhot, Schechter *et al.* (1993); SExtractor, Bertin & Arnouts (1996); StarFinder, Diolaiti *et al.*

(2000)). Recently, the high spatial resolution provided by space images collected with the Hubble Space Telescope in optical bands allowed us new relevant leaps forward in GC studies and software development (HSTphot, Dolphin (2000); EPSF, Anderson *et al.* (2008)). They paved the road for collecting homogeneous, high-quality multi-band images for a significant fraction of Galactic (Marín-Franch *et al.* 2009) and Local Group (LG) (Olszewski *et al.* 1996; Buonanno *et al.* 1999; Rich *et al.* 2005) GCs.

The absolute age of GCs has been studied extensively in the literature (Vandenberg *et al.* 1990; Chaboyer *et al.* 1998), since the main sequence turn-off (MSTO) is the fundamental clock provided by stellar evolution. However, so far this method has been hampered by observational uncertainties. The dominant source of error is the GC distance: the current uncertainties are at least $\approx$5%, this means at least 10% in distance modulus (0.1 mag) and error on absolute age $\geqslant$1 Gyr (Renzini 1991). The error on the reddening and on the chemical composition also contribute to the age uncertainty.

The above problems become even more severe in the NB, since these regions are extremely crowded and heavily reddened. This is why scientists started studying the stellar content of the NB twenty years ago, with the advent of NIR arrays. Early investigations by Philipp *et al.* (1999) and Mezger *et al.* (1999) constrained the contribution of low- and high-mass stars in the dynamical mass (90% vs 10%) and in the K-band flux (6% vs 94%). The young star clusters: Nuclear Star Cluster, Arches and Quintuplet were investigated using space (NICMOS at HST, Figer *et al.* (2002)) and ground-based (SINFONI, Martins *et al.* (2008); Liermann *et al.* (2010a); NACO at VLT, Espinoza *et al.* (2009)) NIR and MIR images.

The metallicity distribution of the NB is also in its infancy, and indeed accurate iron abundances for red and blue supergiants (SG) were only recently provided (Martins *et al.* 2008; Najarro *et al.* 2009; Davies *et al.* 2009). The current estimates suggest a roughly solar iron abundance. This finding has been supported by accurate NIR high-resolution spectra collected with GIANO at TNG (Origlia *et al.* 2015). Note that the above metallicities are systematically more metal-poor than Galactic Cepheids in the inner disk (Genovali *et al.* 2015). This suggests that the NB and the inner disk underwent different chemical-enrichment histories.

The above investigations were complemented by low- and high-resolution NIR spectra collected with ISAAC and with CRIRES at VLT by Ryde & Schultheis (2015). They investigated nine M-type giants located in the NB, and found that they are metal-rich ([Fe/H]=0.11$\pm$0.15). Moreover, they noticed that their α-element abundances are similar to outer-bulge stars, i.e. relatively α-poor. New constraints on the metallicity distribution of the central parsec were provided by Do *et al.* (2015) using low-resolution spectra for a large sample of RG stars (83) in the NB collected with NIFS at Gemini north. They were the first scientists to find evidence of both metal-poor ([Fe/H]$\leqslant$-1) and super metal-rich ([Fe/H]$\geqslant$0.5) stars. These findings open a new path concerning the formation of the NB, i.e. the in situ formation versus the accretion formation.

Note that the NB is, together with the thin disc, the only Galactic region for which we still lack firm identifications of old (t$\geqslant$ 10 Gyr) stellar populations (Bono *et al.* 2015).

2. A new spin on the absolute age of GCs

The absolute age of GCs are mainly based on the comparison between predicted and observed magnitude of main-sequence turn-off stars (MSTO). Pros and cons of optical and NIR CMDs, together with intrinsic and systematic errors have been discussed in many papers (Girardi *et al.* 2002; Salaris & Weiss 2002; Dotter *et al.* 2009; Vandenberg *et al.* 2013; Di Cecco *et al.* 2015; Milone *et al.* 2015; Monelli *et al.* 2015). The same

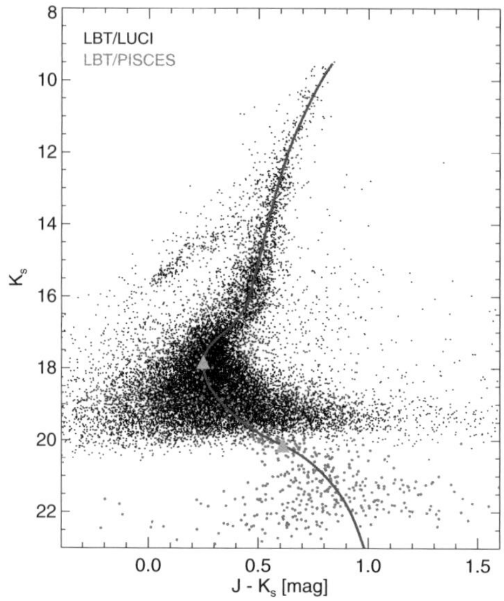

Figure 1. NIR (J,J-K) Color-Magnitude Diagram of the Galactic globular M15. Data were collected seeing limited with LUCI at LBT (black dots) and with the AO system (FLAO) at LBT (red dots) in an off-center field. The red line shows the ridgeline of the cluster, while the bright and faint green triangles mark the main sequence turn-off and the main sequence knee.

applies to the impact that systematics have on relative and absolute ages (Marín-Franch *et al.* 2009; Bono *et al.* 2010; Vandenberg *et al.* 2013).

An independent approach to date GCs, based on a different clock, and in turn on different physics, is the use of the white dwarf cooling sequence. Accurate CMDs provided new firm constraints on the absolute age of GCs (Bedin *et al.* 2005; Hansen *et al.* 2013) and old open cluster (Kalirai *et al.* 2007). The use of the WD luminosity function has solid advantages, when compared with the canonical approach (Salaris *et al.* 2010). Note that deep and accurate NIR photometry of stellar systems is going to play a crucial role, since the WD cooling sequence shows in the faint end, not only a blue (Hansen *et al.* 2004), but also a red turn off (Bono *et al.* 2013). This means new constraints on the absolute age and on the distance of GCs.

A very interesting approach for dating individual and cluster stars is the stellar cosmo-chronology. Its key advantage is that it relies on stellar spectroscopy only, and is therefore independent of errors affecting distance and reddening corrections. Moreover, a small fraction of extremely metal-poor ([Fe/H]$\leqslant$-2.5) stars shows a strong enhancement in neutron-capture elements (r-process) i.e. the channel producing the large majority of heavy elements. Its main problem is the need for accurate measurements of weak, blue lines of ^{232}Th (4019 Å, half-life 14 Gyr) and ^{238}U (3859.6 Å, half-life, 4.5 Gyr) together with the stable s-process element Eu (Frebel & Kratz 2009). This approach has only been applied to a few field stars (Cayrel *et al.* 2001; Sneden *et al.* 2003) and to three red giants (RG) in M15 (Sneden *et al.* 2000). The results are very promising, but still hampered by large errors, indeed U can only be measured using a single optical line.

A new spin on the problem of cluster dating was provided by space and ground-based deep and accurate NIR CMDs. They clearly display a well defined knee along the main sequence (MSK). The bending is due to an atmospheric opacity mechanism: the collisional induced absorption of H2 molecules (Pulone *et al.* 1999; Zoccali *et al.* 2000; Saumon *et al.* 2012). The difference in magnitude or in color between the MSTO and the MSK is a good age indicator, since the MSK is independent of age ($M/M_\odot$ ~0.4-0.5) and only mildly affected by metallicity (Kalirai 2015).

In this context, NIR images collected with the Multiconjugated Adaptive optics Demonstrator (MAD) available at VLT (Marchetti *et al.* (2003)) provided the unique opportunity to fully exploit this technique by using ground-based data (Bono *et al.* 2010). More recently, a modest number of NIR images collected with the AO system available at LBT (First Light Adaptive Optics available at LBT allowed us to constrain with very high accuracy the absolute age of the metal-poor Galactic globular M15. Figure 1 shows the comparison between the data and cluster isochrones, we found an age of 13.3±1.1 Gyr that agrees within the errors with the age based on the MSTO (12.9±2.6 Gyr). However, the error is at least a factor of two smaller than similar evaluations in the literature.

Note that the photometry of crowded stellar fields collected with ground-based AO systems is far from being a trivial issue. The key problem is that the shape of the PSF does depend on the distance from the guide stars used to close the AO loop. Note that this problem changes when moving from the J- to K- band images, and it is also affected by external seeing conditions. The above problems increase when moving from multi-conjugated to single-conjugated AO systems. In order to overcome these problems, different approaches have been suggested, based either on the use of a quadratic change (Bono *et al.* 2010) of the PSF across the FoV, or individual deconvolutions (Schreiber *et al.* 2014), or asymmetric PSF (Fiorentino *et al.* 2014; Monelli *et al.* 2015).

3. Nuclear bulge

The Nuclear Bulge represents the crossroads of several outstanding astrophysical and cosmological open problems. a) There are two competing theories to explain the formation of the GC. In the so-called "in-situ star formation" scenario, the stellar content of the GC builds up with stellar clusters and gas located in its vicinity. In the "accreted" scenario, the stellar content of the NB builds up thanks to the disruption of several GCs due to tidal forces (Capuzzo-Dolcetta & Miocchi 2008) . The former scenario is supported by the presence of the nuclear star cluster and of two very young open clusters: Arches (t~2–2.5 Myr, Figer *et al.* (2002); Najarro *et al.* (2004)) and Quintuplet (t~2.4–5 Myr, Figer *et al.* (1999); Liermann *et al.* (2010b)).

The period (age) range of a few classical Cepheids recently discovered (Matsunaga *et al.* 2011) and their kinematic properties (Matsunaga *et al.* 2015) support the presence of ongoing star formation activity during the last 30 Myra. Dating back to the seminal NIR investigation by Figer *et al.* (2004) there is also evidence of an intermediate-age stellar population (red clump stars). We still solid evidence of an old stellar population in the NB. Two dozen of type II Cepheids were identified by Matsunaga *et al.* (2013). These variables are the progeny of hot HB stars, i.e. stars older than 10 Gyr. However, there is mounting empirical evidence that some of them are in binary systems, and their progenitors appear to have ages of the order of a few Gyrs (Soszyński *et al.* 2011).

Space (HST, SPITZER) and ground based photometric (NAOS/CONICA at VLT) and spectroscopic (KMOS at VLT) data (Schödel *et al.* 2014; Feldmeier 2015; Ghez 2015) improved the analysis of NIR and MIR CMDs of the NB. However, we still lack firm constraints on the old stellar populations. Numerical simulations indicate that a disk-buckling instability drifted the material (stars, gas) present in the pristine Disk into the Bulge (Athanassoula 1992). This prediction is supported by the lack of old stellar tracers (old open cluster, RRL, HB stars) with Disk kinematics.

The radial distribution of RRLs in the Bulge is well known (Soszyński *et al.* 2014; Pietrukowicz *et al.* 2015). However, recent kinematic investigations indicate that RRLs

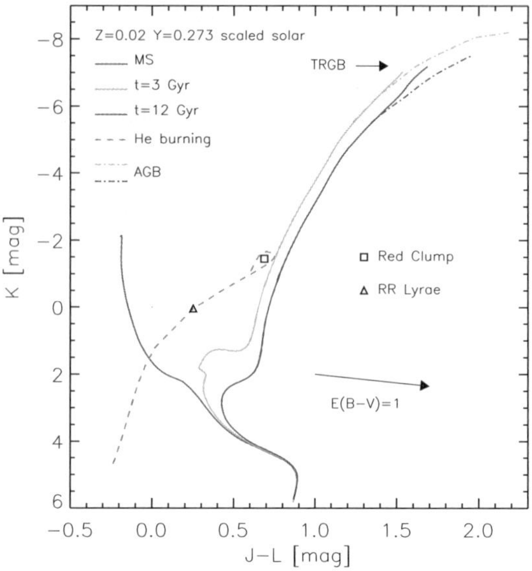

Figure 2. Predicted K,J-L CMD. The isochrones are based on evolutionary tracks (BASTI) computed assuming solar iron abundance (metals, Z=0.02; helium, Y=0.273) and a scaled solar chemical mixture (Pietrinferni *et al.* 2014). The red line displays a young (30 Myr) main sequence (MS), while the green and the blue lines two cluster isochrones for intermediate (3 Gyr) and old (12 Gyr) stellar structures. The purple dashed line shows the helium burning sequence for old and intermediate-age structures, while the dashed–dotted line the Asymptotic Giant Branch (AGB). The black triangle marks the center of the RR Lyrae instability strip, while the black square the Red Clump (RC) stars. The Tip of the Red Giant Branch (TRGB) is marked with a short black arrow. The long black arrow shows the reddening vector for E(B-V)=1 mag.

do not rotate like Bulge giants, indeed their mean radial velocities are vanishing independently of the longitude. Moreover, they have larger velocity dispersions than RGs. This means that RRLs do not follow the cylindrical rotation of Bar/Bulge stellar populations. However, photometric and spectroscopic investigations of RRLs in the Bulge stop at 2 degrees from the NB. This means that we lack information concerning old stellar tracers in the transition between the inner Bulge/Bar and the NB.

4. Paving the road to E-ELT

While waiting for the next generation of ELTs, there are second and third generation instruments at the 8-10m class telescopes, which will pave the way not only for the photometry, but also for the spectroscopy. The MCAO plus four sodium lasers available at Gemini North (GEMS†) provides superb NIR images of Galactic globulars (Turri *et al.* 2015). The Fov is 83x83 arcsec squared with a pixel scale of 0.02 arcsec/pixel and allowed to reach a very good accuracy down to K≈22 mag.

HAWK-I and the new AO Facility at UT4 (GRAAL‡) are also planned to deliver deep NIR images, but on a larger field of view: 7.5 arcmin square. Note that the pixel scale is 0.11 arcsec and is expected to decrease by almost a factor of two the seeing at optical wavelengths. However, the instrument which is going to play a crucial role in this context is the Enhanced Resolution Imager and Spectrograph (ERIS¶) which will

† http://www.gemini.edu/sciops/instruments/gems/
‡ http://www.eso.org/sci/facilities/develop/ao/sys/graal.html
¶ https://www.eso.org/sci/facilities/develop/instruments/eris.html

become available at UT4 (Amico *et al.* 2012; Kuntschner, H. *et al.* 2014). The reason is threefold: a) the field of view (1x1 arcmin square), the pixel scale (0.027 arcsec/px) and the expected performance of the single-conjugate AO system (Strehl ratio larger than 54% in K-band, using both a Natural Guide Star [NGS, R∼12 mag] and a Na Laser Guide star [LGS]) appear very promising for crowded stellar fields. b) The camera is covering a broad wavelength range (1-5 μm), and in turn a very good color sensitivity. c) The unique opportunity to use a NIR integral field unit, SCAO assisted, with a field of view of 8 arcsec square and a medium resolution grating (R∼8,000). This means iron and α-element abundances for evolved cool stars in the Galactic centre and Bulge.

To further constrain the impact of this instrument, Fig.2 shows a K,J-L CMD at solar chemical composition. Theoretical predictions (http://basti.oa-teramo.inaf.it) plotted in this CMD display three interesting features: a) the color sensitivity is 50% larger than classical K,J-K CMDs, and in turn a better sensitivity in the main sequence turn off (MSTO) region of old (K∼3 mag) and intermediate (K∼2 mag) stellar ages. b) The occurrence of the MSK at K∼5 and J-L∼0.9 mag. c) The most important old (RR Lyrae, K∼0 mag) and intermediate (RC, K∼-1.5 mag) age stellar tracers that display half magnitude difference in color.

If we assume that we can routinely reach with ERIS at VLT a limiting magnitude K∼22-23 (10σ), then we could also constrain the age distribution not only in inner/outer Bulge and in low reddening regions of the NB, since the expected apparent magnitude of the MSK, assuming a canonical reddening law (Cardelli *et al.* 1989) is 14.5 (true distance modulus) + 2 (A_K) + 5 (MSK absolute magnitude)∼21.5 mag. Moreover, the opportunity to measure RR Lyrae and RC stars in crowded and reddened central regions of dwarf galaxies in the LG ($\mu \sim 23$ mag).

5. Conclusions and future perspectives

The GCs and the NB are perfect targets for AO observations with this new generation of instruments, due to their high stellar density: *i)* they host plenty of bright stars (R-band < 16-17 mag, blue and red SGs, RGs and Asymptotic Giant Branch stars) that can be used as natural guide stars (NGS) to drive the tip-tilt correction; *ii)* they represent an ideal test-bed for the current on-sky AO performance and, in turn, to investigate our ability in doing accurate photometry/astrometry using these new images where the Point Spread Function (PSF) changes across the FOV. In particular, in the coming era of ELTs, it becomes crucial to understand whether available data reduction software and techniques will allow us to deal with and to exploit current AO data (Fiorentino *et al.* 2014).

An important step forward concerning the stellar content of GCs and of the NB is expected in the near future, once the new generation of extremely large ground-based telescopes (GMT, Giant Magellan Telescope [http://www.gmto.org/]; TMT, Thirty Meter Telescope [http://www.tmt.org/]; E-ELT, European Extremely Large Telescope [http://www.eso.org/sci/facilities/eelt/index.html]) equipped with sophisticated AO systems optimized for the NIR bands, will become available (Greggio *et al.* 2012).

The limiting magnitudes and the volume within which E-ELT will allow us to perform accurate NIR photometry is a quantum jump. Indeed, we can assume limiting magnitudes of J,K∼29–30 mag. There is an unprecedented opportunity to reach with MICADO (http://www.mpe.mpg.de/ir/micado) RR Lyrae and RC stars in the innermost regions of giant elliptical galaxies in nearby galaxy groups (Sculptor, μ ∼20; Centaurus A, μ ∼29; M81 μ ∼28). Moreover, the opportunity to reach TRGB (K∼ -7 mag) and AGB (K∼ -8 mag) stars in Virgo ($\mu \sim 30$), Eridanus (μ ∼32) and Coma (μ ∼35 mag) cluster.

Moreover, the E-ELT intergral field spectrograph HARMONI† and the AO assisted IFUs of the E-ELT multi-object spectrograph (MOSAIC‡) are going to provide a complete spectroscopic analysis (R∼20,000) of the old stellar component (MSTO, $M_K \sim 3$ mag) even in very high-reddened regions of the NB ($A_K \sim 3$ mag, K∼ 21-22). The same applies to galaxies in the LG or in its outskirts (Antlia group, $\mu \sim 26$, Karachentsev & Kashibadze (2005)) using RG and AGB stars. This means a never explored parameter space in the hearts of nearby galaxies. This is a jump in the current understanding of the formation and evolution of galactic bulges, and in particular for their chemical enrichment history. The same applies to the stellar populations in elliptical galaxies for which we only have limited hints from the Galactic bulge and from the dwarf elliptical M32.

Acknowledgements

GB thanks the Japan Society for the Promotion of Science for a research grant (L15518).

References

Amico, P. *et al.* 2012, *SPIE*, 8446, 20
Anderson, J., Sarajedini, A., Bedin, L. R. *et al.* 2008, *AJ*, 135, 2055
Arp, H. C., Baum, W. A., & Sandage, A. R. 1953, *AJ*, 58, 4
Athanassoula, E. 1992, *MNRAS*, 259, 345
Athanassoula, E. 2005, *AIPC*, 804, 333
Bedin, L. R., Cassisi, S., Castelli, F. *et al.* 2005, *MNRAS*, 357, 1038
Bertin, E. & Arnouts, S. 1996, *A&AS*, 117, 393
Bono, G., Genovali, K., Lemasle, B. *et al.* 2015, *ASPC*, 491, 148
Bono, G., Salaris, M., & Gilmozzi, R. 2013, *A&A*, 549, A102
Bono, G., Stetson, P. B., VandenBerg, D. A. *et al.* 2010, *ApJ*, 708, L74
Buonanno, R., Corsi, C. E., Castellani, M. *et al.* 1999, *AJ*, 118, 1671
Buonanno, R. 1989, *ESOC*, 31, 183
Capuzzo-Dolcetta, R. & Miocchi, P. 2008, *ApJ*, 681, 1136
Cardelli, J. A., Clayton, G. C., & Mathis, J. S. 1989, *ApJ*, 345, 245
Cayrel, R., Hill, V., Beers, T. C. *et al.* 2001, *Natur*, 409, 691
Chaboyer, B., Demarque, P., Kernan, P. J. *et al.* 1998, *ApJ*, 494, 96
Davies, B., Origlia, L., Kudritzki, R.-P. *et al.* 2009, *ApJ*, 694, 46
Di Cecco, A., Bono, G., Prada Moroni, P. G. *et al.* 2015, *AJ*, 150, 51
Diolaiti, E., Bendinelli, O., Bonaccini, D. *et al.* 2000, *ASPC*, 216, 623
Do, T., Kerzendorf, W., Winsor, N. *et al.* 2015, *ApJ*, 809, 143
Dolphin, A. E. 2000, *PASP*, 112, 1383
Dotter, A., Kaluzny, J., & Thompson, I. B. 2009, *IAUS*, 258, 171
Espinoza, P., Selman, F. J., & Melnick, J. 2009, *A&A*, 501, 563
Feldmeier, A. 2015, *IAUGA*, 22, 52322
Figer, D. F., Rich, R. M., Kim, S. S. *et al.* 2004, *ApJ*, 601, 319
Figer, D. F., Najarro, F., Gilmore, D. *et al.* 2002, *ApJ*, 581, 258
Figer, D. F., McLean, I. S., & Morris, M. 1999, *ApJ*, 514, 202
Fiorentino, G., Ferraro, I., Iannicola, G. *et al.* 2014, *SPIE*, 9148, 91483U
Frebel, A. & Kratz, K.-L. 2009, *IAUS*, 258, 449
Genovali, K., Lemasle, B., da Silva, R. *et al.* 2015, *A&A*, 580, A17
Ghez, A. 2015, *IAUGA*, 22, 58584
Girardi, L., Bertelli, G., Bressan, A. *et al.* 2002, *A&A*, 391, 195
Greggio, L., Falomo, R., Zaggia, S. *et al.* 2012, *PASP*, 124, 653
Hansen, B. M. S., Kalirai, J. S., Anderson, J. *et al.* 2013, *Natur*, 500, 51

† https://www2.physics.ox.ac.uk/research/visible-and-infrared-instruments/harmoni
‡ http://www.e-elt.nl/index.php?v=mosaic

Hansen, B. M. S., Richer, H. B., Fahlman, G. G. *et al.* 2004, *ApJS*, 155, 551
Kalirai, J. 2015, *IAUGA*, 22, 33338
Kalirai, J. S., Bergeron, P., Hansen, B. M. S. *et al.* 2007, *ApJ*, 671, 748
Karachentsev, I. D. & Kashibadze, O. G. 2005, *astro*, arXiv:astro-ph/0509207
Kristian, J. & Blouke, M. 1982, *SciAm*, 247, 66
Kunder, A., Rich, R. M., Hawkins, K. *et al.* 2015, *ApJ*, 808, L12
Kuntshner, H. *et al.* 2014, *SPIE*, 9147, 1
Liermann, A., Kraus, M., Schnurr, O. *et al.* 2010a, *MNRAS*, 408, L6
Liermann, A., Hamann, W.-R., Oskinova, L. M. *et al.* 2010b, *A&A*, 524, A82
Marchetti, E., Hubin, N. N., Fedrigo, E. *et al.* 2003, *SPIE*, 4839, 317
Marín-Franch, A., Aparicio, A., Piotto, G. *et al.* 2009, *ApJ*, 694, 1498
Martins, F., Hillier, D. J., Paumard, T. *et al.* 2008, *A&A*, 478, 219
Matsunaga, N., Fukue, K., Yamamoto, R. *et al.* 2015, *ApJ*, 799, 46
Matsunaga, N., Feast, M. W., Kawadu, T. *et al.* 2013, *MNRAS*, 429, 385
Matsunaga, N., Kawadu, T., Nishiyama, S. *et al.* 2011, *Natur*, 477, 188
Mezger, P. G., Zylka, R., Philipp, S. *et al.* 1999, *A&A*, 348, 457
Milone, A. P., Marino, A. F., Bedin, L. R. *et al.* 2015, *arXiv*, arXiv:1510.05086
Monelli, M., Testa, V., Bono, G. *et al.* 2015, *ApJ*, 812, 25
Najarro, F., Figer, D. F., Hillier, D. J. *et al.* 2009, *ApJ*, 691, 1816
Najarro, F., Figer, D. F., Hillier, D. J. *et al.* 2004, *ApJ*, 611, L105
Olszewski, E. W., Suntzeff, N. B., & Mateo, M. 1996, *ARA&A*, 34, 511
Origlia, L., Oliva, E., Sanna, N. *et al.* 2015, *arXiv*, arXiv:1510.06870
Philipp, S., Zylka, R., Mezger, P. G. *et al.* 1999, *A&A*, 348, 768
Pietrinferni, A., Molinaro, M., Cassisi, S. *et al.* 2014, *A&C*, 7, 95
Pietrukowicz, P., Kozłowski, S., Skowron, J. *et al.* 2015, *ApJ*, 811, 113
Pulone, L., de Marchi, G., & Paresce, F. 1999, *A&A*, 342, 440
Renzini, A. 1991, *ASIC*, 348, 131
Rich, R. M., Corsi, C. E., Cacciari, C. *et al.* 2005, *AJ*, 129, 2670
Ryde, N. & Schultheis, M. 2015, *A&A*, 573, A14
Salaris, M., Cassisi, S., Pietrinferni, A. *et al.* 2010, *ApJ*, 716, 1241
Salaris, M. & Weiss, A. 2002, *A&A*, 388, 492
Sandage, A. R. 1953, *AJ*, 58, 61
Saumon, D., Marley, M. S., Abel, M. *et al.* 2012, *ApJ*, 750, 74
Schechter, P. L., Mateo, M., & Saha, A. 1993, *PASP*, 105, 1342
Schild, R. & Kent, S. 1981, *SPIE*, 290, 186
Schödel, R., Feldmeier, A., Kunneriath, D. *et al.* 2014, *A&A*, 566, A47
Schreiber, L., Greggio, L., Falomo, R. *et al.* 2014, *MNRAS*, 437, 2966
Sneden, C., Cowan, J. J., Lawler, J. E. *et al.* 2003, *ApJ*, 591, 936
Sneden, C., Johnson, J., Kraft, R. P. *et al.* 2000, *ApJ*, 536, L85
Soszyński, I., Udalski, A., Szymański, M. K. *et al.* 2014, *AcA*, 64, 177
Soszyński, I., Dziembowski, W. A., Udalski, A. *et al.* 2011, *AcA*, 61, 1
Stetson, P. B. 1987, *PASP*, 99, 191
Stolte, A., Hussmann, B., Olczak, C. *et al.* 2015, *yCat*, 357, 80004
Stolte, A. & Brandner, W. 2010, *IAUS*, 266, 123
Turri, P., McConnachie, A. W., Stetson, P. B. *et al.* 2015, *ApJ*, 811, L15
Valenti, E., Zoccali, M., Gonzalez, O. A. *et al.* 2015, *arXiv*, arXiv:1510.07425
Vandenberg, D. A., Brogaard, K., Leaman, R. *et al.* 2013, *ApJ*, 775, 134
Vandenberg, D. A., Bolte, M., & Stetson, P. B. 1996, *ARA&A*, 34, 461
Vandenberg, D. A., Bolte, M., & Stetson, P. B. 1990, *JRASC*, 84, 412
Zoccali, M., Cassisi, S., Frogel, J. A. *et al.* 2000, *ApJ*, 530, 418

Formation, evolution, and survival of massive star clusters
Proceedings IAU Symposium No. 316, 2015
C. Charbonnel & A. Nota, eds.

doi:10.1017/S1743921316007031

The super star cluster driven feedback in ESO338-IG04 and Haro 11

A. Bik[1], G. Östlin[1], V. Menacho[1], A. Adamo[1], M. Hayes[1], J. Melinder[1] and P. Amram[2]

[1]Department of Astronomy, Oskar Klein Centre, Stockholm University, AlbaNova University Centre, 106 91 Stockholm, Sweden
email: arjan.bik@astro.su.se

[2]Aix Marseille Université, CNRS, LAM (Laboratoire dAstrophysique de Marseille), 13388 Marseille, France

Abstract. The stellar content of young massive star clusters emit large amounts of Lyman continuum photons and inject momentum into the inter stellar medium (ISM) by the strong stellar winds of the most massive stars in the cluster. When the most massive stars explode as supernovae, large amounts of mechanical energy are injected in the ISM. A detailed study of the ISM around these massive cluster provides insights on the effect of cluster feedback.

We present high quality integral field spectroscopy taken with VLT/MUSE of two starburst galaxies: ESO 338-IG04 and Haro 11. Both galaxies contain a significant number of super star clusters. The MUSE data provide us with an unprecedented view of the state and kinematics of the ionized gas in the galaxy allowing us to study the effect of stellar feedback on small and large spatial scales. We present our recent results on studying the ISM state of these two galaxies. The data of both galaxies show that the mechanical and ionization feedback of the super star clusters in the galaxy modify the state and kinematics of the ISM substancially by creating highly ionized bubbles around the cluster, making the central part of the galaxy highly ionized. This shows that the HII regions around the individual clusters are density bounded, allowing the ionizing photons to escape and ionize the ISM further out.

Keywords. galaxies: starburst, galaxies: individual: ESO338-IG04, Haro 11, galaxies: ISM, galaxies, star clusters

1. Introduction

Young massive stars and star clusters have a strong impact on the gas in their host galaxy. Young star star clusters, containing many massive stars, emit large amounts of Lyman continuum photons (LyC) able to ionize the ISM. Additionally, momentum is injected by the strong stellar winds of the most massive stars in the cluster. For slightly older clusters, (3 Myrs or older), the most massive stars explode as supernovae enabling the injection of large amounts of mechanical energy into the ISM.

These different feedback mechanisms have different effects on the ISM surrounding the clusters. Optical integral field spectroscopy with MUSE provides the ideal tool for studying affect the clusters have on the ISM in the galaxy. The youngest clusters will ionize their surroundings, resulting in a highly ionized medium around the clusters, traced by the ionization parameter (e.g. Pellegrini *et al.* 2012). The mechanical feedback from stellar winds and supernovae can result in galaxy scale outflows, traced by the kinematics of e.g. the Hα line.

We present optical integral field spectroscopy of two blue compact galaxies, Haro 11 and ESO 338-IG04 (Fig 1). These galaxies are considered local analogues of the high

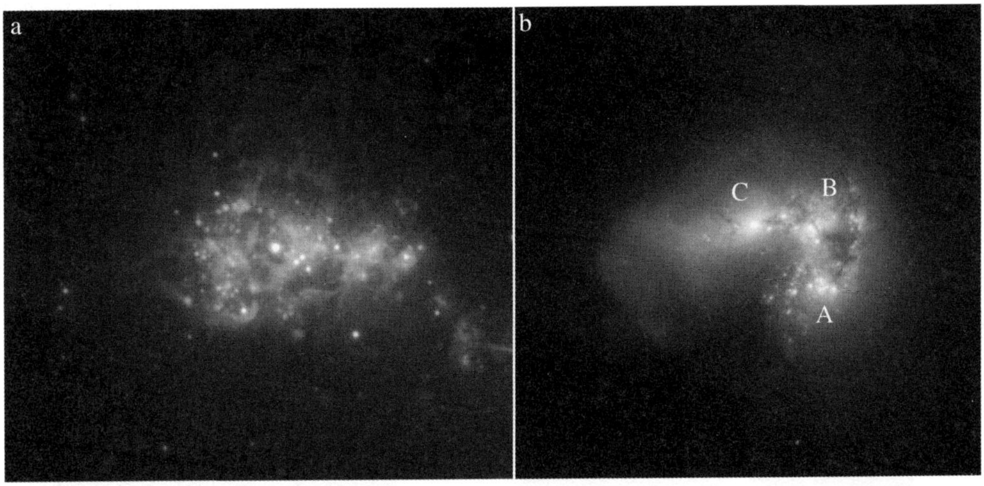

Figure 1. a) Three color-composite of HST observations of ESO 338-IG04. Blue: F140LP (UV), Green: F550M (V-band), Red: FR656N (Hα) The field of view of this image of 20 $\times$ 18 arcsec. **b)** Three color-composite HST observations of Haro 11, Blue: B-band, Green: I-band, Red: K-band. (Taken from Adamo *et al.* 2010. Credit: ESO/ESA/Hubble and NASA). The field of view of this image of 43 $\times$ 43 arcsec.

redshift Lyman break galaxies (Overzier *et al.* 2008). Studying their star formation and ISM properties teaches us about the star formation process in the high redshift universe.

Both galaxies have well studied stellar cluster populations, containing tens of super star clusters (Östlin *et al.* 1998, Östlin *et al.* 2003, Adamo *et al.* 2010, Adamo *et al.* 2011). ESO 338 contains a very young stellar population where the current starburst started $\sim$ 40 years ago (Östlin *et al.* 2003). Several increases in the cluster formation history have been found, the earliest at $\sim$ 20 - 30 Myrs, while the latest increase in cluster formation rate started 10 Myrs ago (Adamo *et al.* 2011). In Haro 11 the star formation is concentrated in 3 knots (Vader *et al.* 1993), each containing tens of young and older super star clusters. These galaxies show a high cluster formation efficiency, suggesting that 50 % of the star formation happens in the form of bound stellar clusters (Adamo *et al.* 2011). These properties make the galaxies ideal targets to study the effect of super star cluster feedback on the ISM of the galaxy.

2. Observations

ESO 338 and Haro 11 were observed with the integral field spectrograph Multi-Unit Spectroscopic Explorer (MUSE; Bacon *et al.* 2010) mounted on the Very Large Telescope (VLT) on Paranal, Chile. The observations of ESO 338 were performed during the first science verification run (Bik *et al.* 2015), the observations of Haro 11 as part of regular service mode observations.

The observations are taken in the no-AO mode in the extended wavelength setting, providing spectra between 4650 and 9350 Å in a field of view of 1 x 1 arcmin with a 0.2 arcsec pixel scale. One pointing, consisting of 4 integrations of 750 seconds, each rotated with 90 degrees, was used to observe ESO 338-IG04. The typical image quality is 0.9 arcsec in the V-band. For Haro 11 two pointings, offset with 30 arcsec east-west were obtained, to cover the large extent of the ionized halo. The image quality here is 0.8 arcsec. The data were reduced using the ESO MUSE pipeline (Weilbacher *et al.* in prep).

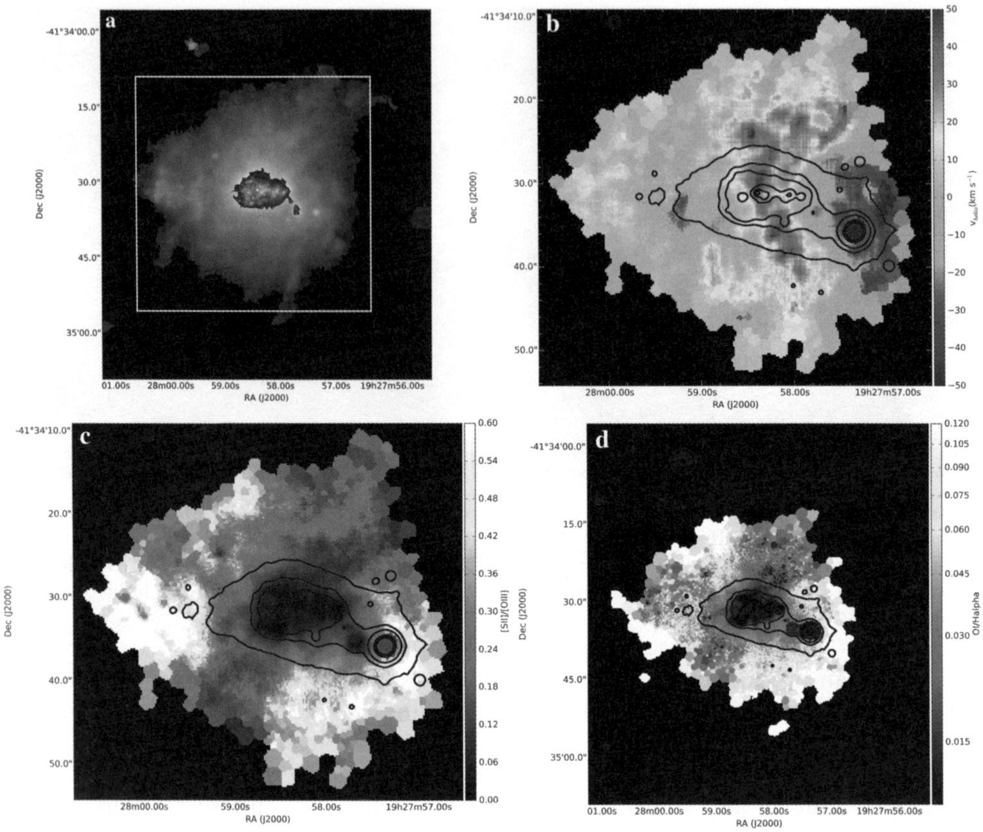

Figure 2. MUSE observations of ESO338-IG04. a) Continuum subtracted Hα with the HST image of Fig 1 super imposed, demonstrating the large halo surrounding ESO 338-IG04. The white box shows the zoomed in portion of the data visible in the other panels. b) The Hα velocity field, indicting large scale redshifted outflows are present. The black contours are contours of continuum emission, derived from stacking the MUSE cube along the wavelength direction. c) The ionization parameter represented by the [SII]/[OIII] line ratios showing the presence of ionization cones. d) The OI/Hα ratio indicting the presence of shocks at the other edges of the halo.

Emission line maps were extracted from the final data by numerically integrating under the line profile. The continuum was subtracted by averaging the continuum red- and blueward of the line. To enhance the low-surface brightness features, we binned the data using the weighted Voronoi tesselation algoritm by Diehl & Statler (2006), which is a generalization of the algoritm developed by Cappellari & Copin (2003). Additionally, the velocity field of the Hα line was calculated by fitting a single Gaussian profile.

3. Results

Figs. 2 and 3 show the results obtained from the MUSE data. Panels (a) show the Hα intensity maps and demonstrate that both galaxies are surrounded by large ionized halos (14.5 kpc for Haro 11 and 4.5 kpc for ESO 338). The velocity map (Figs. 2b and 3b) shows that the velocity field of the ionized gas is not simply tracing the gravitational potential of the galaxy. ESO 338 shows evidence for several redshifted outflows (Bik *et al.* 2015). The velocity map of Haro 11 looks very complicated and most of the gas dynamics is related to the merger event (Östlin *et al.* 2015). A detailed comparison between the Hα intensity

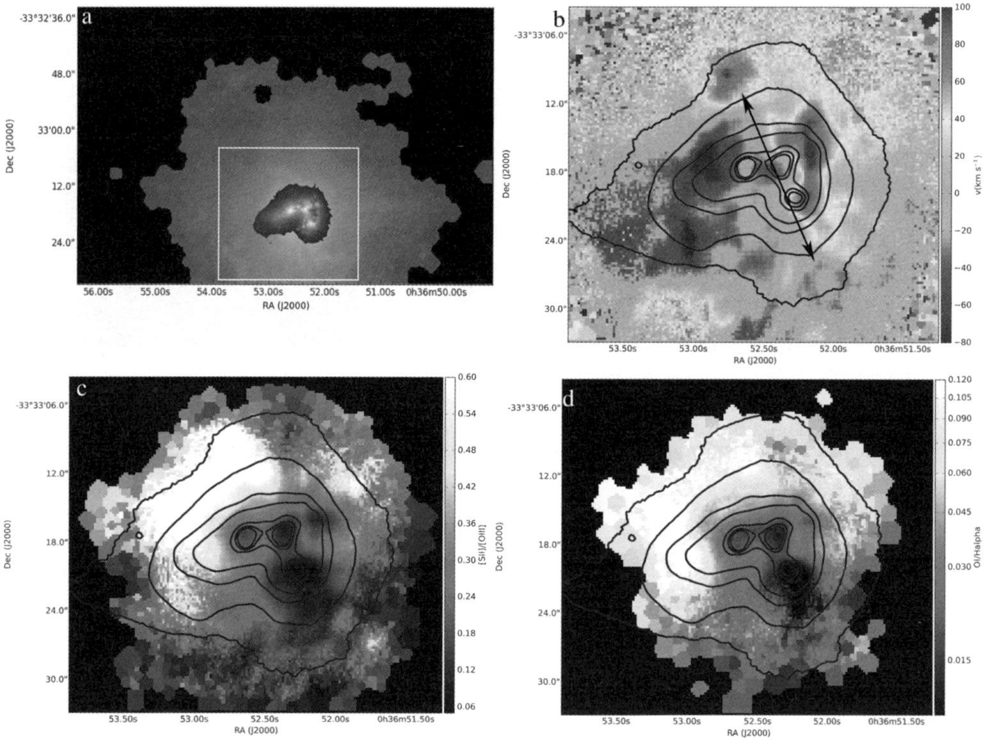

Figure 3. MUSE observations of Haro 11, panels present the same information as in Fig. 1. The arrow in panel b highlights the 2 suspected outflows in Haro 11.

map and velocity map reveals a blue and redshifted outflow candidate (annotated in Fig 3b).

To study the level of ionization in the ISM we calculated the [SII](λ6717 Å+ λ6731 Å)/[OIII]λ5007Å line ratio (Fig. 2c and 3c). The value of this line ratio is related to the ionization parameter and has been calibrated by Pelligrini *et al.* 2012. The lowest values of the line ratios are found towards the location of the young star clusters, consistent with the presence of early O stars in these clusters. However, also on larger scale, a significant fraction of the halos is ionized. In ESO 338 two ionization cones are cleary visible (to the northwest and south east). The ionization parameter map of Haro 11 shows that the Halo is ionized towards the west and south of the galaxy. Only towards the north-east is ISM the neutral. This means that the halos of both galaxies are density bound and facilitate the leakage of LyC photons.

In Fig. 2d and 3d the OIλ6300Å/Hα line ratio is shown, allowing us to discriminate whether the gas is heated by UV photo ionization (low values) or by shocks (high values). As expected from the ionization parameter maps, the location of the clusters show very low values, suggesting the heating by UV photo ionization of the young massive stars. However, where the material becomes more neutral the OI/Hα line ratio suggests the presence of shocks. In ESO 338, the high values are observed towards the outer edges of the halo, while in Haro 11 the north-east side of the halo is strongly shocked.

From the line ratio [SII]λ6717Å/[SII]λ6731Å we calculate the electron density in the halos and the HII regions surrounding the clusters using the calibration of Osterbrock (1989). Fig 4 shows that the halo gas around the galaxies is of low density ($n_e < 100$ cm^{-3}) where the observed line ratio reaches the asymtotic value of the calibration and

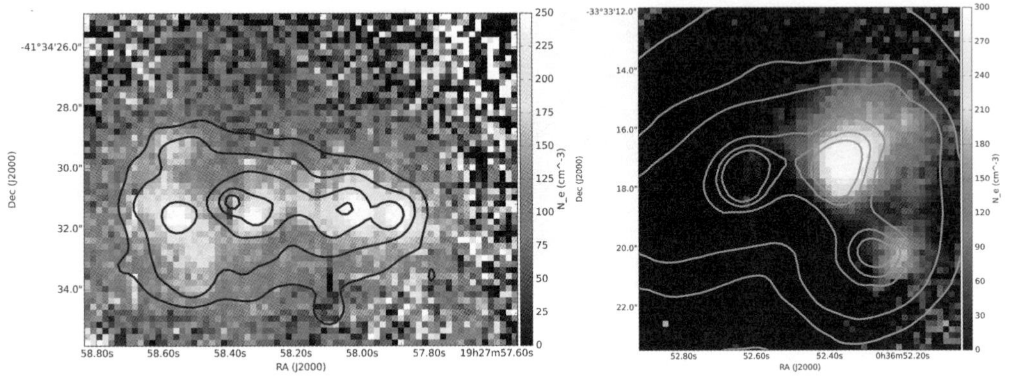

Figure 4. Electron density (n_e) derived from the line ratio maps of [SII]λ6717Å and [SII]λ6731Å. The contours are the same as in Fig. 2*Left:* Electron density map of the inner regions of the halo of ESO 338. The HII regions surrounding the young clusters reach higher densities then the low-density halo gas. *Left:* The electron density map of Haro 11. Knot B shows up most prominent as it contains most of the young cluster (Adamo *et al.* 2010).

cannot be used to derive quantative densities. Towards the HII regions surrounding the clusters in the galaxies, however, we find electron densities of n_e = 200 - 400 cm^{-3}. This suggests that these young clusters are still surrounded by parts of their natal HII region.

4. Discussion

4.1. *LyC and Lyα escape*

The ionization parameter maps show that the halos of both galaxies are density bound over a large part of their halos. The highly ionized cones of ESO 338 as well as the southern and western part of Haro 11 facilitate the escape of LyC photons in the intergalactic medium (see also Zastrow *et al.* 2011 and Zastrow *et al.* 2013). In fact Haro 11 is one of the few nearby galaxies where the LyC continuum radiation has been detected directly (Leitet *et al.* 2011). For ESO 338 an estimate of a LyC escape fraction of 16 % was derived based on analysis of the CII λ1036Å absorption line (Leitet *et al.* 2013).

Comparing the ionization cones and the outflows detected in ESO 338 with the Lyα emission maps of Hayes *et al.* (2007), a spatial correlation was found between the asymmetries in the Lyα map and the location of the outflows, suggesting that the outflows facilitate the escape of Lyα emission (Bik *et al.* 2015).

4.2. *stellar cluster feedback*

The observed properties of the ISM as discussed above can be linked to the star cluster population of ESO 338 and Haro 11. The ionization maps of both galaxies show several small regions of highly ionized gas. Matching their locations with those of the young stellar clusters shows that in the case of ESO 338 these small regions are highly ionized HII regions with high electron density, surrounding the youngest clusters. Östlin *et al.* (2003) derive ages of less than 2 Myrs for those clusters. The data on Haro 11 show a similar picture, all the star forming knots are surrounded by highly ionized, high density HII regions, consistent with them containing young star clusters (Adamo *et al.* 2010), with one exception, however. Knot C is detected in the ionization map, but shows a low electron density, like the rest of the halo around Haro 11. This is consistent with the fact that knot C is the older of the star forming knots, and the ionization might be caused by something other than star formation.

On larger spatial scales, the data show a low-density, partly highly ionized halo around both galaxies. ESO 338 shows clear presence of outflows as well as two ionization cones. Comparing the alignment of the outflows with that of the ionization cones shows that the southern outflow is not aligned with the southern ionization cone, and are probably two unrelated features. LyC photons, capable of ionizing the ISM are only produced by clusters with ages younger than $\sim$ 4 Myrs, while the mechanical energy released by stellar winds and supernovae remains roughly constant for $\sim$ 30 Myrs (Leitherer *et al.* 1999). This means that clusters with ages of $\sim$ 30 Myrs have inject most mechanical energy in the ISM and therefore might be responsible for the observed outflows. The LyC photons responsible for the ionized cones are created by the most recent increase in star formation. These channels might be outflow cones created by the older generation of clusters and might be re-ionized by the younger generation.

5. Conclusion

Both galaxies show that the ISM is significantly altered by the feedback from the super star clusters. Apart from the small HII regions around each young clusters, the ionization makes the galaxy act like a giant HII region expanding into the more neutral outer regions of ISM and creating shocks visible in the OI/Hα line ratio. These results confirm the results of Cormier *et al.* (2012) for Haro 11 derived from mid- and far-infrared fine structure lines. They built up a multi-phase ISM model of Haro 11, consisting of dense, highly ionized HII regions around the individial star clusters. These are surrounded by fragmented photo dissociation regions and embedded in a low-density, low ionized halo. With the MUSE data we confirm this picture and provide the missing two dimensional information on the location of the different ISM phases in the galaxies.

References

&Adamo, A., Östlin,G. Zackrisson, E., *et al.* 2010, *MNRAS*, 407, 870
Adamo, A., Östlin, G., & Zackrisson, E. 2011, *MNRAS*, 417, 1904
Bacon, R., Accardo, M., Adjali, L., *et al.* 2010, *SPIE Conf. Ser.*, 7735, 08
Bik, A. Östlin, G., Hayes. M., Adamo, A., Melinder, J., & Amram, P. 2015, *A&A*, 576, L13
Cappellari, M. & Copin, Y. 2003, *MNRAS*, 342, 345
Cormier, D., Lebouteiller, V., Madden, S. C., *et al.* 2012, *A&A*, 548, 20
Diehl, S. & Statler, T. S. 2006, *MNRAS*, 368, 497
Hayes, M., Östlin, G., Mas-Hesse, J. M., *et al.* 2005, *A&A*, 438, 71
Leitet, E., Bergvall, N., Piskunov, N., & Andersson, B.-G. 2011, *A&A*, 532, A107
Leitet, E., Bergvall, N., Hayes, M., Linné, S., & Zackrisson, E. 2013, *A&A*, 553, A106
Leitherer, C., Schaerer, D., Goldader, J. D., *et al.* 1999, *ApJS*, 123, 3
Osterbrock, D. E. 1989, *Astrophysics of Gaseous Nebulae and Active Galactic Nuclei* (Mill Valley : University Science Books)
Östlin, G., Bergvall, N., & Rönnback, J. 1998, *A&A*, 335, 85
Östlin, G., Marquart, T., Cumming, R. J., *et al.* 2015, *A&A in press*, ArXiv:1508.00541
Östlin, G., Zackrisson, E., Bergvall, N., & Rönnback, J. 2003, *A&A*, 408, 887
Overzier, Roderik A., Heckman, Timothy M., Kauffmann, Guinevere *et al.* 2008, *ApJ*, 677, 37
Pellegrini, E. W., Oey, M. S., Winkler, P. F., *et al.* 2012, *ApJ*, 755, 40
Vader, J. P., Frogel, J. A., Terndrup, D. M., & Heisler, C. A. 1993, *AJ*, 106, 1743
Zastrow, J., Oey, M. S., Veilleux, S., McDonald, M., & Martin, C. L. 2011, *ApJ*, 741, L17
Zastrow, J., Oey, M. S., Veilleux, S., & McDonald, M. 2013, *ApJ*, 779, 76

Formation, evolution, and survival of massive star clusters
Proceedings IAU Symposium No. 316, 2015
C. Charbonnel & A. Nota, eds.

doi:10.1017/S1743921315009412

The Assembly History of the Milky Way Nuclear Star Cluster

A. Feldmeier-Krause[1], N. Neumayer[2], R. Schödel[3], A. Seth[4], P. T. de Zeeuw[1,5], C. J. Walcher[6], N. Lützgendorf[7], M. Kissler-Patig[8], M. Hilker[1] and H. Kuntschner[1]

[1]European Southern Observatory (ESO),
Karl-Schwarzschild-Straße 2, 85748 Garching, Germany
email: afeldmei@eso.org

[2]Max-Planck-Institut für Astronomie,
Königsstuhl 17, 69117 Heidelberg, Germany

[3]Instituto de Astrofísica de Andalucía (IAA)-CSIC,
E-18008 Granada, Spain

[4]Department of Physics and Astronomy, University of Utah,
Salt Lake City, UT 84112, USA

[5]Sterrewacht Leiden,
Leiden University, Postbus 9513, 2300 RA Leiden, The Netherlands

[6]Leibniz-Institut für Astrophysik Potsdam (AIP),
An der Sternwarte 16, 14482 Potsdam, Germany

[7] ESA, Space Science Department,
Keplerlaan 1, 2201 AZ Noordwijk, The Netherlands

[8] Gemini Observatory,
670 N. A'ohoku Place, Hilo, Hawaii, 96720, USA

Abstract. Within the central 10 pc of our Galaxy lies a dense cluster of stars, the nuclear star cluster. This cluster forms a distinct component of our Galaxy. Nuclear star clusters are common objects and are detected in ~75% of nearby galaxies. It is, however, not fully understood how nuclear star clusters form. The Milky Way nuclear star cluster is the closest of its kind. At a distance of only 8 kpc we can spatially resolve its stellar populations and kinematics much better than in external galaxies. This makes the Milky Way nuclear star cluster the perfect local reference object for understanding the structure and assembly history of nuclear star clusters in general. There are of the order of 10^7 stars within the central 10 pc of the Galactic center. Most of these stars are several Gyr old late-type stars. However, there are also more than 100 hot early-type stars in the central parsec of the Milky Way, with ages of only a few Myr. Beyond a projected distance of 0.5 pc of the Galactic center, the density of young stars was largely unknown, since only very few spectroscopic observations existed so far. We covered the central >4 pc^2 (0.75 sq.arcmin) of the Galactic center using the integral-field spectrograph KMOS (VLT). We extracted more than 1,000 spectra from individual stars and identified >20 new early-type stars based on their spectra. We studied the spatial distribution of the different populations and their kinematics to put constraints on the assembly history of the Milky Way nuclear star cluster.

Keywords. Galaxy: center; kinematics; Stars: early-type, emission-line, Wolf-Rayet

1. Introduction

Nuclear star clusters (NSCs) were found in about 75% of low- to intermediate mass galaxies (Carollo *et al.* 1998; Böker *et al.* 2002; den Brok *et al.* 2014; Georgiev & Böker 2014). The typical effective radii of NSCs range from some pc to some tens of parsecs

(Böker *et al.* 2004; Côté *et al.* 2006; Georgiev & Böker 2014). They have masses in the range of 10^5 to 10^8 $M_\odot$ (Walcher *et al.* 2005; Ferrarese *et al.* 2006; Lyubenova *et al.* 2013). The NSC of our own Galaxy has a mass of 3×10^7 $M_\odot$ (Launhardt *et al.* 2002; Feldmeier *et al.* 2014; Schödel *et al.* 2014b) and an effective radius of 4.2 pc (Schödel *et al.* 2014a). Some galaxies contain a supermassive black hole (SMBH) within their NSC. The Milky Way (MW) is such a case, it hosts a black hole of 4.3×10^6 $M_\odot$ (Eckart *et al.* 2002; Ghez *et al.* 2005, 2008; Gillessen *et al.* 2009).

We study the stars of the NSC, in order to understand how and when mass was accreted by the center of the Galaxy, and to learn how the NSC and the SMBH grow. There are two different formation scenarios for NSCs: (1) Gas is accreted from the galactic disk. Then stars are formed from this gas in-situ, in the center of the galaxy (Milosavljević 2004; Pflamm-Altenburg & Kroupa 2009); (2) Stars form in massive star clusters outside the Galactic center. These clusters migrate to the center of the galaxy due to dynamical friction. There they merge and build the NSC (Tremaine *et al.* 1975; Capuzzo-Dolcetta & Miocchi 2008; Gnedin *et al.* 2013; Antonini 2013; Perets & Mastrobuono-Battisti 2014). But it is also possible, and probable, that both scenarios play a role in the formation of NSCs (Neumayer *et al.* 2011; Hartmann *et al.* 2011; De Lorenzi *et al.* 2013; Turner *et al.* 2012).

Once we understand how the MWNSC formed, it can serve as a reference object for understanding the formation of other NSCs. We follow two different paths. One is to study the stellar populations. Different stellar populations can tell us whether all stars formed at the same time, or in different star forming events. Secondly, we study the stellar kinematics. The stellar kinematics contain information about where the stars came from.

2. Young stars in the MWNSC are centrally concentrated

We observed the MWNSC using the K-band Multi-Object-Spectrograph (KMOS, Sharples *et al.* 2013). We observed ~28 pc^2 of the MWNSC, out to the effective radius at 4.2 pc. Here we present first results from the analysis of the central >4 pc^2. We extracted spectra of more than 1,000 stars with a S/N>10. In addition to the spectra, we have photometry in the J, H, and K_S bands (Schödel *et al.* 2010, and in prep.).

Our data set contains spectra of 23 Wolf-Rayet (WR) stars and six narrow-emission line stars (Krabbe *et al.* 1995; Blum *et al.* 2003; Paumard *et al.* 2006; Tanner *et al.* 2006). These stars are young, massive and hot. Nine stellar spectra are featureless and show no spectral lines. These sources are young stars which lose mass through strong stellar winds. The winds interact with the interstellar medium and produce bow shocks (Geballe *et al.* 2006; Viehmann *et al.* 2006; Perger *et al.* 2008; Buchholz *et al.* 2009; Sanchez-Bermudez *et al.* 2014). We classified 76 stars as O/B stars. Their spectra have Brackett γ and He lines. Our data contain 24 new O/B stars, primarily at large radii.

We show the positions of the 114 young early-type stars in Fig. 1. 90% of the young stars are located within a projected distance of 0.5 pc to the central SMBH (Feldmeier-Krause *et al.* 2015). The young stars are less than 10 Myr old (Krabbe *et al.* 1995; Paumard *et al.* 2006; Lu *et al.* 2013).

We derive a lower limit for the total mass of the young star cluster of $M_{total,young}$ = 12,000 $M_\odot$. This result is in excellent agreement with Lu *et al.* (2013), who found $M_{total,young}$ = 15,000 $M_\odot$.

Most of the stars in our data set are cool late-type stars. Previous studies of the central parsec revealed that most of them are several Gyr old red giants (Blum *et al.* 2003; Pfuhl *et al.* 2011), and some are younger red supergiants. We have spectra of 980 such stars. Unlike the young stars, the late-type stars are spread throughout the cluster.

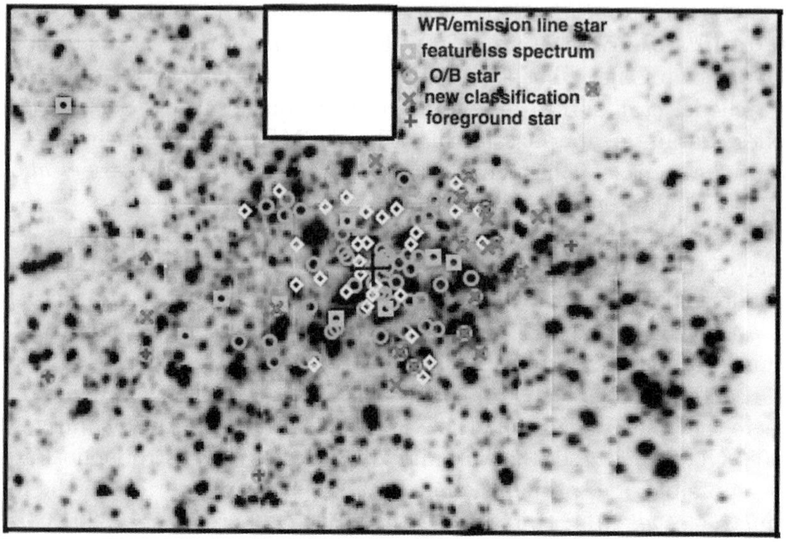

Figure 1. Spatial distribution of early-type stars in the MWNSC. The black box shows the KMOS field of view covering 2.51 × 1.68 pc, the small square in the upper middle was not observed due to the inactive IFU 13. The central black cross denotes the position of Sgr A*. Yellow diamond symbols denote confirmed WR and emission-line stars, cyan squares are stars with featureless spectra, green circles denote O/B stars. Red x-symbols indicate new young star candidates, blue plus-symbols probable foreground stars. Figure adapted from Feldmeier-Krause *et al.* (2015).

The fact that young stars are centrally concentrated suggests that these stars formed in-situ, in the center of the MWNSC. This is evidence against the infalling cluster scenario for these stars. If there was a cluster of young stars that fell towards the center, it must have lost many young stars on its way, due to tidal disruption. It would leave a trail of young stars, but we do not see such a trail. Therefore we favour the in-situ formation of the young stars of the MWNSC.

3. Stellar kinematics of old stars reveal complex structures

The stellar kinematics contain information about the origin of the stars of the MWNSC. We constructed a spectroscopic map of the central 9.5×8 pc of the MWNSC in integrated light using data obtained with the Infrared Spectrometer And Array Camera (ISAAC, Moorwood *et al.* 1998).

We derived the line-of-sight velocity and velocity dispersion of the unresolved faint stars (Feldmeier *et al.* 2014). The maps are shown in Fig. 2. The velocity map on the left panel reveals that the MWNSC rotates in the same sense as the Galaxy. The velocity dispersion map on the right shows an increase towards the central SMBH. We ran axisymmetric Jeans models and derived the total mass of the MWNSC $M_{MWNSC} = 3\times10^7\,M_\odot$ (Feldmeier *et al.* 2014; Schödel *et al.* 2014b). The radius of influence of the black hole is at $2.3^{+2.1}_{-0.7}$ pc (Feldmeier *et al.* 2014).

The velocity map reveals a complex internal structure of the kinematics. There is a misalignment of the kinematic position angle by $9^\circ \pm 3^\circ$ relative to the photometry, which is aligned with the Galactic plane. This misalignment is found at radii from 1 to 4 pc. We further found indications for a substructure rotating perpendicular to the Galactic

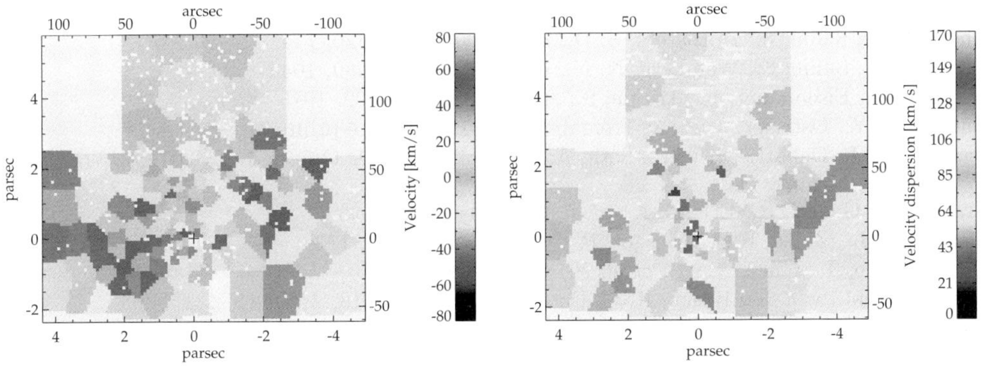

Figure 2. Large-scale kinematic maps of velocity and velocity dispersion (left, and right, respectively) in units of km/s. The coordinates are centred on Sgr A* and along the Galactic plane. The plus sign marks the position of Sgr A*. Figure adapted from Feldmeier *et al.* (2014).

plane at ~0.8 pc distance to the central black hole. This might indicate that the infall of massive star clusters contributes to the formation of NSCs (Feldmeier *et al.* 2014).

4. Summary

We looked at the stellar populations of the central 4 pc^2 of the MWNSC, and we found that young stars are strongly concentrated towards the center. This indicates that they formed there, in the center of the Galaxy. Further, we studied the large scale kinematics on 60 pc^2 of the MWNSC. We used the integrated light of the old stellar populations and found complex internal structures. Those structures may be the remnants of infalling massive star clusters.

There are still several open questions, and also a lot more data to analyse from the outer KMOS fields. We will study the metallicities of the cool, old stars. A metallicity spread as found by Do *et al.* (2015) shows that the late-type stars consist of different populations. We aim to investigate if there are spatial variations or gradients of the stellar population, and if the different stellar populations have different kinematics.

References

Antonini, F. 2013, *ApJ*, 763, 62

Blum, R. D., Ramírez, S. V., Sellgren, K., & Olsen, K. 2003, *ApJ*, 597, 323

Böker, T., Laine, S., van der Marel, R. P., *et al.* 2002, *AJ*, 123, 1389

Böker, T., Sarzi, M., McLaughlin, D. E., *et al.* 2004, *AJ*, 127, 105

Buchholz, R. M., Schödel, R., & Eckart, A. 2009, *A&A*, 499, 483

Capuzzo-Dolcetta, R. & Miocchi, P. 2008, *ApJ*, 681, 1136

Carollo, C. M., Stiavelli, M., & Mack, J. 1998, *AJ*, 116, 68

Côté, P., Piatek, S., Ferrarese, L., *et al.* 2006, *ApJS*, 165, 57

De Lorenzi, F., Hartmann, M., Debattista, V. P., Seth, A. C., & Gerhard, O. 2013, *MNRAS*, 429, 2974

den Brok, M., Peletier, R. F., Seth, A., *et al.* 2014, *MNRAS*, 445, 2385

Do, T., Kerzendorf, W., Winsor, N., *et al.* 2015, *ApJ*, 809, 143

Eckart, A., Genzel, R., Ott, T., & Schödel, R. 2002, *MNRAS*, 331, 917

Feldmeier, A., Neumayer, N., Seth, A., *et al.* 2014, *A&A*, 570, A2

Feldmeier-Krause, A., Neumayer, N., Schödel, R., *et al.* 2015, ArXiv e-prints

Ferrarese, L., Côté, P., Dalla Bontà, E., *et al.* 2006, *ApJ*, 644, L21

Geballe, T. R., Najarro, F., Rigaut, F., & Roy, J.-R. 2006, *ApJ*, 652, 370

Georgiev, I. Y. & Böker, T. 2014, *MNRAS*, 441, 3570
Ghez, A. M., Salim, S., Hornstein, S. D., *et al.* 2005, *ApJ*, 620, 744
Ghez, A. M., Salim, S., Weinberg, N. N., *et al.* 2008, *ApJ*, 689, 1044
Gillessen, S., Eisenhauer, F., Trippe, S., *et al.* 2009, *ApJ*, 692, 1075
Gnedin, O. Y., Ostriker, J. P., & Tremaine, S. 2013, ArXiv e-prints
Hartmann, M., Debattista, V. P., Seth, A., Cappellari, M., & Quinn, T. R. 2011, *MNRAS*, 418, 2697
Krabbe, A., Genzel, R., Eckart, A., *et al.* 1995, *ApJ*, 447, L95
Launhardt, R., Zylka, R., & Mezger, P. G. 2002, *A&A*, 384, 112
Lu, J. R., Do, T., Ghez, A. M., *et al.* 2013, *ApJ*, 764, 155
Lyubenova, M., van den Bosch, R. C. E., Côté, P., *et al.* 2013, *MNRAS*, 431, 3364
Milosavljević, M. 2004, *ApJ*, 605, L13
Moorwood, A., Cuby, J.-G., Biereichel, P., *et al.* 1998, *The Messenger*, 94, 7
Neumayer, N., Walcher, C. J., Andersen, D., *et al.* 2011, *MNRAS*, 413, 1875
Paumard, T., Genzel, R., Martins, F., *et al.* 2006, *ApJ*, 643, 1011
Perets, H. B. & Mastrobuono-Battisti, A. 2014, *ApJ*, 784, L44
Perger, M., Moultaka, J., Eckart, A., *et al.* 2008, *A&A*, 478, 127
Pflamm-Altenburg, J. & Kroupa, P. 2009, *MNRAS*, 397, 488
Pfuhl, O., Fritz, T. K., Zilka, M., *et al.* 2011, *ApJ*, 741, 108
Sanchez-Bermudez, J., Schödel, R., Alberdi, A., *et al.* 2014, *A&A*, 567, A21
Schödel, R., Feldmeier, A., Kunneriath, D., *et al.* 2014a, *A&A*, 566, A47
Schödel, R., Feldmeier, A., Neumayer, N., Meyer, L., & Yelda, S. 2014b, *Classical and Quantum Gravity*, 31, 244007
Schödel, R., Najarro, F., Muzic, K., & Eckart, A. 2010, *A&A*, 511, A18
Sharples, R., Bender, R., Agudo Berbel, A., *et al.* 2013, *The Messenger*, 151, 21
Tanner, A., Figer, D. F., Najarro, F., *et al.* 2006, *ApJ*, 641, 891
Tremaine, S. D., Ostriker, J. P., & Spitzer, Jr., L. 1975, *ApJ*, 196, 407
Turner, M. L., Côté, P., Ferrarese, L., *et al.* 2012, *ApJS*, 203, 5
Viehmann, T., Eckart, A., Schödel, R., Pott, J.-U., & Moultaka, J. 2006, *ApJ*, 642, 861
Walcher, C. J., van der Marel, R. P., McLaughlin, D., *et al.* 2005, *ApJ*, 618, 237

Formation, evolution, and survival of massive star clusters
Proceedings IAU Symposium No. 316, 2015
C. Charbonnel & A. Nota, eds.

doi:10.1017/S1743921315008972

Is the massive star cluster Westerlund 2 double? - A high resolution multi-band survey with the Hubble Space Telescope

Peter Zeidler[1,2], Antonella Nota[2,3], Elena Sabbi[2], Eva K. Grebel[1], Monica Tosi[5], Alceste Z. Bonanos[4], Anna Pasquali[1], Carol Christian[2] and Selma E. de Mink[6]

[1]Astronomisches Rechen-Institut, Zentrum für Astronomie der Universität Heidelberg, Mönchhofstr. 12-14, 69120 Heidelberg, Germany
email: pzeidler@ari.uni-heidelberg.de
[2]Space Telescope Science Institute, 3700 San Martin Drive, Baltimore, MD 21218, USA
[3]ESA, SRE Operations Devision
[4]IAASARS, National Observatory of Athens, GR-15326 Penteli, Greece
[5]INAF - Osservatorio Astronomico di Bologna
[6]Astronomical Institute Anton Pannekoek, Amsterdam University, Science Park 904, 1098 XH, Amsterdam, The Netherlands

Abstract. Westerlund 2 (Wd2) is one of the most massive young star clusters known in the Milky Way. The close proximity (4.16 kpc) to the Sun, and the young age (2.0 Myr) allow us to study star formation in detail at a high spatial resolution. We present results from our recent deep multi-band survey in the optical and near-infrared obtained with the *Hubble* Space Telescope. We demonstrated that, as expected, the region is affected by significant differential reddening with a median value of $E(B-V)_g = 1.87$ mag. The distance was inferred from the dereddened color-magnitude diagrams using Padova isochrones. Analyzing the spatial distribution of stars we found that Wd2 consists of two sub-clumps, namely the main cluster of Westerlund 2 and a less well populated clump located to the North. We estimated the same age of 0.1–2.0 Myr for both sub-clumps, thus we conclude that they are likely coeval.

Keywords. techniques: photometric - stars: early type - stars: pre-main sequence - HII regions - open clusters and associations: individual (Westerlund 2) - infrared: stars

1. Introduction

Westerlund 2 (hereafter Wd2; Westerlund 1961) is one of the most massive ($M \geqslant 10^4\ M_\odot$, Ascenso *et al.* 2007) young (0.1–2.0 Myr) star clusters in the Milky Way (MW). It is located in the Carina-Sagittarius arm $(\alpha, \delta) = (10^h 23^m 58^s.1, -57°45'49'')$(J2000) at a distance of ~ 4.16 kpc from the Sun (Zeidler *et al.* 2015a). It is the central ionizing cluster of the giant HII cloud RCW 49 (Rodgers *et al.* 1960). The estimated molecular cloud mass ranges from $(1.7 \pm 0.8 - 7.5) \times 10^5$ $M_\odot$ (Furukawa *et al.* 2009; Dame 2007, both estimates based on CO submillimeter/milimeter spectroscopy). Moffat *et al.* (1991) estimated that Wd2 hosts more than 80 O-type stars. Due to its close proximity to the Sun and its young age, Wd2 is a perfect target to study young massive star clusters very close to their earliest stages.

We are currently leading a high resolution, multi-band study of Wd2 with the *Hubble* Space Telescope (HST) (PI: A. Nota). We observed Wd2 in four wide-band filters, namely $F555W$, $F814W$, $F125W$, and $F160W$, and two narrow-band filters ($F658N$ and $F128N$) centered on the Hα and Paβ line emission. The optical observations were acquired with the Advanced Camera for Surveys (ACS, Ubeda *et al.* 2012) and the infrared (IR) observations were obtained with the Wide Field Camera 3 in the IR channel (WFC3/IR, Dressel 2012). The field of view (FOV) is $\sim$ 16–20 arcmin^2. Our final photometric catalog comprises 17,121 objects. 2236 point sources are detected in all six filters. The limiting magnitudes are $F555W$ = 26.7 mag, $F814W$ = 25.3 mag, $F125W$ = 23.1 mag, and $F160W$ = 22.6 mag. Our optical data are $\sim$ 5 mag deeper than the photometric data used by Vargas Álvarez *et al.* (2013) obtained with the Wide-Filed Planetary Camera 2 (WFPC2) (Gonzaga 2010). The near-infrared data reach 3–5 mag deeper than those of Ascenso *et al.* (2007). A detailed description of the observations, the data reduction, and the photometric catalog is presented in Zeidler *et al.* (2015a). Our observations were chosen as the 25th anniversary picture for the HST (see Fig. 1).

Our goals are a full census of the stellar population in Wd2 down to 0.1 $M_\odot$, its spatial distribution, its pre-main-sequence (PMS) component, its mass function, and the total cluster and stellar mass. Additionally, we will see if Wd2 already undergoes mass segregation, what is already seen at massive star clusters of this age (e.g. NGC 3603, Pang *et al.* 2013).

2. The gas excess toward Wd2

Wd2 is the central cluster of the giant HII region RCW 49. Due to its location in the Galactic disk it suffers from high differential reddening. Using the Hα and Paβ line emission we created a high resolution, pixel-to-pixel map of the gas excess $E(B-V)_g$. This map gives us the capability of individually deredden the cluster stars assuming that the gas and dust are located between us and Wd2.

2.1. *The pixel-to-pixel gas excess map*

In order to create the color excess map, we removed the continuum emission pixel by pixel. Therefore, we linearly extrapolated the continuum flux at the position of $F658N$ ($F128N$) using the $F555W$ and $F814W$ ($F125W$ and $F160W$) observations. Calzetti *et al.* (1996) provide a relation between the gas excess $E(B-V)_g$, the observed (R_{obs}), and theoretical flux ratio (R_{th}) of Hα and Paβ:

$$E(B-V) = \frac{-\log\left(R_{obs}/R_{int}\right)}{0.4\left[\kappa(\text{H}\alpha) - \kappa(\text{Pa}\beta)\right]}. \tag{2.1}$$

For the theoretical flux ratio R_{int} we derived a value of 17.546 with an assumed electron temperature of 10 000 K and an electron density of 100 cm^{-3} (Osterbrock & Bochkarev 1989). We adopted a normal total-to-selective extinction of $R_V = 3.1$. Using the extinction law of Cardelli *et al.* (1989) we derived $\kappa(\text{H}\alpha) = 2.437$ and $\kappa(\text{Pa}\beta) = 0.8035$.

The median $E(B-V)_g$ is 1.87 mag with a median uncertainty of $\sigma_{E(B-V)} = 0.032$ mag, which is of the order of our photometric uncertainties.

2.2. *Transformation of the gas excess to a stellar excess*

To use the color-excess map for our stellar photometry we needed to transform the gas excess into a stellar excess. Calzetti *et al.* (2000) showed a linear relation between the gas excess $E(B-V)_g$ and the stellar excess $E(B-V)_\star$. Up to now 36 OB-stars were spectroscopically observed in Wd2 (Rauw *et al.* 2004, 2007, 2011; Vargas Álvarez *et al.* 2013). Comparing the theoretical colors (Sparke & Gallagher 2007) observed colors of the stars with spectral type, we can infer the color excess $E(B-V)$. The median values of the gas excess transforms to a stellar excess of $E(B-V)_\star = 1.55$ mag. This procedure is described in detail in Zeidler *et al.* (2015a).

Figure 1. Color composite image of the HST ACS and WFC3/IR data of Wd2, including the *F*125*W* (red), *F*814*W* (green), and *F*555*W* (blue) filters (a color version is published in Zeidler *et al.* (2015a)). North is up, East to the left. The FOV is ∼4 arcmin×4 arcmin. This image was chosen to be the official Hubble 25th anniversary image. Credit: NASA, ESA, the Hubble Heritage Team (STScI/AURA), A. Nota (ESA/STScI), and the Westerlund 2 Science Team.

3. The cluster population

3.1. *The separation of the cluster members*

Due to its location in the Galactic disk, the Wd2 cluster population is contaminated by a high number of foreground objects. In order to get a proper census of the the stellar content in Wd2, as well as its spatial distribution we had to separate the cluster members from the foreground stars. Plotting the $F814W - F160W$ vs. $F814W$ color-magnitude diagram (CMD) of our catalog (see left frame of Fig. 2), we see two populations, a redder cluster member population and a bluer foreground population. The clear gap in their colors gives us the opportunity to separate the two populations with an arbitrary straight line. To show that the blue sequence is foreground, we simulated the MW disk

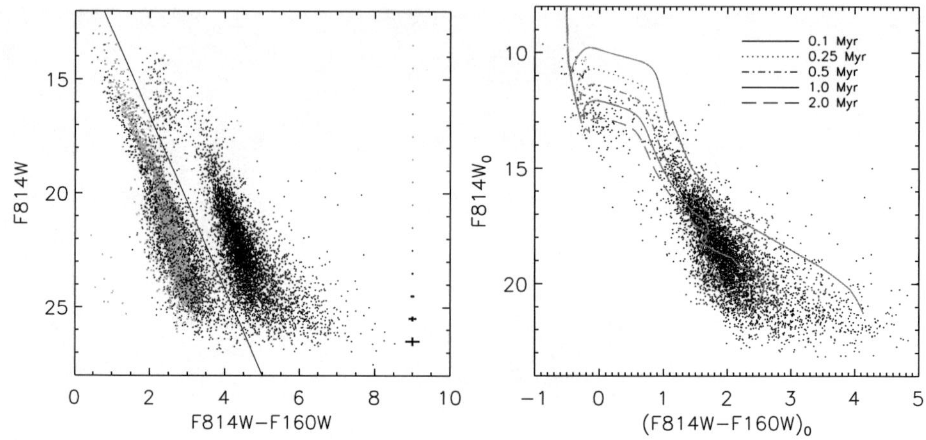

Figure 2. Left: The $F814W$ vs. $(F814W - F160W)$ CMD of the 7,697 point sources from the Wd2 photometric catalog. The points right to the straight line represent likely members of the RCW 49 region (including cluster members). The left sources are probable foreground stars. The gray points indicate the location of the synthetic objects from the Besançon model in the direction of Wd2. The black solid line marks our arbitrary separation between cluster and field stars. **Right:** The dereddened $F814W_0$ vs. $(F814W - F160W)_0$ CMD. Overlayed are the PARSEC 2.1S (Bressan *et al.* 2012) isochrones between 0.1 and 2.0 Myr for $Z = Z_\odot$. The diamonds mark the stars with known spectral type.

with the Besançon model (Robin *et al.* 2003) and indeed the gray points in the left-hand plot of Fig. 2 (representing the model) fall onto the blue sequence.

3.2. *The dereddened CMD of Wd2*

After the selection of likely cluster members, we used our high resolution gas excess map to deredden the photometry of each star. The total-to-selective extinction of $R_V = 3.95$ was determined via distance independent two-color diagrams (TCD, Zeidler *et al.* 2015a). In the right panel of Fig. 2 one can see the dereddened, cluster members in the $F814W_0$ vs. $(F814W - F160W)_0$ CMD with overplotted Padova PARSEC 1.2S (Bressan *et al.* 2012) isochrones in the range of 0.1–2.0 Myr for solar metallicity $Z = 0.0153$. The isochrones extend down to a stellar mass of 0.1 $M_\odot$ Using the turn-on (TO) region we determined a distance for Wd2 of 4.16 kpc in good agreement with Vargas Álvarez *et al.* (2013). The isochrones give an upper age limit of ~ 2 Myr for Wd2.

4. The spatial distribution of the stellar content

After selecting the Wd2 members we analyzed their spatial distribution. In order to compensate for incompleteness effects, we made magnitude cuts, using the luminosity function, where our observations can be considered as mostly complete. We used all objects with $F814W < 22.75$ mag and $F160W < 18.00$ mag. This leaves us with 2610 point sources. We created a 2D density map of the stars (see left-hand plot of Fig. 3) calculating the local surface density within a radius of $r < 10.8''$ from each star. This radius ensures, that we always have enough objects per density bin, yet we are not losing any small scale structures.

We fitted a combination of two 2D Gaussians with a common offset to the smoothed (binsize: 4 px) density map with independently determined peak positions and maximums of the two appearing sub-clumps. We took the 1-σ limits of the two peaks to define the two sub-clumps (thick black contours in Fig. 3), namely the main cluster (MC) and the

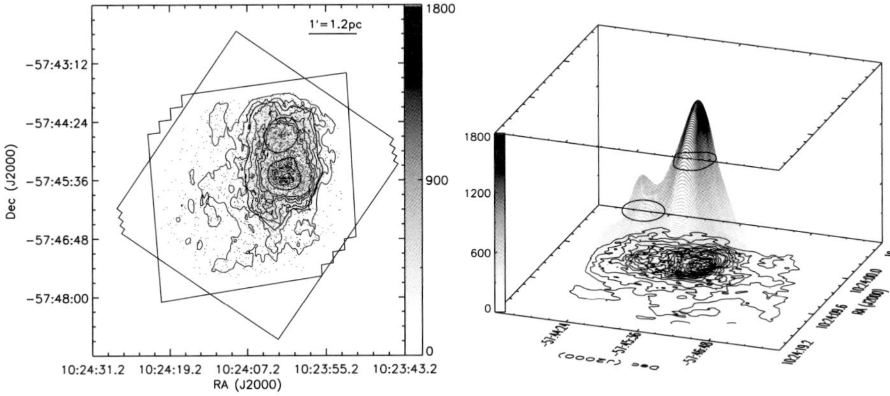

Figure 3. Left: Using the 2610 point sources selected with $F814W < 22.75$ mag and $F160W < 18.00$ mag, we created a 2D plot of the stellar density, taking into account all objects within a radius of $r < 10.8''$ around each star. The thick inner contours mark the two sub-clumps, namely the main cluster and the northern clump. The thick outer contour separates the Wd2 cluster from the periphery. The color-bar indicates the spatial density in arcmin^{-2} **Right:** 3D representation of the smoothed spatial distribution. The z-axis represents the density in arcmin^{-2}.

northern clump (NC). We separated the cluster of Wd2 from the peripheral region of RCW 49 using the 2-σ contour line (outer thick black contour in Fig. 3). In the right-hand plot of Fig. 3 one can see the 3D representation of the smoothed stellar distribution. The peak coordinates for the MC are R.A.=$10^{h}24^{m}01^{s}.65$ and Decl.=$-57°45'34''0$. and for the NC we got R.A.=$10^{h}24^{m}02^{s}.01$ and Decl.=$-57°44'39''2$.. This leads to a distance of the two sub-clumps of $d = 54.2''$ (1.08 pc). The areas of the MC and the NC cover 0.34 arcmin^2 (0.49 pc^2) and 0.39 arcmin^2 (0.56 pc^2), respectively.

The two sub-clumps have the same projected size, but the spatial density of the main cluster is more than a factor two higher. These radii are comparable to the core radii found by Kuhn *et al.* (2014) for 17 cluster regions. The individual CMDs show the same age for both sub-clumps. This leads to the conclusion that both sub-clumps are coeval. For a detailed analysis we refer to Zeidler *et al.* (2015a) and Zeidler *et al.* (2015b).

5. Summary and Conclusions

We presented a detailed analysis of the stellar content of Wd2 using high resolution multi-band observations with the HST. Combing the Hα and Paβ line emissions we created a high-resolution, pixel-to-pixel $E(B-V)_g$ gas excess map. Using the catalog of stars with spectral types, we managed to transform this gas excess to a stellar excess $E(B-V)_\star$. With this we were able to deredden the photometry of the stars in Wd2. To purge our catalog of foreground objects we used the $F814W$ vs. $(F814W - F160W)$ CMD. Two populations, clearly separated by ~ 1 mag (see left-hand plot of Fig. 2) are visible.

Using PARSEC 1.2S (Bressan *et al.* 2012) isochrones and plotting them over the dereddened cluster CMD (see right-hand plot of Fig. 2) we could infer an age of 0.1–2 Myr and a distance of 4.16 kpc for Wd2. In conclusion, Wd2 is of the same age or even younger than other very young star clusters like NGC 3603 (Pang *et al.* 2013) or R136 in the Large Magellanic Cloud (Walborn & Blades 1997; Sabbi *et al.* 2012). It is also younger than Westerlund 1, the most massive known young star cluster in the MW (Clark *et al.* 2005; Gennaro *et al.* 2011; Lim *et al.* 2013).

Fitting two 2D Gaussians to the spatial distributions of cluster members (see Fig. 3) we revealed that Wd2 is composed of two coeval sub-clumps, the MC and the NC having roughly the same spatial extent. The NC contains less than half of the number of objects than the MC. Over-plotting isochrones over the sub-clumps shows that both sub-clumps have the same age of 0.1–2 Myr. Sub-clustering is very common in young clusters (e.g. Kuhn *et al.* 2014). The radii of the sub-clumps are of the same order as the sub-clusters found by Kuhn *et al.* (2014) in 17 star forming regions.

References

Ascenso, J., Alves, J., Beletsky, Y., & Lago, M. T. V. T. 2007, A&A, 466, 137
Bressan, A., Marigo, P., Girardi, L., *et al.* 2012, MNRAS, 427, 127
Calzetti, D., Armus, L., Bohlin, R. C., *et al.* 2000, ApJ, 533, 682
Calzetti, D., Kinney, A. L., & Storchi-Bergmann, T. 1996, ApJ, 458, 132
Cardelli, J. A., Clayton, G. C., & Mathis, J. S. 1989, ApJ, 345, 245
Clark, J. S., Negueruela, I., Crowther, P. A., & Goodwin, S. P. 2005, A&A, 434, 949
Dame, T. M. 2007, ApJ, 665, L163
Dressel, L. 2012, Wide Field Camera 3 Instrument Handbook for Cycle 21 v. 5.0 (Baltimore: STScI)
Furukawa, N., Dawson, J. R., Ohama, A., *et al.* 2009, ApJ, 696, L115
Gennaro, M., Brandner, W., Stolte, A., & Henning, T. 2011, MNRAS, 412, 2469
&Gonzaga, S.and Biretta, J. 2010, in HST WFPC2 Data Handbook, v. 5.0 ,ed., Baltimore, STScI (Baltimore: STScI)
Kuhn, M. A., Feigelson, E. D., Getman, K. V., *et al.* 2014, ApJ, 787, 107
Lim, B., Chun, M.-Y., Sung, H., *et al.* 2013, AJ, 145, 46
Moffat, A. F. J., Shara, M. M., & Potter, M. 1991, AJ, 102, 642
Osterbrock, D. E. & Bochkarev, N. G. 1989, Soviet Ast., 33, 694
Pang, X., Grebel, E. K., Allison, R. J., *et al.* 2013, ApJ, 764, 73
Rauw, G., Manfroid, J., Gosset, E., *et al.* 2007, A&A, 463, 981
Rauw, G., Sana, H., & Nazé, Y. 2011, A&A, 535, A40
Rauw, G., De Becker, M., Nazé, Y., *et al.* 2004, A&A, 420, L9
Robin, A. C., Reylé, C., Derrière, S., & Picaud, S. 2003, A&A, 409, 523
Rodgers, A. W., Campbell, C. T., & Whiteoak, J. B. 1960, MNRAS, 121, 103
Sabbi, E., Lennon, D. J., Gieles, M., *et al.* 2012, ApJ, 754, L37
Sparke, L. S. & Gallagher, III, J. S. 2007, Galaxies in the Universe: An Introduction (Cambridge University Press)
Ubeda *et al.* 2012, Advanced Camera for Surveys Instrument Handbook for Cycle 21 v. 12.0 (Baltimore: STScI)
Vargas Álvarez, C. A., Kobulnicky, H. A., Bradley, D. R., *et al.* 2013, AJ, 145, 125
Walborn, N. R. & Blades, J. C. 1997, ApJS, 112, 457
Westerlund, B. 1961, Arkiv for Astronomi, 2, 419
Zeidler, P., Sabbi, E., Nota, A., *et al.* 2015a, AJ, 150, 78
Zeidler, P., Nota, A., Sabbi, E., *et al.* 2015b, in prep.

Formation, evolution, and survival of massive star clusters
Proceedings IAU Symposium No. 316, 2015
C. Charbonnel & A. Nota, eds.

doi:10.1017/S1743921315010583

Star-Formation Histories of Resolved Galaxies in the Local Volume

Alessandra Aloisi

Space Telescope Science Institute,
3700 San Martin Drive, Baltimore, MD-21218, USA
email: aloisi@stsci.edu

Abstract. Color-magnitude diagrams of resolved stellar populations are the most powerful tool to derive the star formation histories of galaxies. In this review the method to infer star formation histories by comparing synthetic color-magnitude diagrams to deep, high-precision, multi-color photometry is briefly outlined, and our current knowledge of the star formation history of systems within the local Volume is summarized.

Keywords. Local Group, Magellanic Clouds, galaxies: stellar content, galaxies: distances and redshifts, galaxies: evolution

1. Introduction

With the term of resolved galaxies we will refer to galaxies near enough to be resolved into single stars with high-angular resolution (space) telescopes, like the Hubble Space Telescope (HST) which has been a game changer in this field over the past two decades.

Thanks to HST, stellar populations have been resolved in several nearby galaxies, including dwarf and massive galaxies, early- and late-type galaxies, in the Local Group and beyond. Only the filters F555W/F606W (equivalent to Johnson/broadband V) and F814W (equivalent to Johnson-Cousins I) of the HST imaging cameras (e.g., WFC2, ACS/WFC, and WFC3 for $\sim$ 0.05"-0.1" per pixel) have been really exploited so far for this purpose, thanks to their higher sensitivity to both young and old stars.

2. Reconstructing the Star-Formation Histories of Galaxies

A galaxy star-formation history (SFH) is a detailed description of how, when, and where the galaxy formed its stars. SFHs of galaxies can be inferred from the color-magnitude diagrams (CMDs) of their resoved stars by applying the synthetic CMD method. This method consists in simulating a CMD by determining the variation of the star-formation rate (SFR) within the look-back time reached by the available photometry. Theoretical CMDs are created via Monte-Carlo based extractions of single stars from stellar evolution tracks, or isochrones, for a variety of star formation (SF) laws, IMFs, binary fractions, age-metallicity relations, distance, extinction, etc. Observational factors, including photometric errors, incompleteness, and stellar crowding, are also estimated and included in the procedure to fully reproduce an observed CMD. A combination of assumed parameters is acceptable only if the resulting synthetic CMD satisfactorily reproduces all the main features of the observational diagram (Fig. 1, left).

With sufficient investment of HST time, individual stars can be resolved down to the oldest main sequence turn-off (MSTO) in all galaxies of the Local Group and the resulting CMD can be used to infer their SFH over the entire Hubble time. If the oldest MSTO is not reached, then the look-back time depends on which CMD features can be resolved.

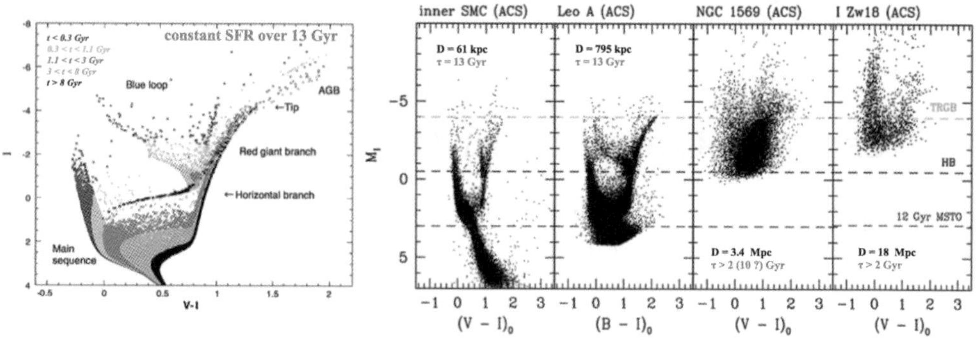

Figure 1. *Left:* Theoretical I vs. $V - I$ CMD for a galaxy that has formed stars constantly over the past 13 Gyr (Tolstoy 2011). Key stellar evolution phases indicating the presence of stellar populations of different ages, are also marked. *Right:* Effect of distance on depth of photometry in the observed CMDs of dwarf galaxies imaged with HST/ACS. The position of key CMD stellar features (indicated as horizontal dashed lines) can be used as indicator of the reachable look-back times and stellar ages within each galaxy (Courtesy: F. Annibali).

The Main Sequence (MS) contains stars with different ages that are nicely spread out in color, and older ages can be reached depending on the depth of the photometry. The Horizontal Branch (HB), an almost horizontal sequence at $I \sim 0$, only contains (metal-poor) stars >10 Gyr old and it is hard to interpret in terms of SFH, except for an indication that ancient stellar populations are present. The Red Giant Branch (RGB) contains stars >1-2 Gyr old, but without further information it is impossible to be quantitative about the SFH because of the age-metallicity degeneracy. The Asymptotic Giant Branch (AGB) above the tip of the Red Giant Branch (TRGB) is a mix of intermediate-age stars. Very young stars in the upper MS and Blue Loops (BLs) are much brighter, and obtaining the SFH over the last 1 Gyr, especially in actively star-forming galaxies, is possible well beyond the Local Group (Fig. 1, right).

3. Star-Formation Histories in the Local Group

The Local Group (LG) contains $\sim$ 50 galaxies within a radius of $\sim$ 1 Mpc, i.e. the distance from us of the most distant galaxies that are still dynamically part of the same group of galaxies of the Milky Way (MW). The LG mass is only dominated by 4 galaxies: the 3 spiral galaxies MW, Andromeda (M31), and Triangulum (M33), and the irregular Large Magellanic Cloud (LMC). All other galaxies, including the Small Magellanic Cloud (SMC), are less massive dwarfs (dEs, dSphs, dSph/dIrrs, and dIrrs), where a dwarf is defined as a galaxy with $M_B > -18$ mag. Resolved stellar population studies of the MW are limited to small nearby regions and star clusters due to our location. All other LG galaxies have been extensively studied and resolved into single stars thanks to their vicinity.

3.1. *Large and Small Magellanic Clouds*

The first global analysis of the resolved stars and SFHs of the LMC (8.5°x 7.5°) and SMC (4°x 4.6°) over their entire history is from the *UBVI* Magellanic Cloud Photometric Survey (MCPS) completed with the 1-m Swope Telescope (0.7" per pixel) at the Las Campanas in Chile (Fig. 2).

Following an initial burst of SF, both the LMC and SMC experienced a quiescent epoch from $\sim$ 10 to 3-5 Gyr ago. SF activity then resumed in both these MW satellites and has proceeded until the current time. The peaks in the SFH of the LMC at 0.5 Gyr and 2-3 Gyr are nearly coincident with similar peaks in the SFH of the SMC, suggesting

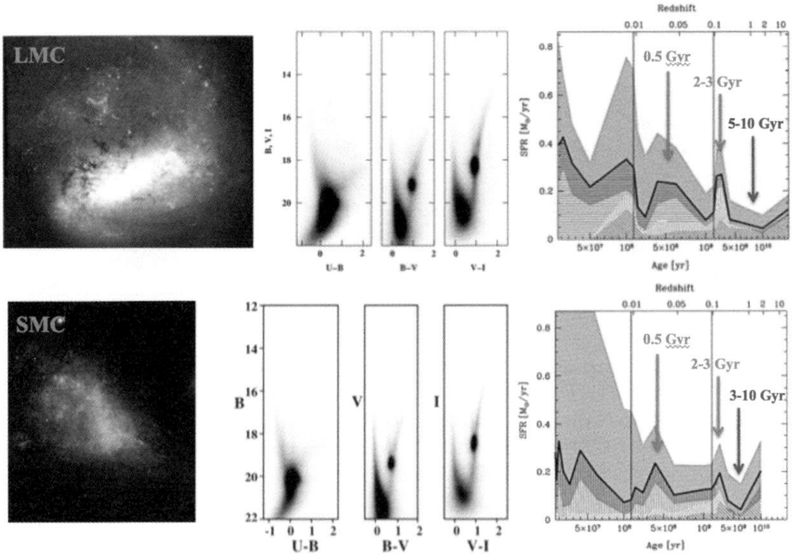

Figure 2. *Top Panels:* Multi-color MCPS image (left), CMDs in the $UBVI$ filters (middle), and global SFH (right) of the LMC. *Bottom Panels:* Same but for the SMC (Harris & Zaritsky 2004, 2009). SFR vs. age is shown as thick black line (right panels); uncertainty on the fit is shown as gray envelope. The metallicity distribution at each age is shown by the mix of colors ($Z = 0.001, 0.0025, 0.004$, and 0.008 $Z_\odot$ in blue, green, orange, and red, respectively).

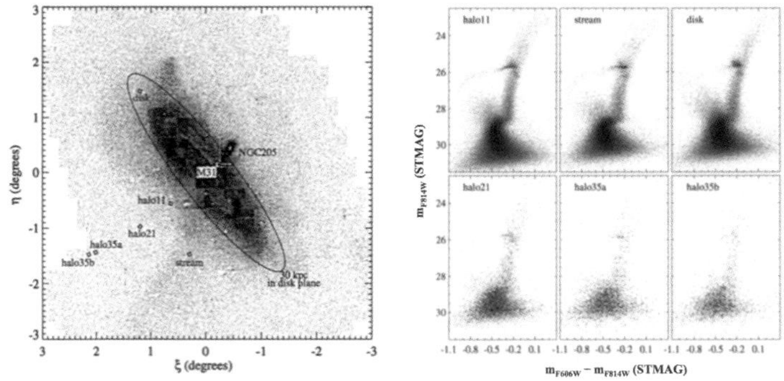

Figure 3. *Left:* M31 stellar density map with the 6 HST fields labelled. An ellipse marks the area within 30 kpc of the galactic center in the inclined disk plane. *Right:* I vs. $V - I$ CMDs of the 6 fields from Brown *et al.* (2009).

a joint history for these galaxies extending back at least several Gyr. These peaks are coincident with past peri-galactic passages with the MW. About 50% of the LMC and SMC mass was assembled 5-8 Gyr ago. The SFH of the field stars in the LMC and SMC is consistent with their cluster formation history (Harris & Zaritsky 2004, 2009).

3.2. *The Andromeda Galaxy (M31)*

Several stellar fields were observed with HST in the halo, disk, and stream of M31 over the past decade (e.g., Brown *et al.* 2009 and references therein; Fig. 3). The 11 kpc halo, the stream, and the disk share an extended SFH. The halo has $\sim 60\%$ of the mass in old (11-13 Gyr) metal-poor ([Fe/H] < -1 dex) stars and $\sim 40\%$ in intermediate-age (6-8 Gyr) metal-rich ([Fe/H] > -0.5 dex) stars likely connected to the extended outer disk (rather than the outer halo). The 11 kpc halo and the stream also have remarkably similar CMDs,

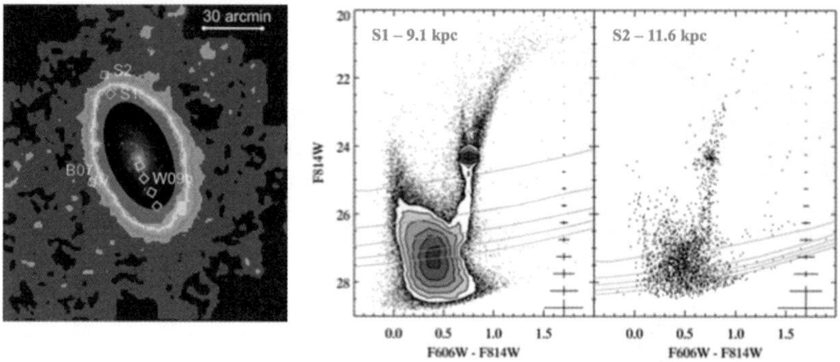

Figure 4. *Left:* Contour map of M33 RGB stellar surface density from the Isaac Newton Telescope Wide Field Camera (INT/WFC) survey with an optical image of the inner disc overlaid. All HST fields observed are labelled, with the two halo fields at 9.1 and 11.6 kpc indicated as S1 and S2. *Right:* I vs. $V - I$ CMD of S1 and S2. Median errors are on the right. Solid gray lines from top to bottom denote 90%, 80%, 70%, 60%, and 50% completeness (Barker *et al.* 2011).

even if the stream is ~ 1 Gyr younger on average. This suggests pollution of the halo by the stream progenitor that has been likely accreted by M31. The disk is instead younger (4-8 Gyr) and more metal-rich ([Fe/H] ~ -0.2 dex). The 21 and 35 kpc halo fields are mostly old, with an age spread around the mean value of ~ 10 and 11 Gyr, respectively, and a wide range in metallicity with a mean [Fe/H] value of ~ -0.7 and -1 dex, respectively. The extended halo of M31 thus cleary shows evidence of hierarchical assembly.

3.3. *The Triangulum Galaxy (M33)*

While several HST observations exist of M33, only a couple of fields (S1 and S2) at 9.1 and 11.6 kpc in the halo of the galaxy along its northern major axis, have been analyzed with the method of the synthetic CMDs to infer their SFH (Barker *et al.* 2011; Fig. 4). The 9.1 kpc halo field resulted on average younger (~ 3 Gyr old) and more metal-rich ([M/H] ~ -0.5 dex) than the 11.6 kpc field which is older (~ 7 Gyr old) and more metal-poor ([M/H] ~ -0.8 dex). The SFH of the 9.1 kpc halo field reveals how M33 inner disk has undergone an inside-out growth.

3.4. *Dwarf Galaxies in the Local Group*

All LG dwarf galaxies have been extensively observed with HST and show different CMDs, with dSphs lacking gas and recent SF, dSph/dIrr containing some gas but not recent SF, and dIrr having gas and undergoing SF. While each galaxy has its unique SFH extending over many Gyr, they all began forming stars ~ 13 Gyr ago. The majority shows no sign of pause due to reionization ~ 12.6 Gyr ago ($z \sim 6$). The SFHs of LG dwarfs can be used to validate galaxy formation and evolution models (Weisz *et al.* 2014; Fig. 5).

4. Star-Formation Histories beyond the Local Group

Going beyond the LG means going to distances > 1 Mpc where resolution of single stars becomes more challenging. While spirals are included in the LG, going beyond means to find other types of galaxies. The closest giant elliptical is Centaurus A (NGC 5128) in the Centaurus group at $D = 3.8$ Mpc. Active galaxies including galaxies undergoing strong SF like, e.g., starbursts and blue compact dwarf galaxies (BCDs), are also only found beyond the LG. The closest starburst is NGC 1569 at $D = 3.4$ Mpc in the IC 342 group, and the closest most metal-poor BCD is DDO 68 in a void at $D = 12.1$ Mpc (Fig. 6).

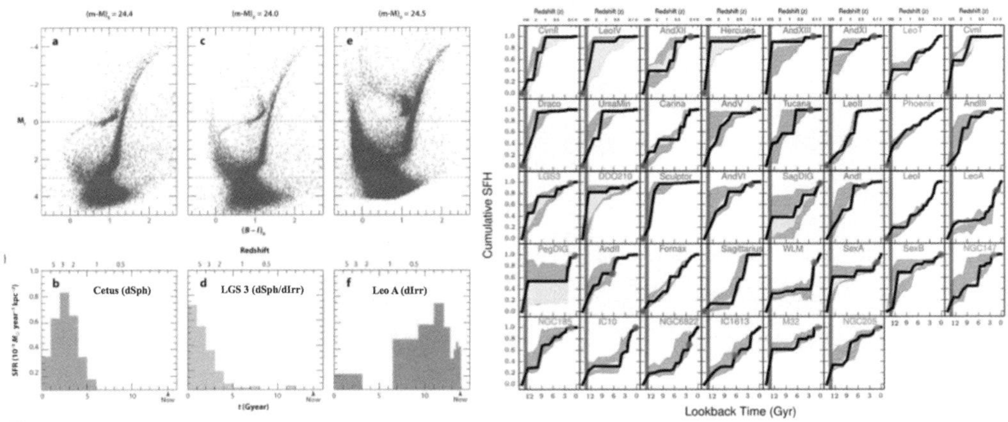

Figure 5. *Left:* HST/ACS I vs. $B - I$ CMDs and SFHs for 3 LG dwarf galaxies from the LCID project (Tolstoy, Hill, & Tosi 2009). *Right:* HST/WFPC2 cumulative SFHs in 38 LG dwarfs sorted by increasing luminosity (Weisz *et al.* 2014). Solid black lines are best-fit SFHs, yellow and gray envelopes represent 1σ uncertainties from random and systematic errors. Colored dots are approximate MSTO age of the CMD used to measure the SFH. Dwarfs are color-coded by morphological type (dSphs = red; dIrrs = blue; dTrans = orange; dEs = green). Red bands indicate the epoch of re-ionization. The majority of LG dwarfs did not form the bulk of their stellar mass prior to reionization, except for And V, And VI, And XIII, Hercules, and Leo IV.

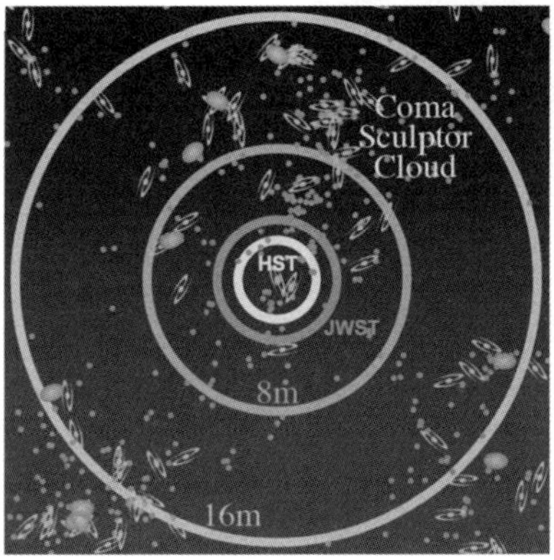

Figure 6. Map of nearby galaxies color-coded by morphological type (Blue = large spiral, Orange = large elliptical, Olive = dwarf galaxy). Circles show how far away a $L_{\odot}$ star can be detected at S/N=5 in two bands in a total exposure of 100 hours (Yellow = 2.4-m (HST), Red = 6.5-m (JWST), Green = 8-m, Purple = 16-m space telescopes). Spirals, ellipticals, and dwarfs come into view beyond 1 Mpc, corresponding to the yellow circle (Postman *et al.* 2010).

4.1. *The closest Giant Elliptical Galaxy: Centaurus A (NGC 5128)*

At D = 3.8 Mpc, NGC 5128 can only be resolved into single stars with HST. Several halo fields were observed with WFPC2 down to the RGBT. The deepest field located at $\sim$ 37 kpc from the galaxy center, was instead observed with ACS/WFC and its CMD revealed the Red Clump (RC) and HB in a giant E/SO galaxy for the first time (Rejkuba *et al.* 2005; Fig. 7). Combining the metallicity-sensitive color of the RGB with the age- and metallicity-sensitive features of the AGB and RC bumps, an average age of $\sim$ 8 Gyr was inferred. An average metallicity of $[M/H] \sim 0.6$ dex indicates that the halo is metal-rich all the way out. No SFH was inferred from the observed CMDs of this galaxy.

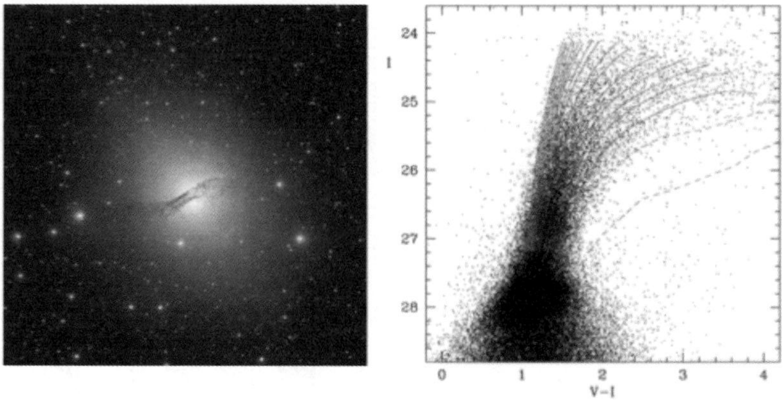

Figure 7. *Left:* Multi-color image of NGC 5128 obtained with the Wide Field Imager attached to the MPG/ESO 2.2-m telescope at the La Silla Observatory in Chile (Credit: ESO). *Right:* I vs. $V - I$ CMD of the deepest halo field in NGC 5128 located at 37 kpc (outside of image bottom left). The CMD displays a wide RGB, an AGB, and a RC; solid and dashed lines are various theoretical models at different metallicities (Rejkuba *et al.* 2005).

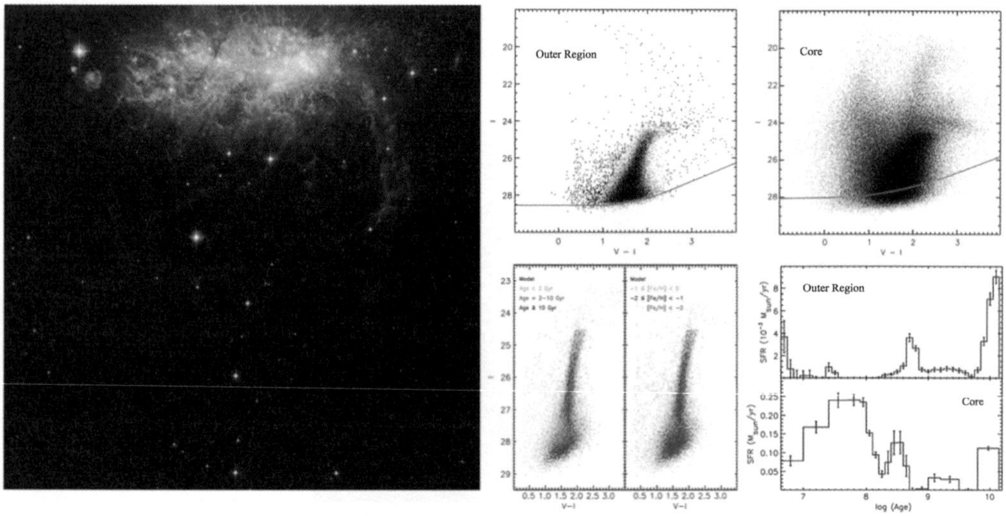

Figure 8. *Left:* ACS/WFC multi-color image of NGC 1569. *Top Middle and Right:* I vs. $V - I$ CMD for the outer region (halo) and the core. 20% completeness is shown as solid line. Features from young, massive, MS and evolved stars dominate the core. Only old RGB stars are visible in the outer region. *Bottom Middle:* Synthetic CMD for the outer region of NGC 1569. Different colors indicate separate ages (left) and metallicities (right). *Bottom Right:* SFR vs. age for the outer (Grocholski *et al.* 2012) and core-dominated (McQuinn *et al.* 2010) regions.

4.2. *The closest (Dwarf Irregular) Starburst Galaxy: NGC 1569*

At $D = 3.4$ Mpc, NGC 1569 is a dwarf irregular with a metallicity of ~ 0.25 $Z_{\odot}$ currently undergoing a strong starburst. Its halo was resolved into single stars with ACS/WFC down to the RC/HB by our group (Grocholski *et al.* 2012; Fig. 8). This CDM stellar feature was used to disentangle the age-metallicity degeneracy of the RGB in the simulations. The observed RGB width required a spread in metallicity (in addition to a spread in ages) similar to that of the MW halo. The oldest strongest SF episode at 10 Gyr accounts for most of the mass. A relatively low level of SF persisted between 10 and 1 Gyr, until the onset of a SF burst at $\sim$ 0.3-0.7 Gyr. The new distance and radial velocity

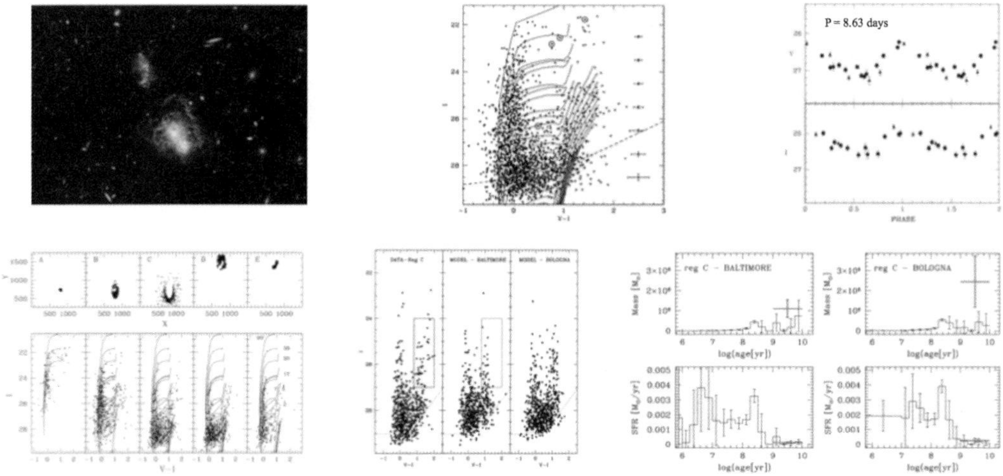

Figure 9. *Top Left:* ACS/WFC multi-color image of I Zw 18. *Top Middle:* I vs. $V-I$ CMD with stellar evolution tracks (RGB tracks in red). Median errors are on the right. Dashed curve is 50% completeness. TRGB gives $D = 18 \pm 2$ Mpc. Four Cepheids (blue circles) give $D = 19 \pm 2$ Mpc. *Top Right:* V and I light curves (corrected for foreground extinction) for the faintest 8.63 day Cepheid. *Bottom Left:* Spatial distribution of stars in regions A-E and corresponding CMDs with stellar evolution tracks for 60, 30, 20, 12, 7, 5, 2, 1, 0.8, and 0.6 $M_\odot$. *Bottom Middle:* Observed and synthetic CMDs for region C obtained with two different codes (Baltimore and Bologna). Dotted curve is 20% completeness. Box indicates TP-AGB stars. *Bottom Right:* Stellar mass and SFR vs. age for region C with Baltimore and Bologna codes. Thick red line at oldest ages indicates the values obtained using only one 1-10 Gyr bin.

suggest that NGC 1569 moved through the IC 342 group over the past few Gyrs The interaction with the group galaxy UGCA 92 may be driving the most recent burst ($<$ 10 Myr old).

4.3. *The Famous Most Metal-Poor BCD at the Edge of the Local Volume: I Zw 18*

I Zw 18 is the farthest galaxy ever been resolved into single stars with HST in the Local Volume ($\sim$ 20 Mpc). With the lowest metal abundance ever observed ($\sim$ 1/50 $Z_\odot$), it is considered the closest analog to primordial galaxies. ACS/WFC time-series observations were obtained by our group, and faint RGB stars that had been previously elusive, were resolved for the first time. A new distance was inferred via TRGB and Cepheids, indicating that the galaxy is much farther away at $\sim$ 18 Mpc (Aloisi *et al.* 2007; Fig. 9). Simulations confirm that I Zw 18 started forming stars $\sim$ 13 Gyr ago (Annibali *et al.* 2013).

4.4. *The Closest Newly Discovered Most Metal-Poor BCD : DDO 68*

DDO 68 is a BCD recently classified as metal-poor as I Zw 18, but a factor of 2-3 closer. Deep ACS/WFC observations by our group down to the RGB give a new distance of $D = 12.1$ Mpc, making the galaxy a few Mpc farther away than previous estimates. The CMDs of regions 1-4 show a gradient in stellar populations, with young stars mostly in the center, and old ($>$ 1 Gyr) RGB stars everywhere. A substantial population of $<$ 10 Myr old stars is in regions 2 and 3 where most of the H II regions are also located.

Particularly interesting is region 3 due to the presence of H II regions on a stream of stars suggesting accretion. Synthetic CMDs indicate a very high costant SF activity bewteen $\sim$ 10 Myr and 1 Gyr. A more recent, stronger, and still active burst is also detected at $\sim$ 1 Myr. 55% of the total mass was produced $>$ 1 Gyr ago (Sacchi *et al.* 2016, in

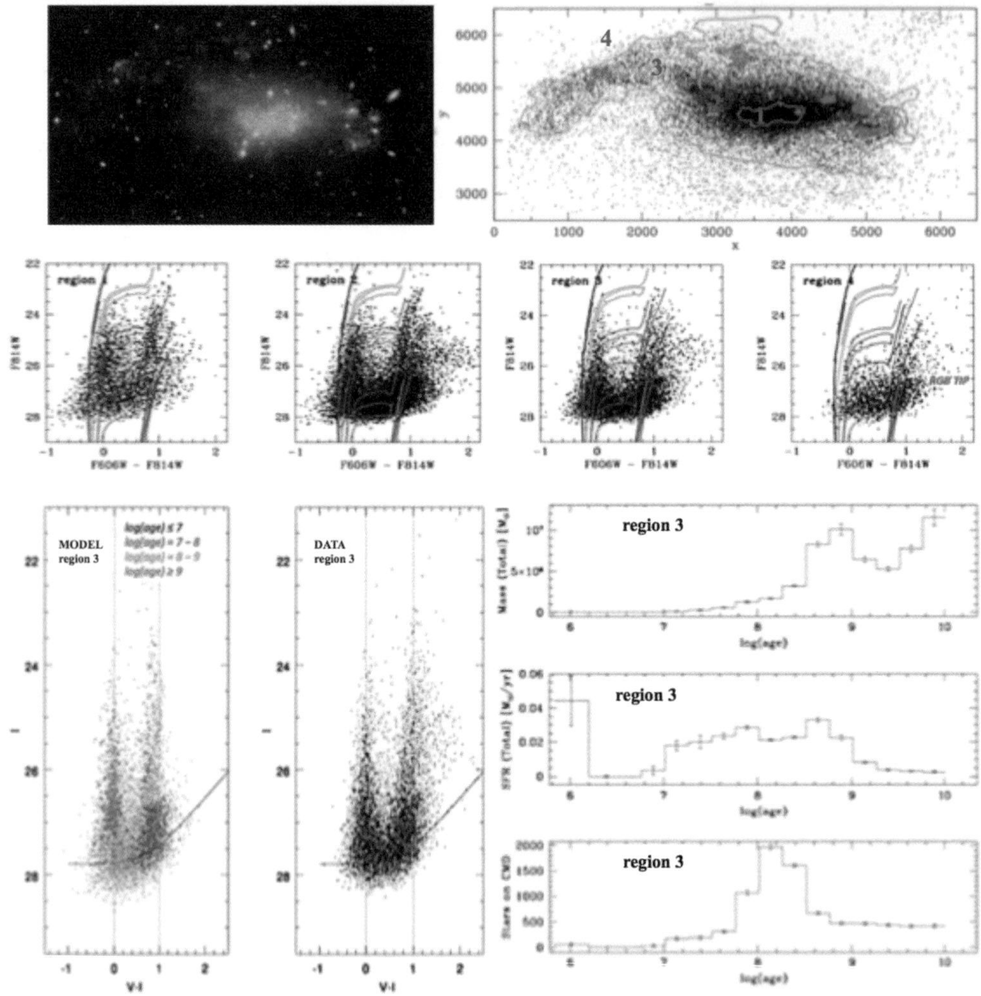

Figure 10. *Top Left:* ACS/WFC multi-color image of DDO 68. *Top Right:* Map of stars photometrized in V and I. Cyan contours define regions 1-4. Magenta contours indicate H II regions from $H\alpha$ imaging. *Middle:* I vs. $V - I$ CMDs of the four regions with isochrones overlaid for 10, 20, 50, 100, 200 Myr, 2 and 13 Gyr (top to bottom). *Bottom Left:* Observed and synthetic CMD of region 3. *Bottom Right:* Stellar mass, SFR, and number of stars vs. age for region 3.

preparation; Fig. 10). A low-mass H I companion of DDO 68 in the direction of the stream of stars of region 3, has been very recently located at $\sim$ 42 kpc (Cannon *et al.* 2014).

5. Summary and Conclusions

The SFH of resolved galaxies is a powerful tool to understand their formation and evolution, including interactions with the surrounding environment. All galaxies resolved into single stars with HST within the Local Volume have revealed stars as old as the look-back times probed by the data. The SFHs of LG dwarf galaxies with similar morphology vary widely, suggesting that many factors are at play in shaping the evolution of these objects. Studies of the halo in the closest spirals and ellipticals support the hierarchical galaxy formation scenario, while the SFH of active dwarf galaxies outside the LG suggest that interactions play a key role in enhancing their present SF activity.

References

Aloisi, A., Clementini, G., Tosi, M., Annibali, F., Contreras, R., Fiorentino, G., *et al.* 2007, *ApJ* (Letters), 667, 151

Annibali, F., Cignoni, M., Tosi, M., van der Marel, R. P., Aloisi, A., & Fiorentino, G. 2013, *AJ*, 150, 143

Barker, M. K., Ferguson, A. M. N., Cole, A. A., Ibata, R., Irwin, M., Lewis, G. F., *et al.* 2011, *MNRAS*, 410, 504

Brown, T. M., Smith, E., Ferguson, H. C., Guhathakurta, P., Kalirai, J. S., Kimble, R. A., *et al.* 2009, *ApJ* (Supplement), 184, 152

Cannon, J. M., Johnson, M., McQuinn, K. B. W., Alfvin, E. D., Bailin, J., Ford, H. A., *et al.* 2014, *ApJ*, 787, 1

Grocholski, A. J., van der Marel, R. P., Aloisi, A., Annibali, F., Greggio, L., & Tosi, M. 2012, *AJ*, 143, 117

Harris, J. & Zaritsky, D. 2004, *AJ*, 127, 1531

Harris, J. & Zaritsky, D. 2009, *AJ*, 138, 1243

McQuinn, K. B. W., Skillman, E. D., Cannon, J. M., Dalcanton, J., Dolphin, A., Hidalgo-Rodríguez, S., *et al.* 2010, *ApJ*, 721, 297

Postman, M., Brown, T., Sembach, K., Giavalisco, M., Traub, W., Stapelfeldt, K., Calzetti, D., Oegerle, W., *et al.* 2010, *Proc. of SPIE*, 7731, 77312K-1

Rejkuba, M., Greggio, L., Harris, W. E., Harris, G. L. H., & Peng, E. W. 2005, *ApJ*, 631, 262

Tolstoy, E. 2011, *Science*, 333, 176

Tolstoy, E., Hill, V., & Tosi, M. 2009, *ARAA*, 47, 371

Weisz, D. R., Dolphin, A. E., Skillman, E. D., Holtzman, J., Gilbert, K. M., Dalcanton, J. J., *et al.* 2014, *ApJ*, 789, 148

Formation, evolution, and survival of massive star clusters
Proceedings IAU Symposium No. 316, 2015
C. Charbonnel & A. Nota, eds.

doi:10.1017/S1743921315010510

The SUNBIRD survey: characterizing the super star cluster populations of intensely star-forming galaxies

Zara Randriamanakoto[1,2] **and Petri Väisänen**[2]

[1] Astronomy Department, University of Cape Town
Private Bag X3, Rondebosch 7701, South Africa
email: zara@saao.ac.za

[2] South African Astronomical Observatory
P.O. Box 9, 7935 Observatory, Cape Town, South Africa
email: petri@saao.ac.za

Abstract. Super star clusters (SSCs) represent the youngest and most massive form of known gravitationally bound star clusters in the Universe. They are born abundantly in environments that trigger strong and violent star formation. We investigate the properties of these massive SSCs in a sample of 42 nearby starbursts and luminous infrared galaxies. The targets form the sample of the SUperNovae and starBursts in the InfraReD (SUNBIRD) survey that were imaged using near-infrared (NIR) K-band adaptive optics mounted on the Gemini/NIRI and the VLT/NaCo instruments. Results from *i)* the fitted power-laws to the SSC K-band luminosity functions, *ii)* the NIR brightest star cluster magnitude – star formation rate (SFR) relation and *iii)* the star cluster age and mass distributions have shown the importance of studying SSC host galaxies with high SFR levels to determine the role of the galactic environments in the star cluster formation, evolution and disruption mechanisms.

Keywords. galaxies: interactions, galaxies: individual: IC 694, NGC 3690, galaxies: star clusters, infrared: galaxies

1. Introduction

Our current understanding of star formation activity in extragalactic sources has greatly increased since the commissioning of the IRAS satellite. With a strong infrared emission ranging in between $10^{11} - 10^{12}\,L_{\odot}$, luminous infrared galaxies (LIRGs) have been the subject of various studies. It has been found that the majority of these galaxies are interacting systems that harbor intense starburst activities in their nuclear regions (Sanders & Mirabel 1996). Such environments are ideal for the birth of the most massive star clusters in the local Universe. Also referred to as super star clusters (SSCs), these extreme form of star clusters are believed to contain vital information regarding the physical conditions under which strong and massive star formation occurs (e.g. Portegies Zwart *et al.* 2010). It has also been observed that active star-forming galaxies with ultra-dense giant molecular clouds (GMCs) favor the formation of the most massive SSCs ($\gtrsim 10^5\,M_{\odot}$) unlike the quiescent galaxies where the cluster masses average around $\sim 10^4 M_{\odot}$. However, the early stages of evolution as well as the role of the environment in shaping the cluster initial mass functions (CIMFs) are not yet fully understood (e.g. Adamo & Bastian 2015). The present work is designed to provide more insight into the universality of the CIMF and to pursue further investigations on whether external factors play an important role during cluster formation and disruption mechanisms.

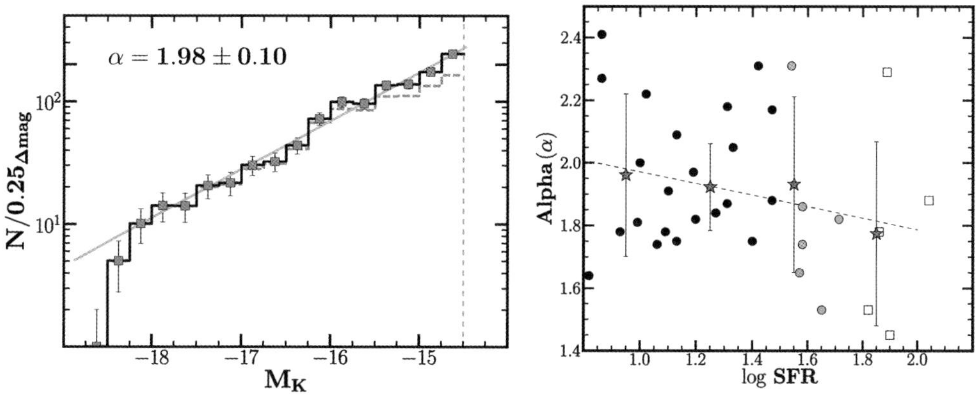

Figure 1. *Left:* K-band SSC LF of the sample, except the data from IRAS 19115−2124. The single power-law fit of the data points is represented by the solid line while the vertical line marks a completeness level of −14.5 mag. *Right:* The power-law slope α plotted against log-SFR. The targets with SFR $> 60\,M_\odot\,yr^{-1}$ are labelled as open squares, whereas the SFR $\leqslant 60\,M_\odot\,yr^{-1}$ targets are marked as cirlces: those with SFR $\leqslant 30\,M_\odot\,yr^{-1}$ are black and those in between 30 and 60 $M_\odot\,yr^{-1}$ are grey. A constant binning of the slopes results in values labelled as stars, and the dashed line represents a linear fit to these new points.

2. The survey

SSC properties of 42 nearby IR-bright galaxies, including LIRGs with extreme starburst activities are investigated in this work. The targets form a representative statistical sample of strongly star-forming galaxies. They are also part of an ongoing survey dubbed SUNBIRD (SUperNovae and starBursts in the InfraReD) to search for dust-obscured core-collapse SNe using NIR AO imaging mounted on the Gemini/NIRI and the VLT/NaCo instruments (e.g. Ryder *et al.* 2014, Väisänen *et al.* 2014). They are IRAS galaxies from the flux-limited RBGS catalogue, selected to have IR luminosities uniformly distributed from $\log(L_\odot/L_{IR}) = 10.6$ to 11.9 and distances $D_L \sim 25 - 150\,Mpc$, with $D_L \approx 200$ Mpc for a handful of the most distant galaxies. The targets are all starburst-dominated based on their cool IRAS colors ($f_{25}/f_{60} < 0.2$), though AGN were not excluded a priori. The sample were also selected in such a way that there are complimentary data set available in the Hubble Legacy Archive, most importantly optical imaging. Source extraction, aperture photometry and NIR selection of the SSC candidates are described in Randriamanakoto *et al.* (2013a) and Randriamanakoto *et al.* (2013b).

3. Luminosity functions of the star clusters

We constructed the K-band SSC luminosity functions (LFs) of the sample. Both constant and variable binning result in similar values of the power-law slopes $\alpha = 1.5 - 2.4$. The median and average of the indices are 1.86 ± 0.24 and 1.92, respectively. Out of the 34 constructed CLFs, 65 % have flatter distributions with $\alpha < 2$. Fitting a single power-law function to the LF of the combined dataset down to a $M_K = -14.5$ mag (which includes the SSCs from all the targets except for the most distant target IRAS 19115−2124 which has a significantly different magnitude), gives a slope $\alpha_1^{con} = 1.98$ with a formal uncertainty of 0.10 (left panel of Fig. 1). This is consistent with the average slope. One of the most important results of this work revealed that starburst galaxies have shallower slopes than normal spirals where $\alpha \approx 2.4$ (Whitmore *et al.* 2014). It is suggested that strongly star-forming galaxies such as LIRGs host SSCs that are disrupted in a way depending on their mass or environment. This would then result in a smaller value of the power-law

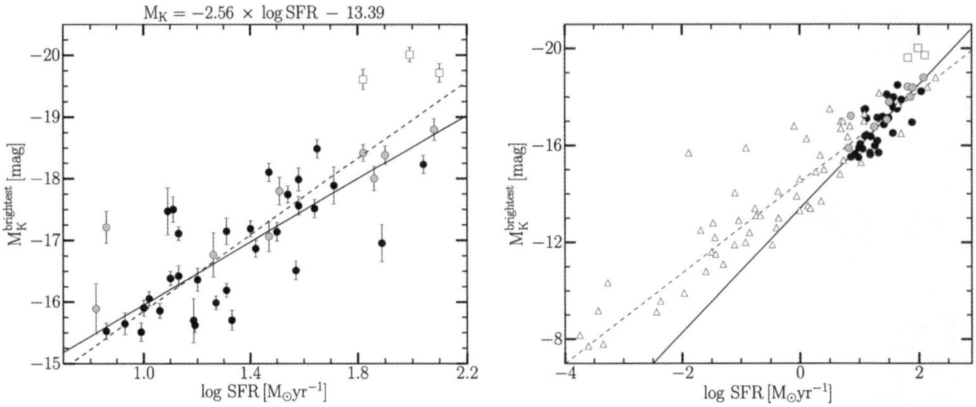

Figure 2. *Left:* A correlation is seen between the NIR brightest star cluster magnitude and the SFR of the galaxy. A weighted linear fit to all the data, including the three most distant targets at $D_L > 150$ Mpc shown as open squares, is represented by the dashed line. The solid line fits the $D_L \leqslant 150$ Mpc targets labeled as circles; those at $D_L \leqslant 100$ Mpc are black and those at $100 < D_L \leqslant 150$ Mpc are grey. *Right:* Another version of the $M^{brightest}$ – SFR relation with an expanded scale including data from Adamo *et al.* 2011 (the triangles) added to the present work (symbols as in the left panel). The solid line results from our best fit and the dashed line is the fit to the optical V-band data after a constant $V - K = 2$ conversion.

index α of the CLF. In fact, a weak correlation has been found between α and the SFR (right panel of Fig. 1). Note, however, that star clusters with different ages were binned together while drawing the LFs. Analyses based on Monte-Carlo simulations and star cluster population in a redshifted Antennae showed that shallower slopes cannot be the result of blending for a target less distant than ≈ 100 Mpc, unless the cluster surface densities are close to confusion limit (Randriamanakoto *et al.* 2013b). Such a limit is quickly reached in case of more distant targets and/or with data taken with poorer resolution. An interesting avenue would be to investigate whether or not the values of α vary with the different merging stages of our targets.

4. The brightest star cluster magnitude - NIR relation

We also used the NIR AO observations of the SUNBIRD survey to check whether one could reproduce a well-known empirical relation that has been established in the optical regime (Bastian 2008). We found a positive correlation between the brightest cluster magnitude and the galaxy SFR (left panel of Fig 2). It is expressed as $M_K^{brightest} = -2.56 \times \log \mathrm{SFR} - 13.39$, considering a distance cutoff $D_L = 150$ Mpc. The slope is within the same uncertainty range to those from optical points with lower SFR levels, assuming a single $V - K$ conversion (right panel of Fig 2). Note that the galaxy IR luminosity was the tracer being used to estimate the corresponding SFR for each target. The conversion eventually assumes that the IR energy is starburst-dominated. We have provided statistical interpretations of the relation: a size-of-sample effect. This simply means that it is more likely for more luminous clusters to form in an environment with extreme star formation activity. However, we also suggested physical interpretations to explain the small scatter seen in the relation. The effects of the galactic environments on cluster formation efficiency Γ or/and possibly other internal cluster effects such as the mass-luminosity relation are likely to play an important role as well (Randriamanakoto *et al.* 2013a). Apart from an estimate of the star cluster physical properties, determining

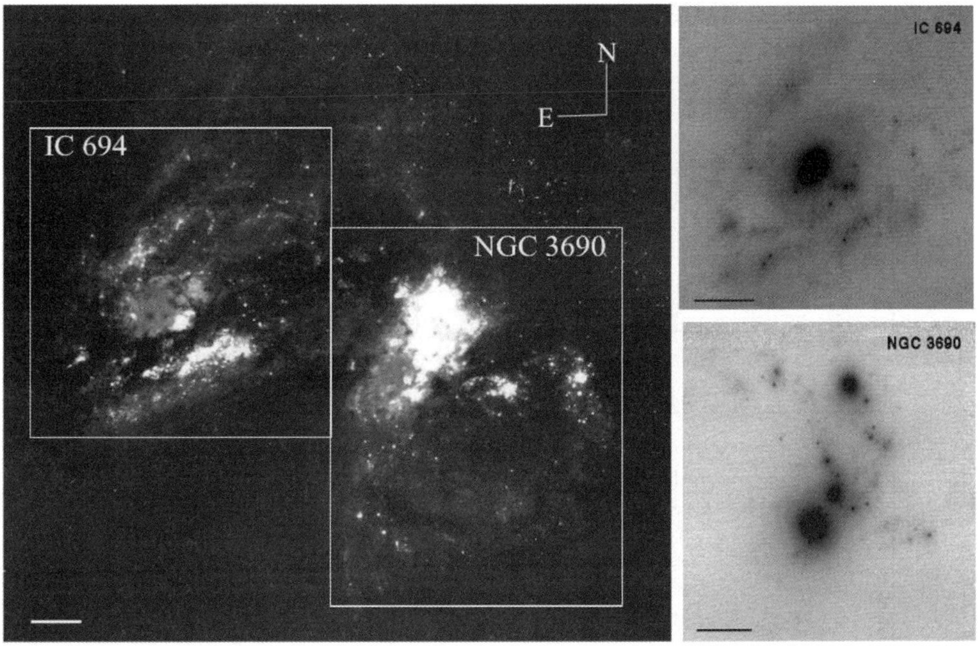

Figure 3. *Left:* WFC3/UVIS three-color image of Arp 299 in a 1.4 by 1.2 arcmin field: F336W (U-band, blue), F438W (B-band, green), and F814W (I-band, red). The boxes represent the FoV of the NIR data. North-East orientation and a length scale of 1 kpc (the horizontal line) are also indicated. *Right:* K-band images of the individual components taken with Gemini/NIRI AO systems in a 22×22 arcsec field: IC 694 (top) and NGC 3690 (bottom). The horizontal lines represent a 1 kpc scale.

the exact roles of the environments and the galaxy reddening would be the next steps required to fully understand the NIR relation.

5. Mass and age modelling of the optically-selected SSCs

NIR AO observations were also combined with HST optical images, whenever available, and new star cluster catalogues were derived based on optical detection. The selection criteria resulted in hundreds of SSCs for each target. We compared `Yggdrasil` SSP models (Zackrisson *et al.* 2011) with the observed SEDs to fit the cluster age, mass, and extinction of starburst-dominated galaxies. A large fraction of the star cluster populations have ages younger than 30 Myr. In the case of Arp 299 (Fig. 3), age-extinction degeneracy could reliably be broken due to inclusion of a UV-filter (F336W). The cluster masses of the interacting systems are generally between $10^4 - 10^8\ M_\odot$ (top panels of Fig 4). Such values are quite massive compared to the mass ranges associated with star clusters hosted in more quiescent environments. Nevertheless, more analyses should be performed to quantify the overestimate of the cluster mass in case of a BIK-fit. The spatial distributions of ages of some cases supported the cruel cradle effect scenario (Kruijssen *et al.* 2011, Väisänen *et al.* 2014) while others indicated the possibility of a mixed population (see middle panels of Fig 4 in the case of Arp 299). At least one of the LIRGs (NGC 3690) has its star cluster population disrupted in a mass-dependent manner. The turnover in the LFs, a truncation of the CMFs at high-mass end (lower right panel of Fig. 5), and the distribution of the cluster frequency (lower panels of Fig 4) serve as a basis for such an argument. Moreover, the cluster formation histories of the LIRG subsample revealed

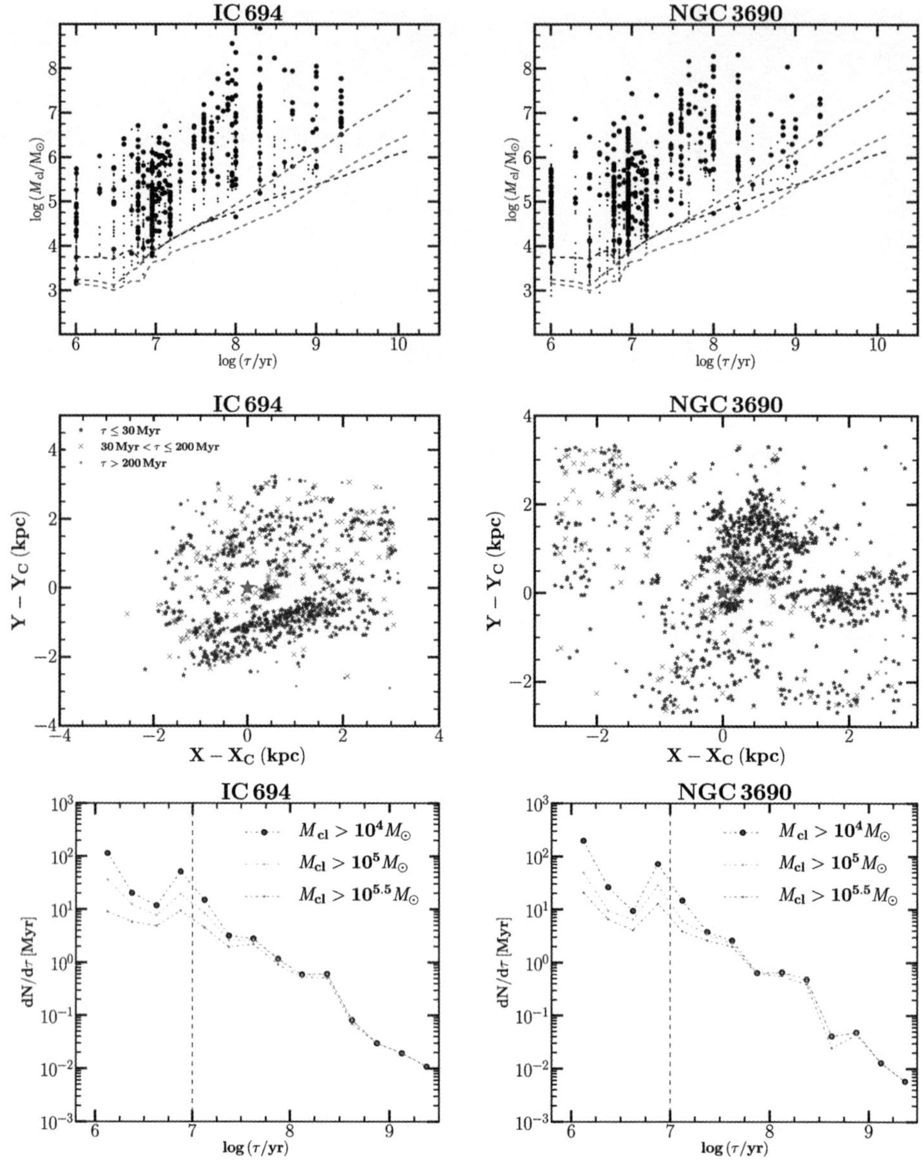

Figure 4. *Top:* The cluster age-mass planes from the χ^2-fit. Black points are the objects with $\sigma \leqslant 0.35$ in all four $UBIK$-images, whereas the dots are those below the error cutoff but in UBI-filters only. The dashed lines denote the photometric detection limits ($\sim 80\,\%$) considering U- (blue), B- (green), and I-band (black) evolutionary tracks. *Middle:* The age spatial distributions of the SSC candidates in the field of Arp 299. The green star represents the position of the galaxy center. The different labels correspond to different age bins. *Lower:* Star cluster frequency $\mathrm{dN}/\mathrm{d}\tau$ as a function of age for Arp 299. Three different lower mass limits were considered to derive various formation rates of the observed clusters.

that more formation of the star clusters are foreseen since the galaxy environments are rich of GMCs necessary for the birth of massive clusters. The next steps of this work would be to correct the cluster mass functions from observational incompleteness and to derive accurate values of Γ on global and local scales.

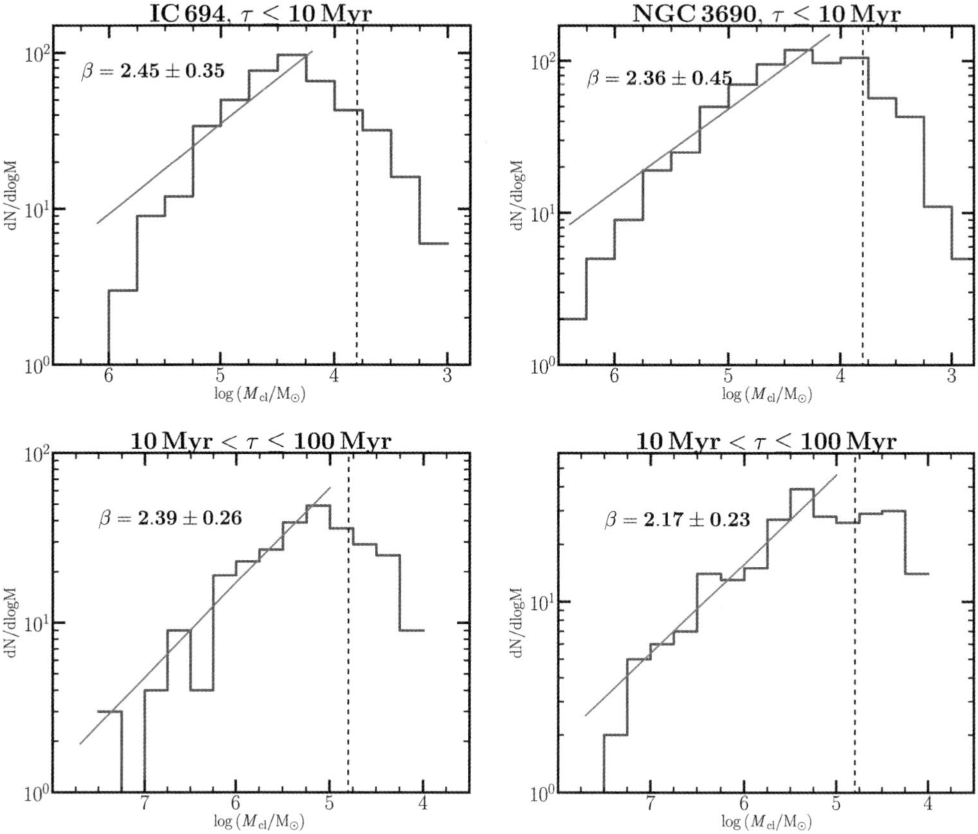

Figure 5. The cluster mass functions of IC 694 (*left*) and NGC 3690 (*right*) plotted with different age bins. The solid lines represent the resulting power-law fit to the high-mass end of the distribution. The dashed lines show approximate cluster mass completeness limits estimated using the age-mass plane. The top panels correspond to the distribution of young clusters with ages less than 10 Myr, and the lower panels for cluster ages between 10 and 100 Myr. The MFs of NGC 3690 seem to have an underlying turnover especially for 10 Myr $< \tau \leqslant$ 100 Myr.

6. Summary & conclusions and ongoing work

Overall, we have shown the capability of NIR AO systems in opening a new window into the understanding of SSC formation, evolution, and disruption mechanisms. It has also been found that interacting systems such as LIRGs offer a totally different environment which can bias the long-term survival chance of its star cluster population. The SUNBIRD survey offers useful data, with the potential of answering some of the most important questions in the field of star cluster research. However, there are still limitations in studying these most massive clusters. New instruments such as the multi-conjugate AO systems mounted on Gemini-South (GeMS/GSAOI) will push down the detection limit and therefore allow the analyses of faint and low-mass star clusters. On the other hand, extending the sample of the survey into a lower SFR baseline would also offer more insights into the star formation process in general. Finally, it will also be interesting to study the kinematics, underlying stellar populations, and the gas of the host galaxies to correlate SSC properties with physical characteristics of the host as a whole in the SUNBIRD sample using the available data from SALT/RSS spectroscopic observations.

References

Adamo, A., *et al.* 2011, *MNRAS*, 417, 1904

Adamo, A., & Bastian, N. 2015, The lifecycle of clusters in galaxies, to appear in *The Birth of Star Clusters*, editor S. W. Stahler, Springer edition, submitted

Bastian, N. 2008, *MNRAS*, 390, 759

Kruijssen, J. M. D., Pelupessy, F. I., Lamers, H. J. G. L. M., *et al.* 2011, *MNRAS*, 414, 1339

Portegies Zwart, S. F., *et al.* 2010, *ARAA*, 48, 431

Randriamanakoto, Z., Escala, A., Väisänen, P., *et al.* 2013a, *ApJ* (Letters), 775, L38

Randriamanakoto, Z., Väisänen, P., Ryder, S., *et al.* 2013b, *MNRAS*, 431, 554

Ryder, S. D., Mattila, S., Kankare, E., & Väisänen, P. 2014, *Proc SPIE*, 9148, 91480D

Sanders, D. B. & Mirabel, I. F. 1996, *ARAA*, 34, 749

Väisänen, P., *et al.* 2014, *arXiv* 1402.2357

Väisänen, P., Barway, S., & Randriamanakoto, Z. 2014, *ApJ* (Letters), 797, LL16

Whitmore, B. C., Chandar, R., Bowers, A. S., *et al.* 2014, *AJ*, 147, 78

Zackrisson, E., Rydberg, C.-E., Schaerer, D., Östlin, G., & Tuli, M. 2011, *ApJ*, 740, 13

Formation, evolution, and survival of massive star clusters
Proceedings IAU Symposium No. 316, 2015
C. Charbonnel & A. Nota, eds.

doi:10.1017/S174392131600870X

Dissecting 30 Doradus: Optical and Near Infrared Star Formation History of the starburst cluster NGC2070 from the Hubble Tarantula Treasury Project

Michele Cignoni[1] **and the HTTP team**

[1]Space Telescope Science Institute, 3700 San Martin Drive, Baltimore, MD, 21218, USA
email: `cignoni@stsci.edu`

Abstract. I will present new results on the star formation history of 30 Doradus in the Large Magellanic Cloud based on the panchromatic imaging survey Hubble Tarantula Treasury Project (HTTP). Here the focus is on the starburst cluster NGC2070. The star formation history is derived by comparing the deepest ever optical and NIR color-magnitude diagrams (CMDs) with state-of-the-art synthetic CMDs generated with the latest PARSEC models, which include all stellar phases from pre-main sequence to post-main sequence. For the first time in this region we are able to measure the star formation using intermediate and low mass stars simultaneously. Our results suggest that NGC2070 experienced a prolonged activity. I will discuss the detailed star formation history, initial mass function and reddening distribution.

Keywords. stellar evolution - star forming region: individual, 30 Doradus, galaxies: stellar content

1. Introduction

Located in the Large Magellanic Cloud (LMC), 30 Doradus is the most active region in the Local Group. The central area of the nebula is dominated by the starburst region NGC 2070, a collective of several dense sub-clusters (Walborn & Blades 1997, Sabbi *et al.* 2012) whose light is dominated by massive OB stars. The most prominent and central of these sub-clusters is the bound super star cluster R 136.

The proximity of the LMC allows to study 30 Doradus with the Hubble Space Telescope (HST) on scales down to ~ 0.01 pc, resolving stars down to the sub-solar regime. However, most of high resolution studies are limited to a few bands and cover only small patches of the 30 Doradus complex. The Hubble Tarantula Treasury Project (HTTP; Sabbi *et al.* 2013, Sabbi *et al.* 2015, submitted), a multi-filter HST survey covering a 14 arcmin $\times$ 12 arcmin wide area around 30 Doradus at high resolution, fills this gap.

In this work we explore the star formation (SF) during the last 50 Myr for the central 40 pc of 30 Doradus, a.k.a. the starburst region NGC 2070, by comparing the HTTP color-magnitude diagrams (CMDs) with synthetic CMDs. These simulations incorporate the latest (V.1.2S) set of PAdova and TRieste Stellar Evolution Code (PARSEC) isochrones (see Bressan *et al.* 2012 and Tang *et al.* 2014), the first theoretical library to include homogeneously all stellar phases from pre-main sequence (PMS) to post-main sequence for all masses between 0.1 and 350 $M_{\odot}$. For the first time, we derive the history of the region using low and intermediate mass stars. In particular, we exploit magnitudes and colors of the PMS Turn-Ons (see e.g. Mayne 2010, Cignoni *et al.* 2010), i.e. the CMD loci where the PMS phase joins to the MS.

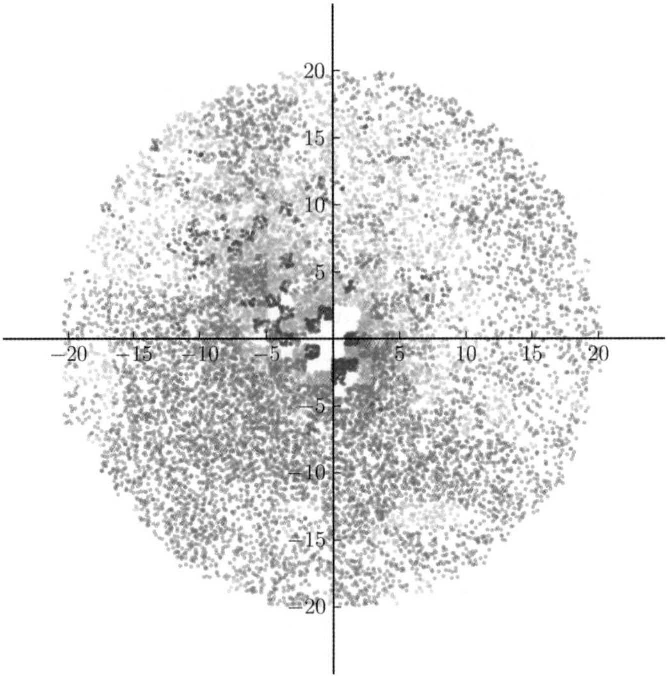

Figure 1. Sub-regions of NGC 2070 where the average completeness is 50% for F555W > 24 (A1; red dots), 23 <F555W< 24 (A2; orange dots) and 22 <F555W< 23 (A3; blue dots).

We use both the optical F555W vs F555W−F775W and the near infrared (NIR) F110W vs F110W−F160W CMDs of the HTTP sample, which offer complementary advantages in terms of higher spatial resolution (the optical) and lower reddening sensitivity (the NIR).

2. Biases

In trying to recover the SFH of NGC 2070 we faced three major problems: 1) membership errors due to field interlopers from the LMC field that mimic older populations; 2) extreme crowding conditions that exacerbate the incompleteness, shortening the reachable look-back time; 3) high level of differential reddening that spreads and dims the CMD, blending together young PMS stars and older MS stars, which introduces further age ambiguities. Here in the following there are our solutions.

Field contamination: A typical approach is to decontaminate the cluster by subtracting the star-counts from a reference field. However, given reddening and incompleteness variations across the whole 30 Doradus, it is impossible to find another direction replicating the observational conditions of NGC 2070. Our solution was to use a field (within the HTTP field of view; see Section 5.1 in Cignoni *et al.* 2015 for details), resembling as much as possible the LMC field *we would observe without* NGC 2070, which means low or negligible SF activity in the last 50 Myr, minimal differential reddening and high completeness down to F555W$\sim$ 24. This reference LMC field was then artificially corrected for the more severe incompleteness and photometric errors of NGC 2070. The resulting field differs only in terms of normalization and differential reddening from the real LMC field contaminating NGC 2070. Finally, reddening and normalization of this field were tuned together with the SFH (see the next Section) of NGC 2070 until an adequate match of NGC 2070's CMD was found;

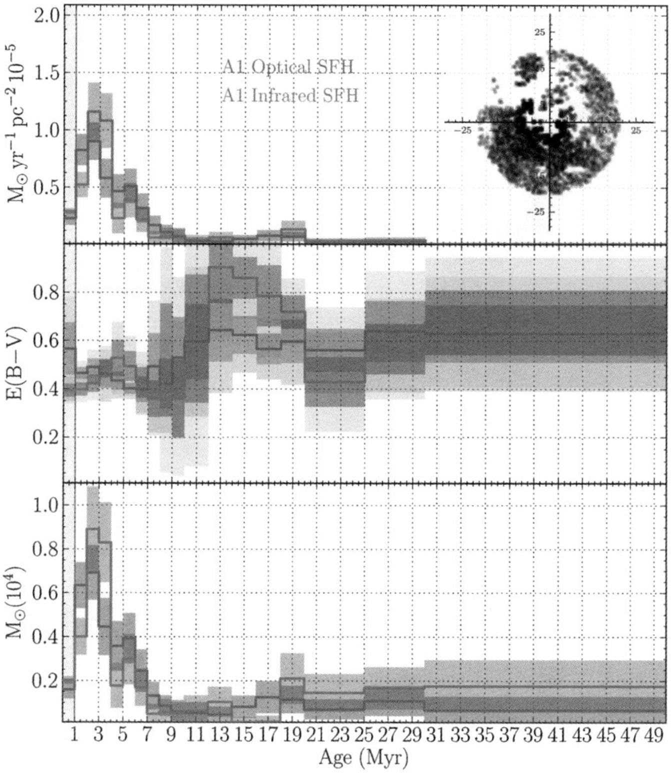

Figure 2. Recovered SFH (top panel), reddening distribution (middle panel) and mass distribution (bottom panel) for the sub-region A1. The optical and NIR solution are plotted in green and red colors, respectively. The bin showing the SF rate in the 0 to 1 Myr range is marked in light grey to indicate that systematic uncertainties may be comparable to or are greater than the formal errors (see text). The inset panel shows the distribution of stars in sub-region A1.

Incompleteness: To test the level of incompleteness of our photometric data and have reliable estimates of photometric errors, we ran extensive artificial star experiments. The experiments consist of adding "fake" sources (5 million) for each of the eight pass-bands, modeled with the point spread function used in the photometric analysis of the frames, onto the actual images. We then applied the same source detection routines used for our science images to the fields containing the combined actual images and the fake sources. The results of the test (difference between input and output fake magnitudes, incompleteness) were then convolved with the stellar models;

Differential reddening: The area is strongly affected by differential reddening. Moreover, we do not know where populations of different ages are located *inside* NGC 2070. For example, young massive stars may have carved the gas with their ionizing flux, hence lowering the extinction. On the other hand, high levels of SF could be sustained only where the gas density is higher, which would lead to the opposite situation. To take into account all these effect, reddening is a free parameter and can vary as a function of age.

3. The SFH procedure

In our procedure, the most likely SFH is the one whose synthetic CMD minimizes the residuals from the observational counterpart. The first step is to create an extensive

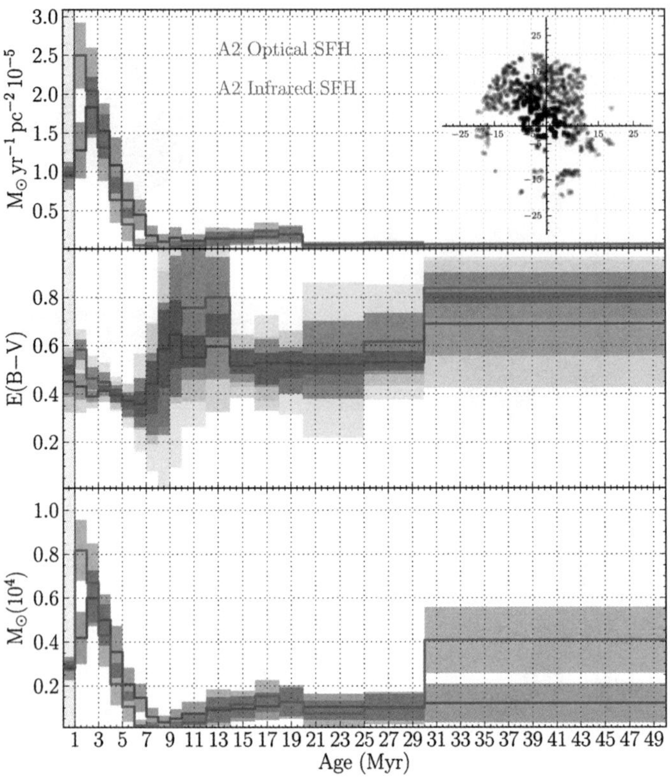

Figure 3. Same as Fig. 2 but for the sub-region A2.

library of basic synthetic CMDs generated from stellar models. Each basic $CMD_{j,k}$ is a "fuzzy" isochrone, with duration Δt and fixed reddening E(B-V), populated through a classical Monte Carlo procedure: 1) synthetic masses and ages are extracted from a Kroupa (2001) initial mass function (IMF) and the j-th SF step, respectively; and 2) synthetic masses and ages are converted to absolute synthetic magnitudes and colors by using a fine grid of isochrones. For our calculations we used the latest PARSEC isochrones for the metallicity Z=0.008; 3) a 30% of synthetic stars is randomly chosen to have a companion. The masses of companions are extracted from the same IMF and their flux is added to the flux of the primaries; 4) The absolute synthetic photometry is put at the distance of the LMC ($(m-M)_0 = 18.5$; see e.g. Panagia *et al.* 1991) and reddened with the k-th reddening E(B-V). To produce realistic simulations, all basic CMDs are degraded with photometric errors and incompleteness as estimated from artificial star tests.

As done previously (Cignoni *et al.* 2006, Cignoni *et al.* 2011), we parameterize a generic CMD as a linear superposition of the basic CMDs. The explored ages range from now up to 50 Myr ago. Since older isochrones tend to be more tightly packed in the CMD than younger isochrones, the duration of each age step increases with age from 1 Myr (at the present time) to 20 Myr (50 Myr ago). The reddening is allowed to vary between 0 and 1 mag with a step of 0.05 mag, while the extinction coefficients R_λ are taken from De Marchi & Panagia (2014). In total we deal with 380 parameters, divided into 18 steps in age (plus one parameter controlling field contamination) and 20 steps in reddening. To explore this wide parameter space we have combined a well tested genetic algorithm

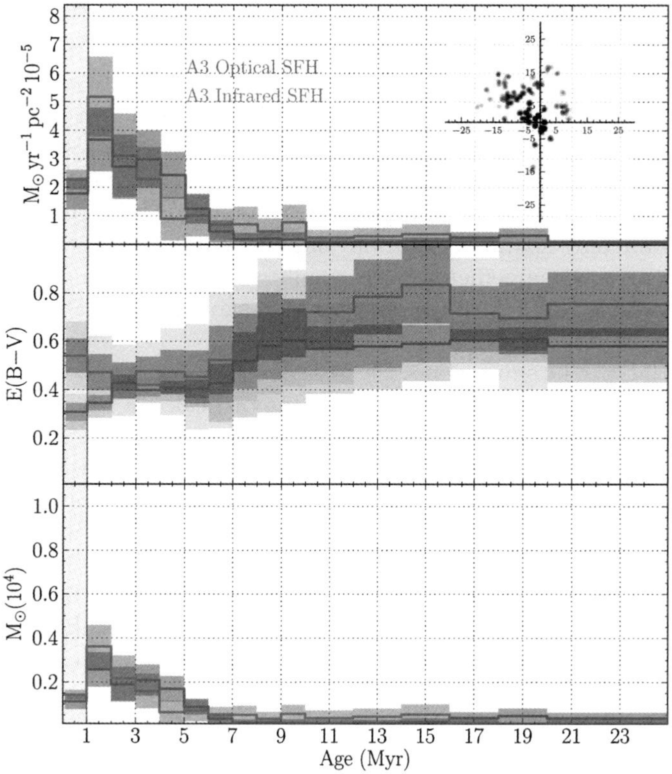

Figure 4. Same as Fig. 2 but for sub-region A3.

(GA), Pikaia†, with a local search routine (see Cignoni *et al.* 2015 for details). As shown in various papers (see, e.g., Ng *et al.* 2002, Aparicio & Hidalgo 2009, Small *et al.* 2013), GAs allow to find global optimum more efficiently than a local search alone. In our approach the synergy of the GA and a local search combines the advantages of both worlds.

Which part of the CMD is used? Not all of the CMD is used to recover the SFH. Our analysis is limited on the bright end by the saturation magnitude and on the faint end by the 50% completeness magnitude level. Nonetheless, even with these conservative selections, the large variations of completeness with magnitude in some regions force us to derive the SFH in sub-regions within which the completeness is reasonably uniform. Fig. 1 shows our selection: sub-regions of NGC 2070 where the completeness in F555W is 50% for F555W > 24 (A1), 23 <F555W< 24 (A2) and 22 <F555W< 23 (A3) are indicated in red, orange and blue, respectively. The stars found in both the F555W and F775W filters inside these sub-regions are used to recover the optical SFHs. Overall, our sub-regions trace the stellar density, from the lowest (A1) to the highest (A3).

This analysis leaves out the very center of NGC 2070, a ≈ 3 pc region mostly represented by R 136.

† Routine developed at the High Altitude Observatory, and available in the public domain http://www.hao.ucar.edu/modeling/pikaia.php.

4. Best SFH for NGC 2070

Figures 2 and 3 show the recovered SF rate (in $M_\odot\,yr^{-1}\,pc^{-2}$; top panel), reddening (middle panel) and stellar mass (bottom panel) as a function of time for sub-regions A1 and A2 respectively, as predicted using the optical (green curve) and NIR (red curve) CMDs. Uncertainties are the quadrature sum of a statistical error (obtained by bootstrapping the data and re-deriving the solutions) and systematic error (obtained by re-deriving the solutions with different age-binnings and CMD binning scheme).

Despite the different spatial resolution and reddening sensitivity, optical and NIR predictions are in good agreement. Both results predict that mild SF activity started throughout the whole NGC 2070 region ≈ 20 Myr ago, and that about 7-8 Myr ago the birth rate accelerated.

Focusing on the specific sub-regions, the SFH of sub-region A1 shows a mild bimodality, with a minor peak around 5-7 Myr ago and a major peak 1-4 Myr ago (1-3 Myr ago in the optical SFH, 2-4 Myr ago in the NIR one). The measured reddening distribution anti-correlates with SF: the average E(B−V) is ≈ 0.6 for stars with ages older than 7 Myr, when the SF activity was lower, and ≈ 0.4 for stars of younger ages, when the SF was stronger. Optical and NIR reddening derivations agree within errors, both in confirming the presence of notable differential reddening.

The SF in sub-region A2 (Fig. 3) is about two times higher than in sub-region A1. The most important feature in the A2 SF is a pronounced peak in both optical and NIR solutions 1-3 Myr ago. Like in sub-region A1, the reddening of sub-region A2 is anti-correlated with the SF activity.

Fig. 4 shows the SFH in the sub-region A3 over the last 25 Myr (the crowding conditions prevented to investigate earlier activity). The peak activity in both the optical and NIR solution is more than two times higher than in A2, and at odds with other sub-regions, is clearly located between 1 and 2 Myr ago (more pronounced in the NIR solution).

5. Conclusions

We summarize here our main conclusions:

SFH: NGC 2070 experienced prolonged activity. We identify three major events in the history of this star forming region: 1) ≈ 20 Myr ago - This epoch demarcates the commencement of the first significant period of SF; 2) 7 Myr ago the SF accelerated throughout the entire region; 3) 1-3 Myr ago the activity reached a peak. In this time range, the SF moves from the periphery to the central regions. Our innermost sub-region (A3) shows a maximum activity 1-2 Myr ago;

Reddening: we find an average E(B−V)≈ 0.4 mag for the young population (<10 Myr old), and ≈ 0.6 mag for the old one (>10 Myr old);

IMF: Except the innermost few pc, where the incompleteness is too severe to allow firm conclusions, a Kroupa IMF in the range $0.5 - 7\,M_\odot$ is compatible with the data.

References

Aparicio, A. & Hidalgo, S. L. 2009, *AJ*, 138, 558
Bressan, A., Marigo, P., Girardi, L., *et al.* 2012, *MNRAS*, 427, 127
Cignoni, M., Degl'Innocenti, S., Prada Moroni, P., & Shore, S. N. 2006, *A&A*, 459, 783
Cignoni, M., Tosi, M., Sabbi, E., *et al.* 2010, *ApJL*, 712, L63
Cignoni, M., Tosi, M., Sabbi, E., Nota, A., & Gallagher, J. S. 2011, *AJ*, 141, 31
Cignoni, M., Sabbi, E., van der Marel, R. P., *et al.* 2015, *arXiv:1505.04799*
De Marchi, G. & Panagia, N. 2014, *MNRAS*, 445, 93

Hunter, D. A., Shaya, E. J., Holtzman, J. A., *et al.* 1995, *ApJ*, 448, 179
Kroupa, P. 2001, *MNRAS*, 322, 231
Mayne, N. J. 2010, *MNRAS*, 408, 1409
Ng, Y. K., Brogt, E., Chiosi, C., & Bertelli, G. 2002, *A&A*, 392, 1129
Panagia, N., Gilmozzi, R., Macchetto, F., Adorf, H.-M., & Kirshner, R. P. 1991, *ApJL*, 380, L23
Sabbi, E., Lennon, D. J., Gieles, M., *et al.* 2012, *ApJL*, 754, LL37
Sabbi, E., Anderson, J., Lennon, D. J., *et al.* 2013, *AJ*, 146, 53
Small, E. E., Bersier, D., & Salaris, M. 2013, *MNRAS*, 428, 763
Tang, J., Bressan, A., Rosenfield, P., *et al.* 2014, *MNRAS*, 445, 4287
Walborn, N. R. & Blades, J. C. 1997, *ApJS*, 112, 457

Formation, evolution, and survival of massive star clusters
Proceedings IAU Symposium No. 316, 2015
C. Charbonnel & A. Nota, eds.

doi:10.1017/S1743921316007018

Nuclear Star Clusters

Nadine Neumayer
Max-Planck Institute for Astronomy, Königstuhl 17, 69117 Heidelberg, Germany
email: neumayer@mpia.de

Abstract. The centers of galaxies host two distinct, compact components: massive black holes and nuclear star clusters. Nuclear star clusters are the densest stellar systems in the universe, with masses of $\sim 10^7 M_\odot$ and sizes of ∼5pc. They are almost ubiquitous at the centres of nearby galaxies with masses similar to, or lower than the Milky Way. Their occurrence both in spirals and dwarf elliptical galaxies appears to be a strong function of total galaxy light or mass. Nucleation fractions are up to 100% for total galaxy magnitudes of $M_B = -19$mag or total galaxy luminosities of about $L_B = 10^{10} L_\odot$ and falling nucleation fractions for both smaller and higher galaxy masses. Although nuclear star clusters are so common, their formation mechanisms are still under debate. The two main formation scenarios proposed are the infall and subsequent merging of star clusters and the in-situ formation of stars at the center of a galaxy. Here, I review the state-of-the-art of nuclear star cluster observations concerning their structure, stellar populations and kinematics. These observations are used to constrain the proposed formation scenarios for nuclear star clusters. Constraints from observations show, that likely both cluster infall and in-situ star formation are at work. The relative importance of these two mechanisms is still subject of investigation.

Keywords. galaxies: nuclei, star clusters, black holes

1. Frequency

Nuclear star clusters (NSCs) are almost ubiquitous at the centres of nearby low mass galaxies (similar to or lower mass than the Milky Way). For very late type, spiral galaxies the fraction of galaxies with a NSC (nucleation fraction) is above 75% (Böker et al. 2002, Georgiev & Böker 2014, see right panel of Fig. 1). For spirals with bulges the central regions are complex due to the presence dust and ongoing star formation. Of these galaxies only 55% are found to host a nuclear star cluster (Carollo et al. 1998). However, due to dust and star formation the detection of the nucleus is often difficult and the true nucleation fraction may be higher. For dwarf ellipticals and S0s at least 66% host a NSC with an increase in nucleation fraction to almost 100% for galaxies with $M_B = -19$ (Côté et al. 2006, den Brok *et al.* 2014, see left panel of Fig. 1). On the other hand, it appears that the most massive ellipticals do not host nuclear star clusters, in fact, their central surface brightness shows a cored distribution. This transition to not having a nuclear cluster appears at about $M_B = -20.5$ (Côté et al. 2006) and is probably linked to the presence of black holes. The most massive ellipticals are built by mergers of galaxies. During the process of merging the black holes drive out the stars from the centre due to three body encounters. This scenario has recently been proposed by Antonini *et al.* 2015.

2. Formation Scenarios

Although nuclear star clusters are so common in lower mass galaxies, their formation scenarios are still under debate. Two main scenarios were proposed to explain the formation and evolution of nuclear star clusters: 1) The infall and subsequent merging of star clusters and 2) the in-situ formation of stars at the center of a galaxy.

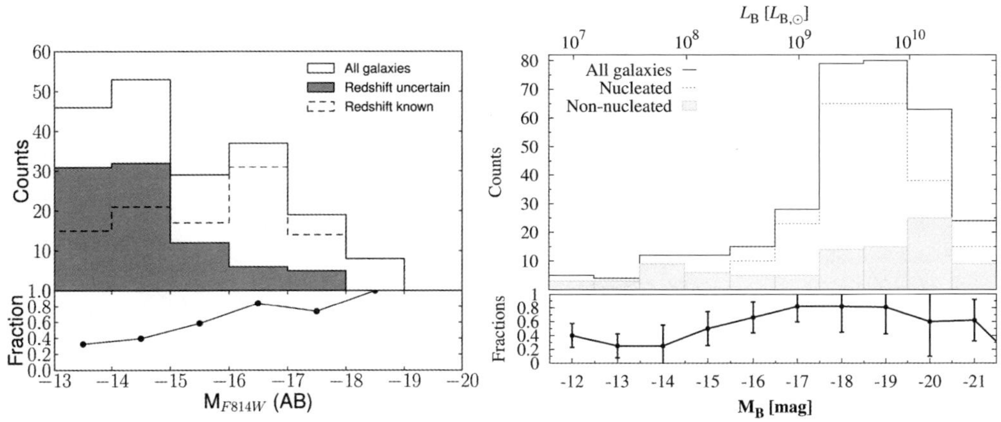

Figure 1. The fraction of galaxies that host a nuclear star cluster is presented as a function of galaxy total magnitude. On the left side for elliptical galaxies (from den Brok *et al.* 2014), and on the right side for spiral galaxies (from Georgiev & Böker 2014).

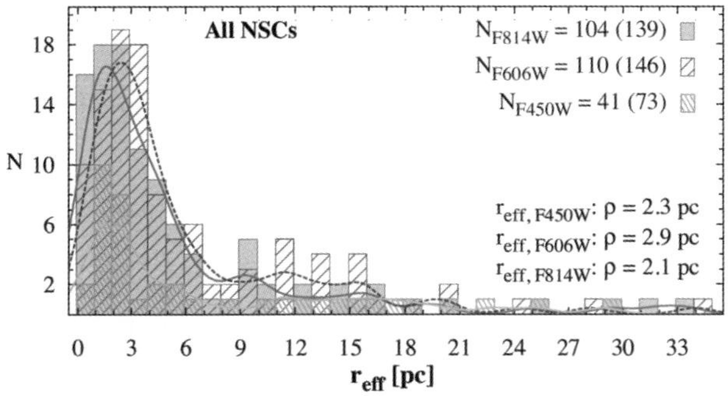

Figure 2. Distribution of effective radii for nuclear star clusters in spiral galaxies (taken from Georgiev & Böker 2014). The typical size of a NSC is between 3 – 5pc.

Theoretical studies for the cluster merging scenario use mainly N-body or semi-analytical approaches (Tremaine *et al.* 1975, Capuzzo-Dolcetta 1993, Lotz *et al.* 2004, Agarwal & Milosavljević 2011, Gnedin *et al.* 2014, Antonini 2013, Arca-Sedda & Capuzzo-Dolcetta 2014). The main support for the in-situ formation is the on-going star formation at the center of galaxies as well as inefficient dynamical friction in bringing the globular clusters to the center of a galaxy (Milosavljević 2004, Schinnerer *et al.* 2008, Pflamm-Altenburg & Kroupa 2009, Seth *et al.* 2008).

Many authors have advocated that the formation is likely a combination of both processes (Hartmann et al. 2011, Neumayer et al. 2011, Turner *et al.* 2012, De Lorenzi et al. 2013, Feldmeier *et al.* 2014, den Brok *et al.* 2014). To test the relative importance of these processes requires observations that spatially resolve the structure, stellar populations, and kinematics of nuclear star clusters.

3. Structural Properties

Observational studies with the Hubble Space Telescope find that the half-light radii of NSCs are typically 3 – 5pc (Böker et al. 2002, Georgiev & Böker 2014, see Fig. 2). Due to these small sizes, their detection requires very high spatial resolution observations, making the HST crucial for systematic searches. The sizes of NSCs appear to vary with the wavelength at which the observations are taken. NSCs seem to be more compact in bluer bands and more extended in redder filters (Georgiev & Böker 2014, Carson *et al.* 2015;

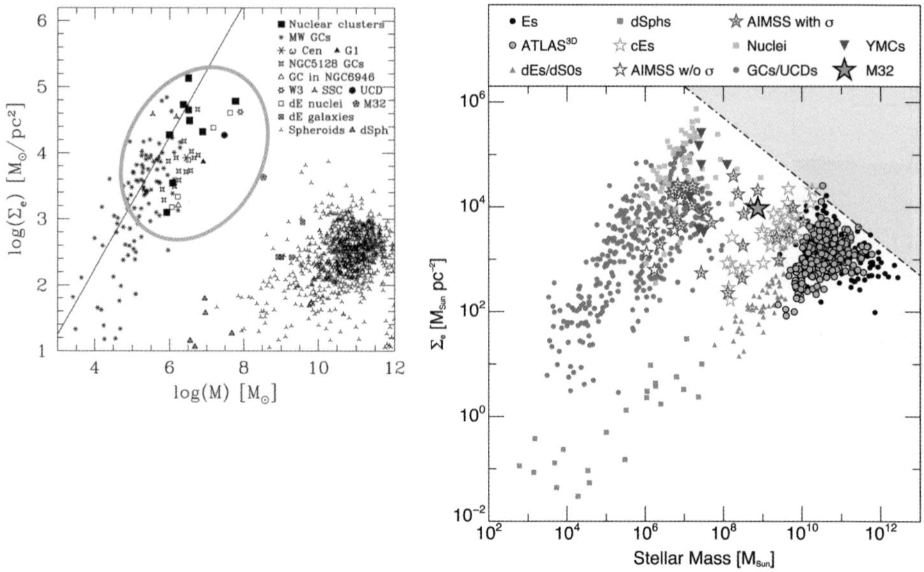

Figure 3. The surface mass density of stellar systems (star clusters and galaxies) plotted against their total mass. NSCs lie at the top of the star cluster sequence and are very different from bulges. Figures are taken from Walcher et al. 2005 and Norris *et al.* 2014.

this can be explained by the presence of young stars that are centrally concentrated, see below in Section 4).

Comparing the properties of nuclear star clusters with Milky Way globular clusters, one finds that they are similar in size, but are on average 4mag brighter (Böker et al. 2004) and hence more massive. Since they have similar half light radii as globular clusters but are more massive, they are also considerably denser than globular clusters. Nuclear star clusters are in fact the densest stellar systems in the universe (Walcher et al. 2005, Misgeld & Hilker 2011, Norris *et al.* 2014). Figure 3 shows the surface mass density of stellar systems as a function of the system's stellar mass. Nuclear star clusters clearly lie at the top of the star cluster sequence (top left of Fig. 3) and are structurally very different from bulges that lie towards the right of the diagrams.

4. Stellar Populations

Nuclear star clusters truly occupy the centers of galaxies, both photometrically but also kinematically (Böker et al. 2002, Neumayer et al. 2011, respectively), and it may be this special location at the bottom of the potential well of the galaxies, that causes the star formation history of NSCs to be rather complex. Several studies have shown that NSCs have multiple stellar populations both in late type (Walcher et al. 2006, Seth et al. 2006, Rossa et al. 2006) and also early type galaxies (Seth et al. 2010, Lyubenova *et al.* 2013). These spectroscopic studies analysed the integrated stellar population properties of nuclear star clusters. Looking at the resolved distribution of stars inside NSCs one finds the following: in the most nearby example of a NSC, at the center of the Milky Way, Feldmeier-Krause *et al.* 2015 showed that the population of young stars (of a few Myrs) is very centrally concentrated (confirming and extending the work of e.g Bartko *et al.* 2010, Do *et al.* 2013, Støstad *et al.* 2015). The young stars are confined to a radius of about 0.5pc around the central black hole, while the distribution of the older stellar population has an effective radius of 4.2pc (Schödel *et al.* 2014).

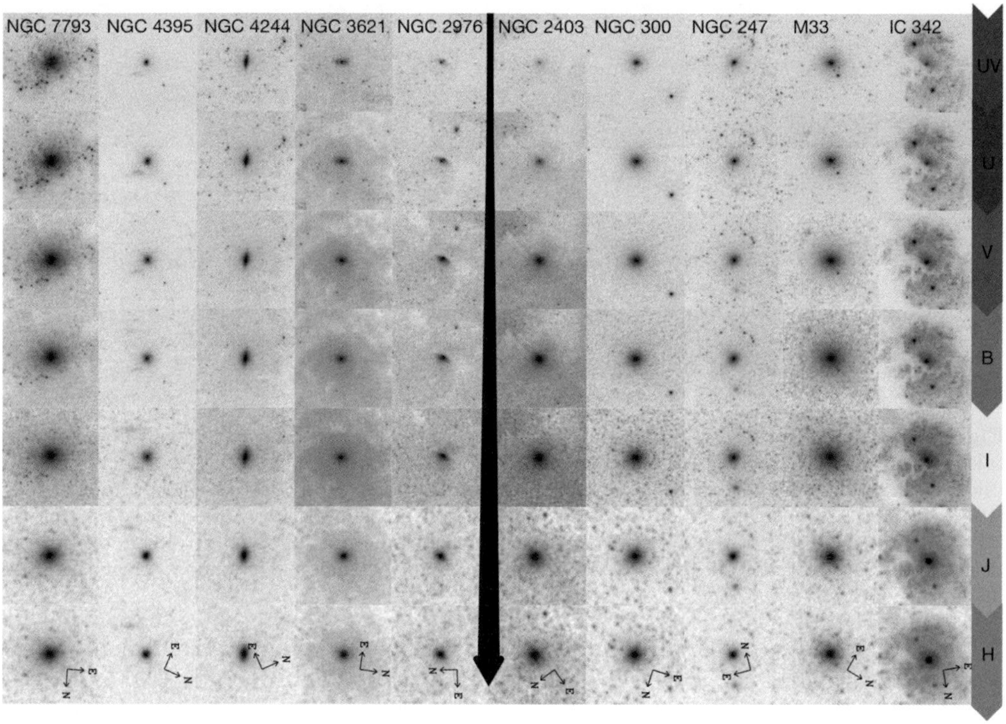

Figure 4. Images of nearby nuclear star clusters in spiral galaxies taken with HST/WFC3 in 7 filters (from Carson *et al.* 2015). The effective radius of the NSCs is more compact in the blue filter while in the redder filters the clusters appear more extended. It appears that the young, blue stars are more centrally concentrated than the old, red stars.

For a sample of 10 nearby NSCs observed with WFC3/HST in seven filters (see Fig. 4) Carson *et al.* 2015 show that the effective radius of the light distribution in the blue filters is more compact while in the redder filters the clusters appear more extended. This hints towards the same finding of young stars being more centrally concentrated, and being formed in-situ at the very center. Moreover, the nucleus of a galaxy appears to be more metal-rich and younger than the surrounding galaxy (Koleva et al. 2011), and the abundance ratios [α/Fe] show that NSCs are more metal enriched than globular clusters (GCs) (Evstigneeva et al. 2007). This finding suggests that NSCs cannot solely be the merger product of GCs, but need some gas for recurrent star formation and enrichment. This finding is also supported by recent kinematical studies (Hartmann et al. 2011, De Lorenzi et al. 2013), where cluster infall alone cannot explain the dynamical state of the NSC.

5. Kinematics

Recent studies of the kinematics of NSCs with integral-field spectroscopy show that the cluster as a whole rotates (Seth *et al.* 2008, Seth et al. 2010, Lyubenova *et al.* 2013, Feldmeier *et al.* 2014, see Figure 5). Combined with the superb spatial resolution of adaptive-optics, the 2D velocity maps resolve stellar and gas kinematics down to a few parsecs on physical scales. In addition, due to the extremely high central stellar density in NSCs, it becomes possible to pick-up kinematic signatures for massive black holes inside NSCs (Seth et al. 2010, Lyubenova *et al.* 2013, den Brok *et al.* 2015, Neumayer *et al.* in prep).

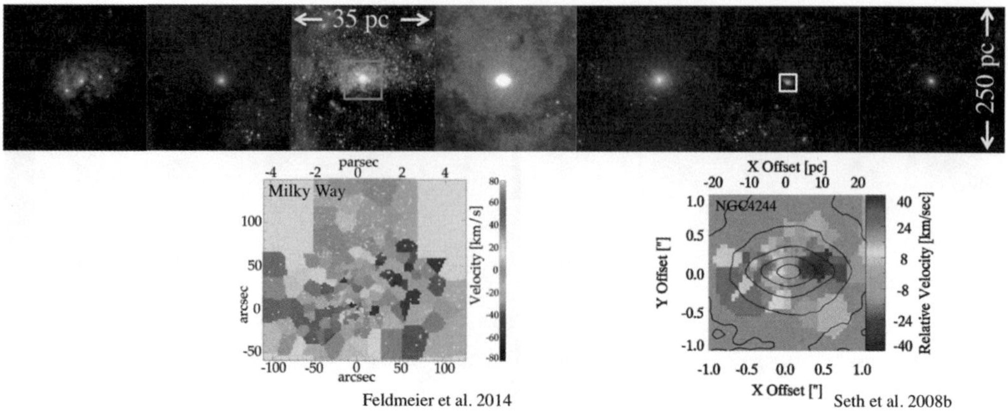

Figure 5. Family portrait of nuclear star clusters in nearby galaxies. Almost all color images taken from Carson *et al.* 2015, except the third from the left: zoom into the Milky Way NSC, taken with Spitzer/IRAC; Stolovy *et al.* 2006). The images are all about 250 pc on the side (taken with HST/WFC3; Carson *et al.* 2015). The bottom shows the radial velocity maps for the nuclear star cluster of the Milky Way and the nearby edge-on galaxy NGC 4244 (taken from Feldmeier *et al.* 2014 and Seth *et al.* 2008). Both edge-on systems clearly show strong rotation in the nuclear cluster.

Recently, Feldmeier *et al.* (2014) observed the NSC of the Milky Way in a long-slit drift scan and constructed the first velocity map of the Milky Way NSC (see Fig. 5, and the chapter of Feldmeier-Krause *et al.* in this edition). This shows that the Milky Way NSC rotates at a maximum velocity of about $\pm 60 km/s$ (comparable to the rotation velocities observed in other NSCs), and its major axis is misaligned with the Galactic Plane by about 9 deg. Moreover, it shows complex kinematic substructures that could be left over coherent structures of infalling star clusters. The comparison of numerical simulations (e.g. Perets & Mastrobuono-Battisti 2014) to the observed maps will help to understand the possible origin of these substructures.

6. Summary - Constraints on Formation Scenarios

The centers of galaxies host two distinct, compact components: massive black holes and nuclear star clusters. Unlike black holes, nuclear star clusters provide a visible record of the accretion of stars and gas into the center of a galaxy. Studying their stellar populations, structure and kinematics allow us to disentangle their formation history and more generally that of galactic nuclei. The two formation scenarios proposed for nuclear star clusters are 1) star cluster infall to the center of a galaxy and subsequent merging; 2) in-situ star formation at the center of a galaxy and build-up of a massive and dense NSC. The observations summarised here can help to set constraints on these formation scenarios in the following way:

NSCs are very common: This observational fact suggests that both formation scenarios are at work to form and grow NSCs.

NSCs have the highest stellar densities in the universe: This observation clearly supports the in-situ formation scenario, since a merger of two globular clusters will not result in a NSC that is denser than the GCs were before. To make the NSC denser requires gas and star formation to dissipate away energy and angular momentum.

NSCs have complex star formation histories: This finding support both formation scenarios.

NSCs rotate: They are thus either built-up from gas and newly formed young stars that fall in from the galactic disk, or from star clusters that come preferentially from within the galactic disk and retain part of their angular momentum. Thus this observable supports both scenarios.

Young stars sit at the center: This observation clearly speaks for in-situ star formation. Especially for the case of the Milky Way nuclear star cluster it became obvious that the infall of young stars from outside the cluster cannot explain the population of very young stars at the very center. These stars must form there.

This list summarises that there are observations supporting both formation scenarios. It is likely that the formation and growth of nuclear star clusters is governed by both star cluster infall and in-situ star formation. Their relative importance is still subject of investigation.

Acknowledgements

I thank Corinne Charbonnel and Antonella Nota for organising this interesting symposium and the IAU for financial support. I thank Iskren Georgiev for providing Fig 2, based on his catalogue of nuclear star clusters.

References

Agarwal, M. & Milosavljević, M. 2011, *ApJ*, 729, 35
Antonini, F. 2013, *ApJ*, 763, 62
Antonini, F., Barausse, E., & Silk, J. 2015, *ApJLetters*, 806, L8
Arca-Sedda, M. & Capuzzo-Dolcetta, R. 2014, *MNRAS*, 444, 3738
Bartko, H., Martins, F., Trippe, S., *et al.* 2010, *ApJ*, 708, 834
Böker, T., Laine, S., van der Marel, R. P., *et al.* 2002, *AJ*, 123, 1389
Böker, T., Sarzi, M., McLaughlin, D. E., *et al.* 2004, *AJ*, 127, 105
Capuzzo-Dolcetta, R. 1993, *ApJ*, 415, 616
Carollo, C. M., Stiavelli, M., & Mack, J. 1998, *AJ*, 116, 68
Carson, D. J., Barth, A. J., Seth, A. C., *et al.* 2015, *AJ*, 149, 170
Côté, P., Piatek, S., Ferrarese, L., *et al.* 2006, *ApJS*, 165, 57
De Lorenzi, F., Hartmann, M., Debattista, V., *et al.* 2013, *MNRAS*, 429, 2974
Do, T., Lu, J. R., Ghez, A. M., *et al.* 2013, *ApJ*, 764, 154
den Brok, M., Peletier, R. F., Seth, A., *et al.* 2014, *MNRAS*, 445, 2385
den Brok, M., Seth, A. C., Barth, A. J., *et al.* 2015, *ApJ*, 809, 101
Evstigneeva, E. A., Gregg, M. D., Drinkwater, M. J., & Hilker, M. 2007, *AJ*, 133, 1722
Feldmeier, A., Neumayer, N., Seth, A., *et al.* 2014, *A&A*, 570, A2
Feldmeier-Krause, A., Neumayer, N., Schödel, R., *et al.* 2015, *A&A*, 584, A2
Georgiev, I. Y & Böker, T. 2014, *MNRAS*, 441, 3570
Gnedin, O. Y., Ostriker, J. P., & Tremaine, S. 2014, *ApJ*, 785, 71
Hartmann, M., Debattista, V. P., Seth, A., *et al.* 2011, *MNRAS*, 418, 2697
Koleva, M., Prugniel, P., de Rijcke, S., & Zeilinger, W. W. 2011, *MNRAS*, 417, 1643
Lotz, J. M., Miller, B. W., & Ferguson, H. C. 2004, *ApJ*, 613, 262
Lyubenova, M., van den Bosch, R. C. E., Côté, P., *et al.* 2013, *MNRAS*, 431, 3364
Milosavljević, M. 2004, *ApJL* 605, L13
Misgeld, I. & Hilker, M. 2011, *MNRAS*, 414, 3699
Neumayer, N., Walcher, C. J., Andersen, D., *et al.* 2011, *MNRAS*, 413, 1875
Neumayer, N. & Walcher, C. J. 2012, *Advances in Astronomy*, 2012,
Norris, M. A., Kannappan, S. J., Forbes, D. A., *et al.* 2014, *MNRAS*, 443, 1151
Perets, H. B. & Mastrobuono-Battisti, A. 2014, *ApJL*, 784, L44
Pflamm-Altenburg, J. & Kroupa, P. 2009, *MNRAS*, 397, 488
Rossa, J., van der Marel, R. P., Böker, T., *et al.* 2006, *AJ*, 132, 1074
Schinnerer, E., Böker, T., Meier, D. S., & Calzetti, D. 2008, *ApJL*, 684, L21

Schödel, R., Feldmeier, A., Kunneriath, D., *et al.* 2014, *A&A*, 566, A47
Seth, A. C., Dalcanton, J. J., Hodge, P. W., & Debattista, V. P. 2006, *AJ*, 132, 2539
Seth, A. C., Blum, R. D., Bastian, N., *et al.* 2008, *ApJ*, 687, 997
Seth, A. C., Cappellari, M., Neumayer, N., *et al.* 2010, *ApJ*, 714, 713
Seth, A. C., van den Bosch, R., Mieske, S., *et al.* 2014, *Nature*, 513, 398
Stolovy, S., Ramirez, S., Arendt, R. G., *et al.* 2006, Journal of Physics Conference Series, 54, 176
Støstad, M., Do, T., Murray, N., *et al.* 2015, *ApJ*, 808, 106
Tremaine, S. D., Ostriker, J. P., & Spitzer, L., Jr. 1975, *ApJ*, 196, 407
Turner, M. L., Côté, P., Ferrarese, L., *et al.* 2012, *ApJS*, 203, 5
Walcher, C. J., Böker, T., Charlot, S., *et al.* 2006, *ApJ*, 649, 692
Walcher, C. J., van der Marel, R. P., McLaughlin, D., *et al.* 2005, *ApJ*, 618, 237

Formation, evolution, and survival of massive star clusters
Proceedings IAU Symposium No. 316, 2015
C. Charbonnel & A. Nota, eds.

doi:10.1017/S1743921315010509

Massive Star Clusters in Dwarf Galaxies

Søren S. Larsen

Department of Astrophysics/IMAPP, Radboud University,
Postbus 9010, NL-6500GL Nijmegen, the Netherlands
email: s.larsen@astro.ru.nl

Abstract. Dwarf galaxies can have very high globular cluster specific frequencies, and the GCs are in general significantly more metal-poor than the bulk of the field stars. In some dwarfs, such as Fornax, WLM, and IKN, the fraction of metal-poor stars that belong to GCs can be as high as 20%–25%, an order of magnitude higher than the 1%–2% typical of GCs in halos of larger galaxies. Given that chemical abundance anomalies appear to be present also in GCs in dwarf galaxies, this implies severe difficulties for self-enrichment scenarios that require GCs to have lost a large fraction of their initial masses. More generally, the number of metal-poor field stars in these galaxies is today less than what would originally have been present in the form of low-mass clusters if the initial cluster mass function was a power-law extending down to low masses. This may imply that the initial GC mass function in these dwarf galaxies was significantly more top-heavy than typically observed in present-day star forming environments.

Keywords. Globular clusters; dwarf galaxies

1. Introduction: globular clusters and halo stars in different environments

It is well-known that the specific frequency of globular clusters, S_N, varies significantly from one galaxy to another†. The overall variation of S_N vs. host galaxy M_V is U-shaped, with a broad minimum between $M_V \sim -18$ and $M_V \sim -20$ where S_N is typically of order unity, but reaching $S_N > 20$–30 for some cD galaxies and dwarf galaxies (e.g., Peng *et al.* 2008; Harris *et al.* 2013). This trend may be driven partly by dynamical effects that lead to more efficient cluster disruption at intermediate host galaxy masses (e.g., Mieske *et al.* 2014) and partly by variations in the "efficiency" of GC formation, relative to field stars (Peng *et al.* 2008; Georgiev *et al.* 2010). Interestingly, the trend of S_N vs. host galaxy mass resembles that of total (dark matter) halo mass vs. stellar mass inferred from abundance matching (e.g. Behroozi *et al.* 2010), suggesting that the mechanisms that drive the two may be related.

In addition to this global relation, there are also variations in GC specific frequency *within* galaxies. This is illustrated, for example, by the difference between the integrated colours of early-type galaxies and the mean colours of their GC populations, with the integrated light being redder (Forte *et al.* 1981; Larsen *et al.* 2001; Forbes & Forte 2001). Direct comparisons of the metallicity distributions for halo RGB stars and GCs confirm that the ratio of GCs to halo stars increases towards lower metallicities (Harris *et al.* 2007). In the Milky Way, GCs account for 1%–2% of the mass of the stellar halo (e.g. Larsen *et al.* 2014a). By decomposing the integrated light of NGC 1399 and M87 into metal-poor and metal-rich components, Forte *et al.* (2007) found that metal-poor GCs account for about 0.9% and 2.1% of the metal-poor stars in these galaxies, which is comparable to the corresponding ratio in the Milky Way. In the Large Magellanic Cloud,

† $S_N \equiv N_{\rm GC} \times 10^{0.4 \times (M_V + 15)}$ for a galaxy of absolute magnitude M_V that hosts $N_{\rm GC}$ GCs

the integrated luminosity of the halo has been estimated to be about $M_V = -15.2$ (Kinman *et al.* 1991), while the old GCs have a total absolute magnitude of $M_V = -10.75$ (Suntzeff *et al.* 1992), or about 1.7% of the total halo luminosity. While these numbers are based on somewhat heterogeneous data and analyses, they do seem to suggest that GCs in general account for a fairly constant fraction of 1%–2% of the total stellar halo mass in most galaxies, albeit probably with some real variation.

It is interesting to compare these numbers with the corresponding fractions of stellar mass in GCs in dwarf galaxies. Within the Local Group, the highest GC specific frequency is found in the Fornax dwarf spheroidal galaxy, which has $M_V \approx -13.2$ and 5 GCs, i.e. $S_N = 26$. The total stellar mass formed over the lifetime of the Fornax dSph is estimated to be $6 \times 10^7\, M_\odot$ (Coleman & de Jong 2008). The total luminosity of the five GCs is $L_V \approx 3.6 \times 10^5\, L_{V,\odot}$ (Webbink 1985), which corresponds to a current mass of $\sim 7.2 \times 10^5\, M_\odot$, assuming $M/L_V = 2\, M_\odot/L_{V,\odot}$ for metal-poor GCs (Strader *et al.* 2009). [We note that dynamical measurements suggest a somewhat higher mass-to-light ratio of $M/L_V \sim 3.5\, M_\odot/L_{V,\odot}$ for the Fornax GCs (Dubath *et al.* 1992; Larsen *et al.* 2012b).]. About half of the initial cluster mass has been lost due to stellar evolution (e.g. Bruzual & Charlot 2003), so the initial stellar mass of the clusters would have been $\sim 1.4 \times 10^6\, M_\odot$ (neglecting dynamical evolution), or about 2% of the total stellar mass formed in Fornax. However, as in larger galaxies, the metallicity distributions of the GCs and the field stars in Fornax differ significantly: whereas four of the five GCs have [Fe/H] < -2 (Larsen *et al.* 2012a), the field star metallicity distribution peaks at [Fe/H] ≈ -1 (Battalia *et al.* 2006; Kirby *et al.* 2011). Leaving out the slightly more metal-rich cluster Fornax 4, the remaining four GCs would have had a total initial mass (again neglecting dynamical evolution) of about $1.2 \times 10^6\, M_\odot$. Only about 5% of the RGB stars in Fornax have [Fe/H] < -2; scaling the total stellar mass accordingly yields $\sim 3 \times 10^6\, M_\odot$, so it is clear that the GCs account for a much larger fraction of the metal-poor stars in Fornax compared to, e.g., the LMC or the Milky Way. A more accurate estimate needs to take into account the lifetimes of RGB stars as a function of metallicity and age and therefore depends on the somewhat uncertain star formation history of the Fornax dSph, but Larsen *et al.* (2012a) estimated that 20%–25% of the metal-poor stars in the Fornax dSph currently belong to the four GCs.

What about other dwarf galaxies? Within the Local Group, the single GC in the WLM galaxy appears to account for a similar high fraction of the metal-poor stars there (Larsen *et al.* 2014a). In their study of GCs in dwarf galaxies beyond the Local Group, Georgiev *et al.* (2010) found several other cases of very high specific frequencies, including the *IKN* dwarf galaxy (a member of the M81 group). These authors quote a visual magnitude of $M_V = -11.5$ for IKN, and with 5 GCs the specific frequency is then an astounding $S_N = 125$. The total integrated luminosity of the GCs corresponds to $M_V = -9.3$, so the GCs then account for about 12% of the *total* luminosity of IKN. From photometry of RGB stars, the field star metallicity distribution appears similar to that in Fornax, with a broad peak around [Fe/H] ~ -1 and only a few stars below [Fe/H] ~ -2 (Lianou *et al.* 2010; Tudorica *et al.* 2015), so the ratio of metal-poor GC stars vs. field stars at low metallicities in IKN is likely to be even higher than in Fornax. It is possible, however, that the luminosity of the IKN galaxy has been significantly underestimated as deep images show it to be very extended (Okamoto *et al.* 2015).

Before drawing sweeping conclusions about the clusters, the possibility that these dwarf galaxies may have lost significant fractions of their metal-poor field stars (albeit while holding on to the GCs) must be considered. Neither of them is obviously involved in interactions with any other nearby galaxies, and WLM in particular is located on the outskirts of the Local Group, more than 900 kpc from both the Milky Way and M31. The possible mass loss history of the Fornax dwarf has been studied in considerable detail and from analysis of its luminosity profile, which drops to very low levels well within the tidal radius, Peñarrubia *et al.* (2009) concluded that Fornax is unlikely to have lost any field stars. More recently, from modelling of the orbit of Fornax around the Milky Way, Battaglia *et al.* (2015) reached a similar conclusion and found that essentially no stellar mass has been lost even for the most eccentric orbits allowed by current observational constraints. IKN is less well studied, but in the stellar density maps of Okamoto *et al.* (2015) the galaxy has a rather regular, round shape without hints of tidal tails.

In summary, GCs can account for a much higher fraction of the metal-poor stars in dwarf galaxies (up to 20%–25% or more), compared to larger galaxies where the corresponding fraction is typically on the order of 1%–2%. It appears unlikely that this difference can be attributed simply to a preferential loss of metal-poor field stars in the dwarfs, so it must reflect differences in the efficiency of cluster formation and/or disruption in the different environments. We will now proceed to examine some of the consequences.

2. Implications for the dynamical evolution of clusters and cluster systems

The high GC- to field star ratios have potentially important implications for the amount of mass that GCs could have lost during their lifetimes, either via "traditional" dynamical mass loss mechanisms such as two-body relaxation and tidal shocks, or as required in models that seek to explain the presence of chemical abundance anomalies in GCs via self-pollution from AGB or massive stars.

2.1. *Multiple populations and the mass budget problem*

It is, by now, well established that a large fraction of the stars in globular clusters have "anomalous" abundances of many light elements. Compared to stars in the field, some elements are enhanced in these stars (e.g., He, N, Na) while others are depleted (e.g., C, O). Far less well established is the origin of these anomalies, although most current scenarios attribute them to some form of self-enrichment within the clusters with an initial "pristine" population of stars (with abundances similar to those seen in field stars) and an "enriched" population. Heavier (α- and Fe-peak) elements do not generally vary appreciably between the pristine and enriched populations, and the process responsible for producing the abundance variations is generally thought to be H-burning at high temperatures ($T > 20 \times 10^6$ K; Langer *et al.* 1993). Two main classes of polluters have been proposed: massive AGB stars undergoing hot bottom burning (Ventura *et al.* 2001; D'Antona & Ventura 2007) or massive main sequence stars, either fast-rotating (Prantzos & Charbonnel 2006; Decressin *et al.* 2007) or in binaries (de Mink *et al.* 2009).

A common challenge in self-enrichment scenarios is to explain the observed large fractions of enriched stars ($> 50\%$). A simple example illustrates this point: in the AGB model, the amount of enriched AGB ejecta produced by the pristine population amounts to about $\beta \approx 5\%$ of the initial stellar mass of the cluster for a Kroupa-like IMF (D'Ercole *et al.* 2008). Thus, even if the enriched gas is turned into new stars with an efficiency of 100%, it would only be sufficient to form an enriched population with an initial mass of $f_{\rm en} = \beta = 5\%$ of the initial mass of the pristine population. The ratio of pristine vs. enriched low-mass stars would then be 100:5, which is in stark contrast to the observed ratios of 50:50 or even 30:70 (Bastian & Lardo 2015). One way out is to postulate that a large fraction of pristine stars were subsequently lost from the clusters; in this basic example $\sim 95\%$ of the pristine stars would have to be lost in order to achieve an equal number of pristine and enriched stars. For every 10 low-mass stars left in the cluster (5 enriched and 5 pristine), there should thus be about 95 ejected stars in the field; in other words, at most $\sim 10\%$ of the stars that were originally formed in GCs should still be cluster members. There is an evident tension between this calculation and the observation that 20%–25% of the metal-poor stars in some dwarf galaxies are members of GCs.

The above calculation is clearly simplistic, and modifications to the AGB scenario are possible that make the numbers look somewhat more favourable. As pointed out in the introduction, about half of the initial mass will have been lost due to stellar evolution after a Hubble time for a standard IMF. If both populations have similar IMFs, they will lose the same fraction of their mass and the final pristine:enriched ratio will be unaffected. However, in the extreme case that the enriched population consists *only* of low-mass stars that are still alive after a Hubble time, the mass budget problem is reduced by a factor of two. Dilution of the polluted ejecta (e.g., D'Ercole *et al.* 2010) with additional pristine gas may help further. On the other hand, other implicit assumptions made above seem rather unrealistic: 1) star formation efficiencies are likely to be less than 100%, 2) the calculation assumes that, while a large fraction of the pristine stars are lost, *no* enriched stars escape from the cluster, 3) the comparison with the dwarf galaxies makes no allowance for stars with similar metallicities formed elsewhere, e.g., in the "field" or in disrupted (low-mass) clusters. We shall return to this last issue below. One could also imagine invoking a *top-heavy* IMF for the *pristine* population, which would have testable consequences for the production of stellar remnants (Dabringhausen *et al.* 2009). While the example above is carried out for the AGB scenario, the "fast rotating massive stars" scenario has similar difficulties, at least in its basic form (but see Charbonnel *et al.* 2014 for possible modifications that may accommodate the GC/field ratios in dwarf galaxies).

Apart from the specific constraints on GC mass loss from dwarf galaxies, there are many other difficulties with self-enrichment scenarios for the origin of abundance anomalies in GCs. We emphasize that there is currently no satisfactory theory, but refer the reader elsewhere for details (e.g., Bastian & Lardo 2015; Bastian, these proceedings).

The above discussion would be void if GCs in dwarf galaxies do not host multiple populations. However, it appears that they do: spectroscopy of a few RGB stars in the Fornax GCs already hinted at the [Na/O] anticorrelation (Letarte *et al.* 2006), and photometry has shown that they are also likely to exhibit substantial [N/Fe] spreads, comparable to those in Galactic GCs (Larsen *et al.* 2014b).

2.2. *Dissolution of low-mass clusters*

The considerations in the previous paragraph pertain specifically to the origin of chemical abundance anomalies in GCs. However, normal dynamical processes such as two-body relaxation and tidal shocks are expected to lead to substantial mass loss from GCs over a Hubble time, in any case. Indeed, such dynamical evolution has often been invoked as a mechanism suitable for transforming the power-law mass functions observed in young cluster populations into the mass functions observed in old GC systems, the latter having a characteristic break around $2 \times 10^5\, M_\odot$ that shows up as a peak when plotted in logarithmic units (e.g., Fall & Zhang 2001; Jordán *et al.* 2007; Kruijssen & Portegies Zwart 2009).

Jordán *et al.* (2007) show that an excellent fit to the GC mass function is provided by the "evolved Schechter function", given as

$$\frac{\mathrm{d}N}{\mathrm{d}M} \propto (M + \Delta M)^{-2} \times \exp\left(-\frac{M + \Delta M}{M^\star}\right) \tag{2.1}$$

where ΔM is the (average) amount of mass lost per cluster, here assumed to be independent of the cluster mass, which is likely to be an oversimplification (e.g. Kruijssen &

Portegies Zwart 2009). $M^\star$ is an exponential cut-off mass to account for the steepening of the MF at high masses that is also seen in young cluster systems (Larsen 2009; Portegies Zwart *et al.* 2010). For $\Delta M = 0$, the function is simply a power-law with an exponential truncation, and if mass loss has occurred we see that the MF is flat for $M < \Delta M$, as observed in old GC systems. A fit to the Milky Way GC system yields $M^\star = 8 \times 10^5 M_\odot$ and $\Delta M = 2.5 \times 10^5 M_\odot$, and the same parameters also fit the combined GC mass function for dwarf galaxies in the Local Group well. The amount of mass lost from the cluster system can now be found by integrating Eq. (2.1) over all masses for $\Delta M = 0$ and $\Delta M = 2.5 \times 10^5 M_\odot$. The result depends on the lower limit adopted for the initial mass distribution; for lower limits of 100 $M_\odot$ and 5000 $M_\odot$ (and adopting an upper present-day mass limit of $10^6 M_\odot$), the ratios of the initial to current masses are $M_{\rm init}/M_{\rm current} = 24$ and 13, respectively. This calculation takes does not account for stellar evolution, which would again increase the ratios by about a factor of two. However, for the sake of comparison with present-day stellar populations, we can ignore stellar mass loss as this will have affected the clusters and the field populations about equally. From a more detailed calculation, Kruijssen & Portegies Zwart (2009) found factors of 64 and 39 for the same lower mass limits, which do include stellar evolution.

The large fractions of dynamical mass loss calculated above may, once again, be compared with the present-day fractions of metal-poor stars in dwarf galaxies that are GC members. As before, far more stars should have been lost from clusters than those that are observed today in the field. Thus, while it is already difficult to accommodate the expected amount of mass loss, either from standard dynamical evolution or as required by simple self-enrichment scenarios separately, the problem becomes even more acute if both were to have occurred.

2.3. *Infant mortality and the efficiency of cluster formation*

Finally, we comment on the issue of "infant mortality". This term is generally used to refer to an early phase of cluster disruption during which a large fraction of the cluster population is dissolved, more or less independently of mass. The notion dates back at least to Lada & Lada (1991,2003), who found that the birth rate of young embedded clusters appears to exceed that of older open clusters by at least an order of magnitude. Subsequent studies have found similar results when comparing compact stellar groupings in external galaxies with older clusters (e.g., Chandar *et al.* 2010). Other studies, however, have questioned the need for an early phase of "infant mortality", arguing that a large fraction of the stellar aggregates present at young ages are not true clusters, but rather unbound associations (Bastian *et al.* 2012).

At any rate, it seems clear that only a relatively minor fraction of stars in present-day star forming regions form within clusters that remain bound on long time scales. The formation efficiency of bound clusters appears to correlate with star formation rate density (Larsen & Richtler 2000; Goddard *et al.* 2010; Silva-Villa & Larsen 2011) and is typically 1%–10% at the current epoch, although it may have been significantly higher in the early Universe when galaxies were more gas rich (Kruijssen 2012).

Again, the observations of GCs in dwarf galaxies put strong constraints on the efficiency of cluster formation and/or "infant mortality" in these environments. Even if no other cluster disruption mechanisms have been at work, a very large fraction of the early star formation must have occurred in the few, massive star clusters that are still surviving to the present day. It is curious that this situation seems to have changed fairly abruptly for metallicities [Fe/H] > -2.

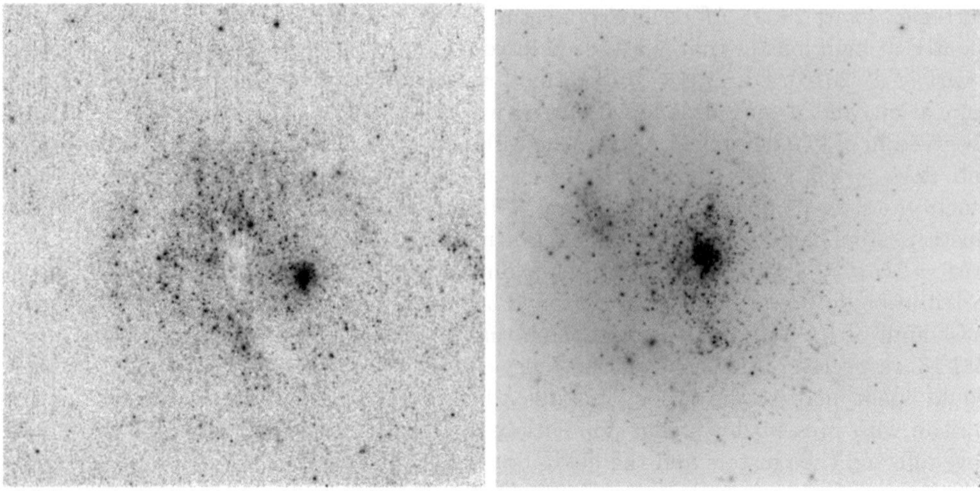

Figure 1. HST/PC F555W images of a star forming complex in the spiral galaxy NGC 6946 (left) and the dwarf galaxy NGC 1705 (right). The dominant object in each panel is a massive star cluster, which is surrounded by numerous fainter clusters and field stars.

2.4. *Young massive clusters in dwarf galaxies*

In the context of dynamical evolution, it is important to note that any direct information about the low-mass end of the initial cluster MF has now been lost, as clusters with initial masses $M < \Delta M$ have disrupted completely. It is, therefore, not strictly necessary to require that a large population of low-mass clusters were present initially, and we can speculate about other possibilities — although such speculations should, preferably, be guided by observational evidence.

As already noted, young cluster populations in the present-day Universe generally do appear to follow power-law MFs down to low masses (Elmegreen & Efremov 1997; Portegies Zwart *et al.* 2010; Adamo & Bastian 2015). The best constraints come from galaxies with rich cluster populations, such as mergers and other starbursts, or large spirals with high star formation rates, although the more limited data available for star-forming dwarfs suggest that many of them also have power-law like cluster mass functions (Billett *et al.* 2002; Seth *et al.* 2004; Cook *et al.* 2012). However, there are some notable exceptions to this statement, in particular the dwarfs NGC 1569 and NGC 1705 which are both dominated by one or two very luminous, young clusters (O'Connell *et al.* 1994). These clusters have estimated masses in the range $8 \times 10^5 \, M_\odot$ to $1.4 \times 10^6 \, M_\odot$ and ages between 5×10^6 years and 16×10^6 years (Larsen *et al.* 2011).

Figure 1 shows HST/PC images of the central regions of NGC 1705 and a stellar complex surrounding a young massive cluster in the spiral galaxy NGC 6946. The two galaxies are at similar distances, 5.1 Mpc for NGC 1705 (Tosi *et al.* 2001) and 5.9 Mpc for NGC 6946 (Karachentsev *et al.* 2000), so the scales are comparable. The corresponding field-of-view is $0.9 \times 0.9 \, \mathrm{kpc}^2$ and $1 \times 1 \, \mathrm{kpc}^2$. The similarity of the two regions is striking. Both clusters have masses of about $10^6 \, M_\odot$ and ages of 10–15 Myr and dominate their immediate surroundings. The cluster in NGC 6946 accounts for about 17% of the V-band light of the surrounding stellar complex (Larsen *et al.* 2002), while that in NGC 1705 appears to have formed as part of a starburst that produced only $10^6 \, M_\odot$ of stars, excluding those in the cluster (Annibali *et al.* 2009). Indeed, there are no other clusters of comparable brightness and mass in NGC 1705, although the cluster mass function integrated over all of NGC 6946 is consistent with a Schechter-like function.

Perhaps one should imagine the Fornax dwarf and others like it, during the earliest stages of their assembly histories, as collections of localised starbursts like those in Fig. 1. Within each region, the conditions may have been conducive to forming massive clusters, for example because of high pressures and densities (Johnson *et al.* 2015). This situation may have applied more generally in the fragments that merged to form galaxy halos, at least those that were large enough to form massive clusters. In this context, it has been suggested that the WLM galaxy, with its single massive GC, may be a surviving analogue of these fragments (Elmegreen *et al.* 2012), although its GC is more metal-poor than typical halo GCs in the Milky Way (Larsen *et al.* 2014a).

3. Summary and Conclusions

High GC specific frequencies appear to be a common feature in dwarf galaxies. It should, of course, be kept in mind that one GC in a galaxy of absolute magnitude $M_V = -15$ already yields $S_N \equiv 1$, so even a single GC will cause a high S_N in faint dwarfs such as Fornax or IKN. The study by Georgiev *et al.* (2010) also includes several dwarf galaxies where no GCs have been detected. However, populations of 5 GCs in galaxies like Fornax and IKN seem unlikely to be simply statistical outliers. In any case, we have seen that these high S_N values, combined with the tendency for GCs to be significantly more metal-poor on average than the majority of field stars in their parent galaxies, put tight constraints on the amount of cluster disruption that could have occurred in these galaxies. First, it is difficult to accommodate the large amounts of mass loss required in basic self-enrichment scenarios for the origin of multiple stellar populations in GCs. However, a more general problem is that if the current GC mass function is the result of dynamical evolution from an initial power-law function similar to that observed in young cluster populations, then this should have produced a larger number of field stars than is currently observed, unless the low-mass clusters were preferentially more metal-rich than the surviving GCs. The possibility that the GC mass function was top-heavy already from the outset may thus warrant serious consideration. It appears that a large fraction of the early star formation in these dwarfs may have occurred in the globular clusters that survive to this day.

Acknowledgements

I am grateful to the Leids Kerkhoven-Bosscha Fonds (LKBF) and the IAU for travel grants that made it possible for me to attend this symposium.

References

Adamo, A. & Bastian, N. 2015, in: *The Lifecycle of Clusters in Galaxies*, ed. S. Stahler, Springer
Annibali, F., Tosi, M., Monelli, M., *et al.* 2009, *AJ*, 138, 169
Bastian, N., Adamo, A., Gieles, M., *et al.* 2012, *MNRAS*, 419, 2606
Bastian, N. & Lardo, C. 2015, *MNRAS*, 453, 357
Battaglia, G., Tolstoy, E., Helmi, A., *et al.* 2006, *A&A*, 459, 423
Battaglia, G., Sollima, A., & Nipoti, C. 2015, *MNRAS*, in press (arXiv:1509.02368)
Behroozi, P. S., Conroy, C., & Wechsler, R. H. 2010, *ApJ*, 717, 379
Billett, O. H., Hunter, D. A., & Elmegreen, B. G. 2002, *AJ*, 123, 1454
Bruzual, G. & Charlot, S. 2003, *MNRAS*, 344, 1000
Chandar, R., Whitmore, B. C., Kim, H., *et al.* 2010, *ApJ*, 719, 966
Charbonnel, C., Chantereau, W., Krause, M., *et al.* 2014, *A&A*, 569, L6
Coleman, M. G. & de Jong, J. T. A. 2008, *ApJ*, 685, 933
Cook, D. O., Seth, A. C., Dale, D. A., *et al.* 2012, *ApJ*, 751, 100
Dabringhausen, J., Kroupa, P., & Baumgardt, H. 2009, *MNRAS*, 394, 1529

D'Antona, F. & Ventura, P. 2007, *MNRAS*, 379, 1431
D'Ercole, A., Vesperini, E., D'Antona, F., *et al.* 2008, *MNRAS*, 391, 825
D'Ercole, A., D'Antona, F., Ventura, P., *et al.* 2010, *MNRAS*, 407, 854
Decressin, T., Charbonnel, C., & Meynet, G. 2007, *A&A*, 475, 859
de Mink, S. E., Pols, O. R., Langer, N., & Izzard, R. G. 2009, *A&A*, 507, L1
Dubath, P., Meylan, G., & Mayor, M. 1992, *ApJ*, 400, 510
Elmegreen, B. G. & Efremov, Y. N. 1997, *ApJ*, 480, 235
Elmegreen, B. G., Malhotra, S., & Rhoads, J. 2012, *ApJ*, 757, 9
Fall, S. M. & Zhang, Q. 2001, *ApJ*, 561, 751
Forbes, D. A. & Forte, J. C. 2001, *MNRAS*, 322, 257
Forte, J. C., Strom, S. E., & Strom, K. M. 1981, *ApJ*, 245, L9
Forte, J. C., Faifer, F., & Geisler, D. 2007, *MNRAS*, 382, 1947
Georgiev, I. Y., Puzia, T. H., Goudfrooij, P., & Hilker, M. 2010, *MNRAS*, 406, 1967
Goddard, Q. E., Bastian, N., & Kennicutt, R. C. 2010, *MNRAS*, 405, 857
Harris, W. E., Harris, G. L. H., Layden, A. C., & Wehner, E. M. H. 2007, *ApJ*, 666, 903
Harris, W. E., Harris, G. L. H., & Alessi, M. 2013, *ApJ*, 772, 82
Johnson, K. E., Leroy, A. K., Indebetouw, R. *et al.* 2015, *ApJ*, 806, 35
Jorcán, A., McLaughlin, D. E., & Côté, P., *et al.* 2007, *171*, 101
Karachentsev, I. D., Sharina, M. E., & Huchtmeier, W. K. 2000, *A&A*, 362, 544
Kinman, T. D., Stryker, L. L., Hesser, J. E., *et al.* 1991, *PASP*, 103, 1279
Kirby, E. N., Lanfranchi, G. A., Simon, J. D. *et al.* 2011, *ApJ*, 727, 78
Kruijssen, J. M. D. & Portegies Zwart, S. F. 2009, *ApJ*, 698, L158
Kruijssen, J. M. D. 2012, *MNRAS*, 426, 3008
Lada, C. J. & Lada, E. A. 1991, in: *The formation and evolution of star clusters*, ed. K. Janes, ASP: San Francisco
Lada, C. J. & Lada, E. A. 2003, *ARA&A*, 41, 57
Langer, G. E., Hoffman, R., & Sneden, C. 1993, *PASP*, 105, 301
Larsen, S. S. & Richtler, T. 2000, *A&A*, 354, 836
Larsen, S. S. 2009, *A&A*, 494, 539
Larsen, S. S., Brodie, J. P., Huchra, J. P., *et al.* 2001, *AJ*, 121, 2974
Larsen, S. S., Efremov, Y. N., Elmegreen, B. G., *et al.* 2002, *ApJ*, 567, 896
Larsen, S. S., de Mink, S. E., Eldridge, J. J., *et al.* 2011, *A&A*, 532, A147
Larsen, S. S., Strader, J., & Brodie, J. P. 2012a, *A&A*, 544, L14
Larsen, S. S., Brodie, J. P., & Strader, J. 2012b, *A&A*, 546, A53
Larsen, S. S., Brodie, J. P., Forbes, D. A., & Strader, J. 2014, *A&A*, 565, A98
Larsen, S. S., Brodie, J. P., Grundahl, F., & Strader, J. 2014, *ApJ*, 797, 15
Letarte, B., Hill, V., Jablonka, P., *et al.* 2006, *A&A*, 453, 547
Lianou, S., Grebel, E. K., & Koch, A. 2010, *A&A*, 521, A43
Mieske, S., Küpper, A. H. W., & Brockamp, M. 2014, *A&A*, 565, L6
O'Connell, R. W., Gallagher, J. S., & Hunter, D. A. 1994, *ApJ*, 433, 65
Okamoto, S., Arimoto, N., Ferguson, A. M. N., *et al.* 2015, *ApJL*, L1
Peñarrubia, J., Navarro, J. F., McConnachie, A. W., & Martin, N. F. 2009, *ApJ*, 698, 222
Peng, E. W., Jordán, A., Côté, P., *et al.* 2008, *ApJ*, 681, 197
Portegies Zwart, S. F., McMillan, S. L. W., & Gieles, M. 2010, *ARA&A*, 48, 431
Prantzos, N. & Charbonnel, C. 2006, *A&A*, 458, 135
Seth, A., Olsen, K., Miller, B., *et al.* 2004, *AJ*, 127, 798
Silva-Villa, E. & Larsen, S. S. 2011, *A&A*, 529, A25
Strader, J., Smith, G. H., Larsen, S. S., *et al.* 2009, *AJ*, 138, 547
Suntzeff, N. B., Schommer, R. A., Olszewski, E. W., & Walker, A. R. 1992, *AJ*, 104, 1743
Tosi, M., Sabbi, E., Bellazzini, M., *et al.* 2001, *AJ*, 122, 1271
Tudorica, A., Georgiev, I. Y., & Chies Santos, A. L. 2015, *A&A*, 581, A84
Ventura, P., D'Antona, F., Mazzitelli, I., & Gratton, R. 2001, *ApJ*, 550, L65
Webbink, R. F. 1985, in: *Dynamics of star clusters*, IAU Symp. 113, eds. J. Goodman & P. Hut, Dordrecht

Formation, evolution, and survival of massive star clusters
Proceedings IAU Symposium No. 316, 2015
C. Charbonnel & A. Nota, eds.

doi:10.1017/S1743921315009126

Stellar population properties of the most massive globular clusters and ultra-compact dwarf galaxies of the Fornax cluster

Michael Hilker[1]

[1]European Southern Observatory, Karl-Schwarzschild-Str. 2,
D-85748, Garching bei München, Germany
email: mhilker@eso.org

Abstract. Most ultra-compact dwarf galaxies (UCDs) and very massive globular clusters reside in nearby galaxy clusters or around nearby giant galaxies. Due to their distance (>4 Mpc) and compactness ($r_{\rm eff}$ < 100 pc) they are barely resolved, and thus it is difficult to obtain their internal properties. Here I present our most recent attempts to constrain the mass function, stellar content and dynamical state of UCDs in the Fornax cluster. Thanks to radial velocity membership assignment of ~950 globular clusters (GCs) and UCDs in the core of Fornax, the shape of their mass function is well constrained. It is consistent with the 'standard' Gaussian mass function of GCs. Our recent simulations on the disruption process of nucleated dwarf galaxies in cluster environments showed that ~40% of the most massive UCDs should originate from nuclear star clusters. Some Fornax UCDs actually show evidence for this scenario, as revealed by extended low surface brightness disks around them and onsets of tidal tails. Multi-band UV to optical imaging as well as low to medium resolution spectroscopy revealed that there exist UCDs with youngish ages, (sub-)solar [α/Fe] abundances, and probably He-enriched populations.

Keywords. galaxies: star clusters, galaxies: abundances, galaxies: clusters: individual (Fornax)

1. Introduction

Most of the massive globular clusters (GCs) of our Milky Way show evidence of multiple stellar populations with differences in their light element abundances (see contributions of Piotto and Milone in this volume). A few GCs even exhibit spreads in iron abundance and probably age. Those are nuclear star cluster candidates whose host galaxies were disrupted during the assembly history of the Milky Way.

In galaxy clusters, disruption of low mass, nucleated galaxies was very common in the past. Indeed, in the Virgo and Fornax clusters there exists a large population of very massive and compact star cluster-like objects, called ultra-compact dwarf galaxies (UCDs). It was early speculated that they might be either "*very bright GCs*" or lower mass examples of "*compact ellipticals like M 32*", or they might "*represent the nuclei of disolved dE,Ns*" (Hilker *et al.* 1999). In the recent years, more and more UCDs have been discovered, some very compact, some rather diffuse, filling up the previously unpopulated region between star clusters and dwarf galaxies in the luminosity-size plane (see figure 1). Different authors apply different definitions for UCDs, either invoking a lower luminosity/mass cut or a lower size cut. Since those definitions are not physically motivated, I consider as 'UCDs' all objects in the red box shown in figure 1. In general, UCDs are characterized by an upper envelope in the mass-size relation and enhanced dynamical mass-to-light ratios (Mieske *et al.* 2008b), roughly occurring at $> 2 \times 10^6 M_\odot$.

If one postulates that the complex GCs ω Centauri and M54 in the Milky Way and G1

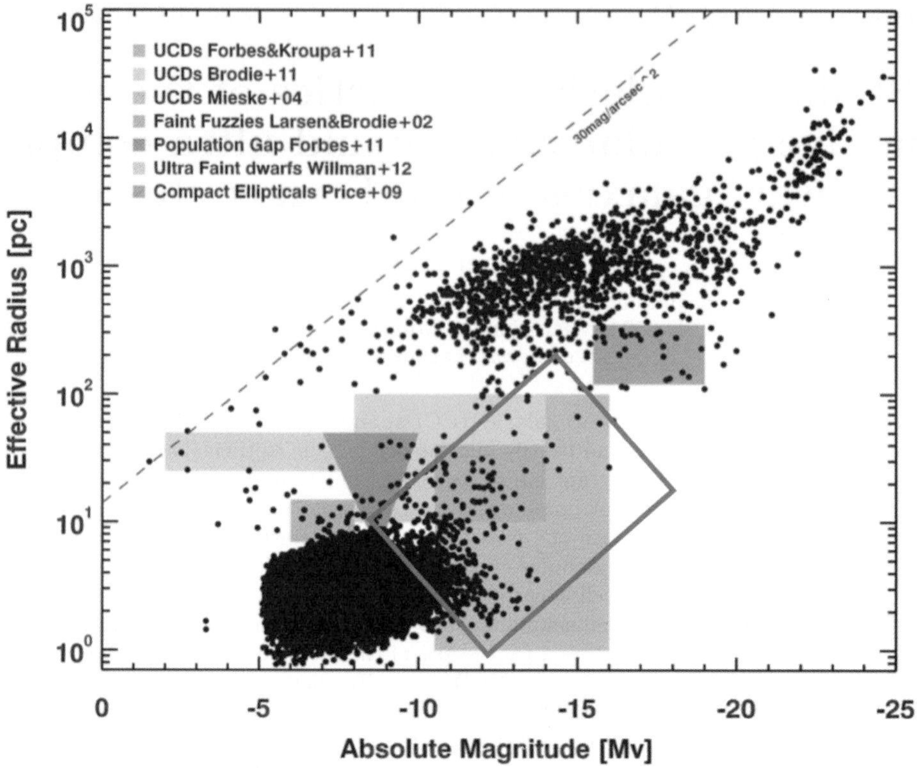

Figure 1. Luminosity-size plane of early-type stellar systems. The compilation is mainly based on similar samples shown in Misgeld & Hilker (2011) and Norris *et al.* (2014). Coloured areas show different definitions of compact objects. Their references are given in the legend. The red rectangle encompasses the objects of interest for this contribution. The dashed line indicates the surface brightness limit of 30 mag/arcsec^2.

in Andromeda are low-mass UCDs, one would expect that UCDs in general should also have complex star formation and chemical enrichment histories. However, due to the large distance (most are located beyond 4 Mpc) and thus unresolved nature of UCDs, multiple stellar populations in them are very difficult to detect and quantify. Most quantities have to be derived from images and spectra of their unresolved stellar populations. The determination of spatially resolved properties of UCDs requires space-based imaging and adaptive optics assisted spectroscopy.

2. The most massive globular clusters in the Fornax cluster

The Fornax cluster has a very well studied GC and UCD population. GC counts from photometric surveys revealed that there exist $\sim 6,450 \pm 700$ GCs within 83 kpc projected distance around the central cluster galaxy NGC 1399 (Dirsch *et al.* 2003). Within 300 kpc of NGC 1399 the GC number counts increase to $\sim 11,100 \pm 2,400$ GCs (Gregg *et al.* 2009, derived from the data of Bassino *et al.* 2006). The number of radial velocity confirmed GCs and UCDs around NGC 1399 is ~950 (Schuberth *et al.* 2010, Hilker, in prep.). Within 300 kpc of NGC 1399 the sample of confirmed UCDs with masses above $5 \times 10^6 M_\odot$ is rather complete and the photometrically selected GCs are well sampled below that limit. Thus, a mass function of UCDs and GCs can be constructed with high confidence (see also contribution of Schulz in this volume). The shape of that

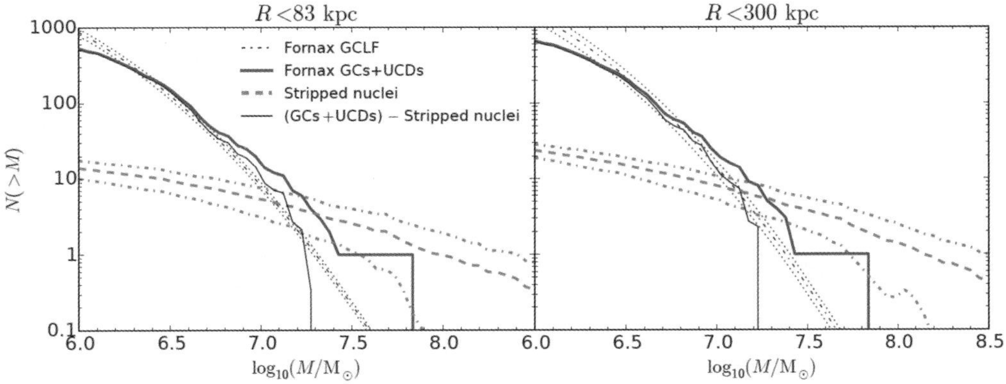

Figure 2. Cumulative mass function of simulated stripped nuclei in Fornax-like clusters within a projected distance of 83 (left panel) and 300 kpc (right panel) from the central galaxy (dashed line with the standard deviation between clusters shown by dash-dotted lines) compared with GCs/UCDs in the Fornax cluster (thick solid line). The thin dash-dotted line shows the integrated Fornax GC luminosity function (GCLF) with the standard deviation given by dotted lines. The thin solid line shows the mean for the simulations subtracted from the Fornax GCLF. Figure taken from Pfeffer *et al.* (2015).

mass function was already presented in Hilker (2009), where we showed that the mass function above $3 \times 10^5 M_{\odot}$ can be approximated with two power laws with slope exponents of -1.9 and -2.7 for objects below and above $2 \times 10^6 M_{\odot}$, respectively. This shape is broadly consistent with the 'standard' (Gaussian) mass function of GCs. Thus, from this point of view, UCDs seem to be consistent with being drawn from the GC mass function (Mieske *et al.* 2012). However, the most massive UCD in Fornax, UCD3, exhibits a compact core and an extended stellar envelope of ~90 pc size (Evstigneeva *et al.* 2008), pointing rather to a remnant of a disrupted, nucleated dwarf galaxy.

We asked ourselves how many UCDs that originated from nuclei of disrupted galaxies are expected to contribute to the mass function in a Fornax cluster-like environment? We conducted cosmological simulations combined with a semi-analytic galaxy formation model and empirical prescriptions for the nucleation fraction and nuclei to galaxy mass fraction to identify stripped nuclei of disrupted satellite galaxies (Pfeffer *et al.* 2014). In figure 2 we show the cumulative mass function of the simulated stripped nuclei and compare it with the mass function of observed GCs/UCDs with masses $>10^6 M_{\odot}$. As one can see, stripped nuclei can only explain the number of very massive UCDs. For masses larger than $10^7 M_{\odot}$ we predict stripped nuclei account for ~40% of GCs/UCDs. For masses between 10^6 and $10^7 M_{\odot}$ the stripped nuclei account for only ~2.5% of GCs/UCDs. Thus, the majority of lower mass UCDs should be of star cluster origin, either formed as very massive genuine globular cluster or being the result of merged super star cluster complexes (e.g., Fellhauer & Kroupa 2002, Brüns *et al.* 2009).

Indeed, there exist young massive star clusters that occupy the same space in the mass-size plane as UCDs (Kissler-Patig *et al.* 2006). Thus, mass and size alone are not indicative of the origin of UCDs.

3. Internal properties of ultra-compact dwarf galaxies

The nature of UCDs can be revealed, or at least constrained, by analysing the spatially resolved surface brightness profiles and internal kinematics and by studying the properties of their unresolved stellar population. A supermassive black hole (SMBH) and/or and

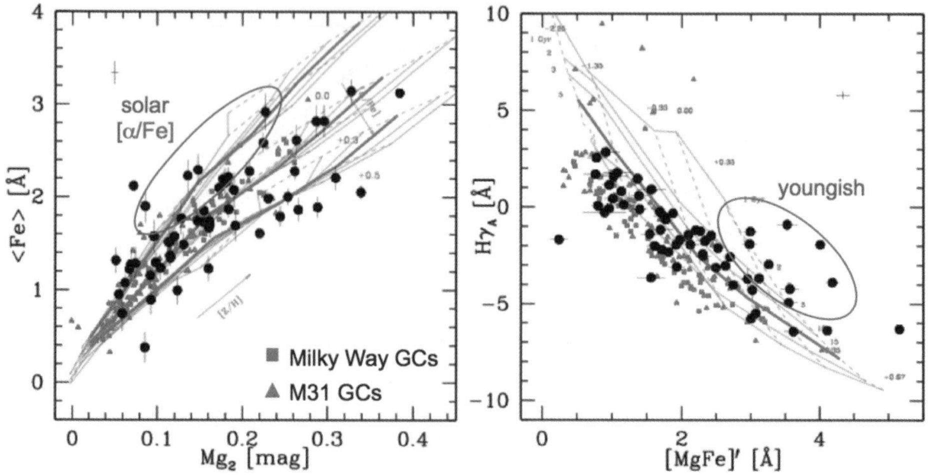

Figure 3. Lick index measurements of ~50 UCDs and massive GCs in the core of the Fornax cluster (black dots) compared to Milky Way and M31 GCs (as indicated). *Left panel*: Mg_2 versus iron index, which allows to differentiate [α/Fe] abundances. *Right panel*: Metallicity index [MgFe]' versus the Balmer index H_{γ_A}, which breaks the age-metallicity degeneracy. The model grids are based on SSP models from Thomas *et al.* (2003). Objects with (sub-)solar [α/Fe] values and young ages are encircled by green ellipses.

extended star formation history would favour their origin as stripped nuclei of a dwarf galaxy. A high metallicity, i.e. not obeying the mass-metallicity relation of nuclei, would rather point to a star cluster origin in an enriched (maybe merger-induced) environment.

Evstigneeva *et al.* (2008) revealed positive colour gradients in the resolved light profiles of some UCDs in Fornax and Virgo. A blue nucleus and a red envelope are consistent with the colour differences of nuclei and stellar body in nucleated dwarf ellipticals (Lotz *et al.* 2004), where nuclei are formed from merging, metal-poor GCs.

In a recent study of the structural composition and clustering properties of 97 UCDs in the halo of NGC 1399, we found evidence for faint stellar envelopes around several UCDs with effective radii of up to 90 pc (Voggel *et al.* 2015). One particularly extended UCD shows clear signs of tidal tails extending out to ~350 pc. We also detected, in a statistical sense, a local overdensity of blue GCs on scales of $\leqslant$ 1 kpc around UCDs. These could either be remnant GCs of a formerly rich GC system around a disrupted dwarf galaxy, or surviving star clusters of a merged super star cluster complex (e.g., Brüns *et al.* 2009).

Combining GALEX NUV/FUV with optical photometry, Mieske *et al.* (2008) showed that the location of several Fornax UCDs and massive GCs in the (FUV-NUV) versus (NUV-V) plane can only be explained with He-enhanced SSP models. This might be a hint to multiple populations in these GCs/UCDs, since He enhancement is also observed in the second generation stars of multi-population Galactic GCs.

Deep VLT/FORS2 spectroscopy on ~50 bright GCs ($M_V < -9$ mag) allowed us to measure Lick indices (Hilker, Puzia, *et al.*, in prep.). In Fig. 3 we show two planes of Lick indices that constrain the [α/Fe] abundances and break the age-metallicity degeneracy, respectively. Besides old GCs with enhanced [α/Fe] abundances, there also exist objects with (sub-solar) [α/Fe] values and some metal-rich intermediate-age GCs/UCDs (2-7 Gyr). Those might have had their origin in dwarf galaxies with low star formation rates and/or as nuclear star clusters with recurrent star formation episodes.

The most convincing case of a stripped nuclei origin of a UCD is the discovery of a SMBH in M60-UCD1, one of the most massive and densest UCDs in the Virgo cluster

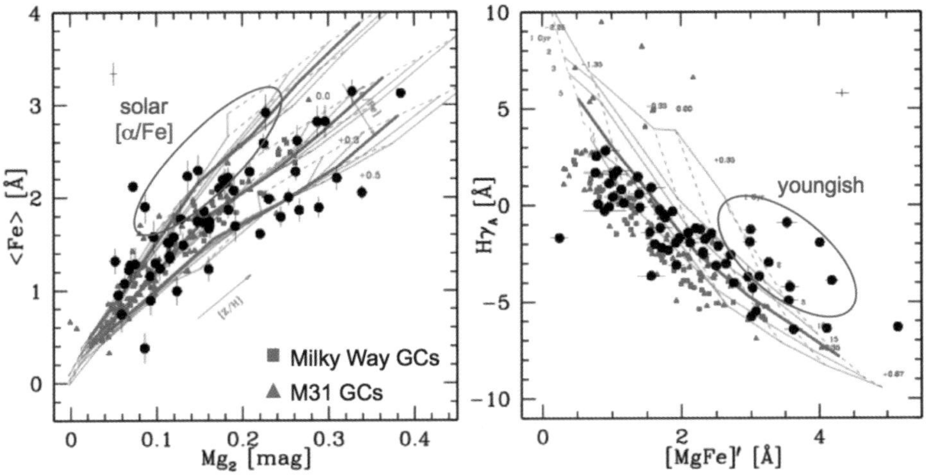

Figure 4. Predicted masses of central black holes in stripped nuclei as function of stripped nucleus mass. The mean and 1-σ confidence interval for the simulations (grey dots) are given by the solid and dashed lines, respectively, using bin sizes of 100 objects. The typical 1-σ confidence interval in MBH is 0.5 dex. For comparison, we also show the black hole mass of M60-UCD1 (Seth *et al.* 2014), the inferred black hole masses of UCDs assuming elevated mass-to-light ratios are due to central black holes (Mieske *et al.* 2013) and the limits for central black holes in the GCs ω Cen and G1 (Lützgendorf *et al.* 2013). For the observed UCDs and GCs, objects with implied black hole masses of zero are given by triangles at the bottom of the figure. Figure taken from Pfeffer *et al.* (2015).

(Seth *et al.* 2014). The SMBH comprises 15% of the UCD's total mass. In our simulations of stripped nuclei (Pfeffer *et al.* 2015) we predict the masses of their central black holes (BHs) as function of stripped nuclei mass (see figure 4). The SMBH mass of M60-UCD1 falls on top of the simulations, and, interestingly, also the inferred BH masses of UCDs with elevated mass-to-light ratios (Mieske *et al.* 2013) are very much consistent with the simulations. More AO-assisted spectroscopic observations are underway to prove or disprove the SMBH hypothesis for UCDs with high dynamical mass-to-light ratios.

4. Summary and outlook

Our general conclusions from the findings on UCDs and massive GCs in the Fornax cluster presented in this contribution can be summarized as follows:

- UCDs are defined through an upper envelope in the mass-size relation and enhanced dynamical mass-to-light ratios – *roughly occurring at* $> 2 \times 10^6 M_\odot$.
- UCDs share properties of nuclei as well as young massive star clusters, e.g. the mass-size relation. *They are a mixed bag of objects.*
- The mass function of massive GCs/UCDs in Fornax shows a break at $\sim 2 \times 10^6 M_\odot$. There is no excess of UCD-mass objects. *Thus, they are mostly of 'star cluster origin'.*
- Stripping of nucleated galaxies in a cosmological framework cannot explain the total number of UCDs, *but* ~*40% of UCDs with a mass* $> 10^7 M_\odot$ *are compatible with stripped nuclei.*
- M60-UCD1 has a SMBH, which makes up 15% of the UCD's mass. *This is direct evidence for the stripping scenario as valid channel to form UCDs.*
- Extended stellar envelopes and overdensities of star clusters around UCDs might hint at the accretion of nucleated dwarf galaxies or at a dissolving super star clusters that were formed from merged star cluster complexes.

Distinguishing stripped nuclei from super star clusters or massive GCs is a difficult task. Many internal properties of UCDs can be explained by both scenarios. More measurements of SMBHs in UCDs, derivation of their star formation histories and light element abundances are needed to unveil the nature of individual UCDs, and thus build up statistically meaningful samples for UCDs of different flavours.

References

Bassino, L. P., Faifer, F. R., Forte, J. C., Dirsch, B., Richtler, T., Geisler, D., & Schuberth, Y. 2006, *A&A*, 451, 789

Brüns, R. C., Kroupa, P., & Fellhauer, M. 2009, *ApJ*, 702, 1268

Brodie, J. P., Romanowsky, A. J., Strader, J., & Forbes, D. A. 2011, *AJ*, 142, 199

Dirsch, B., Richtler, T., Geisler, D., Forte, J. C., Bassino, L. P., & Gieren, W. P. 2003, *AJ*, 125, 1908

Evstigneeva, E. A., Drinkwater, M. J., Peng, C. Y., Hilker, M., De Propris, R., *et al.* 2008 *AJ*, 136, 461

Fellhauer, M. & Kroupa, P. 2002, *MNRAS*, 330, 642

Forbes, D. A., & Kroupa, P. 2011 *PASA*, 28, 77

Forbes, D. A., Pota, V., Usher, C., Starder, J., Romanowsky, A. J., *et al.* 2013 *MNRAS*, 435, L6

Gregg, M. D., Drinkwater, M. J., Evstigneeva, E., Jurek, R., Karick, A. M., Phillipps, S., Bridges, T., Jones, J. B., Bekki, K., & Couch, W. J. 2009 *AJ*, 137, 498

Hilker, M., Infante, L., Vieira, G., Kissler-Patig, M., & Richtler, T. 1999, *A&AS*, 134, 75

Hilker, M. 2009, in: S. Roeser (ed.), *Reviews in Modern Astronomy: Formation and Evolution of Cosmic Structures*, Vol. 21, (Wiley-VCH Verlag GmbH & Co. KGaA), p. 199

Kissler-Patig, M., Jordán, A., & Bastian, N. 2006 *A&A*, 448, 1031

Larsen, S. S., & Brodie, J. P. 2002 *AJ*, 123, 1488

Lotz, J. M., Miller, B. W., & Ferguson, H. C. *ApJ*, 613, 262

Lützgendorf, N., Kissler-Patig, M., Neumayer, N., Baumgardt, H., Noyola, E., *et al.* 2013 *A&A*, 555,26

Mieske, S., Hilker,M., & Infante, L. 2004 *A&A*, 418, 445

Mieske, S., Hilker,M., Bomans, D. J., Rey, S.-C., Kim, S., Yoon, S.-J., & Chung, C. 2008 *A&A*, 489, 1023

Mieske, S., Hilker,M., Jordán, A., Infante, L., Kissler-Patig, M., *et al.* 2008 *A&A*, 487, 921

Mieske, S., Hilker, M., & Misgeld, I. 2012 *A&A*, 537, 3

Mieske, S., Frank, M. J., Baumgardt, H., Lützgendorf, N., Neumayer, N., & Hilker, M. 2013 *A&A*, 558, 14

Misgeld, I. & Hilker, M. 2011, *MNRAS*, 414, 3699

Norris, M. A., Kannappan, S. J., Forbes, D. A.; Romanowsky, A. J., Brodie, J. P., *et al.* 2014 *MNRAS*, 443, 1151

Pfeffer, J., Griffen, B. F., Baumgardt, H., & Hilker, M. 2014, *MNRAS*, 444, 3670

Pfeffer, J., Hilker, M., & Baumgardt, H. 2015 *MNRAS*, submitted

Schuberth, Y., Richtler, T., Hilker, M., Dirsch, B., Bassino, L. P., Romanowsky, A. J., & Infante, L. 2010 *A&A*, 513, 52

Seth, A. C., van den Bosch, R., Mieske, S., Baumgardt, H., & Brok, M. Den, *et al.* 2014, *Nature*, 513, 398

Thomas, D., Maraston, C. & Bender, R. 2003 *MNRAS*, 339, 897

Voggel, K., Hilker, M., & Richtler, T. 2015 *A&A*, accepted

Willman, B., & Strader, J. 2012 *AJ*, 144, 76

Formation, evolution, and survival of massive star clusters
Proceedings IAU Symposium No. 316, 2015
C. Charbonnel & A. Nota, eds.

doi:10.1017/S1743921315010662

Mapping out the origins of compact stellar systems

Aaron J. Romanowsky,[1,2] Jean P. Brodie,[2] and the SAGES[3]

[1]Department of Physics & Astronomy, San José State University,
San Jose, CA 95192, USA
email: aaron.romanowsky@sjsu.edu

[2]University of California Observatories,
1156 High Street, Santa Cruz,CA 95064, USA
email: jbrodie@ucsc.edu

[3]http://sages.ucolick.org

Abstract. We present a suite of extragalactic explorations of the origins and nature of globular clusters (GCs) and ultra-compact dwarfs (UCDs), and the connections between them. An example of GC metallicity bimodality is shown to reflect underlying, distinct metal-poor and metal-rich stellar halo populations. Metallicity-matching methods are used to trace the birth sites and epochs of GCs in giant E/S0s, pointing to clumpy disk galaxies at $z \sim 3$ for the metal-rich GCs, and to a combination of accreted and in-situ formation modes at $z \sim$ 5–6 for the metal-poor GCs. An increasingly diverse zoo of compact stellar systems is being discovered, including objects that bridge the gaps between UCDs and faint fuzzies, and between UCDs and compact ellipticals. Many of these have properties pointing to origins as the stripped nuclei of larger galaxies, and a smoking-gun example is presented of an ω Cen-like star cluster embedded in a tidal stream.

Keywords. galaxies: star clusters – galaxies: nuclei – galaxies: halos

1. Introduction

Globular star clusters (GCs) are spectacular residents of our Galaxy's halo and beyond, yet there are long standing mysteries as to how, when, and where they formed. These mysteries expanded with the recent discoveries of ultra-compact dwarfs (UCDs), which began to bridge the classical gap between galaxies and star clusters and raised questions of how to differentiate between these two classes of stellar system. While studies of GCs in the Milky Way (MW) are focusing on remarkable clues about their origins from multiple stellar populations, another important window is to examine the extragalactic populations of GCs and UCDs – to provide additional context, statistics, and opportunities for witnessing formation in action.

Here we discuss briefly three areas of novel discovery and insight emerging recently from extragalactic studies. The first is new constraints on GC origins in a cosmological context (§2), the second is more extensive mapping of the parameter space of compact stellar systems (§3), and the third concerns observations of tidally stripped nuclei (§4).

2. Globular cluster bimodality and cosmological origins

GCs are thought to trace ancient phases of star and galaxy formation, but many of the particulars are murky. The MW provides important clues through the *bimodality* of its GC system, which consists of a roughly spherical, non-rotating, metal-poor "halo" component, and a flattened, rotating, metal-rich, "bulge/thick-disk" component. This

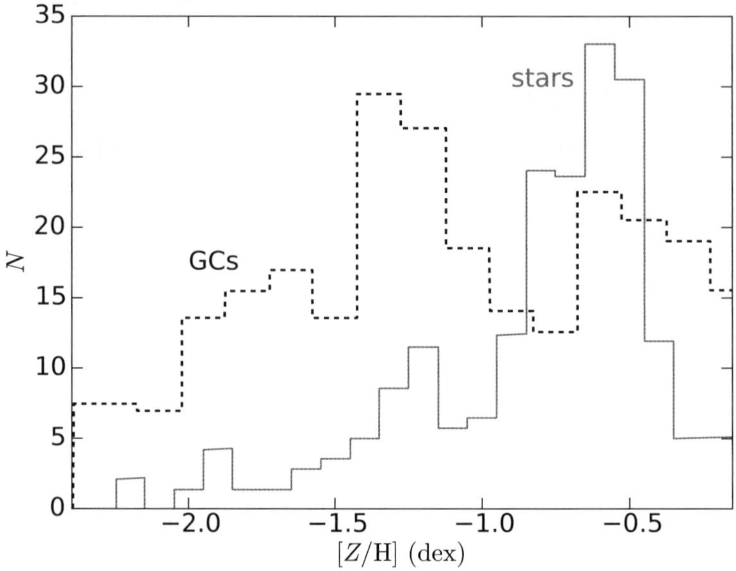

Figure 1. Metallicity distributions of GCs (black dashed histogram) and halo stars (green solid histogram) around the S0 galaxy NGC 3115 (Peacock *et al.* 2015). The stars are from deep *HST* imaging of the RGB at projected galactocentric radii of $\sim$ 50–60 kpc, and the GCs from Subaru/Suprime-Cam photometry at comparable radii (spectroscopic metallicities at smaller radii show similar results). The stellar histograms have been arbitrarily renormalized for comparison to the GCs. Peaks are apparent in both tracers at $[Z/\mathrm{H}] \sim -1.3$ and -0.6, providing the first demonstration beyond the MW that GC bimodality is reflected in the underlying stellar halo.

configuration suggests a early, chaotic assembly through infalling dwarf galaxies (e.g., Searle & Zinn 1978), followed by intense star-forming activity in the main part of the Galaxy. A similar pattern is seen in other galaxies, including the giant ellipticals and lenticulars (E/S0s), which commonly show color bimodality (blue and red GCs; e.g., Zepf & Ashman 1993; Larsen *et al.* 2001). Despite some controversies about interpreting the colors, bimodality in metallicity has now been confirmed through spectroscopic efforts with Keck/DEIMOS as part of the SLUGGS survey† (Brodie *et al.* 2012, 2014; Usher *et al.* 2012), with supporting evidence from kinematics (Pota *et al.* 2013).

Adding to this picture is recent, deep *Hubble Space Telescope* (*HST*) imaging in the outer halo of a nearby S0 galaxy, which detected the elusive second peak of metal-poor red-giant-branch (RGB) stars corresponding to the metal-poor GC peak (Figure 1; Peacock *et al.* 2015). This observation supports the view that blue GC subpopulations trace underlying metal-poor stellar halo populations and that GC bimodality implies two distinct, fundamental modes of early star formation. The different relative peak heights require a higher GC formation efficiency or survival rate in the metal-poor population, for reasons that remain undetermined (see Harris & Harris 2002).

A clearer view of GC origins could be obtained through estimates of their ages, but even in the MW, the best current methods are still uncertain at the $\sim$ 2 Gyr level – which is the difference between forming at $z \sim 3$ and $z \sim 20$. This conundrum motivates an alternative approach where GC metallicities Z are used as a proxy for age, after recognizing that metals build up in galaxies like clockwork (Shapiro *et al.* 2010; Spitler 2010). Thus the GCs can be connected to the observed galaxy mass–metallicity–redshift relation in order to thereby infer both their epochs and their sites of formation.

† `http://sluggs.ucolick.org`

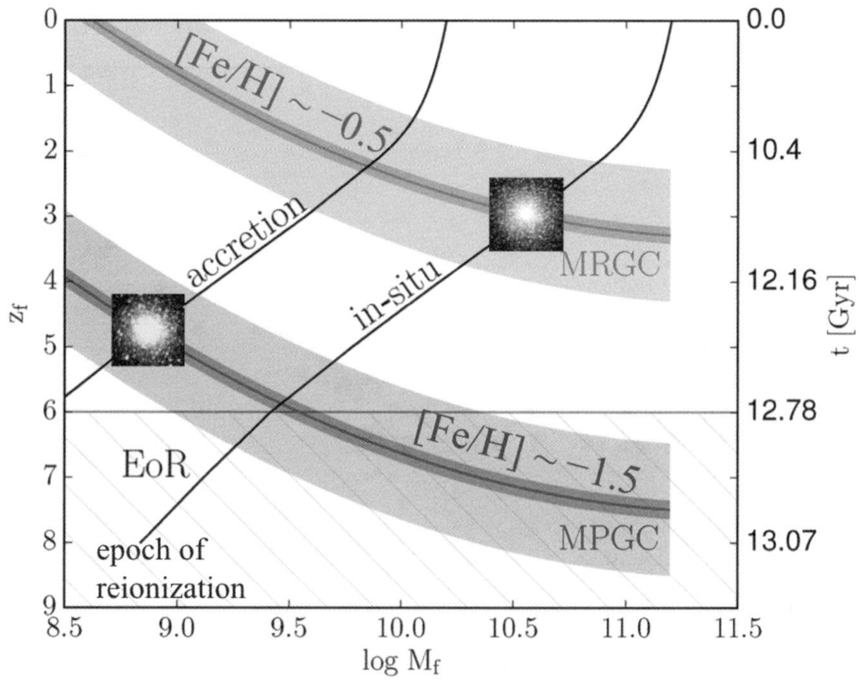

Figure 2. Formation space of GCs around E/S0 galaxies: redshift vs. host galaxy stellar mass. The colored bands show possible solutions for the metal-poor and metal-rich GC subpopulations, bsaed on their measured metallicities. The near-orthogonal curves show example growth tracks, both for the main galaxy and its satellites. Metal-rich GCs are inferred to form at typically $z \sim 3$ in high-mass galaxies, and metal-poor GCs at $z \sim 5$ in dwarfs.

We have applied this "metallicity matching" method to GC spectroscopic data for a sample of E/S0s from SLUGGS (Forbes *et al.* 2015). As Figure 2 shows, for each GC subpopulation, there is a range of possible average formation epochs: for example, the metal-poor GCs could have been formed very early in massive galaxies, or later in dwarf galaxies. This degeneracy is broken by invoking plausible growth curves for the GC host galaxies: one curve for the main progenitor galaxy, and one for a typical 1:10 mass-ratio accreted satellite. Assuming that the metal-rich GCs are a predominantly in situ population, they appear to have formed in the range $z \sim 2$–4. If the metal-poor GCs are primarily accreted, then they formed at $z \sim 4$–5.

This relatively recent birth date for the metal-poor GCs would make them a post-reionization population – an important conclusion since it has been proposed that reionization either produced a hiatus between the two GC subpopulations, or was caused by the GCs themselves (e.g., Beasley *et al.* 2002; Katz & Ricotti 2013). On the other hand, there are significant caveats. One is that the galactic metallicity relation is not well constrained at redshifts above $z \sim 4$. Another is that there may have also been an earlier, in situ phase of metal-poor GC formation in the main galaxy, which is reflected in some problems of crafting a self-consistent pure-accretion model (e.g., Spitler *et al.* 2012)).

The inferred metal-rich GC birth date is more secure, and raises intriguing possibilities for further progress since the Universe at $z \sim 3$ is becoming well studied, as the "high noon" of cosmic star formation. We can now see directly where and how the intense bursts of star formation activity occurred that must have formed the bulk of the metal-rich GC populations. The answer is remarkable: rather than the gas-rich major mergers that dominated the thinking about GC and galaxy formation for many years, high-z

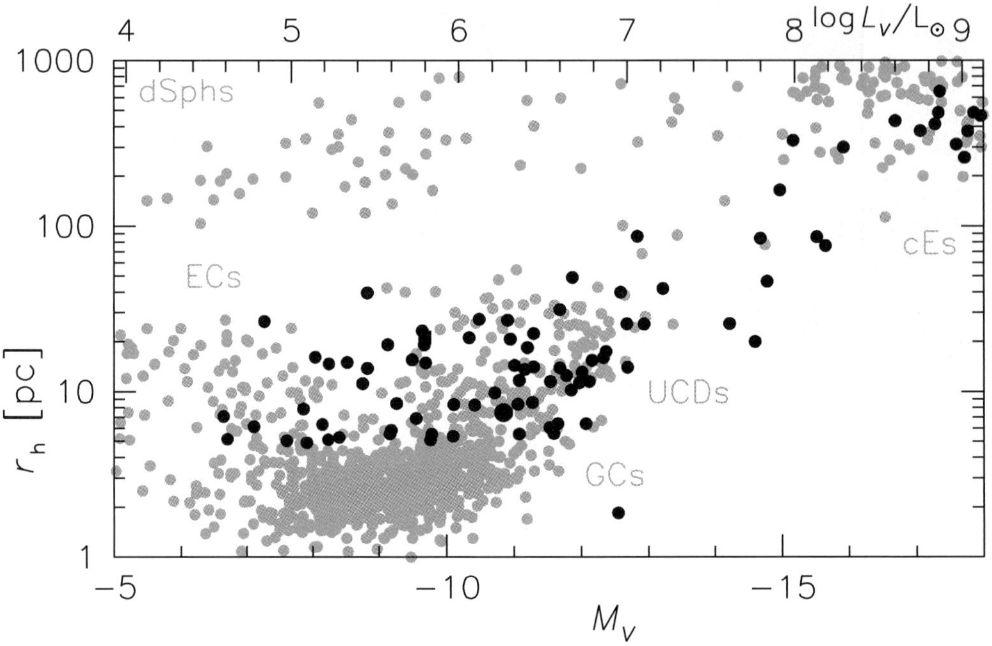

Figure 3. Half-light radius vs. V-band absolute magnitude for distance-confirmed hot stellar sytems. Larger red circles mark objects discovered in the past 2 years. The black square and circle show NGC 2419 and ω Cen, respectively.

star formation is seen to concentrate in extended, turbulent, gas-rich galactic disks (e.g., Genzel *et al.* 2008). These disks host supergiant clumps of gas and young stars that seem natural spawning sites for GCs (Shapiro *et al.* 2012; Kruijssen 2015).

This consolidation of observations at low and high redshifts motivates a new generation of theoretical models for the assembly of GC systems in a cosmological context. The models should include violent disk instabilities (e.g., Porter *et al.* 2014), and should respect the full range of chemo-dynamical constraints: not only the numbers, ages, and metallicities of GCs (Tonini 2013; Li & Gnedin 2014), but also their spatial and orbital distributions – which have so far posed severe theoretical challenges (see Brodie *et al.* 2014, Section 5.4, for more discussion).

3. Compact stellar systems: don't mind the gaps

UCDs were discovered serendipitously through spectroscopic surveys (Hilker *et al.* 1999; Drinkwater *et al.* 2000). They appear similar to GCs but typically have larger sizes (half-light radii $r_{\rm h} \sim$ 10–40 pc rather than $\sim$ 2–4 pc) and higher luminosities (up to $\sim 10^7 L_\odot$). Their formation mechanisms may include merging super star clusters and tidal stripping of galactic nuclei. However, their distributions in basic parameter space and in galactic environments are not clear. Two major efforts to map out the properties of UCDs more systematically include SLUGGS (§2) and the Archive of Intermediate Mass Stellar Systems project (AIMSS; Norris *et al.* 2014).

These surveys revealed that UCDs are *not* synonymous with being very luminous: the family of large-$r_{\rm h}$ objects also extends to lower luminosities and meets up with the "faint fuzzies" or extended clusters (Brodie *et al.* 2011; Forbes *et al.* 2013) This expansion of classification boundaries means that the unusual MW object NGC 2419 is the nearest

Figure 4. Image of the "plume" section of the tidal stream around NGC 3628, from Subaru/Suprime-Cam (20 × 7 kpc image section, *gri* color composite). A central compact stellar cluster is visible, with properties similar to ω Cen.

UCD. At the other end of the spectrum, record-breaking dense stellar systems have been found (Strader *et al.* 2013; Sandoval *et al.* 2015), with these novel extremes of high- and low-surface brightness demonstrating the pivotal role of selection effects in shaping our conceptions of stellar systems. An additional uncharted area of parameter space is the gap between UCDs and compact ellipticals, which is now seen to be filled in with cE–UCD transition objects (see size–luminosity diagram in Figure 3†).

This steadily growing menagerie of diverse compact stellar systems compounds the need to decipher their interrelations and origins. The UCDs probably include more than one underlying subpopulation, and recent cosmological models conclude that most of them are unusually large star clusters, with only a minority being stripped nuclei (Pfeffer *et al.* 2014). Tests of these claims are needed from in-depth observational studies of the UCDs. Some cases of stripped nuclei are obvious, based on an overmassive central black hole, or on an extended star formation history (Seth *et al.* 2014; Norris *et al.* 2015). Metallicity is another useful diagnostic, as cEs and the highest-mass UCDs were found to be relatively metal-rich, in a probable reflection of stripped nuclei origins (Janz *et al.* 2016). Further hereditary clues may come from kinematics and dynamics (Brodie *et al.* 2011; Strader *et al.* 2011; Zhang *et al.* 2015), and from more detailed abundances, such as the very high nitrogen and sodium levels seen in some massive UCDs (Strader *et al.* 2013; Sandoval *et al.* 2015).

4. Ultra-compact dwarfs in formation

Besides the clues to the past of UCDs, it is also possible to witness ongoing UCD formation by tidal stripping. The nearby "Hamburger" spiral galaxy NGC 3628 boasts a tidal stream that appears to contain the nucleus of the disrupting dwarf galaxy (Figure 4; Jennings *et al.* 2015). This object has a luminosity of $\sim 10^6 L_\odot$, a size of ~ 10 pc, and a metallicity of $[Z/\mathrm{H}] \sim -0.8$ dex (from Keck/ESI and Large Binocular Telescope / MODS spectroscopy). It seems destined to become a massive stellar system like ω Cen. Other, similar examples have also been found recently (e.g., Foster *et al.* 2014), and raise the possibility of providing empirical information about the rates and dynamics of GC/UCD formation by tidal stripping.

† Compilation maintained at `http://sages.ucolick.org/spectral_database.html`

Acknowledgements

Supported by National Science Foundation awards AST-1109878, AST-1211995, AST-1515084, and AST-1518294.

References

Beasley, M. A., Baugh, C. M., Forbes, D. A., Sharples, R. M., & Frenk, C. S. 2002, *MNRAS*, 333, 383
Brodie, J. P., Romanowsky, A. J., Strader, J., & Forbes, D. A. 2011, *AJ*, 142, 199
Brodie, J. P., Usher, C., Conroy, C., *et al.* 2012, *ApJ* (Letters), 759, L33
Brodie, J. P., Romanowsky, A. J., Strader, J., *et al.* 2014, *ApJ*, 796, 52
Drinkwater, M. J., Jones, J. B., Gregg, M. D., & Phillipps, S. 2000, *PASA*, 17, 227
Forbes, D. A., Pota, V., Usher, C., *et al.* 2013, *MNRAS*, 435, L6
Forbes, D. A., Pastorello, N., Romanowsky, A. J., *et al.* 2015, *MNRAS*, 452, 1045
Foster, C., Lux, H., Romanowsky, A. J., *et al.* 2014, *MNRAS*, 442, 3544
Genzel, R., Burkert, A., Bouché, N., *et al.* 2008, *ApJ*, 687, 59
Harris, W. E. & Harris, G. L. H. 2002, *AJ*, 123, 3108
Hilker, M., Infante, L., Vieira, G., Kissler-Patig, M., & Richtler, T. 1999, *A&AS*, 134, 75
Janz, J., Norris, M. A., Forbes, D. A., *et al.* 2016, *MNRAS*, in press, arXiv:1511.03264
Jennings, Z. G., Romanowsky, A. J., Brodie, J. P., *et al.* 2015, *ApJ* (Letters), 812, L10
Katz, H. & Ricotti, M. 2013, *MNRAS*, 432, 3250
Kruijssen, J. M. D. 2015, *MNRAS*, 454, 1658
Larsen, S. S., Brodie, J. P., Huchra, J. P., Forbes, D. A., & Grillmair, C. J. 2001, *ApJ*, 121, 2974
Li, H. & Gnedin, O. Y. 2014, *ApJ*, 796, 10
Norris, M. A., Kannappan, S. J., Forbes, D. A., *et al.* 2014, *MNRAS*, 443, 1151
Norris, M. A., Escudero, C. G., Faifer, F. R., *et al.* 2015, *MNRAS*, 451, 3615
Peacock, M. B., Strader, J., Romanowsky, A. J., & Brodie, J. P. 2015, *ApJ*, 800, 13
Pfeffer, J., Griffen, B. F., Baumgardt, H., & Hilker, M. 2014, *MNRAS*, 444, 3670
Porter, L. A., Somerville, R. S., Primack, J. R., & Johansson, P. H. 2014, *MNRAS*, 444, 942
Pota, V., Forbes, D. A., Romanowsky, A. J., *et al.* 2013, *MNRAS*, 428, 389
Sandoval, M. A., Vo, R. P., Romanowsky, A. J., *et al.* 2015, *ApJ* (Letters), 808, L32
Searle, L. & Zinn, R. 1978, *ApJ*, 225, 357
Seth, A. C., van den Bosch, R., Mieske, S., *et al.* 2014, *Nature*, 513, 398
Shapiro, K. L., Genzel, R., & Förster Schreiber, N. M. 2010, *MNRAS*, 403, L36
Spitler, L. R. 2010, *MNRAS*, 406, 1125
Spitler, L. R., Romanowsky, A. J., Diemand, J., *et al.* 2012, *MNRAS*, 423, 2177
Strader, J., Romanowsky, A. J., Brodie, J. P., *et al.* 2011, *ApJS*, 197, 33
Strader, J., Seth, A. C., Forbes, D. A., *et al.* 2013, *ApJ* (Letters), 775, L6
Tonini, C. 2013, *ApJ*, 762, 39
Usher, C., Forbes, D. A., Brodie, J. P., *et al.* 2012, *MNRAS*, 426, 1475
Zepf, S. E. & Ashman, K. M. 1993, *MNRAS*, 264, 611
Zhang, H.-X., Peng, E. W., Côté, P., *et al.* 2015, *ApJ*, 802, 30

Formation, evolution, and survival of massive star clusters
Proceedings IAU Symposium No. 316, 2015
C. Charbonnel & A. Nota, eds.

doi:10.1017/S1743921315009035

Massive star clusters in high-redshift star-forming galaxies seen at a 100 pc scale thanks to strong gravitational lensing

Miroslava Dessauges-Zavadsky[1], Antonio Cava[1], Valentina Tamburello[2], Daniel Schaerer[1], Lucio Mayer[2], Johan Richard[3] and Pablo G. Pérez González[4]

[1]Observatoire de Genève, Université de Genève, 51 Chemin des Maillettes, 1290 Versoix, Switzerland

[2] Center for Theoretical Astrophysics and Cosmology, Institute for Computational Science, University of Zurich, Winterthurerstrasse 190, 8057 Zurich, Switzerland

[3] CRAL, Observatoire de Lyon, Université Lyon 1, 9 Avenue Ch. André, 69561 Saint Genis Laval Cedex, France

[4] Departamento de Astrofísica y Ciencias de la Atmósfera, Facultad de CC. Físicas, Universidad Complutense de Madrid, 28040 Madrid, Spain

Abstract. High-resolution imaging reveals clumpy morphologies among $z = 1 - 3$ galaxies. Most of these galaxies are dominated by disk rotation, which led to conclude that the observed clumps are generated from disk fragmentation due to gravitational instability. Despite the kpc-scale resolution attained by the most advanced facilities and numerical simulations, these clumps are barely resolved at $z > 1$. Thanks to the stretching and magnification power provided by gravitational lensing, we reach the sub-kpc resolving power to unveil their physics. From our literature compilation of data, we show that without lensing there is a bias toward clumps with high masses and sizes. The high-redshift clumps identified in lensed galaxies have stellar masses 2 orders of magnitude lower and a median size of 250 pc. They resemble local star clusters observed in the most intensively star-forming galaxies. The clump masses and sizes observed in lensed galaxies agree with new simulations, which show that the Toomre instability criterion overestimates the clump masses by a factor of $5 - 6$.

Keywords. Gravitational lensing, galaxies: evolution, galaxies: high-redshift

1. Introduction

Deep rest-frame UV HST observations show that $z = 1 - 3$ galaxies, at the peak of the cosmic star formation activity, do not follow the classical Hubble classification: they are mostly irregular and clumpy. Six morphological types have been proposed, among which 5 show prominent clumps or massive star clusters (Elmegreen *et al.* 2005, 2009). While these clumps have been initially associated with interactions and mergers, evidence has been accumulating that only a minority of high-redshift galaxies (about 30%) are in the process of merging. Kinematical galactic internal structure suggests that a substantial fraction of $z \sim 1 - 2$ star-forming galaxies, even those with clumpy morphologies, are dominated by ordered disk rotation, as observationally constrained by Hα line IFU spectroscopy (Genzel *et al.* 2006; Förster Schreiber *et al.* 2009; Wisnioski *et al.* 2015).

These high-redshift disks are very different from their counterparts in the local Universe. They are characterized by high velocity dispersions, low rotation velocity to dispersion ratios, high molecular gas fractions, and high star formation rates (e.g., Wisnioski *et al.* 2015; Dessauges-Zavadsky *et al.* 2015; Schreiber *et al.* 2015). These disks thus

Table 1. Physical properties of simulated clumps.

	Recent simulations by Tamburello *et al.* (2015)	Previous simulations by Ceverino *et al.* 2010
Clump sizes (r)	$100 - 250$ pc	$0.5 - 1$ kpc
Stellar masses ($M_{*\rm clump}$)	$10^7 - 10^8$ $M_\odot$	$10^8 - 10^9$ $M_\odot$
Contribution to the bulge	Negligible	Significant

appear to be highly turbulent, thick, gas-rich, and strongly star-forming. In this context, the dominant interpretation of the origin of clumps, as gravitationally bound structures, is their production by disk fragmentation resulting from gravitational instability. It is debated whether, due to clump interactions and dynamical friction against the surrounding disk, clumps migrate toward the centre of their host galaxy and coalesce into a young bulge, or are disrupted by tidal forces or stellar feedback to form part of a thick disk.

2. New prescriptions for the gravitational fragmentation

The gravitational fragmentation of gas-rich, turbulent galactic disks into giant clumps has been addressed by various numerical simulations (e.g., Bournaud *et al.* 2009; Dekel *et al.* 2009; Ceverino *et al.* 2010). Tamburello *et al.* (2015, hereafter T15) revisit the large-scale fragmentation of galactic disks. Their simulations study isolated galaxies with a wide range of stellar masses, $M_* \sim 10^{10} - 10^{11}$ $M_\odot$, and initial conditions based on the ARGO cosmological hydrodynamical simulations. Their mass resolution is as high as 10^4 $M_\odot$ and they use an efficient stellar feedback model that has been shown to produce realistic spiral galaxies at low redshift, with stellar masses in agreement with constraints (Guedes *et al.* 2011). They investigate, in parallel, the Toomre, Jeans, and Boley *et al.* (2010) gravitational instability criteria. Their main findings are that the clumps that result from fragmentation using the Boley *et al.* instability criterion are smaller both in size and mass (see Table 1). The Toomre mass overestimates the clump masses by a factor of $5 - 6$. Some clumps do still reach a stellar mass as large as $M_{*\rm clump} \simeq 10^9$ $M_\odot$ 0.5 Gyr after formation, but those clumps result from multiple clump-clump mergers and do not reflect the mass scale set by fragmentation (see also Madelker *et al.* in prep.). The largest clumps are found in the most massive galaxies, because these galaxies produce more small clumps and host more mergers that can boost the individual clump mass.

3. Observational constraints on high-redshift clumps

Two approaches have been used to obtain observational constraints on $z > 1$ clumps. The first approach focusses on the study of field galaxies with high-resolution HST imaging and adaptive optic IFU spectroscopy. This approach has the sensitivity as a first limitation, which implies a bias toward galaxies with high star formation rates (SFR) and hence physical conditions not observed in the local Universe. A second limitation is the spatial resolution, in general larger than 1 kpc at $z > 1$ (Genzel *et al.* 2011; Förster Schreiber *et al.* 2011; Guo *et al.* 2012; Wisnioski *et al.* 2012; Swinbank *et al.* 2012). The second approach aims to study gravitationally lensed galaxies. This approach has the following advantage. Strong lensing leads to the stretching of the source image, that conserving the surface brightness produces an amplification in the total observed flux proportional to the stretching factor. This implies the release of the sensitivity limit and enables to study fainter galaxies with lower SFR. More importantly, it also releases the limit on the spatial resolution, such that sub-kpc scales down to 100 pc become reachable (Jones *et al.* 2010; Livermore *et al.* 2012, 2015; Adamo *et al.* 2013; Wuyts *et al.* 2014). For

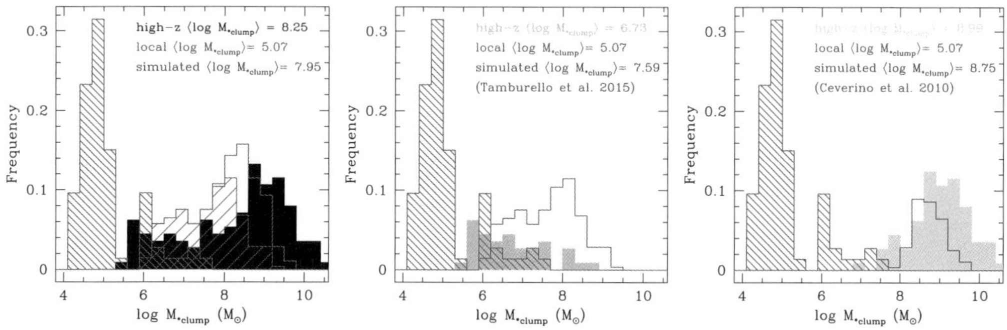

Figure 1. *Left:* Histograms of the stellar mass distribution of high-redshift clumps (filled black), local star clusters (hatched red), and simulated clumps (open blue for Ceverino *et al.* (2010) and hatched blue for T15). *Middle:* Same, but for the high-redshift clumps in "lensed" galaxies (filled green). *Right:* Same, but for the high-redshift clumps in "field" galaxies (filled cyan).

both field and lensed galaxies, the common physical parameters derived for high-redshift clumps are: their size, distance to the centre, stellar mass, SFR, and age.

3.1. *Stellar mass distribution*

In Fig. 1 we compare the stellar mass distributions of high-redshift clumps, local star clusters, and simulated clumps (using our compilation from the literature, exhaustive for $z > 1$ clumps). As shown in the left panel, the high-redshift clumps have on average a factor of 1000 larger masses than local cluster complexes, whereas their masses globally are in line with the clump masses predicted by simulations, except for the higher mass end. The observed high-$M_{*\mathrm{clump}}$ peak is not even reproduced by the T15 simulations. Interestingly, if we plot the clumps identified in lensed galaxies and those studied in field galaxies separately (middle and right panels), we observe that their stellar mass distributions are clearly different, with their means $\langle M_{*\mathrm{clump}}\rangle_{\mathrm{lensed}} = 5.4 \times 10^6$ $\mathrm{M}_\odot$ and $\langle M_{*\mathrm{clump}}\rangle_{\mathrm{field}} = 9.8 \times 10^8$ $\mathrm{M}_\odot$ diverging by 2 orders of magnitude. The $M_{*\mathrm{clump}}$ in lensed galaxies reach the low mass domain found in the recent T15 simulations†. This is not the case of the $M_{*\mathrm{clump}}$ in field galaxies that better agree with the older Ceverino *et al.* (2010) simulations. This shows a clear bias toward the high $M_{*\mathrm{clump}}$ in field galaxies. As a result, the physical properties of clumps as measured in field galaxies may not be representative of the intrinsic properties of high-redshift star clusters because of possible blends of clumps due to the scarce spatial resolution. When spatially resolving the clumps in lensed galaxies, they still have a factor of 100 larger masses than the bulk of local star cluster complexes, but they resemble complexes in the most intensively star-forming galaxies in the local Universe, like the Antennae and NGC 6946.

3.2. *Scaling relations: $M_{*\mathrm{clump}}$ – size and SFR_{clump} – size*

In the local galaxies, the stellar mass of the star cluster complexes is found to scale with their size: $M_{*\mathrm{clump}} \propto r^2$, in a similar way as for the giant molecular clouds (GMCs) (Bastian *et al.* 2005; Bolatto *et al.* 2008). This relation suggests a constant gas surface density. In Fig. 2 we show the stellar masses of high-redshift clumps, local star clusters, and simulated clumps as a function of their sizes. Interestingly, most clumps analysed in lensed galaxies follow the $M_{*\mathrm{clump}}$ – size relation of local GMCs. This suggests that their formation process might be very close to the hierarchical process that governs the formation of cluster complexes in the local Universe. On the other hand, all the other

† The high mass domain is under-represented, because lensing mostly selects galaxies in the lower end of the stellar mass distribution, while in T15 there is the same number of runs with massive and moderate galaxies.

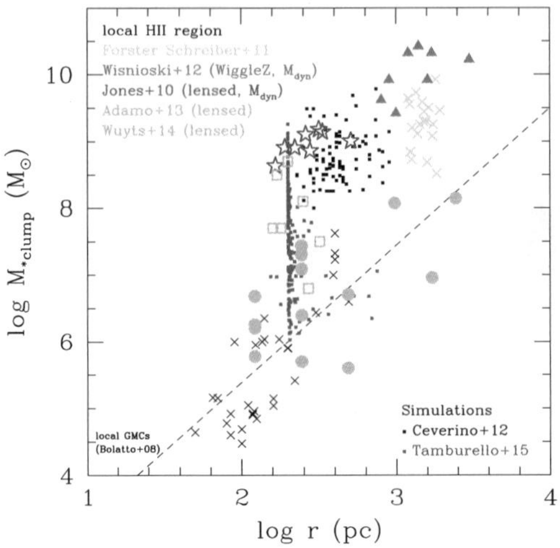

Figure 2. Clump stellar masses as a function of their sizes.

high-redshift clumps observed in field galaxies sit above this relation. This could be due first to the spatial resolution effects, as discussed in Sect. 3.1, that blend clumps to fewer and more massive. The second explanation could be that galaxies at $z > 2$ have disks with larger gas surface densities capable of forming more massive clumps, an effect to be combined with the detection limits that also favour the selection of more massive host galaxies with clumps forming in more extreme environments. This may induce a vertical shift of the $M_{*\rm clump}$ – size relation for the high-redshift clumps (see Adamo *et al.* 2013).

This shift is observed in the $\rm SFR_{clump}$ – size relation, the second scaling relation of local star clusters: $\rm SFR_{clump} \propto r^2$, which suggests a constant surface brightness. When comparing the high-redshift star clusters with the local ones, we observe that the former have surface brightnesses up to 100 times higher than the latter. The difference is too strong to be fully explained by resolution and sensitivity effects. Livermore *et al.* (2012, 2015) suggested that the clump surface brightness or the $\rm SFR_{clump}$ – size relation evolves with redshift. They built up samples of Hα luminosity measurements for clumps in lensed galaxies for three redshift intervals ($1 < z < 1.5$, $1.5 < z < 3$, $z > 3$), and find that the characteristic luminosity of the clump luminosity function clearly increases with redshift.

3.3. *Contribution of clumps to host galaxies*

When analysing the contribution in stellar mass of individual clumps to the host galaxy, we find that it ranges from $< 1\%$ to 50%. The upper end of this contribution is certainly overestimated, since in the field galaxies the observed clumps very likely are blends of several individual clumps. The T15 simulations indeed predict a maximum contribution in stellar mass of the individual clumps to the host of at most 20%. The contribution of the sum of all clumps to stellar mass of the host peaks at $\sim 20\%$, in very good agreement with the T15 simulations. These numbers show that the contribution in stellar mass of both the individual clumps and the sum of all clumps in one host galaxy is relatively small to have an important impact on the bulge growth.

3.4. *Clump survival, growth, and migration*

Since the first observations of clumps in high-redshift star-forming galaxies, the fate of these clumps was questioned. Two scenarios were initially invoked: either the clumps are disrupted by stellar feedback and tidal torques, or the clumps migrate toward the galactic centre of their host galaxies due to interactions and dynamical friction. When

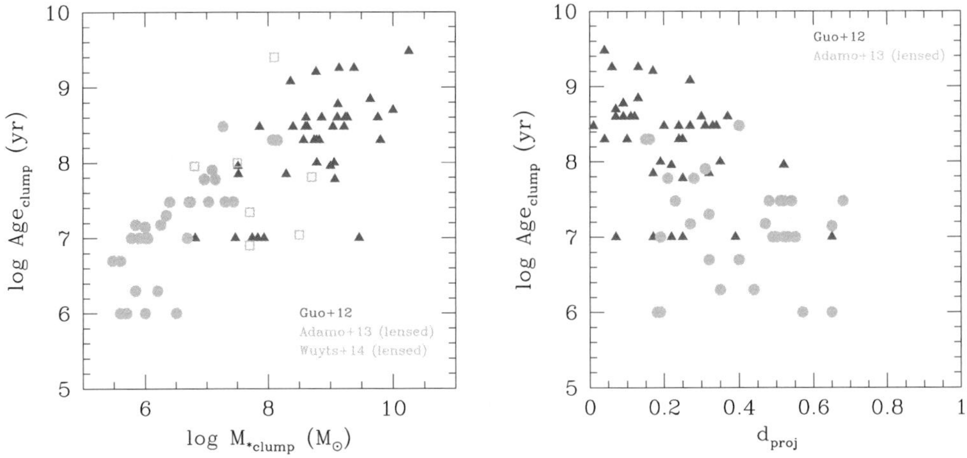

Figure 3. *Left:* Clump ages as a function of their stellar masses. *Right:* Clump ages as a function of their projected distance to the galactic centre of the host galaxies.

estimating the clump ages from the SED modeling, we observe that clumps may reach ages up to 0.5 – 1 Gyr and hence survive over a long timescale. In the left panel of Fig. 3, we show the clump ages as a function of their stellar masses and find that the older clumps are in fact the most massive ones. This agrees with the T15 simulations, where they show that $M_{*\text{clump}}$ grows with time because of clump-clump mergers. Starting with a mean $\langle M_{*\text{clump}} \rangle \simeq 5 \times 10^7$ M$_\odot$ at the time of the clump formation, a mean $\langle M_{*\text{clump}} \rangle \simeq 5 \times 10^8$ M$_\odot$ is reached ~ 0.5 Gyr after formation (see Fig. 4).

In the right panel of Fig. 3, we show the clump ages as a function of their projected distances to the centre of the host galaxies. One observes a trend for the older clumps to be the closer to the galactic centres. This suggests that clumps survive long enough to migrate toward the centre of the hosts. Nevertheless, according to T15 and as seen in Sect. 3.3, these clumps only have a marginal contribution on the galactic bulge growth.

4. Conclusions

It has been shown that a large fraction of star-forming galaxies at $z = 1 - 3$ are found to be disks in rotation. These high-redshift disks are highly turbulent, thick, gas-rich, and strongly star-forming. They induce the formation of clumps or star clusters by disk fragmentation as a result of the gravitational instability. Do these clumps resemble the star cluster complexes observed in the Milky Way and local galaxies?

When analysing the physical properties of the high-redshift clumps observed in field galaxies and the clumps spatially resolved in gravitationally lensed galaxies separately, we find that the stellar masses and sizes of the former are significantly larger than those of local star clusters. On the other hand, the high-redshift clumps in lensed galaxies, although they still have larger stellar masses and sizes, with a mean $\langle M_{*\text{clump}} \rangle = 5.4 \times 10^6$ M$_\odot$ and a median $r = 250$ pc, relative to the bulk of local star clusters, resemble the star complexes in the Antennae and NGC 6946. These results show evidence for a bias toward clumps with high $M_{*\text{clump}}$ and high sizes in field galaxies due to instrumental spatial resolution limitations that blend clumps to fewer and more massive clumps.

Recent simulations by T15 revisit the gravitational fragmentation of turbulent, gas-rich galactic disks and predict smaller clumps in both stellar mass and size. Their findings are in agreement with the physical properties of clumps observed in lensed galaxies.

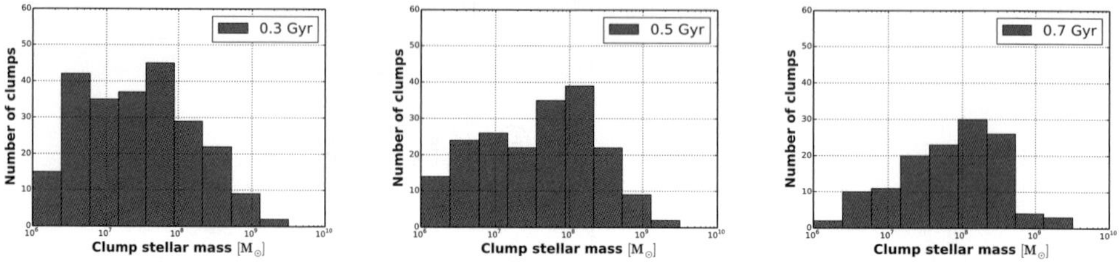

Figure 4. Histograms of the clump stellar mass distribution as a function of time, as derived in the T15 simulations.

They show that the Toomre gravitational instability criterion, so far mostly considered in simulations, overestimates the clump masses by a factor of 5 – 6.

The scaling relations $M_{*\rm clump}$ – size and $\rm SFR_{clump}$ – size empirically determined in star cluster complexes of local galaxies seem to hold for clumps at high redshift, but with a vertical shift well established in the $\rm SFR_{clump}$ – size relation and to be confirmed in the $M_{*\rm clump}$ – size relation. The stellar mass of high-redshift clumps grows with time due to multiple clump-clump mergers. The high-redshift clumps may survive as long as 1 Gyr and migrate toward the galactic centre of their host galaxies. Nevertheless, the clump contribution to the bulge build-up is negligible, with the mean contribution of the sum of all clumps representing $\sim$ 20% of the total stellar mass of the host galaxy.

References

Adamo, A., Östlin, G., Bastian, N., *et al.* 2013,*ApJ*, 766, 105
Bastian, N., Gieles, M., Efremov, Y. N., *et al.* 2005, *A&A*, 443, 79
Bolatto, A. D., Leroy, A. K., Rosolowsky, E., *et al.* 2008, *ApJ*, 686, 948
Boley, A. C., Hayfield, T., Mayer, L., *et al.* 2010, *Icarus*, 207, 509
Bournaud, F., Elmegreen, B. G., & Martig, M. 2009, *ApJ*, 707, L1
Ceverino, D., Dekel, A., & Bournaud, F. 2010, *MNRAS*, 404, 2151
Dekel, A., Sari, R., & Ceverino, D. 2009, *ApJ*, 703, 785
Dessauges-Zavadsky, M., Zamojski, M., Schaerer, D., *et al.*} 2015, *A&A*, 577, 50
Elmegreen, D. M., Elmegreen, B. G., Rubin, D. S., *et al.* 2005, *ApJ*, 631, 85
Elmegreen, B. G., Elmegreen, D. M., Fernandez, M. X., *et al.* 2009, *ApJ*, 692, 12
Förster Schreiber, N. M., Genzel, R., Bouché, N., *et al.* 2009, *ApJ*, 706, 1364
Förster Schreiber, N. M., Shapley, A. E., Genzel, R., *et al.* 2011, *ApJ*, 739, 45
Genzel, R., Tacconi, L. J., Eisenhauer, F., *et al.* 2006, *Nature*, 442, 786
Genzel, R., Newman, S., Jones, T., *et al.* 2011, *ApJ*, 733, 101
Guedes, J., Callegari, S., Madau, P., *et al.* 2011, *ApJ*, 742, 76
Guo, Y., Giavalisco, M., Ferguson, H. C., *et al.* 2012, *ApJ*, 757, 120
Jones, T. A., Swinbank, A. M., Ellis, R. S., *et al.* 2010, *MNRAS*, 404, 1247
Livermore, R. C., Jones, T., Richard, J., *et al.* 2012, *MNRAS*, 427, 688
Livermore, R. C., Jones, T. A., Richard, J., *et al.* 2015, *MNRAS*, 450, 1812
Schreiber, C., Pannella, M., Elbaz, D., *et al.* 2015, *A&A*, 575, 74
Swinbank, M., Smail, I., Sobral, D., *et al.* 2012, *ApJ*, 760, 131
Tamburello, V., Mayer, L., Shen, S., *et al.* 2015, *MNRAS*, 453, 2490
Wisnioski, E., Glazebrook, K., Blake, C., *et al.* 2012, *MNRAS*, 422, 3339
Wisnioski, E., Förster Schreiber, N. M., Wuyts, S., *et al.* 2015, *ApJ*, 799, 209
Wuyts, E., Rigby, J. R., Gladders, M. D., *et al.* 2014, *ApJ*, 781, 61

Formation, evolution, and survival of massive star clusters
Proceedings IAU Symposium No. 316, 2015
C. Charbonnel & A. Nota, eds.

doi:10.1017/S1743921316008759

NGC 346: Looking in the Cradle of a Massive Star Cluster

Dimitrios A. Gouliermis and Sacha Hony

University of Heidelberg, Centre for Astronomy, Institute for Theoretical Astrophysics, Albert-Ueberle-Str. 2, 69120 Heidelberg, Germany
email: gouliermis@uni-heidelberg.de, sacha.hony@free.fr

Abstract. How does a star cluster of more than few 10,000 solar masses form? We present the case of the cluster NGC 346 in the Small Magellanic Cloud, still embedded in its natal star-forming region N66, and we propose a scenario for its formation, based on observations of the rich stellar populations in the region. Young massive clusters host a high fraction of early-type stars, indicating an extremely high star formation efficiency. The Milky Way galaxy hosts several young massive clusters that fill the gap between young low-mass open clusters and old massive globular clusters. Only a handful, though, are young enough to study their formation. Moreover, the investigation of their gaseous natal environments suffers from contamination by the Galactic disk. Young massive clusters are very abundant in distant starburst and interacting galaxies, but the distance of their hosting galaxies do not also allow a detailed analysis of their formation. The Magellanic Clouds, on the other hand, host young massive clusters in a wide range of ages with the youngest being still embedded in their giant HII regions. Hubble Space Telescope imaging of such star-forming complexes provide a stellar sampling with a high dynamic range in stellar masses, allowing the detailed study of star formation at scales typical for molecular clouds. Our cluster analysis on the distribution of newly-born stars in N66 shows that star formation in the region proceeds in a clumpy hierarchical fashion, leading to the formation of both a dominant young massive cluster, hosting about half of the observed pre–main-sequence population, and a self-similar dispersed distribution of the remaining stars. We investigate the correlation between stellar surface density (and star formation rate derived from star-counts) and molecular gas surface density (derived from dust column density) in order to unravel the physical conditions that gave birth to NGC 346. A power law fit to the data yields a steep correlation between these two parameters with a considerable scatter. The fraction of stellar over the total (gas plus young stars) mass is found to be systematically higher within the central 15 pc (where the young massive cluster is located) than outside, which suggests variations in the star formation efficiency within the same star-forming complex. This trend possibly reflects a change of star formation efficiency in N66 between clustered and non-clustered star formation. Our findings suggest that the formation of NGC 346 is the combined result of star formation regulated by turbulence and of early dynamical evolution induced by the gravitational potential of the dense interstellar medium.

Keywords. stars: pre-main-sequence, stars: statistics, H II regions, ISM: structure, ISM: individual: LHA 115-N66, open clusters and associations: individual: NGC 346, Magellanic Clouds

---

1. Introduction

It is generally accepted that most stars form in a clustered mode, i.e., in gravitationally bound concentrations. Star-forming clusters vary in size, mass, and stellar content, from small compact groups of protostars, still embedded in their natal star-forming regions (Lada & Lada 2003), to large Young Massive Clusters (YMCs) that cover more than about 10^4 $M_\odot$ (Portegies Zwart *et al.* 2010). The latter, being fundamental contributors to the stellar mass budget of galaxies, are valuable laboratories for the investigation of

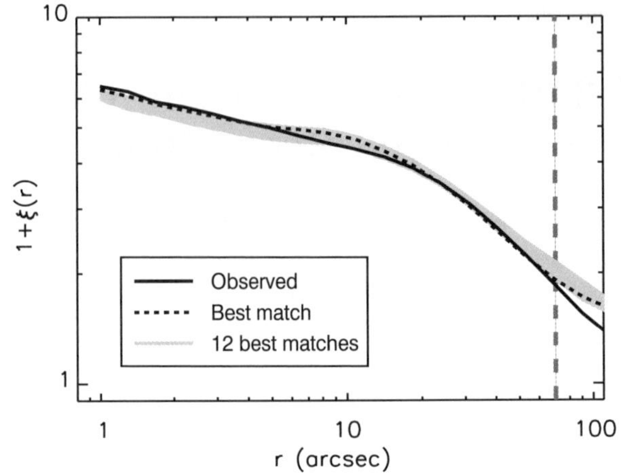

Figure 1. The observed ACF of the young stellar population in N66 (solid black line), and the set of twelve modeled ACFs (light-grey area), which correspond to the best-representative simulated mixed distributions that assume a central compact stellar component and an extended fractal one. The ACF of the best-matching model is plotted with a short dashed line. The vertical dashed line corresponds to the scale beyond which edge-effects hamper the analysis. Plot from Gouliermis *et al.* (2014).

clustered star formation and subsequent evolution. Star clusters themselves are generally not formed in isolation. They are the densest assemblings of larger stellar aggregates and complexes (e.g., Elmegreen 2011), within a hierarchy of stellar structures, which extends up to galactic scales (e.g., Gouliermis *et al.* 2010, 2015).

There is only a handful of YMCs close enough to be resolved into their stellar populations. The study of those located in the Milky Way is limited by the line-of-site contamination by the Galactic disk, while in the Magellanic Clouds deep multi-band photometric surveys allow the study of star formation across the full extent of YMCs natal environments (e.g., Hony *et al.* 2010). In this study we explore the formation of the most massive young stellar cluster in the Small Magellanic Cloud, NGC 346, located in the star-forming complex LHA 115-N66 (in short N66), the brightest HII region in this galaxy. We exploit the exceptionally rich sample of young stars collected with the Hubble Space Telescope (HST) in N66 to address the formation of a YMC from two aspects: (1) The clustering of stars across the length-scale of a giant molecular cloud where the YMC forms (Gouliermis *et al.* 2014), and (2) the relation between the star formation rate and the environmental conditions of the star-forming complex (Hony *et al.* 2015).

2. The Clustering of Young Stars across N66

The young population in NGC 346 consists of low-mass pre–main-sequence stars, identified from their positions in the color-magnitude diagram (e.g., Gouliermis 2012), and high-mass upper–main-sequence stars (with $m_{555} - m_{814} \leqslant 0.0$ mag; $12 \lesssim m_{555} \lesssim 17$ mag), compiling a total sample of 5,150 stars. An age of ~ 5 Myr has been established for these stars by Mokiem *et al.* (2006). We quantify the degree of clustering of the stars by using the *autocorrelation function* (ACF), based on Peebles (1980). For a two-dimensional self-similar distribution this function, $1+\xi$, has a power-law dependency to separation r of the form $1+\xi(r) \propto r^{\eta}$, with the exponent η being related to the 2D fractal dimension as $D_2 = \eta + 2$. The ACF of the young stellar populations of N66 does not follow a single power-law but it has two distinct parts (Fig. 1), and therefore the stellar distribution is not entirely hierarchical.

The broken power-law suggests a complex distribution of stars in N66, which is influenced by the star cluster NGC 346. In order to interpret this behavior we applied numerical simulations of centrally-concentrated stellar clusters, following an Elson, Fall & Freeman (1987) surface density profile, and of three-dimensional fractal stellar distributions, constructed using a box-counting technique. From the comparison of the ACFs of synthetic distributions and their combinations we find that the ACF of N66 is best reproduced as the composite of two distinct spatial distributions; a centrally-condensed cluster with a core radius of ~ 2.5 pc and a profile index $\gamma \simeq 2.27$, and a self-similar stellar distribution with a 3D fractal dimension $D_3 \approx 2.3$ (Fig. 1). About 40% of the total young stellar population belongs to the cluster, while the remaining is spread across the whole extent of the observed star-forming complex. The separation of ~ 6 pc where the ACF power-law breaks indicates the length-scale where the clustering of stars changes from one pattern to the other.

Our findings confirm the appearance of substructure in the region, established in a previous study of ours (Schmeja *et al.* 2009), as the result of hierarchical stellar distribution. The fractal dimension $D_3 \simeq 2.3$, derived for the self-similar stellar component, fits very well to the value established from numerical experiments of supersonic isothermal turbulence (Federrath *et al.* 2009), implying that the observed hierarchy is inherited from the turbulent interstellar gas. On the other hand, numerical simulations find a higher degree of clumping for sink particles formed from turbulent molecular clouds ($D_3 \sim 1.6$; Girichidis *et al.* 2012). The differences in the derived fractal dimensions may reflect discrepancies between the methods used for their measurement, but they may as well demonstrate that *stars have a different spatial distribution than the gas from which they formed.* Even if stars were originally distributed according to a turbulent-driven hierarchy, N-body simulations suggest that a moderate amount of dynamical interactions will partly erase substructure, preventing a star-forming region from retaining a strong signature of the primordial ISM distribution (e.g., Goodwin & Whitworth 2004, Parker *et al.* 2014). The substructure in N66 is unlikely to have formed after the onset of star formation, and it is therefore *the observed fractal dimension is an upper limit to the primordial value.* Dynamics may have facilitated the formation of the YMC within the original self-similar distribution of stars. Kinematic studies in the region will certainly shed light to the importance of dynamical evolution in the formation of NGC 346.

In conclusion, our findings suggest that *star formation in N66 takes place in a clustered, as well as a dispersed mode*, revealed by the bimodal – centrally-condensed and self-similar – clustering of newly-born stars in the region (Gouliermis *et al.* 2014).

3. The Relation between Gas and Young Stars in N66

To understand the formation of NGC 346 in relation to the surrounding interstellar medium (ISM), we examine the ISM properties as a function of the surface density of young stars across N66 (Hony *et al.* 2015). The young stellar sample used for the cluster analysis discussed in the previous section is also used for the derivation of the *surface density of the star formation rate* ($\Sigma_{\rm SFR}$) through star counts. We employ the *"dust method"* to determine the ISM column density, i.e. using the infrared (IR) to submillimetre (submm) dust continuum emission to derive via radiative transfer modeling of its spectral energy distribution (SED) the dust surface density, which we transform into *gas surface density* ($\Sigma_{\rm gas}$). This method has the advantage of not being sensitive to the state of the gas (ionized, neutral or molecular) as long as dust and gas are well mixed.

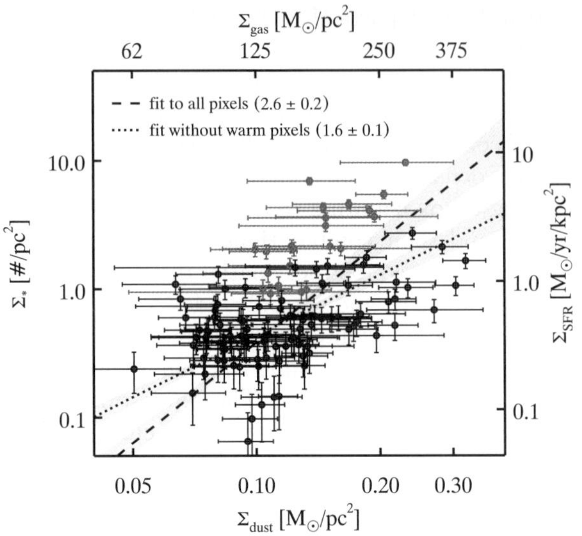

Figure 2. Scatter plot of surface density of young stars versus surface density of dust (and the corresponding SFR and gas surface density) per pixel of 20×20 arcsec^2 (6×6 pc^2) in the star-forming complex N66. The blue points have a strong warm dust component in their SED ($\Sigma_{24\,\mu m}/\Sigma_{250\,\mu m}$ [F_ν/F_ν] > 0.3). The lines-of-sight selected using this color criterion clearly occupy a separate region of the diagram. The dashed and dotted lines are the power-law fits to the data with or without the warm pixels, respectively. The blue areas correspond to the 2σ uncertainties of the best-fitting exponents, which are given in parentheses. There is a clear correlation between Σ_{SFR} and Σ_{gas} with some considerable scatter, suggesting a dependence of the SFR to the density of the gas reservoir. Plot from Hony *et al.* (2015).

This is important because at the age of the young stars in N66 the parental molecular cloud may have been subject to significant photodissociation and using a molecular gas tracer (such as CO) may lead to an incomplete view of the total gas surface densities.

In our analysis, we use IR-to-submm photometric maps from *Spitzer* (SAGE-SMC; Gordon *et al.* 2011) and *Herschel* (HERITAGE; Meixner *et al.* 2013), which we convolve to the lowest available resolution of 20″, corresponding to the SPIRE 250 μm beam. We have also convolved the stellar density map, which has an exceedingly good intrinsic angular resolution, to this effective spatial resolution. After convolution, all maps have been re-projected on the same pixels scheme with pixel size of 20×20 arcsec^2 in order to have spatially independent measurements and to allow pixel-by-pixel SED extraction and comparison between the stellar densities and derived dust surface densities. We thus study the small-scale correlation between newly formed stars and the gas reservoir in 115 pixels (of $\sim$6 pc size), while covering the entire star-forming complex. The conversion of dust to gas surface density was made with the use of a gas-to-dust ratio ($r_{gd} = 1250$), calibrated using gas tracers on large scales. The SFR map is derived from the observed number of young stars in each pixel of the stellar density map using a mass per detected star of 4.3 M$_\odot$, derived using the stellar Initial Mass Function of the region (Sabbi *et al.* 2008), and a star formation duration of 5 Myr (Mokiem *et al.* 2006). This SFR from star-counts is found to be more reliable than indirect SFR tracers (e.g., Hα, 24 μm emission) on parsec scales. We show that these tracers break down on small scales (< 100 pc) by up to a factor of 10.

A correlation between the stellar surface density, $\Sigma_\star$, and the dust surface density, Σ_{dust}, and between the corresponding parameters, Σ_{SFR} and Σ_{gas}, is found with considerable scatter (Fig. 2). A power-law fit to the data yields a steep relation with an exponent

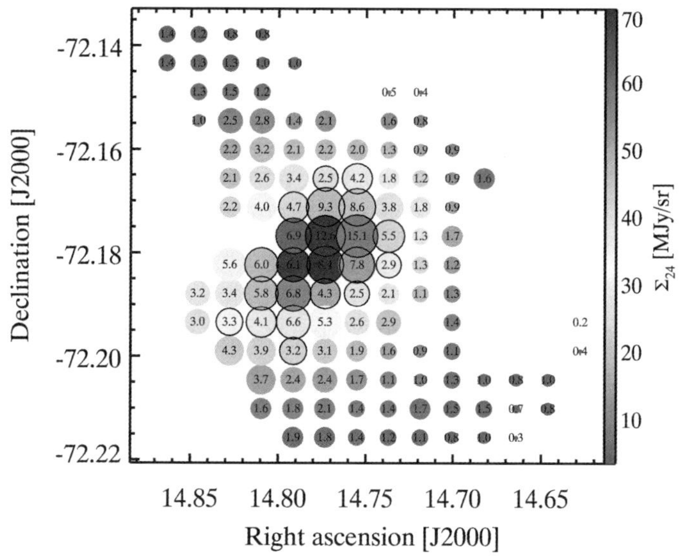

Figure 3. Map of N66 showing the spatial distribution of the derived stellar mass fraction. The size of each point is dependent on $f_\star$ and its color depends on $\Sigma_{24\,\mu m}$. The mass fraction is indicated in each point. Stellar mass fraction clearly peaks on the stellar cluster where the dust is heated to higher temperatures due to the intense radiation field. The warm pixels identified in Fig. 2 are encircled. Plot from Hony *et al.* (2015).

of 2.6 ± 0.2. We find that sight-lines towards the central $\lesssim 15$ pc, where NGC 346 is located, exhibit systematically high values of $\Sigma_\star/\Sigma_{\rm dust}$ by a factor of $\sim$5 compared to the rest of the complex (Hony *et al.* 2015). We investigate the variations in terms of the mass fraction of young stars, $f_\star = \Sigma_\star/\left(\Sigma_\star + \Sigma_{\rm gas}\right)$, across the star-forming complex. The spatial distribution of the observed stellar mass fraction shows that it is systematically high (with a maximum of 15 per cent) within the central 15 pc of the complex, where the YMC is located (Fig. 3). These very high values correspond to very high stellar surface densities, but *not* to low gas column densities. The stellar mass fraction becomes lower outside the central 15 pc, across the remaining surveyed area (66 per cent of the pixels), where it has values systematically below 2 per cent. There is a clear monotonic decrease in $f_\star$ with radial distance from the center, which can be approximated by a power law with exponent -0.7. These variations show also a clear dependence of $f_\star$ on other measurable quantities, such as the surface density of hot dust emission at 24 μm. These findings indicate clearly a higher SFE in the inner 15 pc area, which is dominated by the young stars belonging to the YMC, while the outer parts, characterized by lower SFE, are those dominated by the dispersed self-similar distribution of young stars (Sect. 2).

Our analysis allow us to place these measurements on a Schmidt-Kennicutt (SK) diagram. We find that individual pixels fall systematically above the fiducial SK relation for $\Sigma_{\rm SFR}$ versus $\Sigma_{\rm HI+H_2}$ for integrated disk-galaxies (Kennicutt & Evans 2012) by on average a factor of $\sim$7.

This behavior is consistent with the results by Heiderman *et al.* (2010), who found that the measured SFR lies above the galaxy averages when 'zooming-in' to parsec-scale Galactic star-forming clumps. This is probably caused by less dense gas, which is inefficient in forming stars, and which is included in the galaxy-scale averages but not measured when 'zooming-in' on individual clumps. In our analysis, on the other hand, even though we analyze 6×6 pc^2 sized regions, we cover the complete star-forming complex and find that the entire region (of $\sim$50 pc), being quite active, lies consistently above the SK relation. For N66 one should 'zoom-out' beyond 50 pc to probe this less dense, non-star-forming gas. Indeed, averaging over a larger area (90 pc in radius) we derive a measurement for N66 that lies closer to the SK-relation, but which still remains high by a factor of $\sim$3.

In conclusion, we find an *above-average star formation activity in N66*, and that the observed correlations between stellar and ISM properties reflect *a change in star formation efficiency between clustered and non-clustered star-formation* within the same star-forming complex (Hony *et al.* 2015).

Acknowledgements

Based on observations made with NASA/ESA Hubble Space Telescope, obtained from the data archive at the Space Telescope Science Institute (STScI), Spitzer Space Telescope, *Herschel* and Atacama Pathfinder Experiment (APEX). STScI is operated by the Association of Universities for Research in Astronomy, Inc. under NASA contract NAS 5-26555. *Herschel* is an ESA space observatory with science instruments provided by European-led Principal Investigator consortia and with important participation from NASA. Spitzer Space Telescope is operated by the Jet Propulsion Laboratory, California Institute of Technology under a contract with NASA. APEX is a collaboration between the Max-Planck-Institut für Radioastronomie, the European Southern Observatory, and the Onsala Space Observatory. The authors kindly acknowledge support from the *German Research Foundation* (DFG) through the individual grant GO 1659/3-1 and the collaborative research project SFB881 "The Milky Way System" (subprojects B1, B2, and B5) respectively.

References

Elmegreen B. G., 2011, in: C. Charbonnel & T. Montmerle (eds.), *Star Formation in the Local Universe*, EAS Publications Series Vol. 51 (Cambridge: Cambridge Univ. Press), p. 31
Elson, R. A. W., Fall, S. M., & Freeman, K. C. 1987, *ApJ*, 323, 54
Federrath, C., Klessen, R. S., & Schmidt, W. 2009, *ApJ*, 692, 364
Girichidis, P., Federrath, C., Allison, R., *et al.* 2012, *MNRAS*, 420, 3264
Goodwin, S. P. & Whitworth, A. P. 2004, *A&A*, 413, 929
Gordon, K. D., Meixner, M., Meade, M. R., *et al.* 2011, *AJ*, 142, 102
Gouliermis, D. A. 2012, *Space Sci. Revs*, 169, 1
Gouliermis, D. A., Dolphin A. E., Brandner W., Henning T., 2006, *ApJS*, 166, 549
Gouliermis, D. A., Schmeja, S., Klessen, R. S., *et al.* 2010, *ApJ*, 725, 1717
Gouliermis, D. A., Hony, S., & Klessen, R. S. 2014, *MNRAS*, 439, 3775
Gouliermis, D. A., Thilker, D., Elmegreen, B. G., *et al.* 2015, *MNRAS*, 452, 3508
Heiderman, A., Evans, N. J., II, Allen, L. E., Huard, T., & Heyer, M. 2010, *ApJ*, 723, 1019
Hony, S., Galliano, F., Madden, S. C., *et al.* 2010, *A&A*, 518, L76
Hony, S., Gouliermis, D. A., Galliano, F., *et al.* 2015, *MNRAS*, 448, 1847
Kennicutt, R. C. & Evans, N. J. 2012, *ARAA*, 50, 531
Lada, C. J. & Lada, E. A. 2003, *ARAA*, 41, 57
Meixner, M., Panuzzo, P., Roman-Duval, J., *et al.* 2013, *AJ*, 146, 62
Mokiem, M. R., de Koter, A., Evans, C. J., *et al.* 2006, *A&A*, 456, 1131
Parker, R. J., Wright, N. J., Goodwin, S. P., & Meyer, M. R. 2014, *MNRAS*, 438, 620
Peebles, P. J. E. 1980, *The large-scale structure of the universe* (Princeton Univ. Press)
Portegies Zwart, S. F., McMillan, S. L. W., & Gieles, M. 2010, *ARAA*, 48, 431
Sabbi, E., Sirianni, M., Nota, A., *et al.* 2008, *AJ*, 135, 173
Schmeja, S., Gouliermis, D. A., & Klessen, R. S. 2009, *ApJ*, 694, 367

Formation, evolution, and survival of massive star clusters
Proceedings IAU Symposium No. 316, 2015
C. Charbonnel & A. Nota, eds.

doi:10.1017/S1743921315009047

CHaMP: From Molecular Clouds to Massive Young Clusters

Peter J. Barnes[1,2]

[1]Astronomy Dept., University of Florida, PO Box 112055, Gainesville FL 32611, USA
email: pjb@ufl.edu

[2]School of Science & Technology, University of New England, Armidale NSW 2351, Australia

Abstract. I review the major science outcomes to date of the Galactic Census of High- and Medium-mass Protostars (CHaMP), and also report the latest observational results on this unbiased, uniform sample of massive, cluster-forming molecular clumps, based on new mm-wave and IR data. These clouds represent the vast, subthermally-excited population of clumps predicted by Narayanan *et al.* (2008) to dominate the molecular mass of disk galaxies. Besides confirming their existence, we have presented evidence that these massive clumps probably spend a large fraction (90–95%) of their long lives (possibly up to 100 Myr) in a mostly quiescent, low star formation rate (SFR) state. This is likely ended when a density or internal pressure threshold is crossed, after which vigorous, massive cluster formation consumes the densest gas with a high SFR, dispersing the embedding envelope. New results presented in two other posters at this Symposium include (1) the first analysis of HCN emission from the dense gas (Schap *et al.* 2015), and (2) the first deep photometry of clusters in this sample based on NIR AAT & CTIO data, and on MIR Warm Spitzer IRAC data (Dallilar *et al.* 2015).

Keywords. Star formation, molecular clouds, cluster formation

---

1. Survey Description

The 20°×6° Nanten survey of the Carina Arm of the Milky Way, in the J=1→0 lines of ^{12}CO, ^{13}CO, C^{18}O, and HCO$^+$, was used as the starting point for CHaMP (Barnes *et al.* 2011). We identified 202 massive, dense clouds in the Nanten data, and made more sensitive molecular line maps of the brightest 120 Nanten clouds with the Mopra radio telescope. The Mopra observations resolved these 120 clouds into 303 parsec-scale clumps in the 5× higher-resolution maps, and are in multiple 3mm tracers, including the 4 lines listed above, plus HCN, CN, N$_2$H$^+$, and many others, obtained simultaneously with Mopra's powerful spectrometer.

2. First Molecular Line Results

The first molecular line data that were analysed and published were for the HCO$^+$ line at 89 GHz (Barnes *et al.* 2011). From measurements of basic observed properties of the 303 clumps (such as peak integrated intensity, velocity dispersion, clump size, etc.), we derived a number of physical properties for the entire clump population, such as optical depth, column/surface density, mass, total internal pressure, and virial parameter (among others). From this analysis, a number of unexpected implications have emerged.

For example, the distribution of CHaMP clumps in the mass-density plane shows that virtually all clumps have central densities well below that required to thermalise the HCO$^+$ line, and that only a small fraction of clumps (∼5%) are engaged in vigorous massive star and cluster formation. This confirms that a vast population of subthermally excited massive clumps must dominate the cloud population in the Milky Way's cold ISM, as predicted by Narayanan *et al.*(2008). These mostly quiescent clumps must therefore also survive a very long time (possibly as long as 100 Myr), since the sample is an unbiased population and the actively star-forming clumps are dispersed in a few Myr.

3. First Narrowband NIR Results

CHaMP was designed to be a comprehensive, multi-wavelength survey of massive star formation. Therefore, in addition to the molecular line emission from pre- and protostellar gas, we also obtained complete NIR data on the CHaMP clumps in both broad- and narrowband filters, using the AAT and CTIO Blanco 4m telescopes. The narrowband AAT images of 2μm line emission from CHaMP clumps include Brackett-γ and the $S(1)$ v=1→0 and 2→1 lines of H_2. These data, and the companion broadband images (see below), confirm that most clumps, while massive, are relatively quiescent in their star formation activity (Barnes *et al.* 2013).

Moreover, among the 3mm and NIR lines, only the following pair shows a correlation. HCO^+ is traditionally thought of as a tracer of cold, dense gas, yet it is strongly correlated with Brackett-γ, a tracer of massive young stars. Therefore, not all 'dense gas tracers' trace dense gas! Rather, careful column density calculations show that species such as HCO^+ and HCN trace a combination of column density and excitation.

4. New Molecular Line Results

The HCN line at 88 GHz has hyperfine components which provide a stand-alone opportunity for LTE radiative transfer analysis. The ratios show that simple conversions of integrated intensity to column density are not reliable, and could mask large amounts of quiescent, subthermally-excited, high column density gas, underscoring the conclusions of the mm-wave/NIR comparison (Barnes *et al.* 2013). See the companion paper by Schap *et al.* (2015) for more details.

5. New broadband NIR and deep Warm *Spitzer* photometry

In a companion poster by Dallilar *et al.* (2015), we show preliminary results from an analysis of broadband NIR and deep Warm *Spitzer* photometry, including extinction measures and protostellar content for each of the CHaMP clumps. Again, these show that most clumps have only a small number of embedded protostar candidates, despite being typically quite massive ($\sim$700 $M_\odot$).

6. Implications

ChaMP forces us to revise the prestellar history of molecular clouds, which are much more numerous and have far longer lifetimes than previously realised. This allows ample time, for example, for slow processes like ambipolar diffusion to let the cold gas slip through any magnetic field, which might otherwise provide cloud support against gravity.

References

Barnes, P. J., Yonekura, Y., & Fukui, Y., *et al.* 2011, *ApJS*, 196, 12

Barnes, P. J., Ryder, S. D., & O'Dougherty, S. N., *et al.* 2013, *MNRAS*, 432, 2231

Dallilar, Y., Barnes, P. J., Lada, E. A., & Ryder, S. D. 2015, in IAU Symp. 316, *Formation, Evolution, and Survival of Massive Star Clusters*, C. Charbonnel & A. Nota, eds. (CUP: Cambridge), in press

Narayanan, D., Cox, T. J., & Shirley, Y., *et al.* 2008, *ApJ*, 684, 996

Schap, W., Barnes, P. J., Ginsberg, A., & Ordonez, A. 2015, in IAU Symp. 316, *Formation, Evolution, and Survival of Massive Star Clusters*, C. Charbonnel & A. Nota, eds. (CUP: Cambridge), in press

Formation, evolution, and survival of massive star clusters
Proceedings IAU Symposium No. 316, 2015
C. Charbonnel & A. Nota, eds.

doi:10.1017/S1743921315010777

Kinematic and thermal structure at the onset of high-mass star formation - ISOSS23053

Simon Bihr[1], Henrik Beuther[1], Hendrik Linz[1], Sarah E. Ragan[2], Jochen Tackenberg[1], Rowan J. Smith[3], Thomas Henning[1] and Oliver Krause[1]

[1]Max Planck Institute for Astronomy,
Königstuhl 17, 69117 Heidelberg, Germany
email: name@mpia.de

[2]School of Physics & Astronomy, The University of Leeds,
E.C. Stoner Building, Leeds, LS2 9JT, UK
email: s.ragan@leeds.ac.uk

[3]Jodrell Bank Centre for Astrophysics, University of Manchester,
Oxford Road, Manchester, M13 9PL, UK
email: rowanjsmith.astro@googlemail.com

Abstract. Within the 'earliest phases of star formation' (EPoS) Herschel project, we observed the NH_3 inversion lines of 6 very young high-mass star-forming regions at high spatial resolution (3-5″) with the JVLA. While the Herschel data provide details about the dust continuum, the ammonia data reveal the kinematics and temperatures. Here we present the exciting star-forming clump ISOSS23053 that shows multiple velocity components. We observe a prominent velocity step within the clump, which could be a sign of colliding or converging flows that triggers star formation. Furthermore, we used the JVLA in the C-array configuration to study this source in more detail and we present the first results from the new data. They support the idea of converging flows, as we observe two components in the NH_3(1,1) and (2,2) line, whereas the higher excited NH_3(3,3) line shows one component that links the two lower excited lines.

Keywords. Stars: formation - Stars: massive - ISM: clouds - ISM: kinematics and dynamics

---

1. Introduction

Even though the understanding of high-mass star formation has made tremendous progress over the past decade (e.g., Zinnecker & Yorke 2007; Tan *et al.* 2014), the initial conditions are still poorly constrained. The most prominent candidates for young massive clumps are the infrared dark clouds (IRDCs). With the Herschel bolometers, we encompass the transition from continuum absorption to emission in these objects, but such observations lack any kinematic information. As the formation process of stars is thought to be highly dynamic (e.g., Bonnell et al. 2004; Klessen *et al.* 2005), such information is crucial for constraining star formation theories. Follow-up observations of molecular lines are mandatory for investigating the kinematics of star-forming regions. For this project, we observed the NH_3(1,1) and NH_3(2,2) inversion transitions of ammonia because they probe the temperature and kinematics of dense gas (e.g., Ragan *et al.* 2012b). We selected 6 very young high-mass star-forming clumps from the Herschel EPoS sample (Ragan *et al.* 2012a) to map with the JVLA in the D (most compact) configuration. Within the velocity maps, we find signatures which may be linked to converging flows. The results of this sample are given in Bihr *et al.* (2015). Due to the low spectral

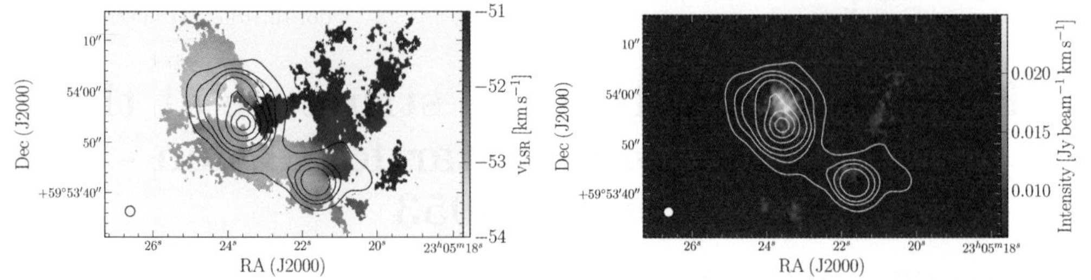

Figure 1. The left panel shows the velocity structure of the NH_3(1,1) line using the JVLA C-array data. The right panel presents the integrated intensity of the higher excited NH_3(3,3) line of the JVLA C-array data. In both panels, the contours show Herschel 70μm emission.

resolution of these early JVLA observations, we were not able to resolve the lines and we could not determine the line width accurately. Hence, we observed two sources of our sample again with the JVLA in C-array to achieve a higher spatial, but also spectral resolution. In this proceeding, we present this data for the clump ISOSS23053.

2. ISOSS23053

The clump ISOSS23053 has a mass of M $\sim$ 500-900$M_\odot$ (Ragan *et al.* 2012a) and shows several indicators of star formation, such as high density dust peaks, spectral signatures of infall and signs of outflows (Birkmann *et al.* 2007). The infrared observations reveal two emission peaks at 70μm (see contours in Fig.1). In contrast to this, the ammonia observations reveal a more clumpy structure and they show a steep velocity step with two components, revealing a difference of $\sim$1.5km s^{-1}. As seen in the left panel of Fig. 1, the velocity step is situated close to the far-IR emission peaks. The exciting question is, if this velocity step traces a converging flow, which triggers the observed star formation activity. Our previous observations (Bihr *et al.* 2015) were not able to answer this question unambiguously, as the spectral and spatial resolution was not sufficient. However, new observations at higher spectral resolution allow us to study this object in great detail. Here, we want to highlight the preliminary analysis of the higher excited NH_3(3,3) line, which is shown in the right panel of Fig.1. We observe a clear detection of this higher excited line at the location of the velocity step, which indicates higher kinetic temperature. In contrast to the two velocity components seen in the NH_3(1,1) and NH_3(2,2) lines, we detect a single component in the NH_3(3,3) line, which spans over both velocity components of the lower excited lines. This is another indicator that we see an interaction between the two velocity components, which may trigger the observed star formation activities.

References

Bihr, S., Beuther, H., Linz, H., *et al.* 2015, A&A, 579, A51
Birkmann, S. M., Krause, O., Hennemann, M., *et al.* 2007, A&A, 474, 883
Bonnell, I. A., Vine, S. G., & Bate, M. R. 2004, MNRAS, 349, 735
Klessen, R. S., Ballesteros-Paredes, J., Vázquez-Semadeni, E., & Durán-Rojas, C. 2005, ApJ, 620, 786
Ragan, S., Henning, T., Krause, O., *et al.* 2012a, A&A, 547, A49
Ragan, S. E., Heitsch, F., Bergin, E. A., & Wilner, D. 2012b, ApJ, 746, 174
Tan, J. C., Beltrán, M. T., Caselli, P., *et al.* 2014, Protostars and Planets VI, 149
Zinnecker, H. & Yorke, H. W. 2007, ARA&A, 45, 481

Formation, evolution, and survival of massive star clusters
Proceedings IAU Symposium No. 316, 2015
C. Charbonnel & A. Nota, eds.

doi:10.1017/S1743921316007079

Fundamental Properties of a Large, Unbiased Sample of Massive, Young, Embedded Star Clusters in the Milky Way

Yigit Dallilar[1], Peter Barnes[1,2], Elizabeth Lada[1], Stuart Ryder[3]

[1]Astronomy Dept., University of Florida, Gainesville FL 32611, USA
email: ydallilar@ufl.edu, pjb@astro.ufl.edu

[2]School of Science & Technology, University of New England, Armidale NSW 2351, Australia

[3]Australian Astronomical Observatory, PO Box 915, North Ryde NSW 1670, Australia

Abstract. In this project, we focus on the analysis of infrared observations of the clumps defined with the Galactic Census of High- and Medium-mass Protostars (ChaMP) (Barnes *et al.* 2011). We derive line of sight infrared extinction values, star counts and protostar candidates around the molecular gas emission obtained with the Mopra telescope. Then, we examine the correlation between radio and infrared properties of the clumps. For this stage of the project, we use the Vela-Carina and 2MASS catalogs to obtain a preliminary understanding of the final results. For the later stages, we will extract infrared photometry from our deep AAT near-IR and *Spitzer* 3.6 and 4.5 μm images. With the final deep photometry results, we will compile the properties of individual clusters.

Keywords. Embedded clusters, Milky Way, extinction, star formation

1. Introduction

The Census of High- and Medium-mass Protostars (CHaMP) is the first large-scale, unbiased, uniform mapping survey of massive molecular clumps in the Milky Way (Barnes *et al.* 2011). Unlike other studies to investigate nearby cloud structures (such as Kong *et al.* 2015), this survey intends to describe a wide-field, unbiased overview of the massive young cluster distribution in a large portion of the Milky Way. This paper focuses on the cluster properties, such as the distribution of protostar candidates and line of sight extinction for each CHaMP clump. Here, we report our preliminary results from CHaMP fields and describe our future plans.

2. Analysis & Preliminary Results

For the preliminary analysis, we exploited the Vela-Carina catalogs†. Source catalogs for each field are obtained to cover the Mopra observations for the ChaMP clumps. For the first step, protostar candidates are identified for each field. We adopt Gutermuth *et al.*'s (2009) infared dereddening procedure. Color excess values from Flaherty *et al.* (2007) are taken as standard values. However, for E_{J-H}/E_{H-K} color excess ratios, the 2MASS near-IR photometry catalog is used and we derived a near infrared extinction vector for each field. With this procedure, we can flag the sources which show a higher infrared excess than the color measurement errors in dereddened [K–3.5] and [3.5–4.6] colors.

† http://irsa.ipac.caltech.edu/data/SPITZER/GLIMPSE/catalogs/VELACAR/

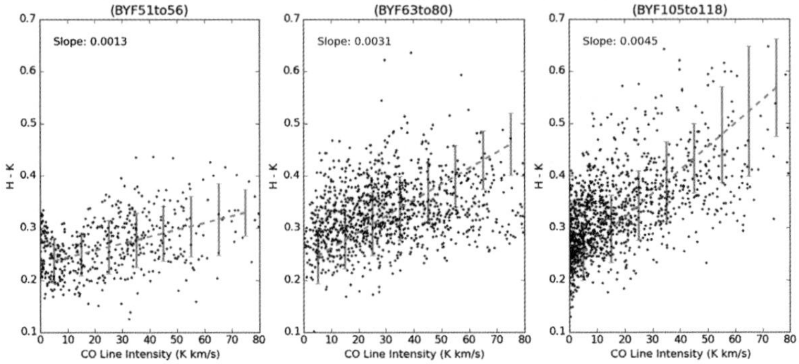

Figure 1. CO integrated line intensity versus $H-K$ color relation, and derived values for a substitute X-factor, for different CHaMP fields. From left to right, figures represents the fields BYF from 51 to 56, 63 to 80, 105 to 118 respectively, with slope values 0.0013, 0.0031, 0.0045.

To obtain clean extinction maps, identified infrared excess sources or protostar candidates are excluded from these catalogs. $H-K$ colors for the rest of the objects are binned into the beam size of the Mopra observations. From these maps, we concluded that regions identified by high CO line intensity are correlated with relatively redder $H-K$ colors. Further, we looked for the pixel by pixel relation between CO total integrated line intensity and $H-K$ colors as a substitute for the column density/extinction relation. This allowed us to compute a linear relation for different fields, and we observed relatively different slopes for various regions. This suggests a clear change in the X-factor over different regions in the Carina Arm of the Milky Way (see figure 1 for three sample fields). If we assume a standard value from dust to gas ratio, $N(H_2)/A_v = 9.4 \times 10^{20} cm^{-2}$ mag^{-1} (Pineda *et al.* 2008), and $H-K$ color excess to extinction conversion, $A_v \sim 16E(H-K)$, slope value of 0.0045 corresponds to $0.7 \times 10^{20} cm^{-2} (Kkm/s)^{-1}$ for the X factor.

3. Conclusion & Future Plans

We demonstrated our preliminary results on the extinction and protostar candidate counts for each CHaMP region exploiting the Vela-Carina catalogs. We derived $H-K$ color maps as a substitute for extinction and investigated its relation to the CO line intensity. Next, we will produce our own source photometry catalog for each region over our deep *Spitzer* images in the 3.5 and 4.6 μm channels and AAT images for the near-IR bands. With this catalog, we will be able to track the extinction for each CHaMP clump instead of big fields. We will obtain the dust to gas ratio to characterise the evolutionary stage of each clump and constrain the clump sizes with a finer resolution on the extinction maps. For this purpose, we will measure the extinction law for each field separately and apply the NICEST method for extinction maps (Lombardi 2008). We intend to compare our results with studies on the nearby molecular cloud regions such as Kong *et al.* (2015).

References

Barnes, P. J., Yonekura, Y., & Fukui, Y. *et al.* 2011, *ApJ*, 196, 12
Flaherty, K. M., Pipher, J. L., & Megeath, S. L. *et al.* 2007, *ApJ*, 663, 1069
Gutermuth, R. A. & Megeath, S. T., Myers *et al.* 2009, *ApJ*, 184, 22
Kong, S., Lada, C. J., & Lada, E. A. *et al.* 2015, *A&A*, 805, 19
Lombardi, M. 2008, *A&A*, 493, 735
Pineda, J. E., Caselli, P., & Goodman, A. A. 2008, *ApJ*, 679, 481

Formation, evolution, and survival of massive star clusters
Proceedings IAU Symposium No. 316, 2015
C. Charbonnel & A. Nota, eds.

doi:10.1017/S1743921315008935

Drama of HII regions: Clustered and Triggered Star Formation

Jin-Zeng Li[1], Jinghua Yuan[1], Hong-Li Liu[1], Yuefang Wu[2] and Ya-Fang Huang[1]

[1]National Astronomical Observatories, Chinese Academy of Sciences, 20A Datun Road, Chaoyang District, Beijing 100012, China
email: ljz@nao.cas.cn

[2]Department of Astronomy, Peking University, 100871 Beijing, China

Abstract. In order to understand the star formation process under the influence of H II regions, we have carried out extensive investigations to well selected star-forming regions which all have been profoundly affected by existing massive O type stars. On the basis of multi-wavelength data from mid-infrared to millimeter collected using *Spitzer*, *Herschel*, and ground based radio telescopes, the physical status of interstellar medium and star formation in these regions have been revealed. In a relatively large infrared dust bubble, active star formation is undergoing and the shell is still expanding. Signs of compressed gas and triggered star formation have been tentatively detected in a relatively small bubble. The dense cores in the Rosette Molecular Complex detected at 1.1 mm using SMA have been speculated to have a likely triggered origin according to their spatial distribution. Although some observational results have been obtained, more efforts are necessary to reach trustworthy conclusions.

Keywords. ISM: bubbles – H II regions – stars: formation – stars: massive

1. N6 – an expanding large bubble with active star formation (Yuan *et al.* 2014)

Although with a diameter of 11 pc, the infrared bubble N6 is still in expansion and actively giving birth to new stars, which have been revealed by the following observational features. A shell structure can be clearly traced by both infrared continuum and molecular line emission (Figure 1). A velocity gradient along the NW-SE orientation for the ring structure has been evidently detected. Among the ten detected molecular clumps, five (A, B, G, H, and I) are located on the border of the ring structure, and four of them show elongations along the shell, consistent with a collect and collapse scenario. Based on the number of YSOs with masses of $3-8\ M_{\odot}$, the lower limits of number and surface density of YSOs inside the bubble are estimated to be 770 and 21 pc^{-2}, respectively, indicating active star formation.

2. G24.134 – compressed gas and triggered star formation (Liu *et al.* 2015)

Based on multi-wavelength investigations, the Galactic bubble G24.136 is found to be compressing its ambient ISM and have the potential to trigger the formation of stars. Dense molecular gas has been detected along the shell structure (Figure 2). The probability distribution function of the column densities can be well fitted using two log-normal functions (Figure 2). Such a feature may be arising from the turbulent cloud and the compressed layers. Six molecular clumps have been found to be along the bubble shell, resembling the scenario of collect and collapse (Figure 2).

3. AFGL 961 – shaping regional core mass function (Li *et al.* 2013)

In the ridge of the Rosette Molecular Complex, the SMA observations have revealed more than 40 pre- and proto-stellar cores which are preferentially located near the massive

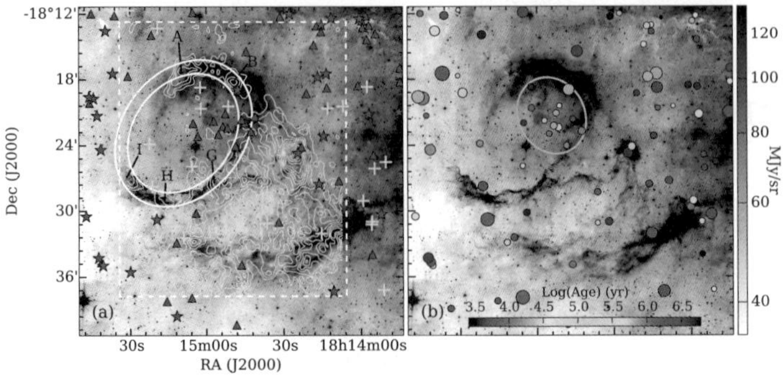

Figure 1. Appearance of the bubble N6 revealed in the infrared (background) and molecular line (contours) emission. Different symbols represents identified YSOs whose ages are color coded and masses are indicated by symbol sizes in the right panel.

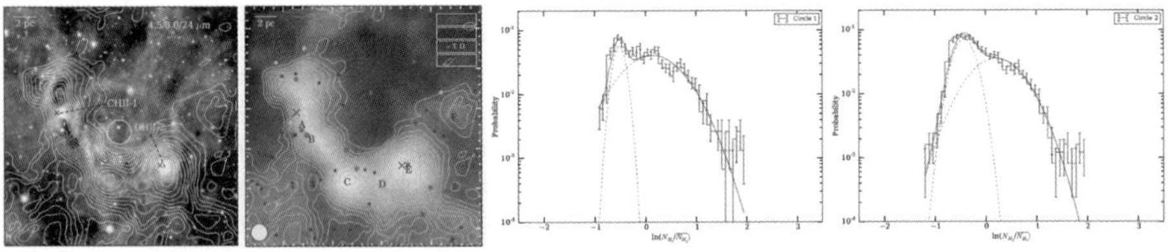

Figure 2. *Left* : Appearance of the bubble G24.136 revealed in the infrared (background) and molecular line (contours) emission. *Right*: Column density PDFs of the selected regions which can be well fitted with the combination of two log-normal distributions.

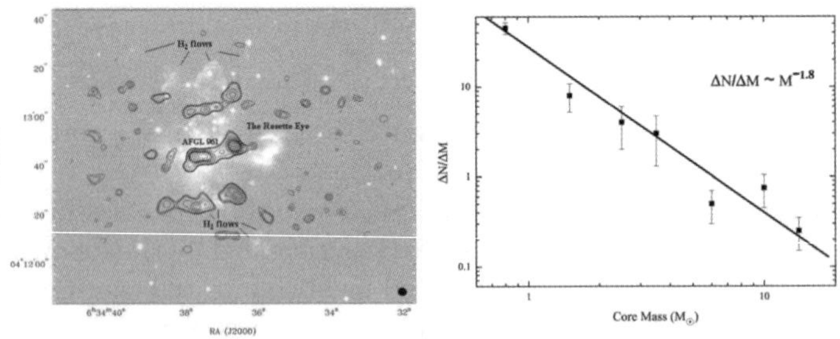

Figure 3. *Left*: Submillimeter contours overlaid onto NTT H_2 $1-0$ S(1) image of the AFGL 961 region. *Right*: Differential core mass function

molecular outflows originating from the central massive binary (Figure 3). The differential core mass function is computed, which yields a somewhat shallower slope of 1.8 as compared to that of the stellar IMF (Figure 3). Dynamical processes from the newly formed massive YSOs could have changed the initial conditions of core or star formation and enhanced the formation of more massive cores near the center of the cluster.

References

Li, J.-Z., Mallamaci, C. C., César Podestà, R., *et al.* 2013, *RAA*, 13, 719
Liu, H.-L., Wu, Y., Li, J., *et al.* 2015, *ApJ*, 798, 30
Yuan, J.-H., Wu, Y., Li, J. Z., & Liu, H. 2014, *ApJ*, 797, 40

Formation, evolution, and survival of massive star clusters
Proceedings IAU Symposium No. 316, 2015
C. Charbonnel & A. Nota, eds.

doi:10.1017/S1743921316007110

HCN Hyperfine Ratio Analysis of Massive Molecular Clumps

William Schap[1], Peter Barnes[1,2], Adam Ginsburg[3] and Antonio Ordonez[1]

[1]Astronomy Dept., University of Florida, PO Box 112055, Gainesville FL 32611, USA
email: wschap@ufl.edu, pjb@ufl.edu, a.ordonez@astro.ufl.edu

[2]School of Science & Technology, University of New England, Armidale NSW 2351, Australia

[3]European Southern Observatory, Garching, Bavaria, Germany
email: aginsbur@eso.org

Abstract. The CHaMP project has identified a uniform sample of 303 massive (20–8000 $M_\odot$), dense (200–30,000 cm^{-3}) molecular clumps in a large sector of the southern Milky Way that includes much of the Carina Arm. These are the kinds of clumps that are likely to be the precursors to IRDCs, large stellar clusters, and massive stars. We report new results of the physical conditions in these clouds based on the J=1→0 emission at 3mm from the HCN molecule. Analysis of the HCN emission from these clumps reveals that the physical conditions in the gas (i.e., the excitation temperature, optical depth, and column density) do not follow the molecular line emissivity in a straightforward way. This means that large fractions of the molecular material involved in massive cluster formation, while not completely "dark", are under-luminous and easily missed in certain studies.

Keywords. Molecular clouds, cluster formation, star formation

1. Data & Analysis

We started with a sample of clumps from the full Mopra HCN data set, BYF 40a–g in Region 6 of CHaMP at an adopted distance of 6.6 kpc (see Fig. 1), and used the same data reduction techniques as described by Barnes *et al.* (2011) for HCO^+. For comparison with the HCO^+ results, we compiled the same observed clump properties in the HCN maps of BYF 40, such as clump positions, sizes, brightness, etc., as shown in Table 1. However, because of the HCN line's hyperfine structure, we used the PySpecKit package (Ginsburg & Mirocha 2000) to perform an LTE analysis of the emission, in order to obtain a self-consistent set of maps of the HCN's optical depth τ, excitation temperature $T_{\rm ex}$, systemic velocity $V_{\rm LSR}$, and velocity dispersion σ_V.

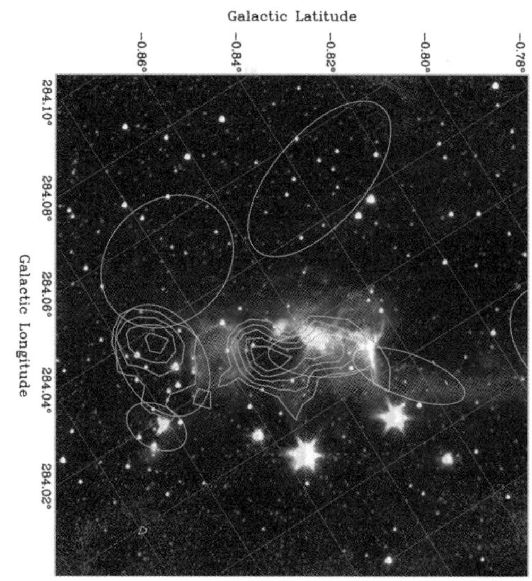

Figure 1. Three-color composite (red = IRAC2, green = IRAC1, blue = Brackett-γ) of the BYF 40 area, with the N_2H^+ contours shown in green and HCO^+ clumps' HPW shown by ellipses. The scale bar is for an assumed distance of 6.6 kpc. Note the mix of molecular and ionised gas in the area.

Table 1. HCN clump properties

Clump	l [deg]	b [deg]	I [K km s^{-1}]	T_{ex} [K]	R [pc]	V_{LSR} [km s^{-1}]	σ_V [km s^{-1}]	N_{HCN} [m^{-2}]
40a	284.012	−0.859	28.80	3.72	0.51	9.02	2.10	8.6e17
40b	284.032	−0.893	10.52	2.04	0.38	9.25	1.66	6.9e17
40c	284.019	−0.903	3.83	1.26	1.21	9.14	1.41	—
40d	283.996	−0.839	4.18	1.27	0.51	8.41	1.85	—
40e	284.056	−0.876	2.39	0.99	1.52	9.62	1.71	1.5e17
40f	284.049	−0.833	1.89	1.14	1.63	8.09	1.41	4.1e17
40g	283.986	−0.816	0.76	0.97	1.33	9.63	1.37	—

Having obtained spatially resolved maps of τ and T_{ex}, we then used

$$N(\mathrm{HCN}) = 9 \times 10^{15}\mathrm{m}^{-2}\, \frac{Q(T_{\mathrm{ex}})e^{E_{\mathrm{u}}/kT_{\mathrm{ex}}}}{(1 - e^{-h\nu/kT_{\mathrm{ex}}})} \int \tau dV \,, \tag{1.1}$$

to compute a spatially resolved map of the HCN column density, including formal error propagation. Illustrative column density values at the clumps' peak emission positions are also given in Table 1.

2. Results

From the computed column densities, we can now compare them with the integrated intensities to explore the relationship (Fig. 2). Neither a linear nor a power-law does a good job of fitting the low-brightness (low-T_{ex}), high-N_{col} (high-τ) points. Using either of these laws to convert from integrated intensity to column density (with an assumed HCN abundance relative to H_2), we obtain a total clump mass around 24,000–37,000 $M_\odot$. Integrating the N_{col} map directly gives a total mass of 57,000 $M_\odot$ instead. Therefore, such laws substantially underestimate the amount of subthermally excited, dense, molecular gas available for (but not yet engaged in) star formation.

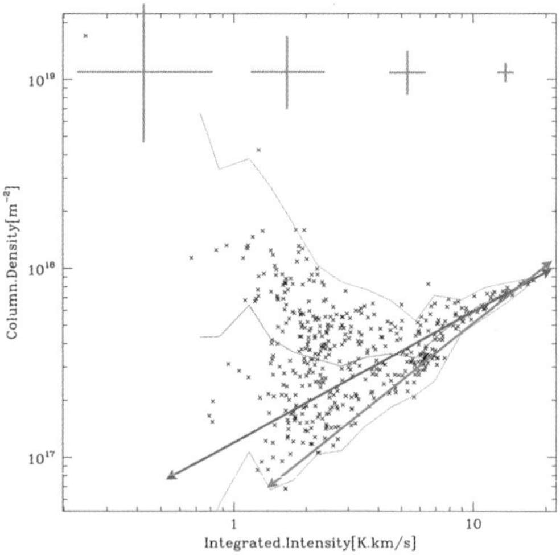

Figure 2. Comparison of HCN integrated intensity with HCN column density, computed by PySpecKit from hyperfine ratio analysis in the BYF 40 area. The straight red line is a least-squares linear fit to the data, while the blue line is a least-squares power-law (index 0.7) fit. The green error bars show typical uncertainties.

3. Implications

The integrated intensity of "dense gas tracers" like HCO^+ (Barnes *et al.* 2013) and now HCN do not trace dense gas, but rather a combination of τ and T_{ex}. Physically-based column density calculations show instead that the molecular mass of dense clumps may be underestimated by substantial factors, which would also imply a correspondingly longer gas depletion timescale, for a given star formation rate. This means that the calibration of Kennicutt-Schmidt-type scaling relations may need to be re-examined.

References

Barnes, P. J., Yonekura, Y., Fukui, Y., *et al.* 2011, *ApJS*, 196, 12

Barnes, P. J., Ryder, S. D., O'Dougherty, S. N., *et al.* 2013, *MNRAS*, 432, 2231

Ginsburg, A. & Mirocha, J. 2000, *Astrophysics Source Code Library*, 1109.001

Formation, evolution, and survival of massive star clusters
Proceedings IAU Symposium No. 316, 2015
C. Charbonnel & A. Nota, eds.

doi:10.1017/S1743921316007043

Massive cold cloud clusters

L. Viktor Toth[1], Sarolta Zahorecz[2,1], Gabor Marton[3], Isabelle Ristorcelli[4,5], Mika Juvela[6], Ludovic Montier[4,5] and Jean-Philippe Bernard[4,5]

[1]Loránd Eötvös University, Pázmány P.s. 1/a, H-1117 Budapest, Hungary

[2]ESO, Karl-Schwarzschild-Str. 2, D-85748 Garching bei München, Germany
email: szahorec@eso.org

[3]Konkoly Observatory, H-1525 Budapest, PO Box 67, Hungary

[4]CNRS, IRAP, 9 Av. colonel Roche, BP 44346, 31028 Toulouse Cedex 4, France

[5]Universite de Toulouse, UPS-OMP, IRAP, 31028 Toulouse Cedex 4, France

[6]Department of Physics, PO Box 64, 00014, University of Helsinki, Finland

Abstract. The all-sky Planck catalogue of Galactic Cold Clumps (PGCC) allows an almost unbiased study of the early phases of star-formation in our Galaxy. Several thousand of the clumps have also distance estimates allowing a determination of mass and density. The nature of Planck cold clumps varies from IRDCs to tiny nearby cold clouds with masses ranging from one to several tens of thousands solar masses. Some of the clumps are embedded in GMCs, others are isolated. Some are close or even very close to OB associations, while others lay far from any UV luminous objects. The small scale clustering of nearby PGCCs was studied with the improved Minimum Spanning Tree (MST) method identifying groups in 3D space, locating also massive cold cloud clusters eg. PGCCMST G210.6-19.5 in LDN 1641.

Keywords. ISM: clouds; ISM: individual LDN 1641, Orion A; ISM: magnetic fields

1. MST analysis of PGCCs

The Planck† survey (Planck Collaboration 2011) provided the first all sky catalogue of cold ISM objects, the Planck Catalogue of Galactic Cold Clumps, (PGCC, Planck Collaboration 2015a) with 13188 Galactic sources. Reliable distance estimates and high quality flux densities were obtained for 3911 PGCCs. We have used the Minimal Spanning Tree method (MST, Cartwright & Whitworth 2004) to identify PGCC clusters. The parameters of MST clusters with more than three members were defined following Cartwright & Whitworth, (2004) and Schmeja & Klessen (2006): cluster radius, R_c, is the distance between the mean position of all cluster members and the most distant sources; convex hull radius, R_h, the radius of a circle with an area equal to the area of the convex hull of the cluster members; elongation measure, $\xi = R_c/R_h$. See an example in Figure 1.

2. Results

Our MST analysis located 137 clusters when applied to the 870 PGCCs in Taurus, Perseus, Auriga and Orion, at a distance of 140-420 pc. LDN 1641 (Orion A main ridge) hosts the most massive cluster. It is located between OriKL and L1641S, and it includes L1641N, NGC1999 and L1641C, see Figure 1. It has a total mass of 1400 $M_\odot$ in 22 PGCCs including PGCC G212.10-19.15, the second most massive PGCC (300 $M_\odot$) within 1 kpc. The cluster members have a median temperature of 12.9 K and a median mass of 48 $M_\odot$.

† http://www.esa.int/Planck

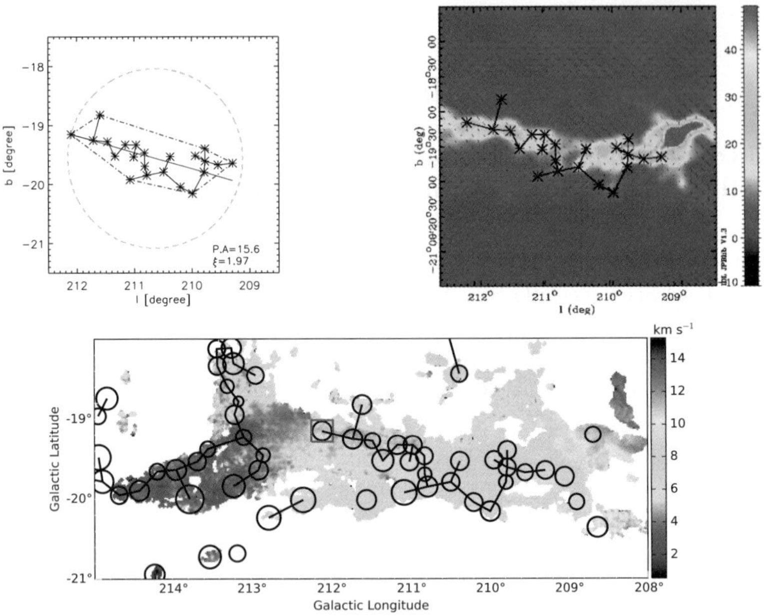

Figure 1. The PGCCMST G210.6-19.5 cluster in L1641, PGCC clumps with reliable distance and flux estimations (marked with asterisks in the top, and circles in the bottom figures), and MST branches (black solid lines). **Top left:** Convex hull (blue dashed-dotted line), cluster elongation direction (red line across), radius (indicated by green dashed circle), position angle and elongation measure are written. **Top right:** B field vectors and MST cluster overlaid on the 353GHz intensity map. **Bottom:** First moment map of the ^{13}CO(2-1) line (Osaka-1.85m survey, Nishimura *et al.* 2015) with PGCC clumps overlaid (circle, size = effective radius). Red square: PGCC G212.10-19.15 the most massive clump. Neighboring MST clusters are also shown.

The magnetic field is perpendicular to the high column density filamentary structure, in good agreement with Planck Collaboration (2015b), i.e. that the B field progressively rotates orthogonal to filaments at higher column densities.

The cluster is divided into two sub-clusters based on the velocity distribution, as seen in Figure 1, the 1st moment map of ^{13}CO(2-1) (Osaka-1.85m Orion Survey, Nishimura *et al.* 2015). The sub-cluster of 9 PGCCs with $b > 210°$ still is the most massive PGCC group in the nearby star forming regions with a total mass of $\sim$620 $M_{\odot}$.

Acknowledgements

This research was partly supported by OTKA grants K101393, NN-11101, 101939, 104607, the EC 7th Framework Programme (FP/2007-2013) under grant agreement No 283393 (RadioNet3), the PECS contract no. 4000109997/13/NL/KML of HSO and ESA, the Academy of Finland grants 250741 and 285769.

References

Cartwright, A. & Whitworth, A. P. 2004, *MNRAS*, 348, 589
Nishimura, A., Tokuda, K., Kimura, K. *et al.* 2015, *ApJS*, 216, 18
Planck Collaboration *et al.* 2011, *A&A*, 536,23
Planck Collaboration *et al.* 2015a, arXiv:1502.01599
Planck Collaboration *et al.* 2015b, arXiv:1502.04123
Schmeja, S. & Klessen, R. S. 2006, *A&A*, 449, 151

Formation, evolution, and survival of massive star clusters
Proceedings IAU Symposium No. 316, 2015
C. Charbonnel & A. Nota, eds.

doi:10.1017/S1743921315008959

Evidence for the non-destruction of the Most Massive Molecular Clouds even after they have given Birth to Massive Star Clusters

J. Zaragoza-Cardiel[1], J. Beckman[2,3,4], J. Font[2,3], A. Camps-Fariña[2,3], B. García-Lorenzo[2,3] and A. Serrano-Borlaff[2,3]

[1]Instituto de Astronomía, Universidad Nacional Autónoma de México, Mexico City, Mexico
email: `javier.zaragoza@astro.unam.mx`

[2]Instituto de Astrofísica de Canarias, C/ Vía Láctea s/n, 38205 La Laguna, Tenerife, Spain

[3]Department of Astrophysics, University of La Laguna, E-38200 La Laguna, Tenerife, Spain

[4]CSIC, 28006 Madrid, Spain

Abstract. We have combined observations of the Antennae galaxies from the radio interferometer ALMA and from the optical interferometer GHαFaS (Galaxy Hα Fabry-Perot System). The two sets of observations have comparable angular and spectral resolutions, enabling us to identify 142 giant molecular clouds (GMCs) and 303 HII regions. We have measured, and compared, their basic physical properties (radius, velocity dispersion, luminosity). We study the mass functions of HII regions and GMCs, where a double population of HII regions due to a double population of GMCs is found. Using the velocity dispersion measurements, we claim that the difference between the regimes is the nature of the dominant binding force. For the regions in the lower mass range, the dominant force is the external pressure, while in the higher mass range it is the internal gravity of the clouds. We have applied the same method to similar observations (GHαFaS & ALMA) of Arp 236 and NGC 1614 finding similar results.

Keywords. Interacting galaxies, star formation, HII regions, GMCs.

---

1. Introduction

The process of star formation enhancement in interacting galaxies is postulated but not very well understood. The study of the kinematics of the star forming regions in galaxy mergers turns out to be crucial to clarify if star formation is indeed enhanced, how the extra-nuclear star forming regions are caused, and to characterize the physical parameters of the gas in these regions. Star formation occurs largely inside molecular clouds, and HII regions trace recent star formation. The study of both types of clouds gives us crucial clues to understand the star formation process. The kinematical properties of these clouds help us to clarify what are the drivers of the turbulence, and the connection with the star formation.

Using ASTRODENDRO (Rosolowsky *et al.* 2008) Zaragoza-Cardiel *et al.* (2014) were able to measure the radius (R), the velocity dispersion (σ_v) and the luminosity for each HII region and GMC observed in the Antennae galaxies. The data used in Zaragoza-Cardiel *et al.* (2014) is already publicly available.† Based on the luminosity functions, and the virial parameter of HII regions and GMCs in the Antennae galaxies, Zaragoza-Cardiel *et al.* (2014) conclude that the most massive GMCs are not dissolved after the birth of massive star clusters, but remain as gravitationally bound HII regions.

† http://vizier.cfa.harvard.edu/viz-bin/VizieR?-source=J/MNRAS/445/1412

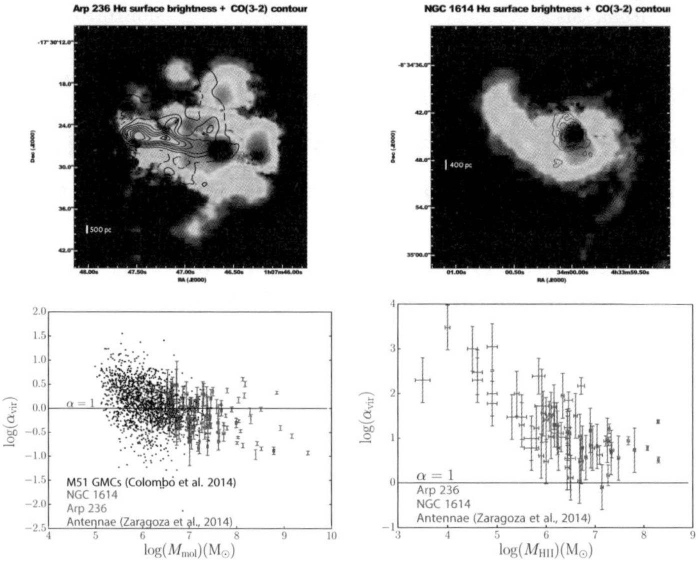

Figure 1. Top left: Arp 236 Hα surface brightness + CO(3-2) contours. Top right: NGC 1614 Hα surface brightness + CO(3-2) contours. Bottom left: Virial parameter, $\alpha_{\rm vir}$, versus molecular gas mass, $M_{\rm mol}$, of GMCs. Bottom right: $\alpha_{\rm vir}$ versus ionized gas mass, $M_{\rm HII}$, of HII regions.

2. HII regions and GMCs in Arp 236 & NGC 1614

We have observed Arp 236, and NGC 1614 with GHαFaS and combined these observations with public ALMA data, Saito *et al.* (2015) for Arp 236, and Sliwa *et al.* (2014) for NGC 1614. The comparison between Hα and CO(3-2) observations is shown in Fig. 1 top. Using ASTRODENDRO we have extracted the properties of HII regions and GMCs in those galaxies.

The virial parameter decreases with the mass of the star forming regions (Fig. 1 bottom), which implies that the more massive the region, the more important is the gravity rather than the external pressure (Bertoldi & McKee 1992). Thus, the massive population of gravitationally bound HII regions is probably due to the massive population of GMCs.

3. Conclusions

We found a population of massive gravitationally bound GMCs and HII regions probably connected based on the luminosity functions and the virial parameter in the Antennae galaxies, Arp 236, and NGC 1614.

References

Bertoldi, F. & McKee, C. F. 1992, *ApJ*, 395, 140
Colombo, D., Hughes, A., Schinnerer, E., *et al.* 2014, *ApJ*, 784, 3
Rosolowsky, E. W., Pineda, J. E., Kauffmann, J., & Goodman, A. A. 2008, *ApJ*, 679, 1338
Saito, T., Iono, D., Yun, M. S., *et al.* 2015, *ApJ*, 803, 60
Sliwa, K., Wilson, C. D., Iono, D., Peck, A., & Matsushita, S. 2014, *ApJL*, 796, L15
Zaragoza-Cardiel, J., Font, J., Beckman, J. E., *et al.* 2014, *MNRAS*, 445, 1412

Formation, evolution, and survival of massive star clusters
Proceedings IAU Symposium No. 316, 2015
C. Charbonnel & A. Nota, eds.

doi:10.1017/S1743921315006833

Do you know the extinction in your young massive cluster?

Guido De Marchi[1], Nino Panagia[2] and Elena Sabbi[2]

[1]ESTEC, 2200 AG Noordwijk, The Netherlands, email: gdemarchi@esa.int

[2]STScI, Baltimore, MD 21218, United States, email: panagia@stsci.edu, sabbi@stsci.edu

Up to ages of $\sim$ 100 Myr, massive clusters are still swamped in large amounts of gas and dust, with considerable and uneven levels of extinction. At the same time, large grains (ices?) produced by type II supernovae profoundly alter the interstellar medium (ISM), thus resulting in extinction properties very different from those of the diffuse ISM. To obtain physically meaningful parameters of stars, from basic luminosities and effective temperatures to masses and ages, we must understand and measure the local extinction law. This problem affects *all* the massive young clusters discussed in his volume.

We have developed a powerful method to unambiguously determine the extinction law in an uniform way across a cluster field, using multi-band photometry of red giant stars belonging to the red clump (RC). In the Large Magellanic Cloud (LMC), with about 20 RC stars per arcmin^2, we can easily derive a solid and self-consistent absolute extinction curve over the entire wavelength range of the photometry. Here, we present the extinction law of the Tarantula nebula (30 Dor) based on thousands of stars observed as part of the Hubble Tarantula Treasury Project (HTTP; Sabbi *et al.* 2013, 2015). Space limitations only allow us to highlight the main results. For more details see De Marchi *et al.* (2015).

From the colour-magnitude diagrams (CMD) we can derive the absolute extinction in all bands towards $\sim$ 3 500 red giants that belong to the RC. The arrows in Fig. 1 indicate the position of the un-reddened RC. The high and variable levels of extinction spread RC stars across the CMD defining tight strips. The slopes of such strips in all bands provide a direct measure of the reddening vector.

In the Tarantula nebula the reddening vector is much steeper than in the Galaxy. While in the diffuse Galactic ISM $R_V = A_V/E(B-V) = 3.1$ (thick blue dashed line in Fig. 2), the Tarantula extinction curve (dots and red line) implies $R_V = 4.5 \pm 0.2$. This value is in excellent agreement with the measurements made inside its core R 136 both from photometry ($R_V = 4.4 \pm 0.2$, black dashed line; De Marchi & Panagia 2014) and spectroscopy ($R_V = 4.4 \pm 0.4$; Maíz Apellániz *et al.* 2014). At optical wavelengths the extinction law is practically parallel to the Galactic law, being shifted to higher values of R by an amount of 0.8 (thin short-dashed line). The difference between the extinction law in 30 Dor and in the Galaxy (green line) is remarkably flat in the optical. This indicates that the extinction curve in the Tarantula is due to dust of the same type as in the Galaxy, but with an extra component that does not depend on wavelength ("grey"). At wavelengths longer than $\sim 1\,\mu$m the contribution of this grey component tapers off as $\lambda^{-1.5}$, like in the Milky Way (dot-dashed line), further suggesting that the nature of the grains is otherwise similar to those in our Galaxy, but with a $\sim$ 2 times higher fraction of large grains. This is consistent with the addition of "fresh" large grains by type II supernova explosions, as recently revealed by the *Herschel* and *ALMA* observations of SN 1987A (e.g. Matsuura *et al.* 2011; Indebetouw *et al.* 2014).

These results leave no doubt that in the Tarantula large grains are more important than in the diffuse ISM in the Galaxy. These properties are not found in the surroundings of the

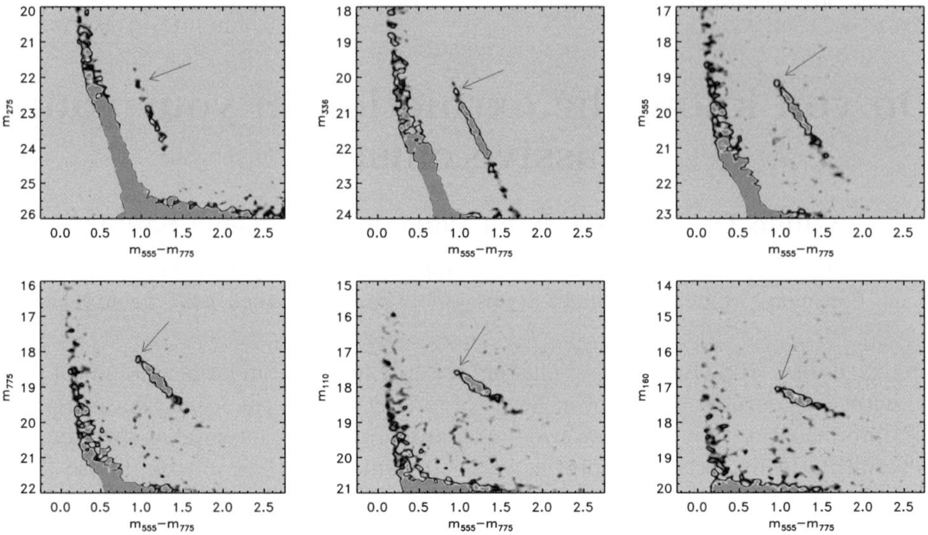

Figure 1. Unsharp masking applied to all CMDs clearly reveals the location of the un-extinguished RC. The reddening vector is obtained through a linear fit to the elongated RC.

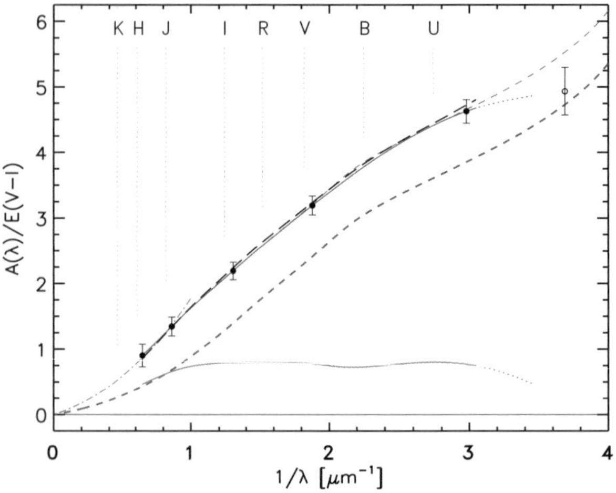

Figure 2. The extinction law of the Tarantula nebula (dots and red line) is compared with the one measured inside the nebula's core (R 136; long-dashed line), and with that of the diffuse Galactic ISM (thick short-dashed line). In the optical domain, the Tarantula and Galactic curves are parallel (the green line is their difference). See the text for details.

Tarantula, so they must result from changes related to the recent/ongoing star formation, implying that the ISM is dynamically altered as large fresh grains are produced and later destroyed. These findings have important implications for the study of the star formation properties in galaxies. Even though these changes might last only some ~ 50 Myr, they affect the HII regions at the peak of their luminoisty, when they are most easily detectable in distant galaxies. Therefore, assuming typical ISM conditions could result in severely underestimated total masses and star formation rates, by a factor of 2 or more.

References

De Marchi, G. & Panagia, N. 2014, MNRAS, 445, 93
De Marchi, G., Panagia, N., Sabbi, E., *et al.* 2015, MNRAS, submitted
Indebetouw, R., *et al.* 2014, ApJ, 782, L2
Maíz Apellániz, J., *et al.* 2014, A&A, 564, A63
Matsuura, M., *et al.* 2011, Sci, 333, 1258
Sabbi, E., *et al.* 2013, AJ, 146, 53
Sabbi, E., *et al.* 2015, ApJS, submitted

Formation, evolution, and survival of massive star clusters
Proceedings IAU Symposium No. 316, 2015
C. Charbonnel & A. Nota, eds.

doi:10.1017/S174392131601187X

Non-thermal emission in Cyg OB2

D. Fenech[1], **J. Morford**[1] **and R. Prinja**[1]

[1]Department of Physics and Astronomy, University College London,
Gower Street, London, WC1E 6BT
email: dmf@star.ucl.ac.uk

Abstract. We report here on the first results from the Cyg OB2 Radio Survey (COBRaS), which is a UCL-led e-MERLIN legacy project to provide a deep-field radio mapping of the Cygnus OB2 association. The project has been awarded a total allocation of 252 hours at C-band (5 GHz) and 42 hours at L-band (1.6 GHz) to image the core of the cluster.

We discuss in particular the presence of non-thermal radio emission at 20 cm (L-band), and its potential as a highly efficient way to identify binaries via single-epoch observations, particularly for colliding-wind binaries. COBRaS data will provide a powerful tool for establishing binary incidence in Cyg OB2, specifically in the difficult intermediate-period range (1–100 yr).

Ultimately, we aim to assemble a substantial and uniquely sensitive radio dataset, which will be exploited to address several fundamentally important areas of stellar astrophysics, including mass-loss, binary frequency, stellar cluster dynamics, and triggered star-formation.

Keywords. radiation mechanisms: nonthermal, binaries: general, stars: early-type, associations: individual: Cyg OB2

1. The Cygnus OB2 Radio Survey (COBRaS)

The Cygnus OB2 association is located at the core of the Galactic Cygnus X region at a distance of 1.45 kpc (Hanson 2003) making it one of the closest young massive stellar clusters. It is known to contain a rich population of massive stars (Setia-Gunawan *et al.* 2003) including 120 ± 20 O-type stars and 2600 ± 400 OB-type stars (knödlseder 2000) and is tremendously rich in both its stellar density and diversity. As one of the most massive clusters in the Milky Way it offers direct comparison to not only massive clusters in general, but also young globular clusters and super star clusters found in high star-forming galaxies. Located behind the great Cygnus rift Cyg OB2 is heavily obscured making this ideal for radio studies.

COBRaS is an e-MERLIN legacy project (P. I. R. Prinja) awarded ~300 hrs of observing time to perform an intensive radio survey of the core of the Cygnus OB2 association in our Galaxy. We will conduct a uniquely probing, targeted deep-field mapping of this young massive cluster with two key science goals. A census of the massive stellar population via the thermal emission from the stellar wind will be used to calculate mass-loss rates and constrain the level of clumping within the wind. In addition, the detection of non-thermal emission will allow a study of the binary population within Cyg OB2.

The COBRaS observations will consist of mosaiced pointings covering approximately 20 square arcmins of the core of the Cyg OB2 cluster including 42 hrs at L-band (1.6 GHz) split over seven pointings with an rms noise of $\sim 15\,\mu$Jy and 252 hrs at C-band (5 GHz) split over 27 pointings with an rms noise of $\sim 3 - 4\,\mu$Jy.

The L-band observations were completed over three observing sessions: 24th Jan. 2014, 11th Apr. 2014 and 25th Apr. 2014. The analysis is ongoing and we present here some preliminary results from the COBRaS project.

2. Colliding winds in binaries

In a binary consisting of two early-type stars (e.g. O+O), the stellar winds collide and electrons are accelerated to relativistic velocities around the stationary shocks in the colliding-wind region (CWR). These CWR produce non-thermal synchrotron emission detectable at radio wavelengths as well as strong x-rays. Such emission has a steep spectral index ($\alpha \sim -0.7$) which can be distinguished from the thermal emission ($\alpha \sim 0.6$) expected from the stellar wind itself.

The e-MERLIN data from COBRaS will allow us to study in detail the individual binary systems and improve our understanding of the physics involved in particle acceleration within the shocks. These observations will also enable a statistical study of the colliding-wind phenomenon and better understand its dependence on stellar and binary parameters.

Accurate knowledge of the binary fraction is crucial for understanding the evolution of stellar populations and is a fundamental requirement for population synthesis models used to study the formation and evolution of stellar clusters. These COBRaS observations will provide single-epoch detection of binary systems enabling direct observation of the binary frequency within Cyg OB2 which is estimated to be ~55% (Kobulnicky *et al.* 2014). In particular, this survey will reveal the intermediate period binaries that are difficult to detect with optical studies.

3. COBRaS L-band detections - non-thermal emission

Approximately one fifth of the L-band observations have been processed and a number of non-thermal sources have been detected in the resulting images. Cyg OB2 #9 is just on example of the well-known binary systems. Cyg OB2 #9 is a highly eccentric O5-5.5I + O3-4III binary system with strong and highly variable radio emission (Blomme *et al.* 2014). Though only recently discovered as a binary, it is one of the most well-studied within the region and has been detected in the COBRaS observations with a flux density of 0.88 mJy.

A number of sources have been observed within the COBRaS data that have no formal identification and limited prior detections. SBHW90 is one example which was observed in a previous radio study at 1.4 GHz and 327 MHz with flux densities of 14.7 and 54 mJy respectively (Setia-Gunawan *et al.* 2003). This suggests SBHW90 may have a steep spectral index which could be from a CWR within a previously unknown binary system. These COBRaS observations provide the first resolved images of SBHW90 which has a flux density of 1.94 mJy at 1.6 GHz, suggesting this source may also be strongly variable.

The COBRaS observations will highlight a population of non-thermal emitting sources within the core of Cyg OB2 including known individual massive stars and binary systems. It is also expected to reveal a number of new sources such as young stellar objects, T-Tauri stars, new binary systems and protoplanetary disks as well as background radio galaxies.

References

Blomme, R., Nazé, Y., Volpi, D., De Becker, M., Prinja, R. K., Pittard, J. M., Parkin, E. R., & Absil, O. 2014, *A&A*, 570, 10

Hanson, M. M. 2003, *ApJ*, 597, 957

Knodlseder, J. 2000, *A&A*, 360, 539

Kobulnicky H. A. *et al.* 2014, *ApJS*, 213, 34

Setia-Gunawan D. Y. A. De Bruyn, A. G., Van der Hucht, K. A., & Williams, P. M. 2003, *ApJS*, 149, 123

Formation, evolution, and survival of massive star clusters
Proceedings IAU Symposium No. 316, 2015
C. Charbonnel & A. Nota, eds.

doi:10.1017/S1743921316007080

Comparing ALMA, VLT, and HST data for Massive, Young Clusters in Grand-Design Spirals

Preben Grosbøl[1] and Horacio Dottori[2]

[1]European Southern Observatory,
Karl-Schwarzschild-Str. 2, D-85378 Garching, Germany
email: pgrosbol@eso.org

[2]Instituto de Física, Univ. Federal do Rio Grande do Sul,
Av. Bento Gonçalves 9500, 91501-970 Porto Alegre, RS, Brazil
email: dottori@ufrgs.br

Abstract. A population of young, massive stellar cluster complexes with near-infrared (NIR) colors indicating high extinction (i.e. Av $\sim 7^{m}$) was identified on HAWK-I/VLT images of several nearby, grand-design spiral galaxies. Models suggest that they are very young cluster complexes still embedded in a dust/gas envelope which will be expelled after 5-7 Myr. This type of very young, embedded clusters are not seen in optical studies using HST data.

A detailed comparison of HST and HAWK-I images was done to better understand the discrepancy between the optical and NIR detection of stellar clusters in nearby galaxies. More than 70% of the NIR clusters are located close to dust lanes which would make an optical detection difficult. A comparison of the ALMA CO(1-0)-map of NGC 4321 and the young, massive clusters shows that 60% of them have CO emission within 2″ indicating a correlation between giant molecular clouds and formation of massive clusters.

Keywords. galaxies: star clusters, galaxies: spiral

1. Introduction

The populations of stellar clusters in 10 nearby, grand-design spiral galaxies were analyzed by Grosbøl & Dottori (2012) using deep, near-infrared (NIR) images observed with HAWK-I/VLT. All the NIR color-color distributions show a characteristic bimodal structure with a main group of sources located near ((H-K), (J-H)) = (0.4, 0.8) corresponding to old clusters with relative low extinction while the other smaller concentration was seen around (0.8, 1.0). Assuming the reddening law by Israel *et al.* (1998), this latter group is consistent with highly extinct, very young clusters.

The (J-K)–K color-magnitude diagrams clearly display the two groups as well separated branches i.e. at (J-K) = 1.2 and 2.0, respectively. The reddening corrected color index Q = (H-K) − 0.84*(J-H) shows a more continuous distribution suggesting that the two branches in (J-K) are due to a change in extinction. The gap between the branches suggests a rapid reduction of extinction in the clusters at an early evolutionary phase as indicated by models of Grosbøl & Dottori (2013).

Optical studies of nearby disk galaxies using HST data (see e.g. Larsen & Richtler (1999), Bastian *et al.* (2012)) do not find a similar population of clusters. This may be due to several reasons such as a) lower spatial resolution of the NIR data as suggested by Bastian *et al.* (2014), b) high attenuation by dust limiting detection at optical wavelengths, or c) such populations only exist in grand-design spirals due to the strong perturbations in their arm regions.

2. Comparison between VLT and HST images

For the 7 galaxies with overlapping HAWK-I/VLT and HST images, the reddest calibrated HST frames(drz) with reasonable quality were selected for comparison. All young sources (i.e. 0.0<Q) with high extinction (i.e. 1.5<(J-K)) were visually checked to see if there were dust in their vicinity. For two galaxies (NGC 1300 and NGC 1566) with open spiral arms and very low inter-arm intensity, some sources appeared to be background galaxies. The low surface brightness of NGC 7424 made the identification of dust on the HST frames rather uncertain. The vast majority (i.e. >70% or 109/145) of the young sources with very red (J-K) colors is located within 2" of dust lanes or patches which would make a detection in blue visual bands very difficult.

3. Comparison between VLT and ALMA CO-map

During the science verification of ALMA, a CO(1-0) line map of NGC 4321 was made in band 3. A total of 34 such clusters with 0.1<Q was found within the area of the ALMA map which detected CO clouds down to a brightness limit of 0.3 Jy/beam km/s. CO emission were detected within 2" for around 60% (i.e. 20/34) of the clusters.

Control samples of clusters with random positions within the map had a flat distribution of distances suggesting that a correlation between the young clusters and CO emission is significant at a 3σ level, even considering the relative shallow map.

4. Conclusions

The comparison of young cluster complexes with very red (J-K) colors on both HST image and K-maps suggests that a majority of them (>70%) is located very close to dust patches. This indicates that the main reason for the lack of this type of sources in HST studies is the high extinction. There may well be a resolution effect which will tend to make very young clusters, observed in larger apertures, redder in (H-K) (e.g. due to nebular Br_γ emission as suggested by Bastian *et al.* (2014)) but this cannot explain the observed increase in (J-H).

Only a larger sample of deep NIR observations of more flocculent spirals will allow to test if the population of very massive, young cluster complexes is unique to grand-design spirals (due to the stronger perturbation in the spiral arms) or not.

Finally, around 60% of the young complexes are found close to CO emission even on the relative shallow ALMA CO(1-0) map suggesting a correlation. Deeper maps would be required for a detailed comparison.

References

Bastian, N., Adamo, A., Schirmer, M., *et al.* 2014, *MNRAS*, 444, 3829
Bastian, N., Konstantopoulos, I. S., Trancho, G., *et al.* 2012, *A&A*, 541, A25
Grosbøl, P. & Dottori, H. 2012, *A&A*, 542, A39
Grosbøl, P. & Dottori, H. 2013, *A&A*, 551, L13
Israel, F. P., van der Werf, P. P., Hawarden, T. G., & Aspin, C. 1998, *A&A*, 336, 433
Larsen, S. S. & Richtler, T. 1999, *A&A*, 345, 59

Formation, evolution, and survival of massive star clusters
Proceedings IAU Symposium No. 316, 2015
C. Charbonnel & A. Nota, eds.

doi:10.1017/S1743921315010686

Exploring the UV excess in star clusters of different mass

Fabiola Hernández-Pérez[1] **Gustavo Bruzual**[2] **and Gladis Magris C.**[3]

[1]Universidad de Los Andes, Departamento de Física, Grupo de Astrofísica Teórica, Mérida, Venezuela
email: fabiola.fchp@gmail.com, fabiolah@ula.ve

[2]Centro de Radioastronomía y Astrofísica (CRyA), Morelia, México
email: g.bruzual@crya.unam.mx

[3]Centro de Investigaciones de Astronomía (CIDA), Mérida, Venezuela
email: magris@cida.gob.ve

Abstract. We compute the expected spectral energy distribution of stellar populations of mass characteristic of star clusters taking into account stochastic fluctuations in the number of stars populating the IMF, and the presence of interacting binary stars in the cluster population. We evaluate under what circumstances the UV excess phenomenon is expected to appear in star clusters of different mass, and which is its most likely source: the stochastic fluctuations, the result of binary interactions, or a mixture of both.

Keywords. EHB stars, UV excess, star clusters.

1. Introduction

Observations of globular clusters (GCs) of M87 from Montes *et al.* (2014) show an UV excess that can not be explained by standard stellar population synthesis (SPS) models (e.g. Bruzual & Charlot (2003), hereafter BC03). It is widely accepted that UV excess could be produced by an Extreme Horizontal Branch (EHB) component, probably due to binary interaction or He enhancement (see Smith *et al.* (2012)).

In this work we propose that stochastic sampling in the IMF could also drastically change the predictions of the models (with and without binary stars), explaining the UV bright GCs.

2. Model and Results

We use a recent implementation of BC03 model that allows a stochastic sampling of the initial mass function (IMF). Our test consists on 221 realizations of the Kroupa IMF, which is filled adding stars with stochastically selected mass, for a total mass of 10^4, 10^5, 10^6 and 10^7 $M_\odot$. In Fig. 1 we present the resulting UVr2CD for a Z=0.004 population at 11 and 12 Gyr, Z=0.004 and Kroupa IMF. As we can see in Fig. 1, fluctuations in the IMF tend to spread the colours predicted by the model. The observations of GCs are covered by the bluest realizations (light gray dots), which are consequence of different numbers of central stars of planetary nebulae (CSPNe). We recall that BC03 model includes CSPNe as the main source of UV radiation in old populations.

In order to evaluate the influence of binary stars in the UV colours, we select some BC03 realizations of the IMF (white squares in left panel of Fig. 1) and use them as input in our SPS model (Hernández-Pérez & Bruzual (2013), hereafter HB13) where EHB stars as product of binary interaction are included. To account for fluctuations in the number

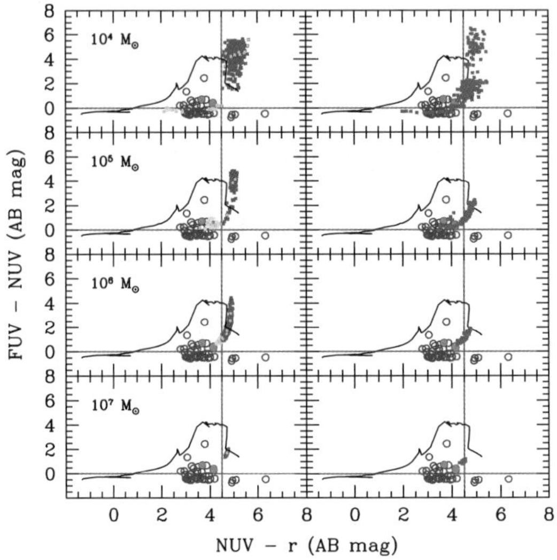

Figure 1. UVr2CD. Squares represent simulated clusters for standard (left panels) and binary (right panels) models. Light gray squares in left panels correspond to simulations with NUV−r < 4.5. Standard model of infinite mass (solid black line) is also shown for comparison. Light gray open circles are M87 GCs from Montes et al 2014 with UV data, black filled triangles are those GCs with good UV photometry. See text for more details

of EHB progenitors, we run ten realizations of the binary related parameters of HB13 model for each white square realization of the IMF, with the same metallicity and ages. The resulting colours are represented as dark gray squares in the right panels of Fig. 1.

In a general way, the cloud of colours predicted by HB13 is similar that BC03, but binaries tend to shift the bulk of the cloud to the blue due to EHB stars. It is clear that observations of GCs, in particular those with good UV photometry (black filled triangles) are reproduced by both models, BC03 and HB13, if stochastic fluctuations in the IMF are included.

An extended analysis, including a consistent GC mass determination, and a more precise analysis of the colour distribution of GC in the UVr2CD, will help us to give a quantitative measure of the relevance of binaries to explain the UV excess in old populations.

3. Conclusions

- Standard model can not explain UV excess observed in GCs of M87.
- Stochastic fluctuations is the dominant source of UV excess predicted by both models for low mass populations.
- For higher masses, the increase in the number of CSPNe (BC03) and EHB (binary) tends to erase stochastic fluctuations.
- Binaries lightly shift the model to the blue in both NUV - r and FUV-NUV colours.

References

Bruzual, G. & Charlot, S. 2003, *MNRAS*, 34, 100 (BC03)
Hernández-Pérez, F. & Bruzual, G. 2013, *MNRAS*, 431, 2612 (HB13)
Montes, M., Acosta-Pulido, J., Prieto, M. & Fernández-Ontiveros, J. 2014, *MNRAS*, 442, 1350
Smith, R., Lucey, J. & Carter D. 2012, *MNRAS*, 421, 2982

Formation, evolution, and survival of massive star clusters
Proceedings IAU Symposium No. 316, 2015
C. Charbonnel & A. Nota, eds.

doi:10.1017/S1743921315010492

Galactic and extragalactic giant molecular clouds and massive star clusters: counting and analysis

A. S. Hojaev[1,2]

[1]National university of Uzbekistan
Vuzgorodok, Tashkent, Uzbekistan

[2]Ulugh Beg Astronomical Institute, Uzbek Academy of Sciences
Astronomicheskaya , 33, Tashkent, 100052, Uzbekistan
email: hojaev@yahoo.com

Abstract. Comparative analysis of GMS and MSC is presented based on data collected on these objects in the Galaxy and other galaxies. The model is proposed of gravitational instability of GMC and MSC system. Some implications are discussed.

Keywords. Galaxy: structure, Galaxy: disk, (Galaxy:) open clusters and associations: general, Galaxy: kinematics and dynamics, ISM: structure, ISM: clouds, galaxies: structure, galaxies: ISM, galaxies: star clusters, surveys

---

1. Introduction

Massive star clusters (MSC), especially super star clusters and young massive star clusters, are unique objects in the Universe. The deep study of MSC together with their predecessors - giant molecular clouds (GMC) resides among the most demanded tasks in modern astronomy. The instrumentation and observing technology existed already permit to find and study GMC and MSC not only within but also outside of the Galaxy. Thus, a few dozens of GMC were scrutinized in the Milky Way (Solomon *et al.* 1987, Grabelsky *et al.* 1988 , Bronfman *et al.* 1989 , Heyer *et al.* 2009, Draine 2011, etc.) and hundreds were revealed in outer galaxies (Wilson & Reid 1991, Taylor *et al.* 2000,Young 2001,Engargiola *et al.* 2003,Rosolowsky *et al.* 2007,Bolatto *et al.* 2008,Pineda *et al.* 2009,Gratier *et al.* 2012,Colombo *et al.* 2014,Kirk *et al.* 2015, etc.). At the same time more than half of thousand MSC including young population was found in the Galaxy and in nearby galaxies (Larsen 1999, McLaughlin & Marel 2005, etc.) Evidently GMC should be closely associated with cold gas-dust clouds, particularly with HI ones and apparently have to play a key-role in MSC forming processes as well as they all together in the kinematics and dynamics of the host galaxy in a whole. Therefore the counting (or census) and survey of the GMC and MSC populations are still important.

2. Results and discussion

In an attempt to address at least some aspects of the physics and evolution of the GMC and MSC, their interplay and impact on the dynamics and evolution of galaxies we drafted a consolidated composite catalogs of GMC that could be observed from the Earth based on the recent data from DB VizieR, SIMBAD at CDS as well as original papers, reports and other publications were used. Catalogs contain the coordinates, sizes, distances, masses as well as some other physical parameters (such as density, temperature, radial velocity, etc.) that are available for the different clouds. They include more than 40 galactic GMC and few hundreds of GMC and super GMC in other galaxies observed in CO lines and detected up to the present. At the same time Larsen (1999) found 551 Young MSC in 21 nearby galaxies, McLaughlin & Marel (2005) identified 153 MSC that can be resolved in the Galaxy, LMC, SMC and the Fornax dwarf spheroidal, Baumgardt *et al.*(2013) reported on 320 MSC in LMC. A preliminary analyses of the data has shown that in case of very young or just emerging MSC they are closely connected with GMS. Older MSC loose the connection with GMC of their origin: the formed MSC, especially with O-B supergiants that blow gas, disrupt surrounding GMC. The lifetime of GMC as well as the star formation efficiency and rate in GMC which are predictable relatively high were estimated by Murray (2011). Using these data the dynamic model of formation and evolution of GMC and MSC system based on gravitational instability develops and numerically calculated. The first results are compared with data of other investigations. The completion and improvement the catalog data are also in progress.

3. Conclusion and implications

The data on GMC and MSC were gathered in composite catalogs and the comparative analysis was made. The model of gravitational instability of GMC and MSC is proposed and numerically computed. As implication of the project we prepared the complex program for future observations of GMC and MSC using the 70 meter Suffa cm-submm radio telescope and other large facilities to cover the lack of observed physical parameters. This project has been supported in part by grant OT-F2-57 funded by the CCSTD at CM of RUz.

References

Solomon, P. M., Rivolo, A. R., Barrett, J., & Yahil, A. 1987, *ApJ*, 319, 730
Grabelsky, D. A., Cohen, R. S., Bronfman, L., & Thaddeus, P. 1988, *ApJ*, 331, 181
Bronfman, L., Nyman, L., & Thaddeus, P. 1989, in: G. Winnewisser & J. T. Armstrong (eds.), *The Physics and Chemistry of Interstellar Molecular Clouds - mm and Sub-mm Observations in Astrophysics*, Lecture Notes in Physics(Berlin: Springer Verlag), 331, 139
Heyer, M., Krawczyk, C., Duval, J.,& Jackson, J. M. 2009, *ApJ*, 699, 1092
Draine, B. T. 2011, *Physics of the Interstellar and Intergalactic Medium* (Princeton University Press), 540
Wilson, C. D. & Reid, I. N. 1991, *ApJ*, 366, L11
Taylor, C., Huttemeister, S., Klein, U.& Greve, A. 2000, *ASP Conf. Ser.*, 215, 51
Young, L. M. 2001, *AJ*, 122, 1747
Engargiola, G., Plambeck, R. L., Rosolowsky, E.& Blitz, L. 2001, *ApJS*, 149, 343
Rosolowsky, E. Keto, E., Matsushita, S.& Willner, S. P. 2007, *ApJ*, 661, 830
Bolatto, A. D., Leroy, A. K.,Rosolowsky, E., Walter, F.& Blitz, L. 2008, *ApJ*, 686, 948
Pineda, J. L. Ott, J., Klein, U., *et al.* 2009, *ApJ*, 703, 736
Gratier, P., Braine, J., Rodriguez-Fernandez, N. J., *et al.* 2012, *A&A*, 542, A108
Colombo, D.,Hughes, A., Schinnerer, E., *et al.* 2014, *ApJ*, 784, 32
Kirk, J. M., Gear, W. K., Fritz, J., *et al.* 2015, *ApJ*, 798, A58, 19
Larsen S. S. 1999, *A&AS*, 139, 393
McLaughlin, D. E.& van der Marel, R. P. 2005, *ApJS*, 161, 304
Murray, N. 2011, *ApJ*, 729 , 133
Baumgardt H., Parmentier G., Anders P.& Grebel E. K. 2013, *MNRAS*, 430, 676

Formation, evolution, and survival of massive star clusters
Proceedings IAU Symposium No. 316, 2015
C. Charbonnel & A. Nota, eds.

doi:10.1017/S1743921316007109

Efficient Selection and Classification of Infrared Excess Emission Stars Based on AKARI and 2MASS Data

Ya-Fang Huang, Jin-Zeng Li, Jing-hua Yuan and Hong-Li Liu

[1]National Astronomical Observatories, Chinese Academy of Sciences,
20A Datun Road, Chaoyang District, Beijing 100012, China
email: huangyf@nao.cas.cn

Abstract. The selection of young stellar objects (YSOs) based on excess emission in the infrared is easily contaminated by post-main-sequence stars and various types of emission line stars with similar properties. We define in this paper stringent criteria for an efficient selection and classification of stellar sources with infrared excess emission based on combined Two Micron All Sky Survey (2MASS) and AKARI colors.

Keywords. stars: pre-main sequence – stars: variables: T Tauri, Herbig Ae/Be

1. Main-sequence Stars and Major Contaminants

First of all, bright dwarfs and giants with known spectral types were selected from the Hipparcos Catalogue and cross-identified with the 2MASS and AKARI Point Source Catalogues to produce the main-sequence and the post-main-sequence tracks (Figure 1(a)). Several kinds of contaminants were then removed based on their distribution in the color-color (C-C) diagrams.

2. Young Stellar Objects

Test sample of Herbig Ae/Be stars and classical T Tauri stars were cross-identified with

Table 1. Source selection criteria defined for various types of YSOs and CBe based on the C-C diagrams compiled based on 2MASS and AKARI.

Source		Defination of the Source Selection Criteria
HAeBe		$(J-H) - 2.3 \times (H-K_S) + 0.81 < 0$
		or $(J-H) - 0.33 \times (H-K_S) - 0.57 < 0$
	and	$(K_S - S9W) + 1.023 \times (S9W - L18W) - 4.523 > 0$
		$S9W - L18W > 1$
	or	$(J-K_S) - 1.9 \times (K_S - S9W) + 2.59 < 0$
CTTS		$-0.3 > (J-H) - 2.3 \times (H-K_S) > -0.81$
		$(J-H) - 0.33 \times (H-K_S) - 0.57 > 0$
	and	$(K_S - S9W) + 1.023 \times (S9W - L18W) - 4.523 > 0$
		$S9W - L18W > 1$
	or	$(J-K_S) - 1.9 \times (K_S - S9W) + 2.59 < 0$
Class I		$(J-H) - 2.3 \times (H-K_S) + 0.31 < 0$
	and	$(K_S - S9W) + 1.023 \times (S9W - L18W) - 7.046 > 0$
		$S9W - L18W > 2$
Class II		$(J-H) - 2.3 \times (H-K_S) + 0.31 < 0$
	and	$4.523 < (K_S - S9W) + 1.023 \times (S9W - L18W) < 7.046$
		$S9W - L18W > 1$
CBe		$(J-H) - 2.3 \times (H-K_S) + 0.3 < 0$
		$(J-H) - 0.33 \times (H-K_S) - 0.57 < 0$
	and	$(K_S - S9W) - 0.978 \times (S9W - L18W) - 0.564 < 0$
	or	$(J-K_S) - 1.9 \times (K_S - S9W) + 2.59 > 0$

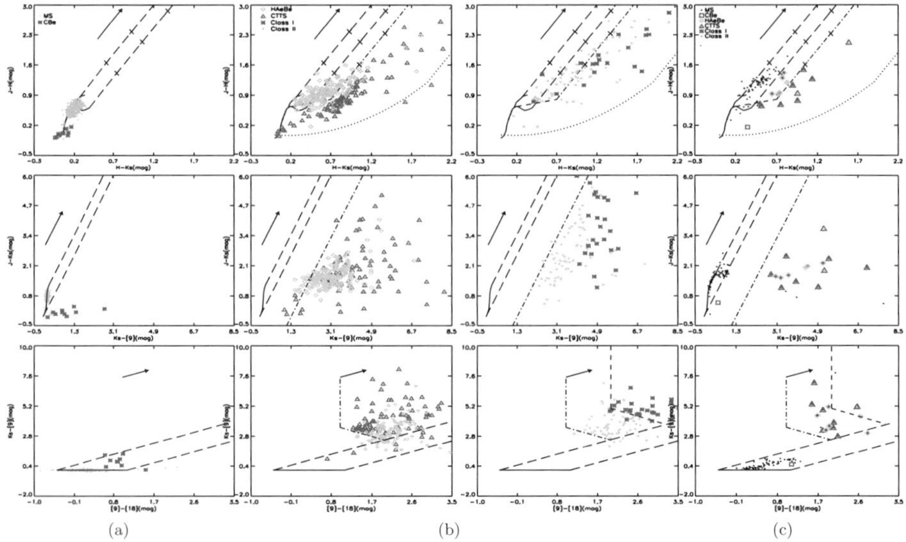

Figure 1. (a) The distribution of the MS and post-MS sources in C-C diagrams. The tracks of them shown in solid lines were compiled from high-precision data from the Hipparcos Main Catalogue. The symbols in the three diagrams are the same. Green dots and purple asterisk represent normal stars and CBe, respectively. Long dashed lines delineate the reddening band for normal stars. The arrow shows a reddening vector of $A_v = 5$ mag Rieke & Lebofsky(1985) in JHK_S and $A_v = 10$ mag in $JK_S[9]$ and $K_S[9][18]$. (b) The distribution of the first test sample of HAeBe stars Thé *et al.*(1994), and CTTSs from Herbig-Bell Catalogue (left), and the second test sample of Class I and Class II sources from Gutermuth *et al.*(2008), Gutermuth *et al.*(2009) (right) in C-C diagrams. Dot-dashed line, short dashed line and dotted line in JHK_S indicate the division line between HAeBe and CTTS, the locus of dereddened TTS Meyer *et al.*(1997) and the locus of dereddened HAeBe Lada & Adams(1992), respectively. Dot-dashed lines in mid-IR C-C diagrams indicate the left boundary of YSOs. Short-dashed line in $K_S[9][18]$ diagram is the division line between Class I and Class II sources. Crosses were over plotted with an interval corresponding to 5 mag of visual extinction in JHK_S diagram. The other lines and arrows are the same as those defined in Figure 1(a). (c) A blind classification of the candidate YSOs with excess emission in the IR toward IC 1396. The lines and arrows are the same as those defined in Figure 1(a).

the 2MASS and AKARI catalogs to define the loci of YSOs with different masses on the C-C diagrams(Figure 1(b)). Well classified Class I and Class II sources were taken as a second test sample to discriminate between various types of YSOs at possibly different evolutionary stages. This helped to define the loci of different types of YSOs and a set of criteria for selecting YSOs based on their colors in the near- and mid-infrared. And then our selection and classification of excess emission sources were summarized in Table 1.

3. Verification of the Sample Selection Criteria

Candidate YSOs toward IC 1396 indicating excess emission in the near-infrared were employed to verify the validity of the new source selection criteria defined based on C-C diagrams compiled with the 2MASS and AKARI data (Figure 1(c)).

References

Gutermuth, R. A., Myers, P. C., Megeath, S. T., *et al.* 2008, , *ApJ* 674, 336
Gutermuth, R. A., Megeath, S. T., Myers, P. C., *et al.* 2009, ApJS, 184, 18
Lada, C. J. & Adams, F. C. 1992, *ApJ*, 393, 278
Meyer, M. R., Calvet, N., & Hillenbrand, L. A. 1997, *AJ*, 114, 288
Rieke, G. H. & Lebofsky, M. J. 1985, *ApJ*, 288, 618
Thé, P. S., de Winter, D., & Pérez, M. R. 1994, A&AS, 104, 315

Formation, evolution, and survival of massive star clusters
Proceedings IAU Symposium No. 316, 2015
C. Charbonnel & A. Nota, eds.

doi:10.1017/S1743921316007055

Strömgren-Hβ photometry of the young open cluster NGC 6913

Nadia Kaltcheva[1], Valeri Golev[2], John Beaver[3], Steven Lund[1] and Michael Briley[4]

[1]Department of Physics and Astronomy, University of Wisconsin Oshkosh, 800 Algoma Blvd., Oshkosh, WI 54901, USA
email: kaltchev@uwosh.edu

[2]Department of Astronomy, Faculty of Physics, St Kliment Ohridski University of Sofia, 5 James Bourchier Blvd., BG-1164 Sofia, Bulgaria
email: valgol@phys.uni-sofia.bg

[3] University of Wisconsin - Fox Valley, 1874 Midway Rd, Menasha, WI, USA
email: john.beaver@uwc.edu

[4] Department of Physics and Astronomy, Appalachian State University, Boone, NC 28608, USA
email: brileymm@appstate.edu

Abstract. We present new CCD $uvby\beta$ photometry of about 50 stars located in the central region of the open cluster NGC 6913. The individual stellar distances and color excesses of established cluster's members are analyzed and new estimate of the cluster's distance is provided.

Keywords. Stellar clusters and associations, $uvby\beta$ photometry, NGC 6913.

1. Introduction

The young open cluster NGC 6913 (M 29) is located in a low-extinction area toward the Great Cygnus Rift. Despite several extensive studies, the distance estimates to the cluster are still controversial, differing by as much as a factor of two. Tifft (1958) suggested two groups at 1.6 and 2.1 kpc. Photoelectric $uvby\beta$ photometry yielded 1.1 kpc, not in agreement with spectroscopically-based estimates (Crawford *et al.* 1977). More recently, utilizing Vilnius, BV and JHK photometry and spectra, Straizys *et al.* (2014) obtained 1.54 kpc. In order to re-evaluate the main cluster parameters, we provide new analysis of the existing $uvby\beta$ data and supplement them with new CCD $uvby\beta$ photometry of the central 15 × 15 arcmin region of NGC 6913. The $uvby\beta$ photometric system (Strömgren 1966, Crawford & Mander 1966) should in general be able to provide independent estimates of reddening, distance, metallicity and age, thus helping to constrain the existing disagreements.

2. Observations

The data were obtained during three consecutive nights in May/June, 2008, on the 0.9 m WIYN (Wisconsin Indiana Yale National Optical Astronomy Observatory) telescope at Kitt Peak National Observatory (KPNO). The telescope was equipped with the S2KB CCD camera, which employs a 2048 × 2048, 21 μ-pixel detector for a field of view of 20.5 arcmin, and a set of KPNO $uvby\beta$ filters.The reduction of the images followed standard procedures for bias subtraction, flat fielding, illumination and scattered light corrections using the IRAF software package. About 20 cluster's stars with $uvby\beta$ photoelectric photometry (Crawford *et al.* 1977) were used as standards. The reduction was carried out in $V, b-y, c_1, m_1$ and β, as outlined by Crawford & Barnes (1970). In order to account for an apparent second-order variations in the point spread function across the field, we investigated various approaches to perform the photometry, including aperture photometry and 2-dimensional fitting with Gauss and Moffat functions. The 2-dimensional Gauss fit was found to provide the least error in the photometric transformations.

3. Results and Discussion

Our analysis of the photoelectric $uvby\beta$ data of Crawford & Barnes (1970) indicates that the bright stars that make the cluster recognizable are located at a distance of 895(+47, −44) pc. The fainter stars in the field seem to be more distant by 500 pc.

The new CCD $uvby\beta$ observations support this conclusion. Up to a limiting magnitude of about 14, we found no B-type stars located closer than 1 kpc. The impression is that the young compact group of bright OB stars is projected onto a background of fainter B-type stars. The obtained distance of 895 $\pm$ 45 pc is significantly smaller than the spectroscopically based 1540 $\pm$ 150 pc (Straizys *et al.* 2014). All these stars are located closer than a distance modulus of 11 mag, which corresponds to 1.6 ± 0.35 kpc (assuming an error of ± 0.5 mag in the distance modulus determination). This disagreement arises most likely from a possible overestimation of the spectroscopic distances(details on such comparisons can be found in Kaltcheva & Golev 2011). The inspection of the photometric diagrams do not imply a peculiar behavior that can significantly affect the $uvby\beta$ photometry-derived distances for the brightest cluster's members.

NGC 6913 is positioned within the boundaries of the Cyg OB1 association, one of the several OB associations within the giant H II region Sh 2-109. This nebulosity is possibly a complex of smaller H II regions overlapping along our line of sight since in this direction we are looking straight down the Orion spur (Sharpless 1959). Recent X-ray emission findings (Uyaniker *et al.* 2001) also suggest that Sh 2-109 consists of overlapping, not physically related, features. A distance less than 1 kpc, as suggested here, would place NGC 6913 closer than other young clusters located within the formal boundaries on the Cyg OB1/OB3 associations, like NGC 6871 and IC 4996, implying overlapping star-forming regions in this direction.

Acknowledgements. This study made use of the NASA Astrophysics Data System, SIMBAD database and WEBDA open cluster database operated at the Institute for Astronomy of the University of Vienna. N.K. acknowledges support from the SNC Endowed Professorship at the University of Wisconsin Oshkosh. This work is supported by the National Science Foundation grant AST-1516932.

References

Crawford, D. L. & Mander, J. 1966, *AJ*, 71, 114

Crawford, D. L. & Barnes, J. V. 1970, *AJ*, 75, 978

Crawford, D. L., Barnes, J. V., & Hill, G. 1977, *AJ*, 82, 606

Kaltcheva, N. & Golev, V. 2011, *Stellar Clusters & Associations*, A RIA Workshop on Gaia Proceedings, p. 299

Sharpless, S. 1959, *ApJ Suppl.*, 4, 257

Straizys, V., Milasius, K., & Boyle, P., *et al.* 2014, *AJ*, 455, 850

Strömgren, B. 1966, *ARAA*, 4, 433

Tifft, W. G. 1958, *AJ*, 63, 127

Uyaniker, B., Fürst, E., Reich, W., Aschenbach, B., & Wielebinski, R. 2001, *A&A*, 371, 675

Formation, evolution, and survival of massive star clusters
Proceedings IAU Symposium No. 316, 2015
C. Charbonnel & A. Nota, eds.

doi:10.1017/S1743921315009102

The Spectral Energy Distribution of the Earliest Phases of Massive Star Formation

Randolf Klein[1], Jennifer Cooper[1], Leslie Looney[2], Thomas Henning[3], Sukanya Chakrabarti[4] and Sachin Shenoy[1]

[1]SOFIA-USRA, NASA Ames Research Center, MS 232-12, PO box 1, Moffet Field, CA 94035
email: rklein@sofia.usra.edu

[2]University of Illinois, Dept. of Astronomy, MC-221, 1002 W. Green St., Urbana, IL 61801

[3]Max-Planck-Institut für Astronomie Königstuhl 17, D-69117 Heidelberg

[4]School of Physics and Astronomy, Rochester Institute of Technology
84 Lomb Memorial Dr., Rochester, NY 14623

Abstract. We have selected cold and massive ($M > 100 M_{\odot}$) cores as candidates for early phases of star formation from millimeter continuum surveys without associations at short wavelengths. We compared the millimeter continuum peak positions with IR and radio catalogs and excluded cores that had sources associated with the cores' peaks. We compiled a list of 173 cores in over 117 regions that are candidates for very early phases of Massive Star Formation (MSF). Now with the Spitzer and Herschel archives, these cores can be characterized further. We are compiling this data set to construct the complete spectral energy distribution (SED) in the mid- and far-infrared with good spatial resolution and broad spectral coverage. This allow us to disentangle the complex regions and model the SED of the deeply embedded protostars/clusters. We present a status report of our efforts: a preview of the IR properties of all cores and their embedded source inferred from a grey body fit to the compiled SEDs.

Keywords. stars: formation, infrared: ISM, ISM: clouds

For our millimeter survey (Klein et al. 2005), we defined the following criteria for cloud cores to be in an early stage of massive star formation (MSF):

(*a*) No radio, IRAS, MIR, NIR sources should be within $10''$ of the core's peak. As a basis for these constraints the NVSS, IRAS, MSX, and 2MASS databases were used.

(*b*) A minimum core mass of $100\,M_{\odot}$. The star formation efficiency can reach values of 50% (Hillenbrand & Hartmann 1998). Thus, to build a $8\,M_{\odot}$ star and its low-mass companions, one needs at least a $130\,M_{\odot}$ core (Miller & Scalo 1979).

Similarly, Sridharan et al. (2005) selected successfully high-massive starless core candidates from their millimeter continuum survey by the lack of mid-infrared (MIR) emission. Using the above criteria, we compiled a source list from the following millimeter surveys in order to get a comprehensive set of cloud cores in an early stage of MSF: Hunter et al. 2000; Sridharan et al. 2002; Faúndez et al. 2004; Klein et al. 2005; Beltrán et al. 2006.

The source list collected from the above surveys is a comprehensive set of massive millimeter cores. This selection results in a list of 173 millimeter cores across the Galaxy allowing a global study of the earliest stage of MSF and their evolution.

We collected for all cores the data available from Herschel (SPIRE and PACS photometry), which provided the data around the peak of the SED. This in turn, allowed to fit a grey-body to the emission by the cold dust in these envelops (using the Herschel and sub-millimeter data points). For the grey-body fits the Rayleigh-Jeans tail has been multiplied by ν^{β} with β being the spectral index of the dust emissivity. For 132 cores, the fits were successful. The top pannel of Fig. 1 shows the distribution of the fit parameters.

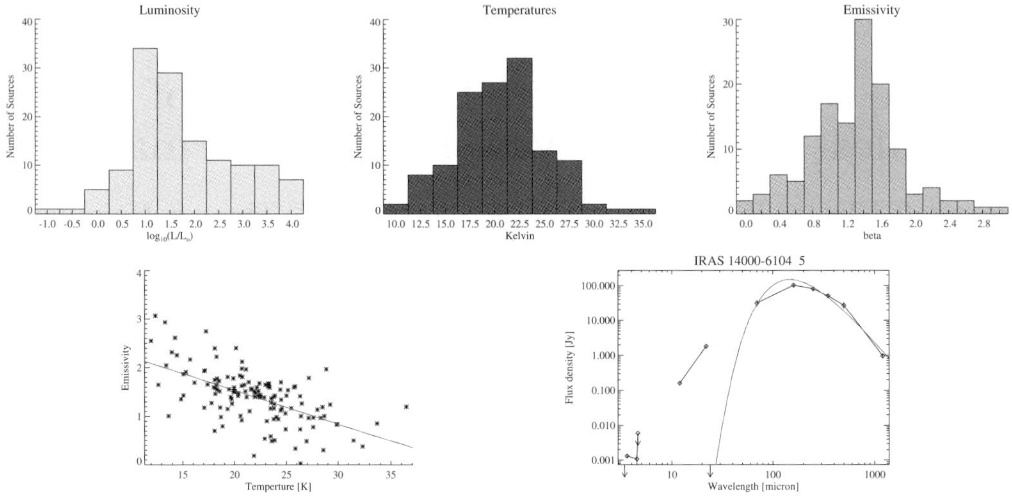

Figure 1. Top: grey-body fit parameter distributions; bottom left: Correlation of dust temperature and emissivity; bottom right: typical SED

Figure 1 shows on the bottom left how the fitted temperature and dust emissivity are related. The data points are scattered, but there is a correlation between the fitted cold dust temperature and the dust emissivity. The red line is a linear fit: $\beta = -0.07T[K]+2.9$.

On the bottom right is the SED of a typical source, i.e. all fit parameters are close to average. The parameters for this source: $L = 41\,L_{\odot}$, $T = 22\,\mathrm{K}$, $\beta = 1.5$. The blue line is the grey-body fit. Only the far-infrared part is fitted.

The Herschel data allowed us a first characterization of the dusty envelopes of the massive protostar candidates with a grey-body fit. While a few sources are quite faint for massive star-forming cores (but maybe they aren't, yet), most cores emit $100\,L_{\odot}$. The fitted cold dust temperature and the dust emissivity are correlated. If temperature is an indication of age (core is heating up with progressing star formation), the change in emissivity could be dust evolution.

The next step will be to model the full SED from the NIR to mm-wavelengths, e.g. with the Robitaille et al. (2006) models, but early tests show that the model library does not contain good fits for the coldest sources. We will also employ the semi-analytic SED models of Chakrabarti & McKee (2005).

Acknowledgement

R.K. acknowledges support throught the NASA ADAP grant NNX11AF95G.

References

Beltrán, M. T., Brand, J., Cesaroni, R., Fontani, F., Pezzuto, S., *et al.* 2006, A&A, 447, 221
Chakrabarti, S. & McKee, C. F. 2005, ApJ, 631, 792
Faúndez, S., Bronfman, L., Garay, G., Chini, R., Nyman, L.-Å., & May, J. 2004, A&A, 426, 97
Hillenbrand, L. A. & Hartmann, L. W. 1998, ApJ, 492, 540
Hunter, T. R., Churchwell, E., Watson, C., Cox, P., Benford, D. J., *et al.* 2000, AJ, 119, 2711
Klein, R., Posselt, B., Schreyer, K., Forbrich, J., & Henning, T. 2005, ApJS, 161, 361
Miller, G. E. & Scalo, J. M. 1979, ApJS, 41, 513
Robitaille, T. P., Whitney, B. A., Indebetouw, R., *et al.* 2006, ApJS, 167, 256
Sridharan, T. K., Beuther, H., Saito, M., Wyrowski, F., & Schilke, P. 2005, ApJ, 634, L57
Sridharan, T. K., Beuther, H., Schilke, P., Menten, K. M., & Wyrowski, F. 2002, ApJ, 566, 931

Formation, evolution, and survival of massive star clusters
Proceedings IAU Symposium No. 316, 2015
C. Charbonnel & A. Nota, eds.

doi:10.1017/S174392131501056X

The Orion Nebula in the Far-Infrared: High-J CO and fine-structure lines mapped by FIFI-LS/SOFIA

Randolf Klein[1], Leslie W. Looney[2], Erin Cox[2], Christian Fischer[3], Christof Iserlohe[3], Alfred Krabbe[3]

[1]SOFIA-USRA, NASA Ames Research Center, MS 232-12, PO box 1, Moffet Field, CA 94035
email: rklein@sofia.usra.edu

[2]University of Illinois, Dept. of Astronomy, MC-221, 1002 W. Green St., Urbana, IL 61801
Deutsches SOFIA Institut, Pfaffenwaldring 29, D-70569 Stuttgart

Abstract. The Orion Nebula is the closest massive star forming region allowing us to study the physical conditions in such a region with high spatial resolution. We used the far infrared integral-field spectrometer, FIFI-LS, on-board the airborne observatory SOFIA to study the atomic and molecular gas in the Orion Nebula at medium spectral resolution.

The large maps obtained with FIFI-LS cover the nebula from the BN/KL-object to the bar in several fine structure lines. They allow us to study the conditions of the photon-dominated region and the interface to the molecular cloud with unprecedented detail.

Another investigation targeted the molecular gas in the BN/KL region of the Orion Nebula, which is stirred up by a violent explosion about 500 years ago. The explosion drives a wide angled molecular outflow. We present maps of several high-J CO observations, allowing us to analyze the heated molecular gas.

Keywords. stars: formation, infrared: ISM, ISM: Orion Nebula

1. Preliminary Results

The observations were taken during the commissioning of FIFI-LS (Klein et al. 2014) on board of NASA's airborne observatory SOFIA (Young et al. 2012) last year and as recent as March 2015. The data reduction and calibration is still under development. Mainly, the atmospheric absorption correction and flux calibration is still in works.

The results of the fine structure line observations of [CII] and [OI] are shown in Fig. 1. Next to the line emission, a FIFI-LS observation also yields the underlying continuum. The continuum emission is peaking in the BN/KL object, where deeply embedded protostars heat the dust (De Buizer et al. 2012). The dust east of the Trapezium is the second peak. The Bar is not very prominent in the dust continuum. However, the [CII] emission clearly traces the Orion Bar, and the north-eastern wall of the cavity, which is cleared out by the trapezium stars. The emission comes from the Photon-Dominated-Region (PDR), where the UV-radiation from the Trapezium Cluster irradiates the surrounding molecular gas.

Figure 1 also shows the [OI] lines, which trace the Bar like the [CII] lines, as these fine-structure lines are the main cooling lines in a PDR. When the flux calibration is done, we can start interpreting the line ratios in term of density, temperature, and excitation conditions. Since the Bar is a nearly edge-on PDR, the width and position of the line emission can indicate how far the UV can penetrate.

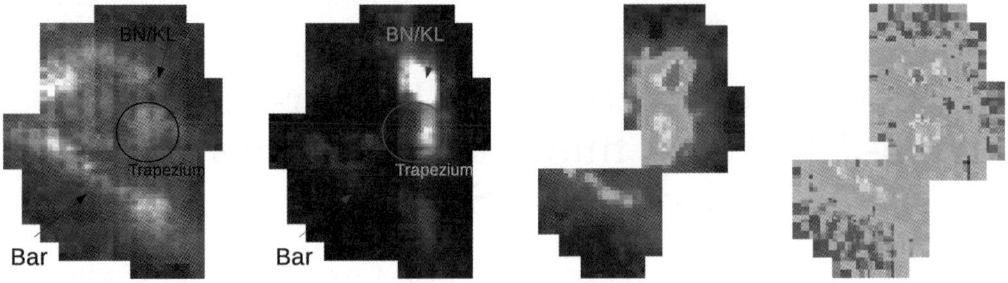

Figure 1. From left to right: [CII]158 μm line and continuum emission, [OI]63 μm and [OI]145 μm line emission

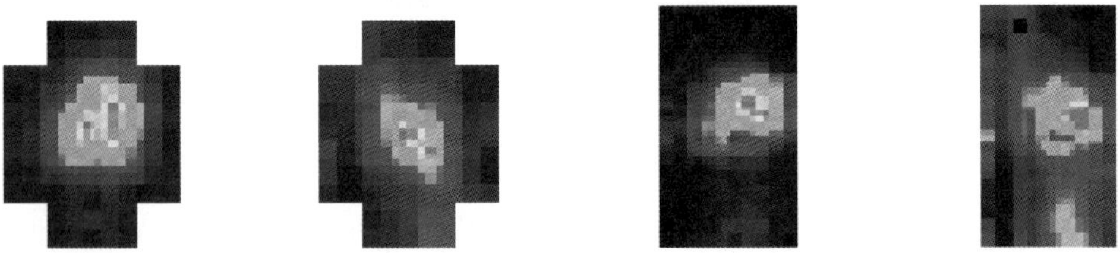

Figure 2. High J-CO from the BN/KL-object lines from left to right: $J = 16 \rightarrow 15$ at 163 μm, continuum at 163 μm, $J = 14 \rightarrow 13$ at 186 μm, continuum at 186 μm

With FIFI-LS, we also detected the high-J CO lines in the Becklin-Neugebauer/ Kleinmann-Low (BN/KL) object reported by Goicoechea et al. (2015a). Figure 2 shows the two lobes of the wide molecular outflow caused by an explosions about 500 years ago.

2. Outlook

The next step will be to calibrate the FIFI-LS data and cross calibrate the results with existing observations like Goicoechea et al. (2015b). After that a quantitative analysis can follow. We will look at the fine structure line intensities and compare them to PDR models like Pound & Wolfire (2008) to drive the physical conditions in the PDRs in the Orion Nebula and how the parental could is affected by the forming stars. Similarly, the study of the warm gas in BN/KL via the high-J CO lines will help to understand the effects of star formation on the surrounding molecular cloud. The large maps by FIFI-LS of this well studied region will demonstrate the mapping capabilities of FIFI-LS on board SOFIA.

References

De Buizer, J. M., Morris, M. R., Becklin, E. E., Zinnecker, H., Herter, T. L., Adams, J. D., Shuping, R. Y., & Vacca, W. D. 2012, ApJ, 749, L23

Goicoechea, J. R., Chavarría, L., Cernicharo, J., Neufeld, D. A., Vavrek, R., Bergin, E. A., Cuadrado, S., Encrenaz, P., Etxaluze, M., Melnick, G. J., & Polehampton, E. 2015a, ApJ, 799, 102

Goicoechea, J. R. *et al.* 2015b, ArXiv e-prints

Klein, R., Beckmann, S., Bryant, A., Colditz, S., Fischer, C., Fumi, F., Geis, N., Hönle, R., Krabbe, A., Looney, L., Poglitsch, A., Raab, W., Rebell, F., & Savage, M. 2014, in Society of Photo-Optical Instrumentation Engineers (SPIE) Conference Series, Vol. 9147, 2

Pound, M. W. & Wolfire, M. G. 2008, in ASP Conference Series, Vol. 394, Astronomical Data Analysis Software and Systems XVII, ed. R. W. Argyle, P. S. Bunclark, & J. R. Lewis, 654

Young, E. T. *et al.* 2012, ApJ, 749, L17

Formation, evolution, and survival of massive star clusters
Proceedings IAU Symposium No. 316, 2015
C. Charbonnel & A. Nota, eds.

doi:10.1017/S1743921315010844

Water Masers Outburst in the Massive Stellar Cluster W49A

Busaba H. Kramer[1,2], Karl M. Menten[1], Tomasz Kamiński[3], Bo Zhang[4], Nimesh A. Patel[5] and Alex Kraus[1]

[1]Max-Plank-Institut für Radioastronomie, Auf dem Hügel 69, D-53121 Bonn, Germany
email: bkramer@mpifr-bonn.mpg.de

[2]National Astronomical Research Institute of Thailand, Chiang Mai 50200, Thailand

[3]European Southern Observatory, Alonso de Cordova 3107 Vitacura, Santiago, Chile

[4]Shanghai Astronomical Observatory, Shanghai 200030, P. R. China

[5]Harvard-Smithsonian Center for Astrophysics, Cambridge, MA 02138, USA

Abstract. We report a multi-wavelength study of a recent major flare ($\sim$ 80,000 Jy at V_{LSR} $\sim$ -98.1 km s^{-1}) of the 22-GHz water maser in W49A. In February 2014, we started monthly monitoring with the Effelsberg 100-m radio telescope. In May 2014, we carried out the nearly simultaneous observations of the 22-GHz transition with selected submillimeter water transitions using the IRAM 30-m telescope (at 183 GHz) and the Atacama Pathfinder Experiment (APEX) 12-m telescope (from 321 to 475 GHz). We have also performed interferometric observations using the NRAO Very Long Baseline Array (VLBA) at 22 GHz and the Submillimeter Array (SMA) at 321 and 325 GHz. One remarkable result is the detection of very high velocity emission features in several transitions. Our data also represent its first detection of the 475-GHz water transition in a star-forming region. Studying these multiple masing transitions in conjunction with theoretical modeling of their excitation not only places strong constraints on the physical conditions of the masing gas but also allows us to study their association with the embedded massive stellar cluster in W49A.

Keywords. masers, stars: formation, ISM: molecules, radio lines: ISM

---

1. Results and Summary

Results from single dish observations are shown in Figure 1. Several high velocity features were present in several transitions. The 475 GHz water maser transition was detected for the first time in a star-forming region. Several features including the high velocity outburst features are coincide.

The global distribution of the masers spots from VLBA observations agrees well with the previous observations over 30 years ago. The north-south arc-like structure observed with VERA (Honma *et al.* 2004) during the major outburst in 2003 is also present. The 2014 outburst feature (V_{LSR} $\sim$ -98.1 kms^{-1}) is located near to the 2003 outburst feature within the north-south arc-like feature. The SMA map of the 321-GHz transition also shows that the peak of high velocity features (-98 to -95 km s^{-1}) is located in the south-east offset to the central low velocity features (0 to 22 km s^{-1}). Further results will be presented elsewhere (Hutawarakorn Kramer *et al.* in prep.)

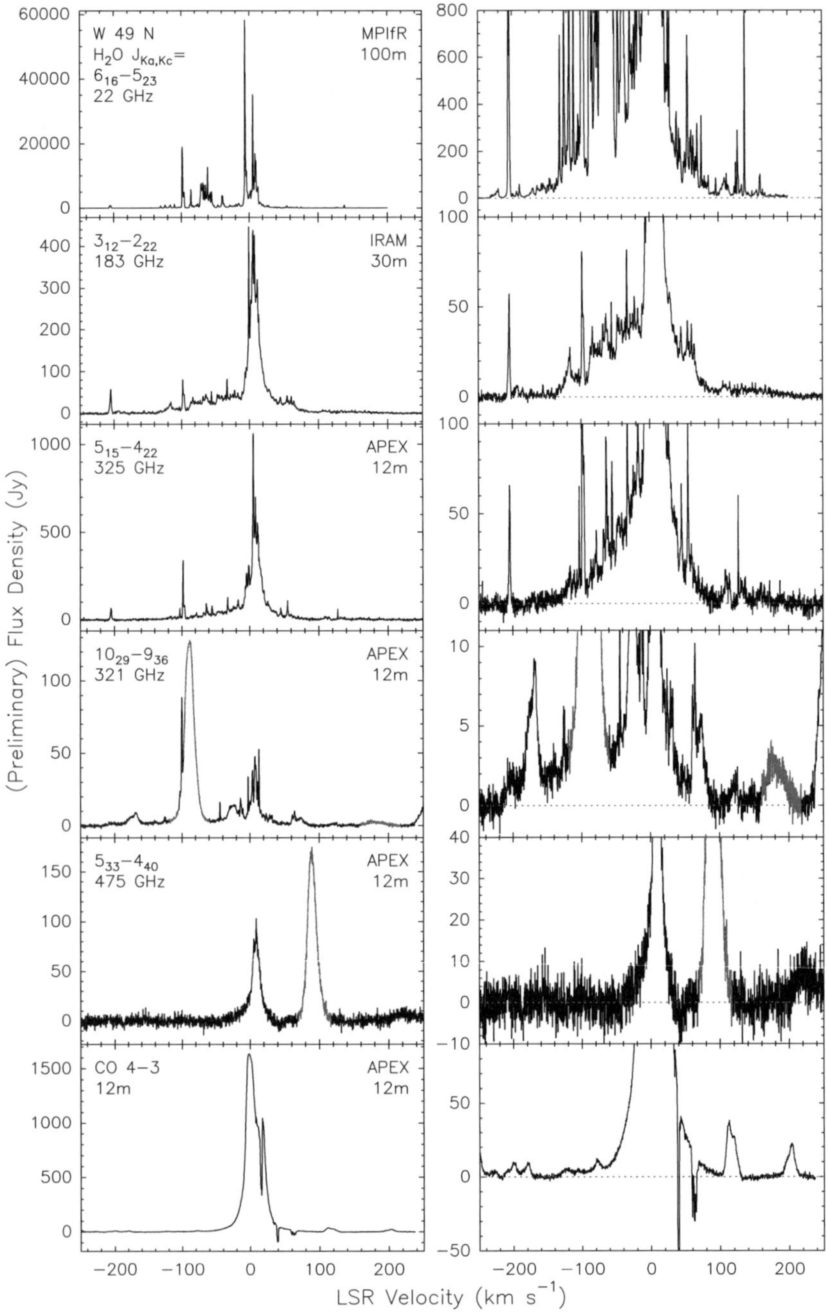

Figure 1. Water and CO *J=4-3* spectra observed toward in W49A during May 2014. Other known molecular lines are indicated in red color.

References

Honma, M., Choi Y. K. & Bushimata, T., *et al.* 2004, *PASJ*, 56, L15
Kramer, B. H., Menten, K. M., Kamiński, T.,*et al.* in prep.

Formation, evolution, and survival of massive star clusters
Proceedings IAU Symposium No. 316, 2015
C. Charbonnel & A. Nota, eds.

doi:10.1017/S1743921315010558

Is there a cluster in the massive star forming region IRAS 20126+4104?

V. A. Montes[1], **Peter Hofner**[1,2], **C. Anderson**[1,2], **and V. Rosero**[1,2]

[1]Physics Department, New Mexico Tech
801 Leroy Pl., Socorro, NM 87801, USA
email: vmontes@nmt.edu

[2]National Radio Astronomy Observatory,
1003 Lopezville Road, Socorro, NM 87801, USA

Abstract. A Chandra X-ray Observatory ACIS-I observation and a 6 cm continuum radio observation with the Karl G. Jansky Very Large Array (VLA) together with a multiwavelength study in infrared (2MASS and Spitzer) and optical (USNO-B1.0) shows an increasing surface density of X-ray sources toward the massive protostar. There are at least 43 YSOs within 1.2 pc distance from the massive protostar. This number is consistent with typical B-type stars clusters (Lada & Lada 2003).

Keywords. Stars: formation, X-ray: stars, Infrared: stars, radio continuum: stars

1. Introduction

Two main competitive theories intend to explain massive star formation: the turbulent core model, which is an extension of the low-mass star formation model (McKee & Tan 2003), and models involving competitive accretion or stellar collisions (Bonnell & Bate 2006). The characterization of the cluster in which massive stars form can help discriminate between the two main scenarios. Until recently it was believed that massive stars were only formed in substantial clusters. However, a recent study of the massive star forming region IRAS 20126+4104 using Spitzer observations by Qiu *et al.* (2008), suggested that the massive protostar was isolated, and the region was showing no obvious cluster. This is a well studied massive protostar located at a distance of 1.7 kpc. This early B-type star is currently forming through disk accretion (Cesaroni *et al.* 2014), has a luminosity of about 10^4 $L_\odot$, and is associated with a bipolar outflow (Cesaroni *et al.* 1997, Lebron *et al.* 2006). Here we adopt a multiwavelength approach to characterize the stellar environment of the IRAS 20126+4104 region combining Chandra X-ray ACIS-I and VLA 6 cm continuum observations with near-infrared (2MASS) data of the region. We detected 150 X-ray sources in the ACIS-I field and 13 radio sources within the $9'.2$ VLA primary beam. Associating X-ray sources with their near-infrared counterparts in the 2MASS catalog, and a color study of those counterparts, allow us to determine the foreground/background contamination, and to investigate the population of young stars in the IRAS 20126+4104 region.

2. Analysis: Contamination and source characteristics

The analysis of the region explicitly explained in Montes *et al.* (2015), was mainly done using the 2MASS counterparts of the X-ray sources. It was found that 88 X-ray sources have near-infrared counterparts and consequently 62 have no 2MASS counterparts. Plotting the positions of the latter sources shows a uniform distribution over the ACIS-I field contrasting with the positional plot of the X-ray sources having near-infrared counterparts which appear clustered around the protostar. The distribution of the sources without 2MASS counterparts is consistent with extragalactic contamination due to AGNs (20 ± 10 sources from Getman *et al.* (2006) simulation) and late-type stars which would be below the 2MASS sensitivity limit at this distance. From the 88 2MASS counterparts, 79 with good photometry quality were used to plot the H-K_s versus J-H color-color diagram showing that the majority of the sources appear to be located in the color space corresponding to WTTS, 3 are in the CTTS locus and one having a large infrared excess which is likely a class I protostar. The 79 counterparts were also used to plot the J versus J-H color-magnitude diagram and compared to the Besançon simulation of Galactic stellar populations (Robin *et al.* 2003) allowing to determine the location of expected foreground and background contamination in the J versus J-H color-magnitude diagram which results in 30 sources. A similar study with optical data using USNO-B1.0 catalog plotting the R versus R-I color-magnitude diagram allows to add 10 more sources to the contamination list of foreground and background sources. This study determines that approximatively 80 X-ray sources are associated with the massive protostar.

3. Discussion

In the previous study of the cluster by Qiu *et al.* (2008), an area of 5'x 5' was observed with the Spitzer Space Telescope and 19 YSOs were identified, leading to the conclusion that no obvious cluster is associated with the massive protostar. In our study, 23 X-ray sources were identified in the same area, with only 3 sources coincident with the sources found by Qiu *et al.* (2008). We also find 7 compact radio sources in this area, with 4 of them having neither X-ray nor Spitzer counterparts. Thus, combining the results of our study and that of Qiu *et al.* (2008), we find 43 YSOs within 1.2 pc radius from the massive protostar. This is likely a lower limit. Comparing it with the compilation of clusters in Lada & Lada (2003, Table 1), this number is consistent with typical embedded clusters about B-type stars. Another hint for the presence of a cluster is the increase of the surface density of X-ray sources toward the massive protostar (Montes *et al.* 2015). In conclusion, the massive protostar IRAS 20126+4104 is associated with about 80 X-ray sources, which are likely lower mass YSOs.

Acknowledgement

P. H. acknowledges support from NSF grant AST-0908901 and CXC award G03-4018A.

References

Bonnell, I. A. & Bate, M. R. 2006, *MNRAS*, 370, 488

Cesaroni, R., Felli, M., Testi, L., Walmsley, C. M., & Olmi, L. 1997, *A&A*, 325, 725

Cesaroni, R., Galli, D., Neri, R., & Walmsley, C. M. 2014, *A&A*, 566, A73

Getman, K. V., Feigelson, E. D., Townsley, L., Broos, P., Garmire, G., & Masahiro Tsujimoto, M. 2006, *ApJS*, 163, 306

Lada, C. J., Lada E. A. 2003, *ARA&A*, 41, 57

Lebron, M., Beuther, H., Schilke, P., & Stanke, Th. 2006, *A&A*, 448, 1037

Mckee, C. F. & Tan, J. C. 2003, *ApJ*, 585, 850

Montes, V. A., Hofner, P., Anderson, C., & Rosero, V. 2006, *ApJS*, 219, 41

Robin, A. C., Reyl, C., Derrire, S., & Picaud, S. 2003, *A&A*, 409, 523

Qiu, K., Zhang, Q., Megeath, T., Gutermuth, R. A., Beuther, H., Shepherd, D. S., Sridharan, T. K., Testi, L., & De Pree, C. G. 2008, *ApJ*, 685, 1005

Formation, evolution, and survival of massive star clusters
Proceedings IAU Symposium No. 316, 2015
C. Charbonnel & A. Nota, eds.

doi:10.1017/S1743921316008899

Westerlund 1: monolithic formation of a starburst cluster

Ignacio Negueruela[1], J. Simon Clark[2], Ben W. Ritchie[2] and Simon P. Goodwin[3]

[1]Departamento de Física, Ingeniería de Sistemas y Teoría de la Señal
Escuela Politécnica Superior, Universidad de Alicante
Carretera de San Vicente del Raspeig s/n, E03690, San Vicente del Raspeig, Alicante, Spain
email: ignacio.negueruela@ua.es

[2]Department of Physics and Astronomy, The Open University, Walton Hall
Milton Keynes MK7 6AA, United Kingdom

[3]Department of Physics and Astronomy, University of Sheffield, Sheffield, S3 7RH, UK

Abstract. Westerlund 1 is in all likelihood the most massive young cluster in the Milky Way, with a mass on the order of $10^5\ M_\odot$. To determine its bulk properties we have made multi-epoch radial velocity measurements for a substantial fraction of its OB stars and evolved supergiants and obtained multi-object spectroscopy of candidate cluster members in its locale. The results of these two studies show that Westerlund 1 is apparently subvirial and appears completely isolated, with hardly any massive star in its vicinity that could be associated with it in terms of distance modulus or radial velocity. The cluster halo does not extend much further than five parsec away from the centre. All these properties are very unusual among starburst clusters in the Local Universe, which tend to form in the context of large star-forming regions.

Keywords. open cluster and associations: Westerlund 1, stars: formation

---

Star-forming regions in the Galaxy and beyond demonstrate hierarchical structure in both time and space, whereby star clusters form in larger complexes within their natal giant molecular clouds. A prime example is the galactic G305 complex, which consists of a number of proto-clusters embedded in the periphery of a wind-blown bubble driven by the young massive clusters Danks 1 and 2 (Clark *et al.* 2004; Davies *et al.* 2012). A further, diffuse population of isolated massive stars is distributed across the complex, with their properties implying that star formation has occurred for several Ma and is still proceeding apace (delineated by multiple deeply-embedded proto-stars/clusters, maser emission and compact H II regions).

However, our data on Westerlund 1 (Wd1) appear to tell a different story. Observations of evolved OB stars imply a highly uniform cohort, evolving continuously from O9 III to B2.5 Ia (e.g. Negueruela *et al.* 2010), consistent with a simple homogeneous, co-eval 5Ma stellar population, also accommodating the simultaneous presence of red supergiants and Wolf-Rayets (Clark *et al.* 2005). The apparent instantaneous star-forming event suggested by the massive stars within Wd1 is also echoed by observations of the main sequence turn-on, implying an age-spread of < 0.4 Ma (Kudryavtseva *et al.* 2012).

So how did Wd1 form? To answer this question we adopted a three-pronged approach:

• A large multi-epoch and -object spectroscopic survey undertaken to identify binary and pulsating stars (see Ritchie *et al.* 2009 and Fig. 1) also allowed us to distinguish those cluster members with constant radial velocity and thus determine the mean radial velocity dispersion of the cluster. This reveals a velocity dispersion of $< 4.6\,\mathrm{km\,s^{-1}}$ (Clark *et al.* 2014 and in prep.), implying that Wd1 is sub-virial despite the fact that, given its

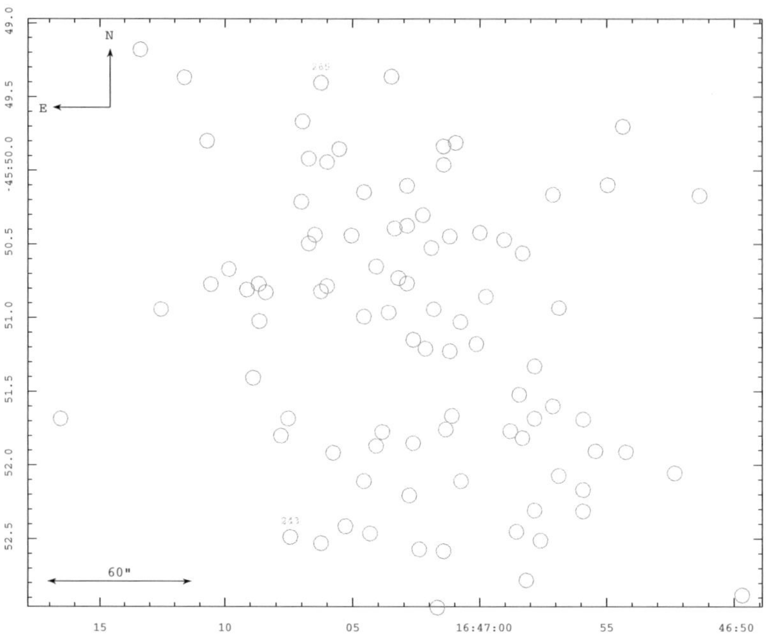

Figure 1. Finding chart for targets observed with FLAMES on the VLT for our radial velocity survey in the central regions of the cluster.

age and radial extent, one would anticipate that it should be dynamically relaxed and in virial equilibrium.

• Optical and near- to mid-IR wide-field imaging (e.g. *Spitzer* GLIMPSE/MIPSGAL) reveals no indication of satellite clusters, remnant giant molecular cloud nor ongoing star formation on spatial scales comparable to other star forming complexes such as G305.

• AAOmega multi-object spectroscopic observations were made of photometrically identified candidate early-type stars in the vicinity of Wd1; these revealed no evidence for a spatially extended "halo population".

Given these observations it appears that Wd1 formed in a single, compact and spatially isolated "starburst". Such a conclusion raises a number of important questions, such as (i) how can one accumulate so much mass in such a compact region?, (ii) why do we currently find Wd1 in a sub-virial state?, and (iii) why does it appear to have formed in a different manner compared to other massive stellar aggregates - does it simply represent an extreme example of a process which yields a continuum of cluster complexes/associations in terms of densities, spatial extents and age-spreads or is a different physical process responsible for its formation? In any event the sub-virial nature of Wd1 implies long-term survivability and so that it may represent the birth of a proto-globular cluster in the local Universe.

References

Clark, J. S. & Porter, J. M. 2014, *A&A*, 565, A90
Clark, J. S., Negueruela, I., Crowther, P. A., & Goodwin, S. P. 2005, *A&A*, 434, 949
Clark, J. S., Ritchie, B. W., Najarro, F., Langer, N., & Negueruela, I. 2014, *A&A*, 565, A90
Davies, B., Clark, J. S., Trombley, C., *et al.* 2012, *MNRAS*, 419, 1871
Kudryavtseva, N., Brandner, W., Gennaro, M., *et al.* 2012, *ApJL*, 750, L44
Negueruela, I., Clark, J. S., & Ritchie, B. W. 2010, *A&A*, 516, A78
Ritchie, B. W., Clark, J. S., Negueruela, I., & Crowther, P. A. 2009, *A&A*, 507, 1585

Formation, evolution, and survival of massive star clusters
Proceedings IAU Symposium No. 316, 2015
C. Charbonnel & A. Nota, eds.

doi:10.1017/S1743921316008711

Mid-infrared [NeII] Imaging of Young Massive Star Clusters Near Galactic Nuclei

Sherry C. C. Yeh[1], **Chao-Wei Tsai**[2], **Thomas R. Geballe**[3] **and Cinthya N. Herrera**[4]

[1]Subaru Telescope, National Astronomical Observatory of Japan
650 North A'ohoku Place, Hilo, Hawaii 96720, USA
email: yeh@naoj.org

[2]Jet Pulsation Laboratory
MS 169-327, 4800 Oak Grove Drive, Pasadena, CA 91109, USA
email: Chao-Wei.Tsai@jpl.nasa.gov

[3]Gemini Observatory, 670 North A'ohoku Place, Hilo, Hawaii 96720, USA
email: TGeballe@gemini.edu

[4]Institut de Radioastronomie Millimétrique
300 rue de la Piscine, Domaine Universitaire, 38406 Saint Martin d'Héres, France
email: herrera@iram.fr

Abstract. We investigate the formation of young massive clusters near the nuclei in NGC 6946, IC 342, Maffei II, and NGC 7714, using ground-based mid-infrared [NeII] imaging. We derive the cluster formation efficiency and cluster mass function, and the results suggest that environmental effects on YMC formation may not be significant.

Keywords. Star Clusters, Nearby Galaxies, Infrared

Young massive clusters (YMCs) are common (> 50%) near galactic nuclei of all Hubble types (Böker *et al.* 2002). Galactic nuclei are ideal test beds to examine the impact of extreme environments on YMC formation, because conventional star formation processes do not necessarily operate in such environments. *Is cluster formation efficiency universal?* Cluster formation efficiency (CFE) is the fraction of stars formed in bound clusters, it is also a key parameter in numerical simulations of star cluster assembly histories. Observationally, empirical CFEs were derived using optically-visible, relatively older clusters (> 10 Myr) (Bastian 2008, Adamo *et al.* 2011). These works suggest that CFE is universal ~10%. Theoretical work (Kruijssen 2012), however, suggests that CFE should depend on the host galaxy's environment, from 1% in low surface gas density ($\Sigma_{\rm gas}$) galaxies, to 70% in very high $\Sigma_{\rm gas}$ environments. *Does cluster mass assembly depend on the environment?* While observations suggest that cluster mass assembly is independent of galactic environments and cluster mass function is universal (Dowell *et al.* 2008), theories suggest otherwise, that high shear environments impede cluster formation (Weidner *et al.* 2010), thus cluster mass function should vary in different environments.

Extremely young YMCs are very obscured even in the near-infrared wavelengths (A_K > 3, Tsai *et al.* 2013). Extinction is almost absent at wavelengths > 10 μm, and mid-infrared observations on a 8-m telescope deliver diffraction limited, sub-arcsecond angular resolution, which is required to resolve the size of YMCs (a few pc) at a distance of few tens of Mpc. Such observations also simultaneously recover extended emission associated with unbound, field star formation, which allows us to derive CFE, while radio interferometric observations cannot recover extended emission. Our work is based on data taken in 2013 and 2014 using Cooled Mid-Infrared Camera and Spectrometer (COMICS) on Subaru Telescope. We mapped the nuclear regions of NGC 6946, IC 342, Maffei II, and NGC

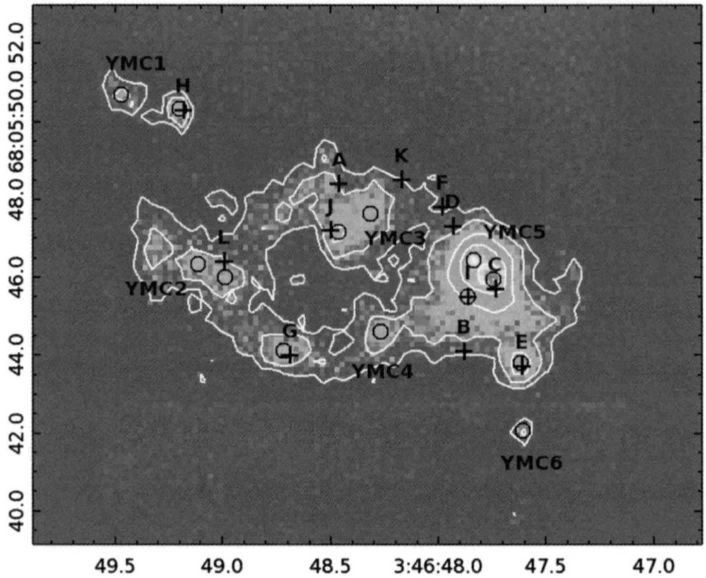

Figure 1. IC 342 [NeII] 12.8 μm image with 0.4" resolution. Lowest contours mark 3-σ detections. YMCs are marked in black circles, and their radio continuum counterparts are marked by crosses. The YMCs have no counterparts in the optical wavelengths.

7714, using the [NeII] 12.81 μm emission line, a direct tracer of massive star formation (Ho & Keto 2007) and much brighter than any hydrogen recombination lines. We detected 30 YMCs with estimated masses of 10^4 to 10^6 $M_\odot$, as well as extended [NeII] emission, which presumably is associated with unbound field star formation. Fig. 1 shows an example of the [NeII] images. *CFE Universality*: The derived CFEs in our sample galaxies are 12% to 22%, which disagree with theoretical predictions (Kruijssen 2012) of CFEs in extreme environments ($> 50\%$), and they crudely agree with the empirical value of 10% found in exposed star clusters (Bastian 2008) in other galactic environments. This may suggest a universal cluster formation efficiency independent of the natal galactic environment. *CMF vs. Galactic Environments*: The CMF derived from our YMC sample has a power law index of -2.2. Taken into account the uncertainties, this agrees the YMC HII region luminosity function index of -2.2 (Tsai *et al.* 2006), and the cluster mass function index of -2 derived from older clusters in other galactic environments (Chandar *et al.* 2010). This perhaps also suggests that the environmental effects of YMC mass assembly is insignificant.

References

Adamo, A., Östlin, G., & Zackrisson, E. 2011, *MNRAS*, 417, 1904
Bastian, N. 2008, *MNRAS*, 390, 759
Böker, T., Laine, S., van der Marel, R. P., Sarzi, M., Rix, H.-W., Ho, L. C., & Shields, J. C. 2002, *ApJ*, 123, 1389
Chandar, R., Fall, S. M., & Whitmore, B. C. 2010 *ApJ*, 711, 1263
Dowell, J. D., Buckalew, B. A., & Tan, J. C. 2008, *ApJ*, 135, 823
Ho, L. C. & Keto, E. 2007, *ApJ*, 658, 314
Kruijssen, J. M. D. 2012 *MNRAS*, 426, 3008
Tsai, C.-W., Turner, J. L., Beck, S. C., Crosthwaite, L. P., Ho, P. T. P., & Meier, D. S. 2006, *AJ*, 132, 2383
Tsai, C.-W., Turner, J. L., Beck, S. C., Meier, D. S., & Wright, S. A. 2013, *ApJ*, 776, 70
Weidner, C., Bonnell, I. A., & Zinnecker, H. 2010, *ApJ*, 724, 1503

Formation, evolution, and survival of massive star clusters
Proceedings IAU Symposium No. 316, 2015
C. Charbonnel & A. Nota, eds.

doi:10.1017/S1743921315010534

Identifying and Characterizing Stellar Clusters Toward the Galactic Anti-Center from Pan-STARRS1 Survey

C.-C. Lin[1,2], W.-P. Chen[2], J. Hou[1], L. Chen[1] and Z. Shao[1]

[1]Shanghai Astronomical Observatory, CAS,
80 Nandan Road, Shanghai 200030, China
email: cclin@shao.ac.cn

[2]Graduate Institute of Astronomy, National Central University,
300 Jhongda Road, Jhongli 32001, Taiwan

Abstract. The current sample of star clusters (SCs) is still incomplete beyond 1.8 kpc. The Pan-STARRS1 with its wide field of views and sensitive cameras provides us an opportunity to identify and characterize SCs as much as possible. By using a star counting algorithm, we obtained a total of 491 stellar density enhancements in a field of $20^\circ \times 20^\circ$ toward the Galactic Anti-Center (GAC), of which 50 are known SCs. The remaining 441 candidates are verified with radii, interstellar reddening, distances, and ages with proper motions and multiple bands photometries. Our results push the completeness of SCs almost near the edge of GAC. With the revised SC sample, the separation between Sagittarius and Perseus arms are about 3.2 ± 0.2 kpc and the widths of the nearby spiral arms—Sagittarius, Orion, and Perseus—with are 1.4 ± 0.1, 1.6 ± 0.1, and 3.3 ± 0.2 kpc, respectively.

Keywords. Galaxy: structure; open clusters and associations: general

1. Motivation and Pan-STARRS1

Some hundreds of thousand star clusters (SCs) are expected to exist on the Galactic plane based on the number of present SCs in the solar neighborhood (Piskunov *et al.* 2006). However, in the past few decades, the catalogs of SCs contain only a few thousand entries and mostly limited in 1.8 kpc (Kharchenko *et al.* 2013). The discrepancy is due partly to dust extinction in the Galactic plane, and partly to lack of comprehensive all-sky searches for distant systems.

Our goal is to conduct a comprehensive search for SCs from Pan-STARRS1 —the Panoramic Survey Telescope And Rapid Response System— which is equipped with a 1.8 m telescope, a 1.4 giga-pixel camera, and multiple optical bands to a depth of about 24 magnitude, about 100 times more wide and sensitive than currently available surveys (Kaiser *et al.* 2010). Here, we present the pilot study by applying a star-counting algorithm (Lin *et al.* 2011) to the Pan-STARRS1 data, which contains about 30 million objects in the Galactic Anti-Center (GAC) with a 400 deg^2 field of view.

2. Identification and Characterization of Star Cluster Candidates

By using a star-counting algorithm, 491 stellar density enhancements were identified toward the GAC with a 400 deg^2. Among them, 49 were matched with known open clusters and one was matched with a globular cluster, Palomar 2. A total of 109 known SCs were located toward the GAC. Except for larger radii, and embedded ones, the detection rate of SCs with radii smaller than $10'$ from the algorithm used was $50/60 \approx 83\%$.

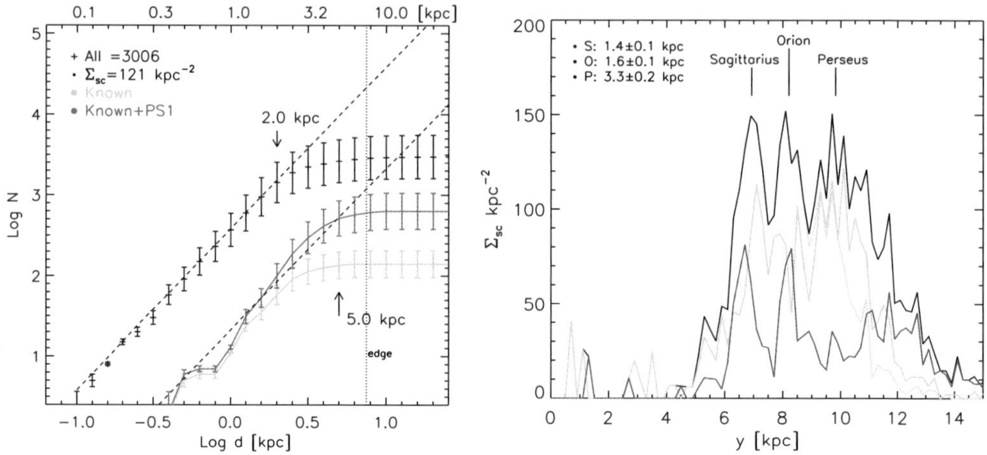

Figure 1. (Left) The cumulative number distribution as a function of the heliocentric distance of the PS1 SC candidates. The purple line shows cumulative number density distribution of 491 candidates combined with known SCs. Poisson errors are marked. (Right) The SCs surface density distribution from Galactic center to GAC. The thick grey line represents known SC sample from Kharchenko *et al.* (2013). The black line demonstrates the revised SC sample. The pink line shows older age of SCs than 100 Myr, while the blue line represents SCs younger than 100 Myr.

The remaining 441 candidates were characterized with radius, reddening, distance, age, and lowest mass members have been implemented on these candidate along with proper motions and multi-band photometry. Compared with known SCs, the resulting reddening values and distances are generally larger because of the optical filters are more sensitive to dust extinction. The age determination for star clusters with red giant members are more trustworthy than those without red giant members.

3. Discussion and Future Works

The revised SC sample is estimated to be complete up to a heliocentric distance of 5 kpc toward the GAC, see Figure 1. Moreover, this sample allows us to estimate the separation between Sagittarius and Perseus arms to be about 3.2 kpc with 0.2 kpc uncertainty and the widths of the nearby spiral arms—Sagittarius, Orion, and Perseus—with the full-width-half-maximum to be 1.4 ± 0.1, 1.6 ± 0.1, and 3.3 ± 0.2 kpc, respectively.

The candidates in the current search still need to be verified to be real SCs or not either by Pan-STARRS1 images inspection or by follow-up observations. The age, in particular, needs to be determined with not only the Padova model with different metallicity, but also the models providing pre-main sequence isochrones. The spectroscopic surveys, such as SDSSIV-APOGEE2 and LAMOST, will provide the radial velocity information for members in SCs so that the 3D motion of SCs in the Milky way can be estimated to trace how SCs interact with the spiral arms or giant molecular clouds.

References

Kaiser, N., Burgett, W., Chambers, K., *et al.* 2010, in SPIE, Vol. 7733
Kharchenko, N. V., Piskunov, A. E., Schilbach, E., *et al.* 2013, A&A, 558, A53
Lin, C. C., Chen, W. P., & Magnier, E. A. 2011, APRIM, V1, 181
Piskunov, A. E., Kharchenko, N. V., Röser, S., *et al.* 2006, A&A, 445, 545
Wang, P. F., Chen, W. P., Lin, C. C., *et al.* 2014, ApJ, 784, 57

Formation, evolution, and survival of massive star clusters
Proceedings IAU Symposium No. 316, 2015
C. Charbonnel & A. Nota, eds.

doi:10.1017/S1743921316008747

A starburst region around $l = 347° - 350°$

A. Marco[1], I. Negueruela[1], M. Monguió[1], C. González-Fernández[2], J. Maíz Apellániz[3], R. Dorda[1] and J. S. Clark[4]

[1]Departamento de Física, Ingeniería de Sistemas y Teoría de la Señal. Escuela Politécnica Superior. University of Alicante. Apdo.99 E-03080. Alicante, Spain
email: amparo.marco@ua.es

[2]Institute of Astronomy, University of Cambridge, Madingley Road, Cambridge CB3 0HA, United Kingdom

[3]Centro de Astrobiología, CSIC-INTA, campus ESAC, apartado postal 78, 28 691 Villanueva de la Cañada, Madrid, Spain

[4]Department of Physics and Astronomy, The Open University, Walton Hall, Milton Keynes, MK7 6AA, United Kingdom

Abstract. Very recently, a number of obscured massive open clusters have been identified in the Milky Way. A very significant fraction of them lie either close to the base of the Scutum Arm or towards Galactic longitude of 350°. We are studying these clusters and their neighbourhoods, finding very good evidence for a major starburst region close to the near tip of the Galactic Long Bar.

Keywords. open cluster and associations:VdBH 222 and Teutsch 85, supergiant, Galactic structure

1. Introduction

In the past few years, several clusters with masses around or above 10 000 $M_{\odot}$ have been discovered in the Milky Way. Clusters rich in red supergiants (RSGs) cover the age range 10 - 25 Ma, while younger clusters are mostly signposted by OB supergiants (SGs), which are much fainter in the infrared, and so, more difficult to detect through the Galactic plane extinction. The clusters rich in RSGs are concentrated towards $l = 26° - 28°$, including two large associations, one connected to Stephenson 2 and the other one to RSGC 3 (Negueruela *et al.* (2011), Negueruela *et al.* (2012) and González-Fernández & Negueruela (2012)). We are currently studying whether these two associations (with an estimated distance of $\sim$ 6 kpc) are separate or part of a single structure. Large-scale star formation there has been confirmed by the parallax distance of 5.5 kpc to the giant star-forming complex W43 (Zhang *et al.* (2014)).

Towards the other tip of the Long Bar, distances are much more difficult to measure. We have found a gap in the extinction towards $l = 350°$, where we find two massive clusters, VdBH 222 and Teutsch 85, in an association containing $\sim$ 50 RSGs. Although our knowledge of Galactic structure seemed to favour a location close to the far end of the Bar, our ongoing spectroscopic studies rule out such a long distance. Even so, these clusters are excellent laboratories for stellar astrophysics, providing a unique opportunity to check the relative fraction of RSGs to blue SGs (BSGs). On average, BSGs outnumber RSGs by a factor $\sim$ 3 in Galactic open clusters (Eggenberger *et al.* (2002)).

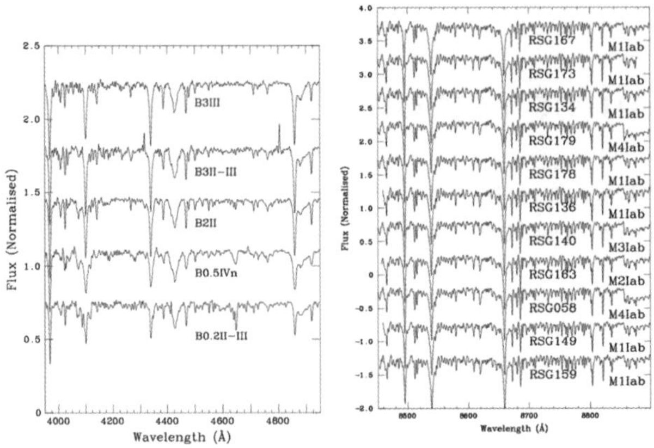

Figure 1. Left. Spectra of the brightest blue members of VdBH 222 and Right. Spectra of red supergiants in and around the open cluster Teutsch 85

2. Results

2.1. *The open cluster VdBH 222*

VdBH 222 (l = 349.13°; b = −00.44°) is a young massive cluster with a likely mass greater than 10 000 $M_\odot$. Its population of massive evolved stars is comparable to that of large associations, such as Per OB1. Marco *et al.* (2014) identified a large population of luminous cool SGs confirmed as cluster members via radial velocity measurements. They found nine RSGs (plus one other candidate) and two yellow SGs and a large population of OB stars. Ten of them are bright enough to be BSGs. The cluster lies behind ∼ 7.5 mag of extinction for the preferred value of $R_V = 2.9$. Isochrone fitting allows a narrow range of ages between 12 and 18 Ma. The cluster radial velocity is compatible with distances of ∼ 6 and ∼ 10 kpc. To characterize with more precision the population of OB stars and limit the value of R_V, we took spectra of the stars in the upper main sequence with FORS2 at the UT1 in the Paranal Observatory in June 2015. These new data place the open cluster at ∼ 6 kpc with an age around 18–20 Ma. We found no BSGs (see Fig. 1), against a complement of 2 yellow and 10 red SGs. This result is at odds with the statistics mentioned above.

2.2. *The open cluster Teutsch 85*

The second open cluster in this starburst region is Teutsch 85. It is located 1.8° (190 pc at 6 kpc) away from VdBH 222. We have identified ∼ 14 RSGs with similar radial velocities within 7′ of the cluster center an important number of RSGs with approximately the same radial velocities in the region connecting the two clusters (see Fig. 1). We are analyzing deep $UBVR$ photometry and FORS2 spectroscopy of this cluster. Based on a preliminary analysis it seems to have an age of 25 − 30 Ma, but only one or two BSGs against 14 RSGs. This result is again at odds with the statistics for the Milky Way.

References

Marco, A., Negueruela, I., González-Fernández, C., *et al.* 2014 *A&A*, 567, A73
Eggenberger, P., Meynet, G., & Maeder, A. 2002, *A&A*, 386, 576
Zhang, B., Moscadelli, L., & Sato, M., Reid M. J., *et al.* 2014, *ApJ*, 781, A89
Negueruela, I., González-Fernández, C.; Marco, A., & Clark, J. S. 2011, *A&A*, 528, A59
Negueruela, I., Marco, A., González-Fernández, C., *et al.* 2012 *A&A*, 547, A15
González-Fernández, C. & Negueruela, I. 2012 *A&A*, 539, A100

Formation, evolution, and survival of massive star clusters
Proceedings IAU Symposium No. 316, 2015
C. Charbonnel & A. Nota, eds.

doi:10.1017/S1743921315008947

Evolved massive stars in W33 and in GMC G23.3−0.3

M. Messineo[1], J. S. Clark[2], D. F. Figer[3], K. M. Menten[1], R.-P. Kudritzki[4], F. Najarro[5], R. M. Rich[6], V. D. Ivanov[7], E. Valenti[7], C. Trombley[3], C.-H. R. Chen[1], B. Davies[8] and J. W. MacKenty[9]

[1]Max-Planck-Institut für Radioastronomie, Auf dem Hügel 69, D-53121 Bonn, Germany,
email: messineo@mpifr-bonn.mpg.de

[2]Department of Physics and Astronomy, The Open University, Walton Hall, Milton Keynes, MK7 6AA, UK

[3]Center for Detectors, Rochester Institute of Technology, 54 Memorial Drive, Rochester, NY 14623, USA

[4]Institute for Astronomy, University of Hawaii, 2680 Woodlawn Drive, Honolulu, HI 96822, USA

[5]Centro de Astrobiologia (CSIC-INTA), Ctra. de Torrejon a Ajalvir km4, E-28850, Torrejonde Ardoz, Madrid, Spain

[6]Physics and Astronomy Building, 430 Portola Plaza, Box 951547, Department of Physics and Astronomy, University of California, Los Angeles, CA 90095-1547, USA

[7]European Southern Observatory, Karl-Schwarzschild-Strasse 2, D-85748 Garching bei Munchen, Germany

[8]Astrophysics Research Institute, Liverpool John Moores University, Twelve Quays House, Egerton Wharf, Birkenhead, Wirral. CH41 1LD, UK

[9]Space Telescope Science Institute, 3700 San Martin Drive, Baltimore, MD, 21218, USA.

Abstract. We conducted infrared spectroscopic observations of bright stars in the direction of the molecular clouds W33 and GMC G23.3−0.3. We compared stellar spectro-photometric distances with parallactic distances to these regions, and we were able to assess the association of the detected massive stars with these molecular complexes. The spatial and temporal distributions of the detected stars enabled us to locate sources of ionizing radiation and to gather precise information on the star formation history of these clouds. The studied clouds present different distributions of massive stars.

Keywords. infrared: stars stars: evolution - supergiants open clusters and associations: general ISM: clouds - ISM: supernova remnants

1. Introduction

Young massive stars and stellar clusters continuously form in the Galactic disk, generate bubbles and HII regions in molecular clouds, and subsequently enrich the interstellar medium via winds and supernovae. We conducted an infrared survey for bright stars in the Galactic giant molecular clouds (GMCs) G23.3−0.3 and W33. A large number of extraordinary sub-clumps/clusters of massive stars were detected. The spatial and temporal distribution of these massive stars yielded information on the star formation history of these clouds.

2. GMC G23.3−0.3

The GMC G23.3−0.3, shown in Fig. 1, is located at the parallactic distance of 4.6 kpc (Brunthaler *et al.* 2009), and is remarkably rich in HII regions and supernova remnants

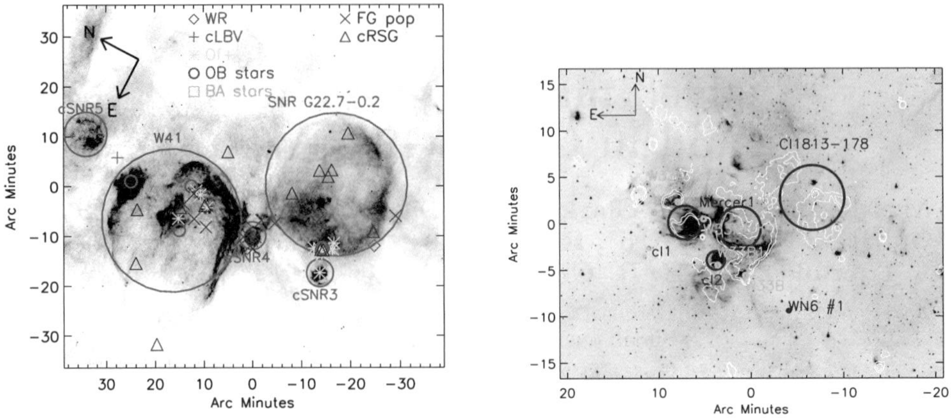

Figure 1. *Left:* 20 cm emission from MAGPIS of GMC G23.3−0.3 centred at RA 18°33′50.″96 and Dec. −08°55′34.″9. Detected massive stars are indicated (e.g., RSGs with triangles, early-type stars with diamonds, plus, and circles). SNRs are indicated with red circles. *Right:* GLIMPSE 8 um image of W33, centred at RA 18°13′51.″88 and Dec. −17°56′17.″5, with contours of 90 cm emission from MAGPIS. Prestellar cores are in green (Immer *et al.* 2014), while candidate stellar clusters are in red.

(SNRs) (Messineo *et al.* 2010, 2014). We discovered a dozen massive O-type stars, one candidate luminous blue variable, and several red supergiants (RSGs). The O-type stars have masses from 25 to 50 $M_{\odot}$ and ages of a few million years, while the RSGs belong to a burst that occurred 20-30 Myr ago. GMC G23.3−0.3 has had one of the longest known history of star formation. We detected massive stars in the cores of SNR W41 and SNR G22.7−0.2 (Fig. 1).

3. W33

The W33 complex (Fig. 1) is located on the Galactic plane at about 12.°9 of longitude and at the parallactic distance of 2.4 kpc (Immer *et al.* 2013). We spectroscopically detected a few evolved O-type stars and one Wolf-Rayet star, but none of the late-type objects has the luminosity of a RSG star. Several dense molecular cores that may harbor proto clusters were recently discovered on the east side of the cloud (Immer *et al.* 2014). We were able to identify episods of sequential star formation: evolved massive stars/clusters that are triggering these new condensations (Messineo *et al.* 2015). Therefore, W33 is characterized by discrete sources and has had at least 3-5 Myr of star formation, which is now propagating from west to east. The massive stellar cluster Cl1813−178 (Messineo *et al.* 2011) is located behind the W33 complex at a distance of about 4.8 kpc.

References

Brunthaler, A., Reid, M. J., Menten, K. M., Zheng, X. W., Moscadelli, L., & Xu, Y. 2009, *ApJ*, 693, 424

Immer, K., Reid, M. J., Menten, K. M., Brunthaler, A., & Dame, T. M. 2013, *A&A*, 553A, 117

Immer, K., Galván-Madrid, R., König, C., Liu, H. B., & Menten, K. M. 2014, *A&A*, 572A, 63

Messineo, M., Figer, D. F., Davies, B., Kudritzki, R. P., Rich, R. M., MacKenty, J., & Trombley, C. 2010, *ApJ*, 708, 1241

Messineo, M., Menten, K. M., Figer, D. F., Davies, B., Clark, J. S., Ivanov, V. D., Kudritzki, R.-P., Rich, R. M., MacKenty, J. W., & Trombley, C. 2014, *A&A*, 569A, 20

Messineo, M., Davies, B., Figer, D. F., Kudritzki, R. P., Valenti, E., Trombley, C., Najarro, F., & Rich, R. M. 2011, *ApJ*, 733, 41

Messineo, M., Clark J. S., Figer, D. F., Kudritzki, R.-P., Najarro, F., Rich, R. M., Menten K. M., Ivanov, V. D., Valenti, E., Trombley, C., Chen C.-H. R., & Davies, B. 2015, *ApJ*, 805, 110

Formation, evolution, and survival of massive star clusters
Proceedings IAU Symposium No. 316, 2015
C. Charbonnel & A. Nota, eds.

doi:10.1017/S1743921315006882

e-MERLIN Radio Continuum Measurements of OB Star Winds in CYG OB2

Jack Morford[1], **Raman Prinja**[1] **and Danielle Fenech**[1]

[1]Dept. of Physics & Astronomy, University College London
Gower St, London, WC1E 6BT, U.K.
email: jmorford@star.ucl.ac.uk

Abstract. Here, we report on the first results from the e-MERLIN Cyg OB2 Radio Survey (COBRaS), which is designed to exploit e-MERLIN's enhanced capabilities to conduct deep-field mapping of the tremendously rich Cyg OB2 association. The project aims to deliver the most detailed radio census of the most massive OB association in the northern hemisphere. There exists considerable evidence for clumping in the winds of hot stars, which has hugely important consequences for mass-loss determinations. The amount of mass lost from a massive star is a crucial parameter required for stellar and cluster evolution models that are paramount to our understanding of the formation and evolution of massive star clusters. Presenting some of the first 20cm (L band) detections of massive stars in Cyg OB2, both accurate mass-loss determinations and constraints upon clumping within their winds are made. These data substantially increase the observational detections of the outer wind of massive stars and in combination with other observations at different wavelengths, COBRaS will greatly advance our knowledge of clumping as a function of radial distance around massive stars.

Keywords. stars: early-type - stars: individual: CYG#12, CYG#7 & A15 - stars: winds - Galaxy: open clusters and associations: individual: Cyg OB2 - Radio: stars.

1. COBRaS: The Cyg OB2 Radio Survey

COBRaS is an e-MERLIN legacy project awarded 42 hours at L-band (1.5GHz) and 252 hours at C-band (5GHz). As an intensive radio survey, the project aims to deliver targeted, high-resolution, deep field maps covering the core of the OB rich stellar association. These observations will push the limits of the instruments capabilities in detecting sub milli-Jansky free-free radio emission from the winds of hot stars. Two leading scientific goals dominate the project. Firstly to detect the thermal free-free emission from the outer wind regions of single massive stars allowing for accurate determinations of their mass-loss rate, $\dot{M}$ from which constraints can be made upon on their small-scale wind structure. Secondly, detections of the non-thermal emission within the region will allow an investigation into T Tauri objects, Young Stellar Objects (YSOs) and the massive star binary population, providing a better estimate of the region's binary fraction.

The Cyg OB2 association is a young 'super-star' cluster ($\sim$ 2-3Myr old) located in the heart of the Cygnus X region and hosting approximately 2600$\pm$400 OB stars (Knödlseder *et al.* 2000). Due to it's 'nearby' galactic location ($\sim$ 1.4-1.7kpc away) and the fact that it suffers from a large visual extinction, $3^m \leqslant A_\nu \leqslant 8^m$ (Guarcello *et al.* 2013), make it ideally suited to studies within the radio.

2. Mass-loss from massive stars

Massive stars provide a pivotal feedback mechanism in re-cycling energy and material back into the ISM via their stellar winds. This feedback arguably provides the raw

Table 1. Positions, spectral types and predicted and observed 20cm fluxes and $\dot{M}$ for 3 stars within the COBRaS L-band pointing's from ~ 35-40 minutes on source.

Object Name	RA (J2000)	DEC (J2000)	Spectral Type	Flux 20cm (μJy) Predicted	Flux 20cm (μJy) Observed	$\dot{M}$(x10^{-6}M$_\odot$yr^{-1}) Predicted	$\dot{M}$(x10^{-6}M$_\odot$yr^{-1}) Observed
Cyg #7	20 33 14.2	41 20 21.5	O3If	20.5	<150	1.745	<7.8
A15	20 31 36.9	40 59 09.1	O7I	16.7	<138	0.915	<4.8
Cyg #12	20 32 40.9	41 14 29.2	B3.5Ia+	1770	341	10	2.9

materials to trigger star formation and undoubtedly plays a vital role in the evolution of stellar clusters. The mass-loss parameter itself is a key determinant of a stars evolution and lifetime, and a very important parameter within stellar, cluster and galactic evolution models. Different mass-loss diagnostics have been found to be discordant with one another and Puls *et al.* 2006 suggests that previously found mass-loss rates have been over-estimated and can be attributed to the effect of wind clumping. This uncertainty in current stellar mass-loss estimates is in urgent need of clarification.

Clumping within a stellar wind is considered as 'small-scale structure' and can be described as small-scale clumps of over-dense material throughout the wind. Originating from density in-homogeneities within the inner wind that re-distribute material into enhanced density clumps, it is thought that the thermal free-free emission detected from these clumps will cause an over-estimate of the stars mass-loss rate. As a ρ^2 diagnostic, the radio thermal continuum emission is highly sensitive to any clumping and for a star of known $\dot{M}$, constraints can be put on the clumping factor, $f_{cl} = <\rho_{cl}^2>/<\rho_{sw}>^2$.

3. Pre-liminary COBRaS L-band results

The L-band (42 hours) observations were completed by April 2014 and currently ~ 1/5 of the dataset has been completely reduced. The preliminary results shown below include the three stars: Cyg #7, A15 and Cyg #12 and table 1 displays information regarding their spectral types, predicted and detected 20cm fluxes and $\dot{M}$.

Unfortunately we received a null detection for both Cyg #7 and A15 and thus only have an upper limit on their 20cm flux and $\dot{M}$ value. Cyg #12 is detected with a 20cm flux of 341μJy, a significantly lower value than previously measured by Clark *et al.* 2012 who published a value of $F_{20cm} = 3270\mu$Jy. The calculated $\dot{M}$'s shown in table 1 have all been calculated through using a clumping factor $f_{cl} = 1$, however using previous $\dot{M}$ or F_{20cm} values taken from the literature we obtain a range of possible mass-loss rates for Cyg #12 giving the constraint 25 $> f_{cl} <$1. This is a loose constraint since the large variability in observed F_{20cm} suggests eruptive behaviour, it is possible that the object is in transition between the Blue Hyper-giant (BHG) and Luminous Blue Variable (LBV) evolutionary phases. It is important to stress that once the L-band dataset has been reduced in it's entirety, our sensitivity level will increase by a factor of ~ 3-4 which will significantly increase the number of single star wind detections at 20cm.

References

Knödlseder, J. 2000, *A&A*, 360, 539

Guarcello, M. G., Drake, J. J., Wright, N. J., Drew, J. E., Gutermuth, R. A., Hora, J. L., Naylor, T., Aldcroft, T., Fruscione, A., García-Alvarez, D., Kashyap, V. L., & King, R. 2013, *ApJ*, 773, 135

Puls, J., Markova, N., & Scuderi, S. 2006, *ArXiv e-Prints*

Clark, J. S., Najarro, F., Negueruela, I., Ritchie, B. W., Urbaneja, M. A., Howarth, I. D, 2012 *A&A*, 541, 145

Formation, evolution, and survival of massive star clusters
Proceedings IAU Symposium No. 316, 2015
C. Charbonnel & A. Nota, eds.

doi:10.1017/S1743921315010923

Environmental effects on stellar populations of star clusters and dwarf galaxies

Stefano Pasetto[1], Mark Cropper[1], Yutaka Fujita[2], Cesare Chiosi[3] and Eva K. Grebel[4]

[1]Mullard Space Science Laboratory, University College London, Holmbury St Mary, Dorking, Surrey RH5 6NT, United Kingdom
email: s.pasetto@ucl.ac.uk

[2]Department of Earth and Space Science, Graduate School of Science, Osaka University,1-1 Machikaneyama-cho, Toyonaka-shi, Osaka 560-0043, Japan

[3]Dept. of Physics & Astronomy "Galileo Galilei", University of Padua, Vicolo dell'Osservatorio, 5, 35141 Padova PD, Italy

[4]Astronomisches Rechen-Institut, Zentrum für Astronomie der Universität Heidelberg, Mönchhofstr 12-14, 69120, Heidelberg, Germany

Abstract. We investigate the competitive role of the different dissipative phenomena acting on the onset of star formation of gravitationally bound systems in an external environment. Ram pressure, Kelvin-Helmholtz and Rayleigh-Taylor instabilities, and tidal forces are accounted for separately in an analytical framework and compared in their role in influencing the star forming regions. We present an analytical criterion to elucidate the dependence of star formation in a spherical stellar system on its surrounding environment. We consider the different signatures of these phenomena in synthetically realized colour-magnitude diagrams (CMDs) of the orbiting system thus investigating the detectability limits of these different effects for future observational projects and their relevance. The developed theoretical framework has direct applications to the cases of massive star clusters, dwarf galaxies in galaxy clusters and dwarf galaxies orbiting our Milky Way system, as well as any primordial gas-rich cluster of stars orbiting within its host galaxy.

Keywords. Star cluster, gas instabilities, star formation, colour magnitude diagrams

1. Non-inertial linear response theory for gas instabilities in spherical coordinates.

We model a time-dependent gravitationally bound system of stars and both molecular and HI gas as a spherical system interacting with the external environment (Pasetto *et al.* 2015). We derived an analytical criterion that depending on the phase space condition $\{\mathbf{x}, \mathbf{v}\}$ of the system orbiting a more massive companion allows the distinguishing of different contributions affecting the star formation $\hat{\gamma}^2(\mathbf{x}, \mathbf{v}; t) > 0$ or its absence $\hat{\gamma}^2(\mathbf{x}, \mathbf{v}; t) < 0$ on colour magnitude diagrams synthetically generated (Fig.1 upper panel). The individual component contributions, Rayleigh-Taylor type $\hat{\gamma}^2_{RT}$, mixed instability term type $\hat{\gamma}^2_{mix}$, not-inertial acceleration of Rayleigh-Taylor type $\hat{\gamma}^2_{a-RT}$, the Kelvin-Helmholtz type term $\hat{\gamma}^2_{KH}$, the inertia term type $\hat{\gamma}^2_{I}$, carry all different dependence on $\{\mathbf{x}, \mathbf{v}\}$ and are worked out analytically (See Fig.1 lower panel).

References

Pasetto S., Cropper M., Fujita Y., Chiosi C., & Grebel E. K., 2015, A&A, 573, A48
Pasetto S., Bertelli G., Grebel E. K., Chiosi C., & Fujita Y., 2012, A&A, 542, A17

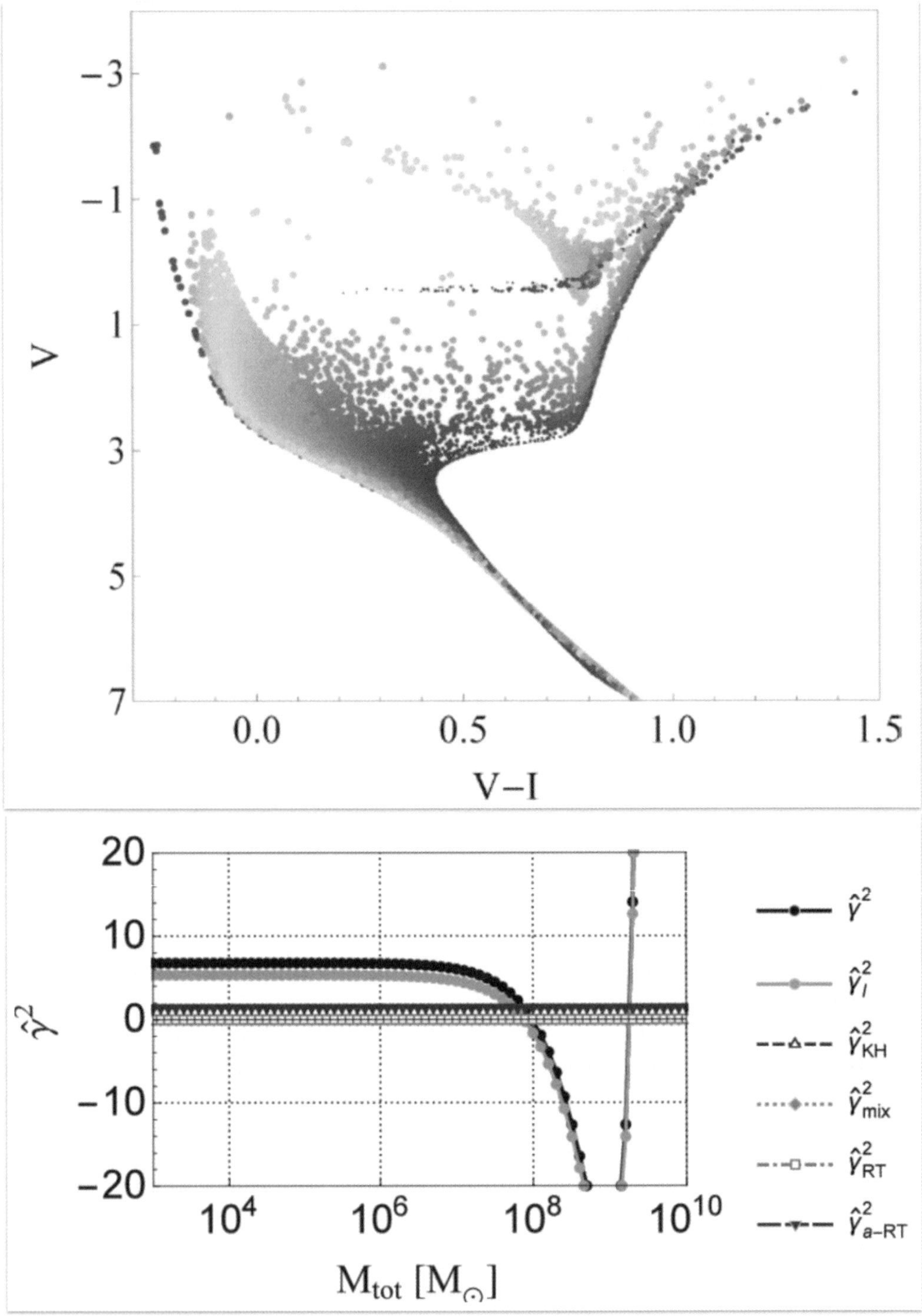

Figure 1. Upper panel: synthetic CMD realized for the simulation of the upper panel (see Pasetto *et al.* 2012 for further details). Lower panel: Instability criteria plotted as function of total mass, gas+stars, of an orbiting object for assigned location within the Milky Way external potential (see Pasetto *et al.* 2015 for further details).

Formation, evolution, and survival of massive star clusters
Proceedings IAU Symposium No. 316, 2015
C. Charbonnel & A. Nota, eds.

doi:10.1017/S1743921316000016

Formation of the twin galactic starburst regions NGC 6357 and NGC 6334

Kazufumi Torii[1], Keisuke, Hasegawa[1], Akio Ohama[1], Hiroaki Yamamoto[1], Kengo Tachihara[1] and Yasuo Fukui[1]

[1]Department of Physics, Nagoya University, Furo-cho, Chikusa-ku, Nagoya, Aichi 464-8601, Japan. email: `torii@a.phys.nagoya-u.ac.jp`

Abstract. New CO J=1–0 observations with NANTEN and NANTEN2 reveal that extensive collisions between two molecular clouds at relative velocity of 15 km s^{-1} triggered the O star formation in the Galactic mini-starbursts NGC 6357 and NGC 6334. Correlated/anti-correlated gas distributions and intermediate velocity features between the two clouds lend support for the cloud-cloud collision scenario. The timescale of the collision and high-mass star formation is as short as less than 0.5 Myrs, suggesting rapid O star formation.

Keywords. ISM: clouds — ISM: molecules — ISM: kinematics and dynamics — stars: formation

---

High-mass star formation is one of the top-priority issues in astrophysics. We have now growing evidence for O star formation triggered by supersonically colliding clouds in massive star clusters and young single O stars (e.g., Furukawa *et al.* 2009; Ohama *et al.* 2010; Torii *et al.* 2011, 2015; Fukui *et al.* 2014, 2015a,b). This suggests collisions play a substantial role in O star formation.

NGC 6357 and NGC 6334 are outstanding twin young HII regions with O stars extensively distributed across ~100 pc at distance 1700 pc (Figure 1a). These large HII regions are rich in molecular gas and are thus reasonable targets to study the mechanism of O star formation. Most recently, based on new NANTEN and NANTEN2 CO J=1–0 observations, we discovered in NGC 6357 and NGC 6334 two molecular clouds are colliding with each other at a relative velocity of ~15 km s^{-1} to trigger O star formation.

The results are summarized in Figure 1. The red-shifted cloud (hereafter the red cloud) has strong CO emission and are elongated along the Galactic plane over 100 pc, showing a good coincidence with NGC 6357 and NGC 6334 and their O-star clusters depicted by crosses (Figure 1b). On the other hand, the blue-shifted cloud (hereafter the blue cloud) is weak in CO, but shows an elongated distribution similar to the red cloud (Figure 1c). The blue cloud toward NGC 6357 shows a good coincidence with the red cloud above 0.8° and nested distributions with the red cloud below 0.8° (black dashed lines). The blue cloud in NGC 6334 also shows a good correlation with the red cloud especially toward the clusters (white dashed lines). These correlated and anti-correlated distributions between the two velocity clouds are usually seen in cloud-cloud collision as predicted in numerical studies (Habe & Ohta 1992; Takahira *et al.* 2014). In addition, in the longitude-velocity diagram in Figure 1d the two clouds are connected by intermediate velocity bridging features, which are similar to that seen in RCW 38 (Fukui *et al.* 2015b). Synthesis observations by Haworth *et al.* (2015) indicate that these "bridging features" are unique characteristics to cloud-cloud collision.

The collisions in NGC 6357 and NGC 6334 are occurring between the two filamentary clouds across ~100 pc, along the Galactic arms. The timescale of the collision can be estimated as a crossing time to be ~0.5 My. The results demonstrate that cloud-cloud collision has a potential to cause extensive O star formation on a Galactic scale. Further

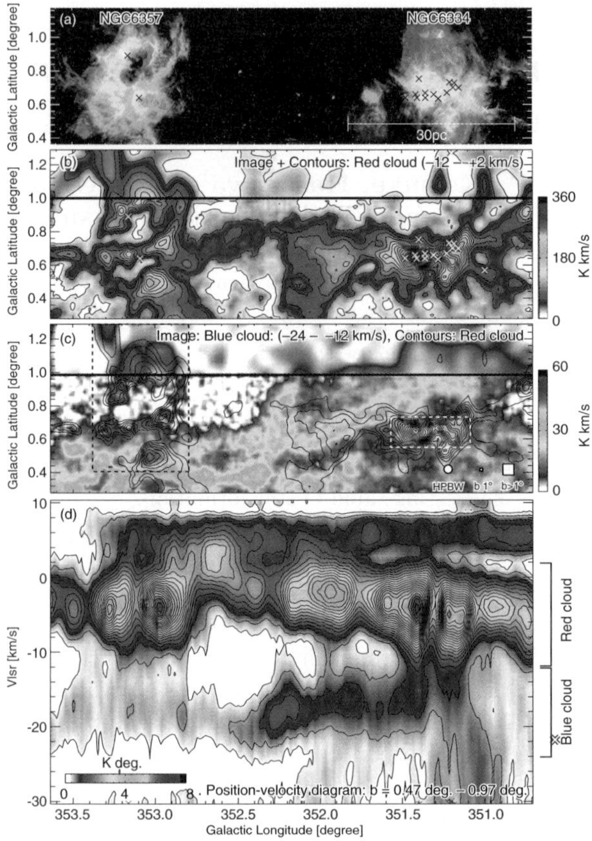

Figure 1. (a) A composite image of NGC 6357 and NGC 6334, where red and green indicate Hα (Parker *et al.* 2005) and Spitzer 8 μm (Benjamin *et al.* 2003), respectively. (b) The red cloud toward the same region in (a). (c) The contour map of the red cloud is superimposed on the image of the blue cloud. The dashes lines indicate the region where correlations of the gas distributions between the two clouds are seen. (d) Longitude-velocity diagram of the two clouds. In panels (a)–(c) crosses indicate the clusters.

molecular studies which cover several 100-pc scales along the arms will be extremely important in order to better understand the role of cloud-cloud collisions in galactic activity.

References

Benjamin, R. A. *et al.* 2013, *PASP*, 115, 953

Fukui, Y. *et al.* 2014, *ApJ*, 780, 36

Fukui, Y. *et al.* 2015a, *ApJ*, 807, L4

Fukui, Y. *et al.* 2015b, ArXiv e-prints, arXiv:1504.05391

Furukawa, N. *et al.* 2009, *ApJ*, 696, L115

Habe, A. & Ohta, K. 1992, *PASJ*, 44, 203

Haworth, T. J. *et al.* 2015, *MNRAS*, 450, 10

Parker, Q. A. *et al.* 2005, *MNRAS*, 362, 689

Takahira, K. *et al.* 2014, *ApJ*, 792, 63

Torii, K. *et al.* 2011, *ApJ*, 738, 46

Torii, K. *et al.* 2015, *ApJ*, 806, 7

Formation, evolution, and survival of massive star clusters
Proceedings IAU Symposium No. 316, 2015
C. Charbonnel & A. Nota, eds.

doi:10.1017/S1743921315006997

N131: A dust bubble was born from the disruption of a gas filament?

Chuan-Peng Zhang

National Astronomical Observatories, CAS, 100012 Beijing, China
email: cpzhang@nao.cas.cn

Abstract. N131 is an infrared dust bubble residing in a molecular filament. We aim to study the formation and fragmentation of this bubble with multi-wavelength dust and gas observations. Towards the bubble N131, we analyzed archival multi-wavelength observations including 3.6, 4.5, 5.8, 8.0, 24, 70, 160, 250, 350, 500 μm, 1.1 mm, and 21 cm. In addition, we performed new observations of CO (2-1), CO (1-0), and ^{13}CO (1-0) with the IRAM 30-m telescope. Multi-wavelength dust and gas observations reveal a ringlike shell with compact fragments, two filamentary structures, and a secondary bubble N131-A. The bubble N131 is a rare object with a large hole at 24 μm and 21 cm in the direction of its center. The dust and gas clumps are compact and might have been compressed at the inner edge of the ringlike shell, while they are extended and might be pre-existing at the outer edge. The column density, excitation temperature, and velocity show a potentially hierarchical distribution from the inner to outer edge of the ringlike shell. We also detected the front and back sides of the secondary bubble N131-A in the direction of its center. The derived Lyman-continuum ionizing photon flux within N131-A is equivalent to an O9.5 star. Based on the above, we suggest that the bubble N131 might be triggered by the strong stellar winds from a group of massive stars inside the bubble. We propose a scenario in which the bubble N131 forms from the disruption of a gas filament by expansion of H II region, strong stellar winds, and fragments under self-gravity.

Keywords. infrared, formation, bubbles, H II regions, clouds.

Based on the intriguing morphological structure (in Fig. 1) and velocity distribution of the N131 studied in Zhang *et al.* (2013) and Zhang *et al.* (2015), we propose that the progenitor of the bubble N131 was a filamentary nebula elongated along the direction of the clumps AD and BC (see the sketch in Fig. 2). It is likely that within the middle of the filamentary nebula, a cluster of massive stars ionized, heated, and blew away the surrounding interstellar medium and compressed the pre-existing clumps. With the expansion of H II region at relatively high temperature, pressure, and a strong stellar winds lasting more than 10^6 yr, the feedback from the massive stars broke up the pre-existing filamentary nebula and separated it into the two unconnected clumps AD and BC. With the expansion of evolving H II region, a ringlike structure is formed. Bodenheimer *et al.* (1979) have simulated a similar scenario. We provide further evidence to back up this argument.

The clumps AD and BC are two unconnected filamentary nebula on the opposite sides of the bubble. Along the direction of the clumps C, B, A, and D in Fig. 1, there is a velocity gradient from -14.5 to -6.5 km s^{-1} (Zhang *et al.* 2013). This velocity coherence indicates that in the past the clumps AD and BC were likely connected together. The velocity gradient may be partly produced by the bubble expansion.

These broken filamentary morphologies, caused by bubble expansion, are rather common in large scale surveys, i.e. the 500 pc filamentary gas wisp from Li *et al.* (2013), and the bubble N39 from Deharveng *et al.* (2010). In addition, the secondary bubble N131-A (Zhang *et al.* 2015) close to the N131 is also embedded in the filamentary nebula (or the clumps AD). The N131-A is expanding outwardly, and compressing the surrounded

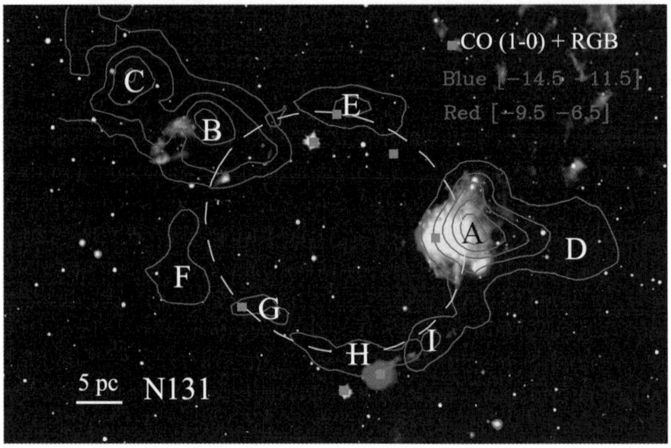

Figure 1. Integrated intensity contours of the CO (1-0) line of the blueshifted and redshifted clouds superimposed on the RGB image with 24 μm (red), 8.0 μm (green), and 4.5 μm (blue). The CO (1-0) data is from Zhang *et al.* (2013). The integration range is from -14.5 to -11.5 km s^{-1} for the blueshifted cloud, and from -9.5 to -6.5 km s^{-1} for the redshifted cloud. The symbols "■", letters (A, B, ..., and I), and ellipse indicate the positions of eight IRAS point sources, nine molecular clumps, and ringlike shell of the bubble, respectively.

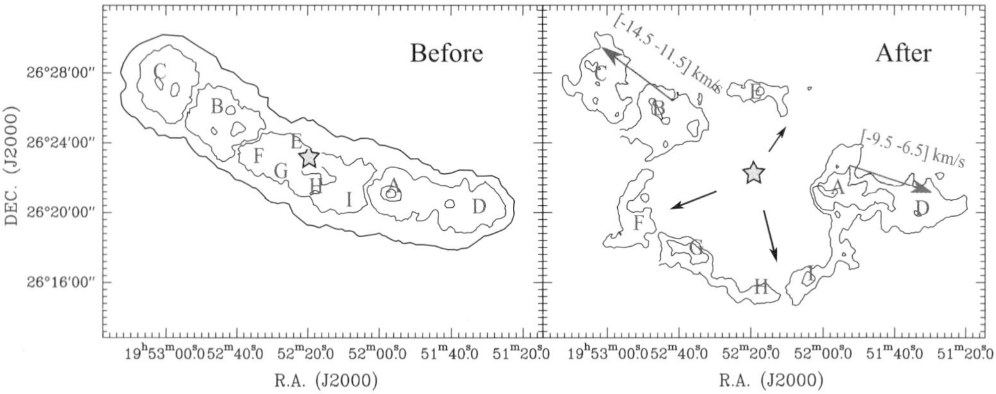

Figure 2. A schematic illustration depicting our target region before and after the formation of the bubble N131. The yellow star indicates a group of ionizing stars, which caused expansion of the shell. In the upper panel, we indicate the proposed pre-existing clumps A, B, ..., and I. In the lower panel, we indicate the observed clumps A, B, ..., and I, and the velocity structure.

material. The N131 in its early stage might have had a similar evolutionary process as the N131-A. With the expansion of the bubble, finally the strong stellar winds helps to form the bubble N131.

To probe the scenario, we also compared the relative masses of these clumps surrounding the bubble N131 with the proposed pre-existing clumps in (Zhang *et al.* 2015). The similar masses between the clump AD, clump BC, and ringlike shell suggests that the bubble might originate from a filamentary nebula in view of mass conservation.

References

Bodenheimer, P., Tenorio-Tagle, G., & Yorke, H. W. 1979, ApJ, 233, 85
Deharveng, L., Schuller, F., Anderson, L. D., *et al.* 2010, A&A, 523, A6
Li, G.-X., Wyrowski, F., Menten, K., & Belloche, A. 2013, A&A, 559, A34
Zhang, C.-P., Li, G.-X., Wyrowski, F., *et al.* 2015, A&A, submitted
Zhang, C.-P., Wang, J.-J., & Xu, J.-L. 2013, A&A, 550, A117

Formation, evolution, and survival of massive star clusters
Proceedings IAU Symposium No. 316, 2015
C. Charbonnel & A. Nota, eds.

doi:10.1017/S1743921315009734

The influence of feedback from massive stars on the formation and emergence of massive clusters

James E. Dale[1,2]

[1]Universitäts–Sternwarte München, Scheinerstr. 1, 81679 München, Germany.
[2]Excellence Cluster 'Universe', Boltzmannstr. 2, 85748 Garching, Germany.
email: dale.james.e@gmail.com

Abstract. Massive star clusters are of fundamental importance both observationally, since they are visible at such great distances, and theoretically, because of their influence on the large–scale ISM. Understanding stellar feedback is a prerequisite for making sense of their formation and early evolution, since feedback influences cluster structure, star formation efficiency, and sets the timescales on which clusters emerge from their parent clouds to become optically visible. I review the progress made in understanding these issues from a numerical perspective.

Keywords. stars: formation, galaxies: star clusters, ISM: kinematics and dynamics

1. Introduction

Throughout most of the history of the Universe, star formation has occurred in a hierarchically–clustered fashion. Dispersal of the more loosely bound majority of stellar structures result in field galactic stellar populations, while the most strongly–bound minority survive as long–lived stellar clusters [Kruijssen (2012a)].

Owing to their rapid evolution, massive stars are mostly to be found in stellar associations, both those destined to be dispersed and those that will remain bound. This has important implications for the evolution of the interstellar medium (ISM), the structure of galaxies and the reionisation of, and distribution of metals within, the early Universe, since the effects of feedback from an evenly–distributed population of sources are very different from those of a clustered population.

As well as their central role in ISM dynamics, clusters are an invaluable observational tool, providing a window into the evolution of the ISM over a large fraction of the age of the Universe [e.g. Adamo *et al.* (2011)]. However, the interpretation of observations of unresolved clusters relies on a detailed understanding of how their spectral energy distributions (SEDs) evolve during the most complex phase of their lives. There are many physical processes contributing to the evolution of both the *intrinsic* and *apparent* brightnesses and SEDs of clusters, including the triggering, termination and overall rate and cadence of star formation, the clearing of gas from the cluster volumes, and the disruption of the parent giant molecular clouds (GMCs). Worryingly (in view of the importance of clusters) these processes are very poorly understood.

What is clear is that feedback is a central plank in a comprehensive model of star formation, and is sufficiently complex that numerical simulations are an indispensable tool in helping to understand it. In particular, the task of explaining how the many different mechanisms interact with one another is best achieved with the aid of computations. This is especially true of primordial, or Population III star formation where at present, we have almost no observational data to guide us.

2. Population III vs. Population II star formation

The very first generation of stars have as yet not been observed (they are one of the principal targets for the forthcoming *James Webb Space Telescope*), but theoretical considerations demonstrate that they formed in a very different fashion from their descendants. While most star formation has occurred through gravitational fragmentation in GMCs within galaxies occupying dark matter haloes, the first stars formed before the existence of either molecular gas or galaxies. Pop III stars formed from baryons accumulating in $\sim 10^6\,M_\odot$ dark–matter minihaloes with virial radii of $\sim 100\,$pc. Without any molecules except H_2, these systems are less able to cool than GMCs, and have higher temperatures and Jeans masses. As a result, Pop III stars were expected to be considerably more massive than present–day objects [e.g. Abel *et al.* (2002), Hosokawa & Omukai (2009)] and, owing to the difficulty of fragmentation, were thought to form alone and not in clusters.

However, recent 3–D simulations with more advanced treatments of thermodynamics find that angular momentum conservation and fragmentation result in the formation of Pop III *clusters* [e.g. Clark *et al.* (2008), Stacy *et al.* (2010), Greif *et al.* (2011), Smith *et al.* (2011)], both increasing the numbers of low–mass stars and limiting the growth of the central objects (fragmentation–induced starvation – Peters *et al.* 2010, Smith *et al.* 2011). This steepening of the Pop III IMF dramatically reduces the quantity of feedback produced per unit stellar mass.

3. Feedback mechanisms through cosmic time

One of the chief difficulties in modelling and understanding feedback is that there are many different mechanisms operating on widely different length– and timescales, which interact with other in a complex and non–linear manner. There has been considerable progress in recent years in understanding the individual mechanisms, as described below. However, no simulation including all of them has yet been attempted.

3.1. *Accretion feedback*

All stars exert feedback as they form via thermal accretion feedback and jets. Thermal feedback derives from the conversion of gravitational potential energy to heat as accreting material falls onto the protostar, and from the Kelvin–Helmholtz contraction of the protostar itself (some models, e.g. [Offner *et al.* (2009a)], also include the energy released by deuterium burning). The infrared radiation released is absorbed by dust and, where the gas–dust mixture is dense enough, couples to the gas. Accretion heating in primordial star formation has been modelled (e.g. [Clark *et al.* (2008), Clark *et al.* (2011), Smith *et al.* (2011, 2012) and has only a minimal effect, heating the gas only modestly and leaving fragmentation largely unchanged. This is partly due to higher background temperatures of primordial gas, but the absence of dust can also disable accretion heating.

On the contrary, accretion heating in present-day star formation results in lower accretion rates and suppression of fragmentation (e.g. [Bate (2009, 2012, 2014), Offner *et al.* (2009a), Urban *et al.* (2010), Krumholz *et al.* (2010, 2011, 2012)].

Jets are also driven by accretion, and their short cooling times allow them to be thought of as depositing chiefly momentum. As far as the author is aware, jets from primordial stars have not been modelled, but their main effects on present–day star formation are to limit accretion onto individual objects and to drive small–scale turbulence (e.g. [Carroll *et al.* (2009), Offner & Arce (2014), Federrath *et al.* (2014)]. Because they are both driven by, and have influence on, the accretion process, the interaction of accretion heating and jets can be complex (e.g. [Krumholz *et al.* (2012), Hansen *et al.* (2012)], lessening each others' effectiveness. This highlights the issue that different feedback mechanisms are not independent and their effects are not necessarily additive.

It might be argued that the range of influence of accretion feedback is too small to be of any consequence for an entire cluster, but this is not the case. [Krumholz *et al.* (2010, 2011, 2012)] show that it is a necessary but not sufficient condition for obtaining the correct IMF robustly. [Bate (2009, 2012), Stamatellos *et al.* (2012)] demonstrate the importance of radiative feedback in particular on the low-mass end of the IMF. Accretion feedback also affects the environments massive stars are born into, in the sense of helping set the star formation efficiency the clouds have achieved when the massive stars ignite.

3.2. *HII regions*

HII regions are the most studied form of stellar feedback. Owing to very low primordial metallicities, Pop III stars had higher surface temperatures, and cooling within their HII regions was less efficient, leading to higher HII region temperatures ($\approx$30 kK as opposed to present–day values of $\approx$10 kK). This higher temperature is sufficient to ionise He and an important observational indicator of Pop III stars is expected to be simultaneous Ly–α and He II emission (e.g. [Johnson, (2009)]. Primordial HII region temperatures were sufficient to expel baryons from minihaloes, allowing large fractions of the ionising photons to escape and play their part in cosmic reionisation (e.g. [Yoshida *et al.* (2007)]. Anisotropic accretion was nevertheless able to continue, as in modern star formation (e.g. [Tan & McKee (2008), Hosokawa *et al.* (2012), Stacey *et al.* (2012)], but the effectiveness of feedback depends strongly on the uncertain shape of the Pop III IMF. Radiative feedback from *massive* Pop III stars suppresses star formation very effectively, but that from normal-mass ones does not (e.g. [Jeon *et al.* (2015)]).

Present-day simulations of feedback have largely concentrated on HIIRs and agreement is emerging that $\sim 10^4$ M$_\odot$ clouds can be severely damaged or destroyed by HII regions on Myr timescales ([Arthur *et al.* (2011), Walch *et al.* (2012), Dale *et al.* (2012), Colín *et al.*(2013)]. However, as suggested by, e.g. [Matzner (2002)] and confirmed by [Dale *et al.* (2012, 2013, 2014)], massive clouds with higher escape velocities are more resilient. This is shown qualitatively in Figure 1, which depicts the effects of $\approx$3 Myr of photoionisation and wind feedback on a series of molecular clouds ranging in mass from 10^4–10^6 M$_\odot$.

3.3. *Stellar winds*

Stellar winds from massive main–sequence stars have received comparatively little attention as feedback agents. Since they are driven by metal lines in stellar atmospheres, winds were absent in Pop III stars, and are generally weaker in stars with low metallicity. The dynamically–relevant quantities are the rates of release of momentum $\dot{p} = \dot{M}v_\infty$ and mechanical energy $L = 0.5\dot{M}v_\infty^2$, which depend on the intrinsic stellar properties $\dot{M}$ (the mass loss rate) and v_∞ (the wind terminal velocity). Both of these in turn depend on the metallicity. At moderately low metallicity ($\gtrsim$ 0.01) [Vink *et al.* (2001)] deduce $\dot{M} \propto Z^{0.7-0.8}$ and [Leitherer *et al.* (1992)] find $v_\infty \propto Z^{0.13}$. [Kudritski (2002)] find that steeper relations apply at lower metallicities.

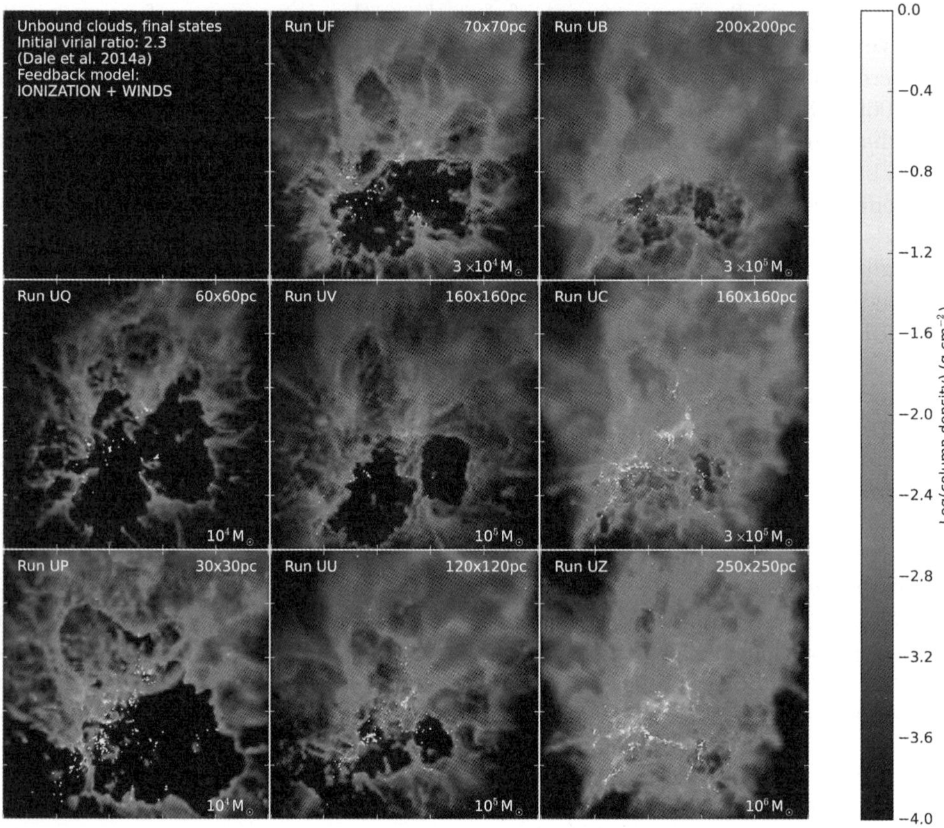

Figure 1. Column–density images of 10^4–$10^6\,M_\odot$ turbulent clouds after ≈ 3 Myr of photoionisation and wind from Dale *et al.* (2014). Image sizes are given in the top right corner, and clouds masses in the bottom right.

The influence of winds on modern star formation is much debated. [Rogers & Pittard (2013)] find that winds open channels through the cloud and leak, so that merely injecting energy is not sufficient to disrupt a cloud – the cloud must retain the energy long enough for a large fraction of it to be converted to bulk kinetic energy. [Rosen *et al.* (2014)] evaluated, for four young systems (30 Dor, M17, Carina, NGC 3603), how much wind energy had been injected, and what fraction had been converted to work, finding that this was generally rather low, $\sim 10\%$, and inferring that much of the energy was lost due to leakage of hot gas. This is in line with a large body of earlier research (e.g. [Garcia–Segura & Mac Low (1995), Oey (1996), Naze *et al.* (2002), Dopita *et al.* (2005)] suggesting that wind bubbles expand slower than spherical models predict.

3.4. *Supernovae*

On the largest scales, supernovae (SNe) are the most important stellar feedback mechanism. Pop III SNe were thought to be unusually violent, since 140–260 $M_\odot$ stars die as pair-instability SNe, releasing 10^{53} erg (e.g. [Heger & Woosley (2002), Whalen *et al.* (2008a), Greif (2015)]. Whether and how often these occur depends on the Pop III IMF, which is still very uncertain. What is clear is that the pre–SN action of photoionisation governs the effectiveness of supernovae – without it, SNe explode in dense material which rapidly quenches them. However, as described above, photoionisation on its own is able to expel baryons from primordial minihaloes. This allows the SNe to enter their radiative phases much later and to expand to much larger distances and distribute metals over much larger volumes (e.g. [Whalen *et al.* (2008b), Jeon *et al.* (2014)].

A similar phenomenon is observed in present–day models, which demonstrate that the environments in which SNe explode completely control their influence. [Iffrig & Hennebelle (2014)] model SN detonations inside and outside clouds and find very different results in terms of how much dense gas survives the explosion. [Walch & Naab (2015)] examined the effect of photoionisation acting before the SN, finding that pre–ionisation results in $\sim$50% more momentum being deposited.

4. Summary

Accretion–driven feedback is likely less effective at early epochs than in present–day star formation, where it plays an important role in moderating star formation rates and fragmentation and is one of the main processes regulating the IMF. Similarly, main sequence winds were absent in primordial star formation owing to the lack of driving metal ions in stellar atmospheres. Their influence on later star formation is still unclear, but both observations and simulations suggest that their effectiveness is strongly reduced by leakage of hot gas from clouds.

Photoionisation has been of central importance at all redshifts, operating on very similar physical principles, modulo the higher HIIR temperatures at high redshift. However, its effectiveness depends strongly on halo/GMC escape velocities, and on the Pop III IMF at high redshifts. While most primordial minihaloes were vulnerable to photoionisation, many present–day GMCs are not and other mechanisms such as radiation pressure need to be invoked to explain the very young, dense gas–free clusters observed by, e.g. [Hollyhead *et al.*(2015)]. This issue is now receiving attention [Skinner & Ostriker (2015)].

Supernovae have likewise been important at all redshifts, albeit in different senses. Today, SNe help regulate the cycling of matter within galaxies, but bottom-up hierarchical structure formation implies that Pop III (and possibly Pop II) star formation was well underway *before* galactic assembly, so that feedback from Pop III (and possibly early Pop II) stars influenced the *formation* of galaxies (e.g. [Bromm & Yoshida, (2011)]. The details of the role of SNe depend, again, on the Pop III IMF, which sets the proportion of PISNe. At all times, the effectiveness of SNe depends very strongly on the earlier-acting feedback processes, especially photoionisation. While Pop III SNe likely played little part in expelling baryons from minihaloes (since ionisation was able to accomplish this alone), present–day SNe may assist in disrupting molecular clouds, although observations of gas–free clusters too young to have hosted SNe imply that this is certainly not always the case.

Simulations suggest that feedback–triggered star formation has been operating throughout much of the Universe's history, although it seems to be of only second–order importance today, and the identification of triggered stars may be very difficult (e.g. [Dale *et al.* (2015b)]. Several primordial calculations suggest that Pop III and Pop II star formation could be contemporaneous and that Pop II formation can be triggered by Pop III stars (e.g. [Pawlik *et al.*, (2013), Jeon *et al.*, (2015)]. Conversely, [Wise *et al.* (2012)] showed that patchy IGM pollution allows Pop III formation to continue in pockets up to z=6. These effects may in principle be easier to deduce observationally.

The interplay between star formation, gas expulsion and cloud disruption is still not well understood. This emergence timescale is important, since clusters are brightest at around the time when they are clearing away residual gas, before the first SNe. Observations (e.g. [Hollyhead *et al.* (2015)] find this timescale to be short, 4Myr even for massive

($> 10^4 M_\odot$) clusters. [Dale *et al.* (2015a)] are able to reproduce such short timescales only for substantially lower–mass ($\sim 10^3 M_\odot$) clusters.

While important observationally, gas expulsion form clusters appears to have only modest dynamical influence. Simulations by [Offner *et al.* (2009b), Girichidis *et al.* (2011), Kruijssen *et al.*, (2012b)] find that stars rapidly decouple from the gas, forming stellar-dominated subsystems largely immune to gas expulsion/cloud dispersal. [Dale *et al.* (2015a)] observed that, even where feedback was able to expel large fractions of the gas from cloud potential wells, the fractions of *stars* unbound were much more modest.

In general, many of the processes that influence the evolution of clusters have done so in a very similar fashion throughout the history of the Universe, and primordial and present-day star formation are not as different as they may at first appear. These different processes interact with one another in a complex and, as yet, only dimly understood fashion, and this topic will continue to be a major focus of star formation research for the foreseeable future.

References

Abel, T., Bryan, G., & Norman, M., 2002, *Science*, 295, 93

Adamo, A., Östlin, G., & Zackrisson, E., 2011, *MNRAS*, 417, 1904

Arthur, S. J., Henney, W. J., Mellema, G., de Colle, F., & Vázquez–Semadeni, E., 2011, *MNRAS*, 414, 1747

Bate, M. R., 2009, *MNRAS*, 392, 1363

Bate, M. R., 2012, *MNRAS*, 419, 3115

Bate, M. R., 2014, *MNRAS*, 442, 285

Bromm, V. & Yoshida, N., 2011, *ARA&A*, 49, 373

Carroll, J. J., Frank. A., Blackman, E. G., Cunningham, A. J., & Quillen, A. C., 2009, *ApJ*, 695, 1376

Clark, P. C., Glover, S. C. O., & Klessen, R. S., 2008, *ApJ*, 672, 757

Clark, P. C., Glover, S. C. O., Smith, R. J., Greif, T. H., Klessen, R. S., & Bromm, V., 2011, *Science*, 331, 1040

Colín, P., V azquez–Semadeni, E., & Gomez, G. C., 2013, *MNRAS*, 435, 1701

Dale, J. E., Ercolano, B., & Bonnell, I. A., 2012, *MNRAS*, 424, 377

Dale, J. E., Ercolano, B., & Bonnell, I. A., 2013, *MNRAS*, 430, 234

Dale, J. E., Ngoumou, J., Ercolano, B., & Bonnell, I. A., 2014, *MNRAS*, 442, 694

Dale, J. E., Ercolano, B., & Bonnell, I. A., 2015, *MNRAS*, 451, 987

Dale, J. E., Haworth, T. H., & Bressert, E., 2015, *MNRAS*, 450, 1199

Dopita, M. A., Groves, B. A., Fischera, J., Sutherland, R. S., Tuffs, R J., Popescu, C. C., Kewley, L. J., Reuland, M., & Leitherer, C., 2005, *ApJ*, 619, 755

Federrath, C., Schrön, M., Banerjee, R., & Klessen, R. S., 2014, *ApJ*, 790, 128

Garcia–Segura, G. & Mac Low, M–M., 1995, *ApJ*, 445, 145

Girichidis, P., Federrath, C., Banerjee, R., & Klessen, R. S., 2011, *MNRAS*, 413, 2741

Greif, T. H., Springel, V., White, S. D. M., Glover, S. C. O., Clark, P. C., Smith, R. J., Klessen, R. S., & Bromm, V., 2011, *ApJ*, 737, 75

Greif, T. H., 2015, *Computational Astrophysics & Cosmology*, 2, 3

Hansen, C. F., Klein, R. I., McKee, C. F., & Fisher, R. T., 2012, *ApJ*, 747, 22

Heger, A. & Woosley, A. E., 2002, *ApJ*, 567, 532

Hollyhead, K., Bastian, N., Adamo, A., Silva–Villa, E., Dale, J., Ryon, J. E., & Gazak, Z., 2015, *MNRAS*, 449, 1106

Hosokawa, T. & Omukai, K., 2009, *ApJ*, 691, 823

Hosokawa, T., Yoshida, N., Omukai, K., & Yorke, H. W., 2012, *ApJ*, 760, L37

Iffrig, O. & Hennebelle, P., 2014, *A&A*, 596, 95

Jeon, M., Pawlik, A. H., Bromm, V., & Milosavljević, M., 2014, *MNRAS*, 444, 3288

Jeon, M., Bromm, V., Pawlik, A. H., & Milosavljević, M., 2015, *MNRAS*, 452, 1152
Johnson, J. L., 2010, *MNRAS*, 404, 1425
Krumholz, M. R., Cunningham, A. J., Klein, R. I., & McKee, C. F., 2010, *ApJ*, 713, 1120
Krumholz, M. R., Klein, R. I., & McKee, C. F., 2011, *ApJ*, 740, 74
Krumholz, M. R., Klein, R. I., & McKee, C. F., 2012, *ApJ*, 754, 71
Kruijssen, J. M. D., 2012, *MNRAS*, 426, 3008
Kruijssen, J. M. D., Maschberger, T., Moeckel, N., Clarke, C. J., Bastian, N., & Bonnell, I. A., 2012, *MNRAS*, 419, 841
Kudritski, R. P., 2002, *ApJ*, 577, 389
Leitherer, C., Robert, C., & Drissen, L., 1992, *ApJ*, 401, 596
Matzner, C. D., 2002, *ApJ*, 566, 302
Nazé, Y., Chu, Y. H., Guerrero, M. A., Oey, M. S., Gruendl, R. S., & Smith, R. C., 2002, *ApJ*, 124, 3325
Oey, M. S., 1996, *ApJ*, 467, 666
Offner, S. S. R., Klein, R. I., McKee, C. F., & Krumholz, M. R., 2009, *ApJ*, 703, 131
Offner, S. S. R., Hansen, C. E., & Krumholz, M. R., 2009, *ApJ*, 704, L124
Offner, S. S. R. & Arce, H. G., 2014, *ApJ*, 784, 61
Pawlik, A, H., Milosavljević, M., & Bromm, V., 2013, *ApJ*, 767, 59
Peters, T., Klessen, R. S., Mac Low, M–M., & Banerjee, R., 2010, *ApJ*, 725, 134
Rogers, H. & Pittard, J. M., 2013, *MNRAS*, 431, 1337
Rosen, A. L., Lopez, L. A., Krumholz, M. R., & Ramirez–Ruiz, E., 2014, *MNRAS*, 442, 2701
Skinner, M. A. & Ostriker, E. C., 2015, *ApJ*, 809, 187
Smith, R. J., Glover, S. C. O., Clark, P. C., Greif, T. H., & Klessen, R. S., 2011, *MNRAS*, 414, 3633
Smith, R. J., Hosokawa, T., Omukai, K., Glover, S. C. O., & Klessen, R. S., 2012, *MNRAS*, 424, 457
Stacy, A., Greif, T. H., & Bromm, V., 2010, *MNRAS*, 403, 45
Stacy, A., Greif, T. H., & Bromm, V., 2012, *MNRAS*, 422, 290
Stamatellos, D., Whitworth, A. P., & Hubber, D. A., 2012, *MNRAS*, 427, 1182
Tan, J. C. & McKee, C. F., 2008, *AIP Conf. Proc.*, 990, 47
Urban, A., Martel, H., & Evans, N. J., 2010, *ApJ*, 710, 1343
Vink, J. S., de Koter, A., & Lamers, H. J. G. L. M., 2001, *A&A*, 369, 574
Walch, S. K., Whitworth, A. P., Bisbas, T. G., Wünsch, R., & Hubber, D. A., 2012, *MNRAS*, 427, 625
Walch, S. K. & Naab, T., 2015, *MNRAS*, 451, 2757
Whalen, D., van Veelen, B., O'Shea, B. W., & Norman, M. L., 2008, *ApJ*, 682, 49
Whalen, D., O'Shea, B. W., Smidt, J., & Norman, M. L., 2008, *ApJ*, 679, 925
Wise, J. H., Turk, M. J., Norman, M. L., & Abel, T., 2012, *ApJ*, 745, 50
Yoshida, N., Oh, S. P., Kitayama, T., & Hernquist, L., 2007, *ApJ*, 663, 687

Formation, evolution, and survival of massive star clusters
Proceedings IAU Symposium No. 316, 2015
C. Charbonnel & A. Nota, eds.

doi:10.1017/S1743921315009394

Galactic flows and the formation of stellar clusters

Romas Smilgys[1] **and Ian A. Bonnell**[1]

[1]Scottish Universities Physics Alliance (SUPA), School of Physics and Astronomy, University of St. Andrews, North Haugh, St. Andrews, Fife KY16 9SS, UK
email: rs202@st-andrews.ac.uk

Abstract. We investigate the formation of stellar clusters from a Galactic scale SPH simulation. The simulation traces star formation over a 5.6 Myr timescale, with local gravitational instabilities resulting in $\sim 10^5$ solar masses of star formation in the form of sink particles. We investigate the time evolution of the physical properties of the forming clusters including their half-mass radii, their energies and the depletion time of the gas. Star formation is driven by the large scale flows which compress the gas to higher densities where self gravity takes over and collapse occurs. We show that the more massive clusters (up to $\sim 2 \times 10^4$ solar masses) gather their material from of order 10 pc due to these large scale motions associated with the spiral arm passage and shock. The bulk of the gas becomes gravitationally bound near 1-2 Myr before sink formation, and in the absence of feedback, significant accretion ongoing on longer timescales. We trace the hierarchical merging process of cluster formation which naturally results in age spreads of order the crossing time of the original region which provides the gas reservoir for the cluster.

Keywords. stars: formation, ISM: clouds, (Galaxy:) open clusters and associations: general

1. Introduction

Star formation is one of the key processes in understanding galaxies evolution and physical conditions how gas are being transformed into stars. Most stars are believed to be formed in clusters (Lada & Lada (2003), Gutermuth *et al.* (2009), Bressert *et al.* (2010)) such that establishing the physical conditions for the formation of stellar clusters is crucial in order to develop a comprehensive theory of star formation. However, physical conditions of cluster formation are not well understood.

Observations of recent years are starting to unveil much more details (i.e. Peretto *et al.* (2013)), but they alone provides only a single snapshot in cluster evolution. This is where numerical simulations comes into their role by providing insight into the physical process of cluster formation. However, most of cluster formation based simulations (Bate *et al.* (2003), Bonnell *et al.* (2003)) used idealised initial conditions, ignoring how such initial conditions arise, or even if they are realistic. So large scale Galactic dynamics are getting ignored in these simulations.

First attempts to study star formation including Galactic scale flows were made by Bonnell *et al.* (2006). Bonnell *et al.* (2013) produced a set of 3 nested simulations (Global disc simulation, high resolution Cloud re-simulation and Gravity re-simulation) to cover wide range of scales. Global disc simulation used entire Galactic disc. Cloud simulation used high density region from Global simulation within about 2 kpc size. And finally Gravity simulation used 250 pc region from Cloud simulation. These simulations shown that there is possible to study star formation on cloud scale without excluding galactic spiral arm passage through ISM. We continue to investigate origin of star formation

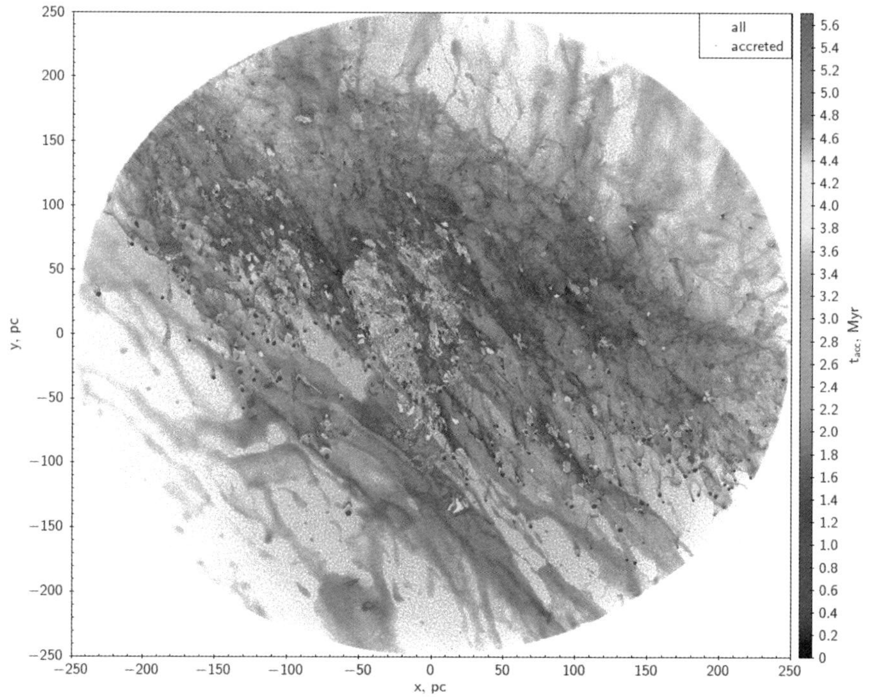

Figure 1. Initial conditions of Bonnell *et al.* (2013) simulation in terms of x-y positions. Grey colour shows where are all gas particles located, while coloured ones show where accreted particles (We denote the gas particles which are involved in star formation as accreted particles) are located. Accreted particles are colour coded by accretion time, which shows when they will be accreted.

between hundreds of parsecs scales down to sub-parsec scales and the formation of stellar clusters. Lagrangian nature of SPH simulations allows easily to follow and recover evolution of cluster forming regions, and to seek where, when and how star formation is triggered.

2. Galactic flows

We start with Bonnell *et al.* (2013) Gravity simulation initial conditions. These conditions were inherited from a set of nested simulations based on larger galactic and spiral arm scales, setting the initial positions and velocities of the SPH particles. We plot all the particles of the Gravity simulation in Figure 1 in greyscale.

The Lagrangian nature of particles in SPH allows us to follow positions, velocities and other parameters over the time. We can thus follow the star formation process, the onset of self-gravity, collapse and accretion over all time steps. We select all these accreted gas particles and plot on the top of all simulation gas particles, and colour coding them by accretion time (Figure 1).

The gas particles in Figure 1 clearly show the presence of the spiral arm, with gas compressed as it exits the spiral arm. Darker grey areas show larger density regions. Gas particles that are subsequently involved in star formation are initially located in the densest parts of the simulation. In the middle is visible a large chunk of accreted gas extending ~ 150 pc from upper to lower side. This is the main star forming region, where 10 massive clusters form in 5.6 Myr.

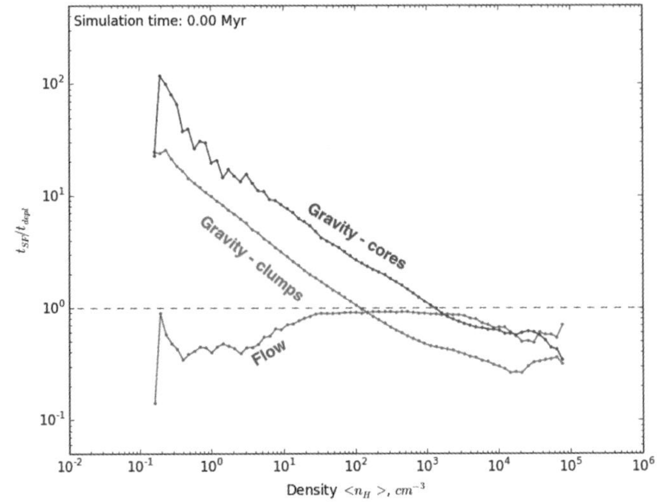

Figure 2. Star formation and depletion time ratio vs gas SPH density diagram at the start of simulation. Star formation can be explained only when this ratio is lower than 1. The diagram shows that at large scales (low densities) star formation is driven by Galactic flows, while on small gravity takes over.

Colours of accreted particles show that early star formation (blue) is concentrated into tiny clumps, which collapse in ~ 1 Myr to form first sinks. These clumps are surrounded by green, yellow and red regions, which are widespread over much larger volumes. This indicates that the gas, which forms stars later, is still more far away from star formation locations than the gas forming stars earlier. So early star formation is likely to be driven by the rapid collapse of local dense regions.

In order to understand what is driving star formation, we compare the actual star formation timescale, the depletion time, against simple predictions for the timescale for star star formation. These include the free fall time of the region depending on the local gas density, the collapse time of the individual star/sink, and the timescale for compression due to the large scale kinematics of the gas flows. We use predicted star formation and depletion time ratio (Figure 2) as a function of gas density to understand star formation triggering mechanisms over such large range of scales. All three predicted star formation times can be expressed in a ratio with depletion time, which is real time in the simulation required for particles to reach the location where they are accreted by a sink, or formed into a new one. The predicted timescale must be less than the actual (depletion) timescale as recorded by the simulation if the associated physics is driving the star formation.

Figure 2 show that gravity drives star formation only at larger densities, while only flow can explain star formation of low densities. In contrast, at lower densities it is the large scale flows from the Galactic shock which drives compression until larger densities are reached where self-gravity can take over.

3. Stellar clusters

Soon after looking at the positions of sinks at the end of the simulation we noticed that sinks tend to be not alone but often forms tight groups (clusters of sinks). These clusters can be used as references for massive star clusters. We wrote an algorithm to detect these clusters on every time step of the simulation. The detection is based on a

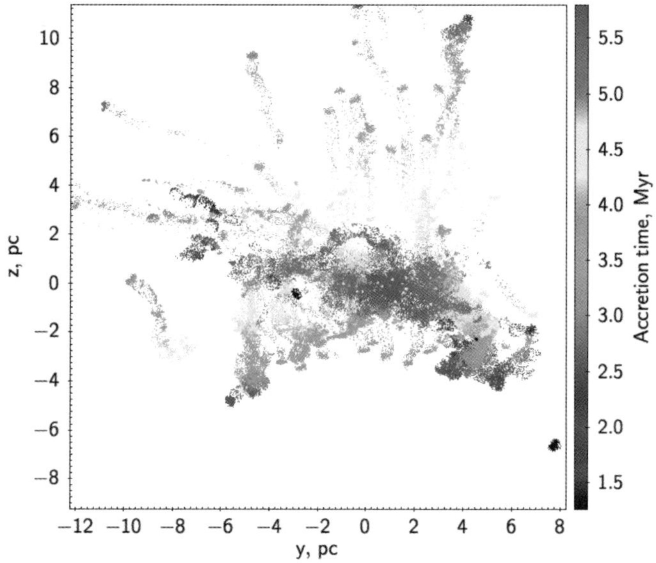

Figure 3. Star formation traces inside most massive cluster in terms of y-z positions. Colours represents when star formation is happening. Diagram shows that early star formation takes place in smaller groups of clumps, which later assemble into final larger cluster.

cluster definition of at least 6 gravitationally bound sinks where members are in at least 1 pc neighbourhood with other cluster members. As a particular case we look at final simulation clusters and trace their formation history backwards in time.

Firstly we look at all sink formation and accretion events in a position-position diagram (Figure 3) of the most massive final cluster. We can trace star formation history and hence how the cluster formed and grew. The final positions of each particle is plotted relative to the cluster centre of mass. The colour represents the time at which accretion occurred. We can clearly see there paths of ongoing accretion pointing towards the centre of cluster. These paths are left by moving sinks, which keeps accreting gas along their way. The early star formation is quite isolated, where the first sinks form (blue colour). At later times sinks are migrating towards the cluster centre while accreting gas along their paths (green and yellow tracks) and finally form 2-3 pc size cluster in the middle. There are also visible sub-grouping in the early to mid star formation, which indicates that sub-clusters are starting to form earlier, and later merge to form larger cluster.

We repeat our cluster finding procedure over the entire field to find also smaller and less massive clusters. Over 10 clusters with masses $>1000\ M_{\odot}$ form in the main star forming region with sizes of 1-2 pc and distances between clusters of $\sim$10 pc. To show the variety of clusters we plot their mass-radius relation (Figure 4), where we use half-mass radii to express cluster sizes. We plot the evolution of mass-radius relation for all time steps. The diagram shows a clear trend that more massive clusters are also larger in size. The earliest clusters form at $\sim$ 2 Myr from the start of the simulation with masses up to 1000 - 2000 $M_{\odot}$. At later times, clusters grow most significantly in their masses while only slightly increasing their sizes (at $\sim$4 Myr we see most massive clusters of $\sim$5000 $M_{\odot}$ and at the end of simulation reaching almost 30000 $M_{\odot}$).

Finally we look at the question of stellar age spreads within clusters. We reconsider star formation traces seen in Figure 3 and plot accretion time versus distance from the cluster mass centre where accretion occurs (Figure 5). Tracks are shown for 3 most massive clusters and coloured by cluster to which they belong. The evolutionary tracks move

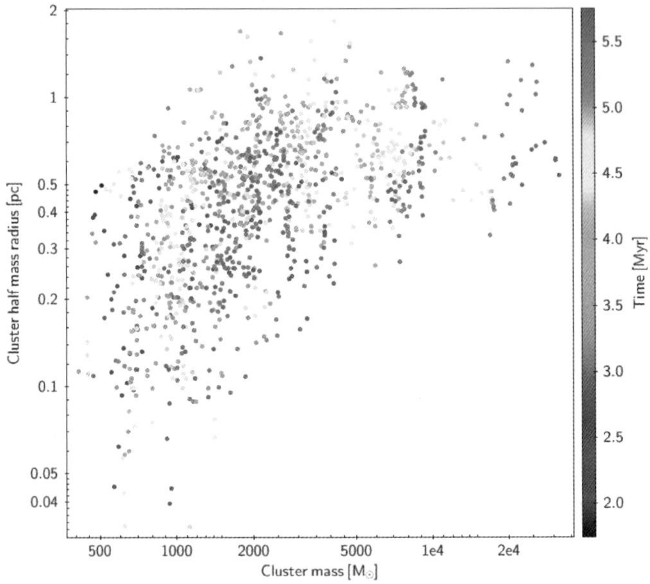

Figure 4. Evolution of mass-radius relation for simulation clusters. Diagram shows that more massive clusters are also larger in size and when time passes clusters grow mostly in mass, while keeping similar sizes.

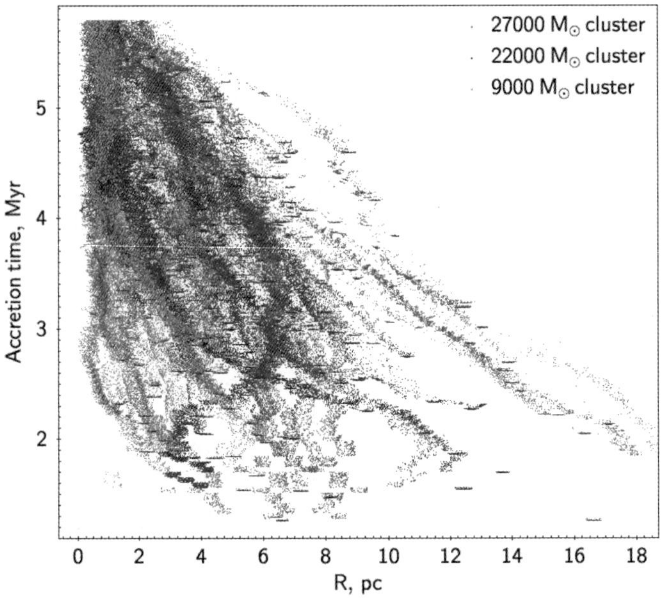

Figure 5. Accretion history of 3 most massive final cluster. Diagram shows that accretion inside clusters is happening continuously over several Myr along sink paths towards cluster centre. This indicates possible age spreads inside cluster, which assembles from smaller clusters.

towards smaller radii over all times due to the hierarchical nature of the fragmentation and subsequent merging. The tracks start at 1.5 - 2 Myr, which is the start of the formation of individual sinks. Sink formation is denoted as strong 0.5 pc wide horizontal bars at the beginning of the tracks. Clusters keep accreting gas all the time up to the end of simulation. However, cluster merging from several sub-clusters indicates potential stellar age spreads up to several Myr.

4. Summary

We analysed data produced by the SPH simulation, which uses Galactic dynamics to see how star formation is triggered and how do stellar clusters form. We found that gravitationally unstable clouds start to collapse and first stars and clusters after 1-2 Myr. The local collapse is followed by the larger region global collapse, creating more massive clusters. Galactic flows are responsible for compressing these clouds and explains star formation from large scales. We found that forming clusters starts to form widespread hierarchal merging make place to form larger clusters. More massive clusters are also larger in size and mass-radius relation show that forming clusters are mostly increasing in mass while sustaining similar sizes over the time. Cluster merging show potential age spread of several Myr.

References

Bate M. R., Bonnell I. A., Bromm V., 2003, *MNRAS*, 339, 577

Bonnell I. A., Bate M. R., Vine S. G, 2003, *MNRAS*, 343, 413

Bonnell I. A., Dobbs C. L., Robitaille T. P., Pringle J. E., 2006, *MNRAS*, 365, 37

Bonnell I. A., Dobbs C. L., Smith R. J., 2013, *MNRAS*, 430, 1790

Bressert, E., Bastian, N., & Gutermuth, R., *et al.*, 2010, *MNRAS*, 409, L54

Gutermuth, R. A., Megeath, S. T., & Myers, P. C., *et al.*, 2009, *ApJS*, 184, 18

Lada, C. J., & Lada, E. A., 2003, *ARAA*, 41, 57

Peretto N., Fuller G. A., Duarte-Cabral A., Avison A., *et al.*, 2013, *A&A*, 555, A112

Formation, evolution, and survival of massive star clusters
Proceedings IAU Symposium No. 316, 2015
C. Charbonnel & A. Nota, eds.

doi:10.1017/S1743921315008984

Stellar feedback from a massive Super Star Cluster in the Antennae merger

Cinthya N. Herrera[1]† **and Francois Boulanger**[2]

[1]National Astronomical Observatory of Japan,
2-21-1 Osawa, Mitaka, Japan
email: `cinthya.herrera@nao.ac.jp`, `herrera@iram.fr` (present address)

[2]Institut d'Astrophyisque Spatiale
Université Paris Sud, Batiment 121, 91405 Orsay, France
email: `francois.boulanger@ias.u-psud.fr`

Abstract. Stellar feedback from massive stars can unbind and disperse large amount of molecular gas, affecting the star formation efficiency. Based on ALMA and VLT observations in the Antennae galaxies we study a massive ($\sim 10^7$ $M_\odot$) and young ($\sim$3 Myr) SSC, B1, associated with compact molecular and ionized emission, which suggests that it is embedded in its parent cloud. However, we found contradictories and puzzling results on the structure and dynamics of the matter around the cluster, indicating that SSC B1 is not embedded in its parent cloud after all. We propose that radiation pressure was highly enhanced at the early stages of the SSC formation, disrupting the parent cloud in <3 Myr. We show evidences of outflowing gas from the parent cloud in the more extended CO gas. Higher angular resolution observations are needed to validate this interpretation and to understand the origin and fate of the component seen to be associated with SSC B1.

Keywords. ISM, galaxies: star clusters, interactions

---

1. Introduction

Star formation is an inefficient process. Observations of giant molecular clouds (GMCs) in the Milky Way suggest that the highest star formation efficiency (SFE) occurs in massive ($>10^4$ $M_\odot$) GMCs reaching only 20% (Murray 2011). Many studies have considered *stellar feedback* as a main energy source to maintain turbulence as star clusters are formed and thus preventing star formation (e.g., Tan *et al.* 2006, Krumholz *et al.* 2006). Several stellar feedback mechanisms have been proposed and analytically quantified (e.g., photoionization and expansion of HII regions, stellar winds, SN explosions, radiation pressure). Eventually, the energy and momentum injected by the newly formed stars will unbind and disperse the matter around them. How quickly this occurs is key to the SFE.

Massive stellar feedback is important to galaxy evolution and star formation history in the universe. Observations of starburst galaxies suggest that $P_{\rm rad}$ is the main driving mechanism of galactic molecular outflows (e.g., Geach *et al.* 2014), dispersing and disrupting large amounts of molecular gas thus affecting the total SFE. At smaller scales, $P_{\rm rad}$ is found to dominate within the shells close to the central SSC in the 30 Dor star forming region (Lopez *et al.* 2011). Numerical simulations also indicate that $P_{\rm rad}$ would be the dominant stellar feedback mechanism close to massive star clusters where the escape velocity is larger than the sound speed of the HII gas (Krumholz & Matzner 2009; Murray *et al.* 2010). However, it is still unclear how this occurs in GMC scales since observational support is insufficient.

† Present address: IRAM, 300 Rue de la Piscine, 38406 Saint Martin d'Hères, France.

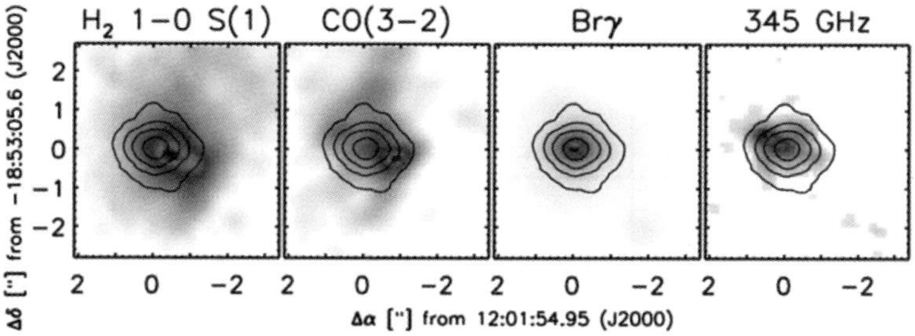

Figure 1. Intensity maps for SGMC 4/5. From the left to right, K-band continuum emission contours overlaid on H_2 1–0 S(1), CO(3–2), Brγ and continuum emission at 345 GHz. The peak emission of the contours marks the position of SSC B1.

Super star clusters (SSCs) are one of the most extreme forms of star formation, being compact (few parsecs size) and massive ($>10^5$ M$_\odot$). The are recurrently found in galaxy interactions, common phenomena in the Universe. SSCs must be ubiquitous in the Universe. SSCs in nearby mergers are ideal sites to study massive stellar feedback. The Antennae galaxy (NGC 4038/39) is a nearby (22 Mpc, Schweizer *et al.* 2008) merger between two gas-rich spiral galaxies that hosts a large number of SSCs (Whitmore *et al.* 1995). In the region where the two disks interact, the overlap region, large amount of molecular gas is observed to be fragmented into Super-Giant Molecular Complexes (SGMCs, Wilson *et al.* 2000), likely the birth places of SSCs. Interferometric observations show that SGMCs are clumpy and present different velocity components (Herrera *et al.* 2012). In this study we focus in the Antennae overlap region, where the most massive ($\geqslant 10^5$ M$_\odot$) and youngest (<10 Myr) SSCs are located (Mengel *et al.* 2005).

2. Observations

We use two datasets on the Antennae overlap region. First, observations obtained with the SINFONI/VLT imager spectrometer facility, in the near-IR K-band which includes Brγ and several H_2 rovibrational lines. Second, observations done with the ALMA interferometer in the Cycle 0 early science process (PI: B. Whitmore). These observations are done in Band 7 (345 GHz), which includes the CO(3–2) line and the dust continuum emission. Detailed information of these datasets can be found in Herrera *et al.* (2012) and Whitmore *et al.* (2014) for the VLT and ALMA observations, respectively.

3. A unique SSC in the Antennae overlap region

We search in our SINFONI field-of-views for SSCs still associated with compact molecular (H_2) and ionized (Brγ) gas emission. In SGMC 4/5 we found a unique SSC, B1, which is also the most massive ($\sim 10^7$ M$_\odot$) SSC in the overlap region. It is a bright target to study how massive SSCs disperse their parent molecular clouds.

Fig. 1 shows the distribution of the H_2, CO, Brγ and continuum emission. While the molecular gas extends across the SGMC, the ionized gas and K-band continuum are symmetrically distributed around the cluster. Fig. 2 presents the CO(3–2) channel maps where we can distinguish two spatially and spectrally separated components. The low velocity emission is mainly associated with SSC B1, while the higher velocity component seems to surround the cluster in a bubble-like shape structure. We hypothesize that the low velocity component traces the parent molecular cloud of SSC B1.

We estimate the mass of the GMC from the continuum emission at 345 GHz. We approximate the spectral dependence on the dust emission with a gray body and assume

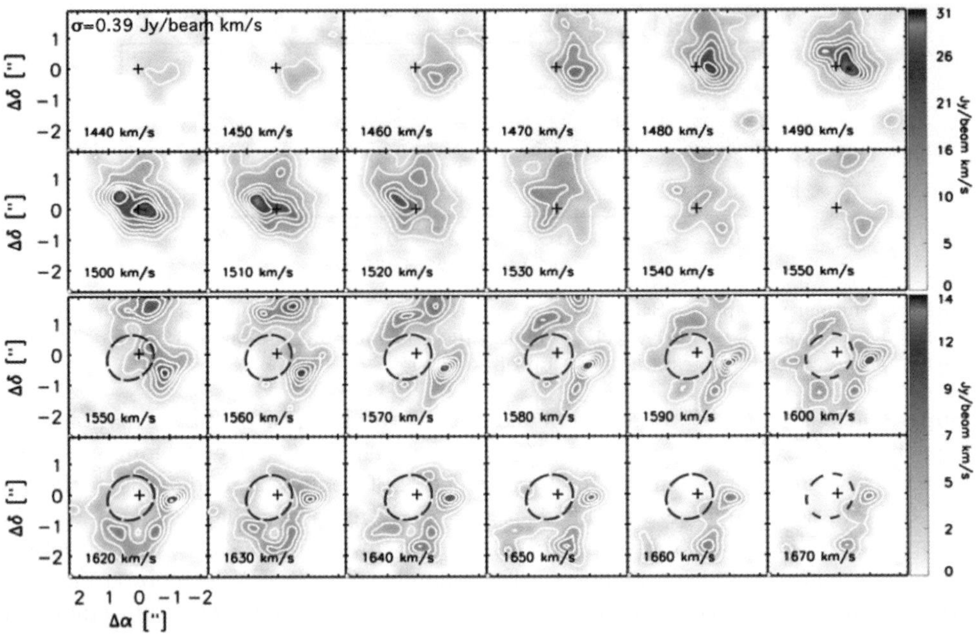

Figure 2. CO(3–2) channel maps of SGMC 4/5. The position of SSC B1 is marked with a cross. Top panel: channel maps from 1440 km s^{-1} to 1550 km s^{-1} by 10 km s^{-1}. Contours start at 10σ with a spacing of 10σ. Bottom panel: CO channel maps from 1550 km s^{-1} to 1670 km s^{-1} by 10 km s^{-1}. Contours start at 5σ with a spacing of 5σ. Offset position are the same as in Fig. 1.

that the dust properties in the Antennae are the same as those in the Galaxy. For a typical dust temperature of 20 K, we derive a GMC mass of 4.6$\times$10^7 M$_\odot$. We also estimate the molecular gas mass from the observed CO emission. Assuming the canonical $X_{\rm CO}$ conversion factor (Bloemen *et al.* 1986), we estimate the molecular mass to be 2.4$\times$10^8 M$_\odot$, 5 times larger than the value estimated from the continuum emission. This excess can be accounted for by a difference on the $X_{\rm CO}$ factor. Zhu *et al.* (2003) found a $X_{\rm CO}$ factor of at least 5 times smaller than the canonical value for the entire overlap region (at 1.5 kpc scale).

4. Physical structure of the matter surrounding SSC B1

We describe the physical structure of the molecular gas surrounding SSC B1 which we associate with the low velocity component of the CO emission.

Gas pressure in the molecular gas. Radiation pressure is the pressure exerted by the stellar radiation over the molecular gas that surrounds the central star cluster. We quantify $P_{\rm rad}$ in the molecular gas surrounding SSC B1 as $P_{\rm rad}=\frac{L_{\rm cl}}{4\pi {\rm R}_{\rm in}^2\,{\rm c}}(1+\tau_{\rm rad})$ (e.g., Murray *et al.* 2010), where $L_{\rm cl}$ is the luminosity of SSC B1, R$_{\rm in}$ is the HII radius and $\tau_{\rm rad}$ accounts for the trapping of the IR photons within the cloud. We compute P$_{\rm rad}$/k$_{\rm B}$=1.1$\times$10^8 (1+$\tau_{\rm rad}$) K cm^{-3}.

We model the HII region surrounding the SSC as an ionized nebula with dust grains. For different values of R$_{\rm in}$, we solve the radiative transfer equation and measure the outward radiation field (χ) and $P_{\rm rad}$ exerted at the surface of the photodissociation region (PDR). We compare our grid with the outputs of the PDR Meudon code (Le Bourlot *et al.* 2012), which predicts molecular line intensities from PDRs. We find that PDR models with $\chi=10^3$–10^4 times the mean radiation field in the solar neighborhood and a gas pressure of about 10^8 K cm^{-3}, are needed to account for the observed $R_{S(1)} \equiv H_2$ 1–0/2–1 S(1) line emission. The χ value is consistent with that estimated by Gilbert *et al.* (2000) from the comparison of PDR models and several near-IR H_2 line intensities from single-slit spectroscopy. The estimated gas pressure of 10^8 K cm^{-3} agrees with the measured molecular gas pressure, supporting low values of $\tau_{\rm rad}$. There is no significant trapping of the IR photons within the cavity of the molecular gas surrounding SSC B1.

Gas pressure in the hot plasma. SSC B1 was detected as a compact X-ray source by Chandra (Zezas *et al.* 2006). From its X-ray luminosity ($L_X = 1.9\times10^{38}$ erg s^{-1}) we find an electron density of 0.6 cm^{-3}. For a temperature of T = 10^7 K, the pressure of the hot plasma is $P_{\rm hot,X}/k_B = 1.9\ n_e\ T = 1.2\times10^7$ K cm^{-3}.

The hot gas pressure is one order of magnitude weaker than the pressure of the molecular gas. This difference indicates that there is a leak of hot gas within the cavity. The parent molecular cloud is not a closed, homogeneous shell around the cluster, but it is already broken. This is consistent with our finding that $\tau_{\rm rad}$ is negligible. This result also implies that the radius estimated for the molecular cloud is not an unique value. The molecular gas surrounding the cluster is clumpy, with molecular clumps located at different distances from the cluster.

5. Outflowing gas

The radiation pressure is not enough to compensate the gravitational attraction exerted by SSC B1 and the mass of the cloud itself. Today, $P_{\rm rad}$ is weaker than the gravitational force and cannot push away the molecular gas associated with SSC B1.

We speculate that at least a significant part of the natal cloud has already been blown away and that the high velocity component is tracing outflowing molecular gas. This gas would have been earlier accelerated by $P_{\rm rad}$ when it was much higher than today. We estimate the momentum of the outflowing gas from the mass in the CO high velocity component and the gas expansion velocity $v_{\rm exp}$ estimated from the line width of this component. To test the hypothesis that this gas was accelerated by $P_{\rm rad}$ over a time scale of $t_{\rm feed}$, we write $L_{\rm cl}/c(1+\tau_{\rm rad})t_{\rm feed} = M_{\rm high}v_{exp}$, where $t_{\rm feed}=R_{\rm high}/v_{\rm exp}$=2 Myr, with $R_{\rm high} \simeq$100 pc the mean size of the bubble-like shape structure observed in this component. We found a high value of $\tau_{\rm rad}$, possible when the parent cloud was not yet disrupted. (In the numerical models by Murray *et al.* (2010), $\tau_{\rm rad}$ reaches values of several tens of units at the very beginning of the disruption of the matter). If we correct $M_{\rm high}$ by the $X_{\rm CO}$ factor estimated by Zhu *et al.* (2003), $\tau_{\rm rad}$ will be ~2. We estimate the outflow rate of the parent molecular cloud to be 18 $M_\odot$ yr^{-1} (using the corrected $X_{\rm CO}$ factor), assuming a SFE for SSC B1 of 5–10%.

The high velocity component is also observed in the H_2 1–0 S(1) emission. Fig. 3 shows the spectra for the v=1–0 and 2–1 S(1) H_2 line emission. The v=2–1 line can be fitted with a single Gaussian curve while v=1–0 requires two components. The right panel shows in solid contours the high velocity component of H_2 overlaid to the CO high velocity component. The spatial extent of both high velocity components are similar.

The $R_{S(1)}$ ratio is commonly used to disentangle between H_2 gas heated by UV radiation or collisions (e.g., Herrera *et al.* 2011). We find $R_{S(1)}$=0.36 and <0.18 for the low and high H_2 velocity components, respectively. The former ratio can be accounted for by UV heating of the gas in PDRs. The latter can be observed towards PDRs with high radiation fields (>10^4) and gas pressure (>10^8 K cm^{-3}). However, since this component is extended and there is no evidence for extended massive stellar population, we favor

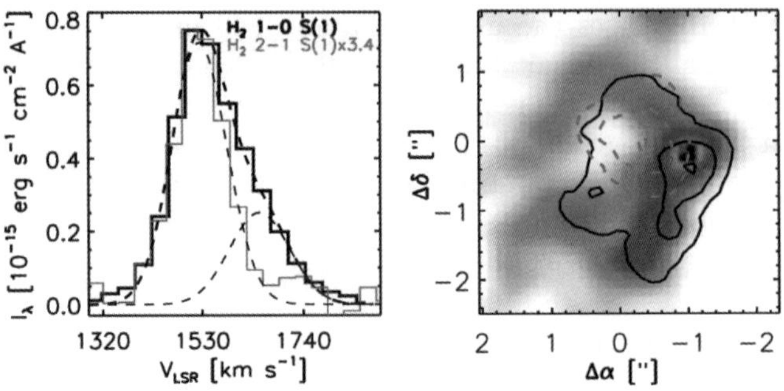

Figure 3. H_2 2–1/1–0 S(1) ratio in the SGMC 4/5. Left panel: spectra for v=1–0 S(1) (v=2–1) H_2 in thick line (thin line, scaled by a factor 3.4). The dashed lines are the gaussian fits for the spectra. Right panel: color image shows integrated intensity of the CO(3–2) high velocity component. Dashed and solid contours are the H_2 emission integrated in 1300–1512 km s^{-1} and 1580–1760 km s^{-1}, respectively. Offset position are the same as in Fig. 2.

shocks as main gas heating mechanism. Values in the range 0.1–0.2 are reproduced in J- and C-shock models (Kristensen 2007) for gas with densities in the range 10^4–10^6 cm^{-3} and shock velocities from 15 to 50 km s^{-1}. This supports the idea that the high velocity component is tracing outflowing gas. Evidence of high velocity ionized gas can be observed in the Brγ line emission. At the velocities of the low velocity component ($<$ 1600 km s^{-1}), the Brγ emission is symmetrical. An excess of ionized gas emission, where the CO(3–2) has a low column density, is observed at velocities of 1600–1700 km s^{-1}, that we interpret as an evidence of escaping ionised gas.

6. Discussion

In our interpretation, today is too late to witness the direct disruption of the natal cloud of SSC B1. For a source of similar characteristics, Murray *et al.* (2010) predicted that the disruption occurs in $\lesssim 1$ Myr. We argue that the outflowing gas from the disruption of the parent cloud is traced by the high velocity component. The bottom panel of Fig. 2 shows the channel maps of this component, which surrounds SSC B1 in a bubble-like shape structure with an inner radius of about 100 pc. Similar gas dynamics have been observed in the starburst galaxies M82 and NGC 253 (Weiss *et al.* 1999; Sakamoto *et al.* 2006), which have been interpreted as expanding super-bubbles.

Uncertainties on different parameters do not significantly impact our interpretation. First, there is an uncertainty on the X_{CO} factor, which implies an uncertainty on the molecular mass obtained from the CO luminosity for both velocity components. Second, there is an uncertainty on the efficiency of the formation of SSC B1. Even though low SFE (5%–10%) are common in Galactic molecular clouds, larger values have also been reported towards massive star forming regions. We have been conservatives by applying the commonly used values for both factors. By doing this, the results on the high velocity component are coherent indicating that most of this gas would be outflowing gas from the parent molecular cloud disruption. The non-detection of red-shifted gas towards the line of sight of the cluster, as typically observed in expanding shells, can be explained by the presence of the low velocity component. This component is very dense and it may be hiding the red-shifted outflowing gas. The origin and fate of the gas which is the closest to SSC B1, the low velocity component, is not clear since we cannot resolve structures smaller than 50 pc. In any case, the results on this component are somehow contradictories and thus difficult to interpret. Its virial parameter indicate that it is self-bound. However, the cluster escape velocity (at the distance of the internal cavity which is determined by the Brγ size) is 49 km s^{-1} while the observed CO line width is 59 km s^{-1}, yielding an expansion velocity of at most 30 km s^{-1}. Another source of confusion is given by a star forming region likely happening very close to SSC B1. Brogan *et al.* (2010) detected a water maser where the dust continuum (Fig. 1, peak emission at the position [0.5,0.4]), HCO+(4–3) line, and mid-IR continuum emission (Snijders *et al.* 2006) peak. The velocities of the H_2O maser and the HCO+(4–3) line coincide. The no detection of Brγ, K-band continuum nor H_2 emission suggest that this source is an extremely young star forming region.

If our interpretation is not correct, we do not really understand how the cluster will get rid of its surrounding matter to match the observed values of SFE. This component is still very massive, comparable with the mass of the cluster, and even more it is self-bound. Further observations are needed to quantify and test our proposed scenario. Specifically, higher resolution observations will reveal the spatial distribution of the gas very close to the cluster, i.e. the gas observed in the low velocity component.

References

Bloemen, J. B. G., Strong, A. W., Blitz, L., Cohen, R. S., & Dame, T. M., *et al.* 1986, *A&A*, 154, 25

Brogan, C., Johnson, K., & Darling, J. 2010, *ApJ*, 716, L51

Geach, J. E., Hickox, R. C., Diamond-Stanic, A. M., Krips, M., Rudnick, G. H., Tremonti, C. A., Sell, P. H., Coil, A. L., & Moustakas, J. 2014, *Nature*, 516, 68

Gilbert, A. M., Graham, J. R., McLean, I. S., Becklin, E. E., Figer, D. F., Larkin, J. E., Levenson, N. A., Teplitz, H. I., & Wilcox, M. K. 2000, *ApJ*, 533, L57

Herrera, C. N., Boulanger, F., & Nesvadba, N. P. H. 2011, *A&A*, 534, A138

Herrera, C. N., Boulanger, F., Nesvadba, N. P. H., & Falgarone, E. 2012, *A&A*, 538, L9

Kristensen, L. E. 2007, PhD thesis, LERMA, Observatoire de Paris-Meudon

Krumholz, M. R. & Matzner, C. D. 2009, *ApJ*, 703, 1352

Krumholz, M. R., Matzner, C. D., & McKee, C. F. 2006, *ApJ*, 653, 361

Le Bourlot, J., Le Petit, F., Pinto, C., Roueff, E., & Roy, F. 2012, *A&A*, 541,A76

Lopez, L. A., Krumholz, M. R., Bolatto, A. D., Prochaska, J. X., & Ramirez-Ruiz, E. 2011, *ApJ*, 731, 91

Mengel, S., Lehnert, M. D., Thatte, N., & Genzel, R. 2005, *A&A*, 443, 41

Murray, N. 2011, *ApJ*, 729, 133

Murray, N., Quataert, E., & Thompson, T. A. 2010, *ApJ*, 709, 191

Sakamoto, K., Ho, P. T. P., Iono, D., Keto, E. R., Mao, R.-Q., Matsushita, S., Peck, A. B., Wiedner, M. C., Wilner, D. J., & Zhao, J.-H. 2006, *ApJ*, 636, 685

Snijders, L., van der Werf, P. P., Brandl, B. R., Mengel, S., Schaerer, D., & Wang, Z. 2006, *ApJ*, 648, L25

Schweizer, F., Burns, C. R., Madore, B. F., Mager, V. A., Phillips, M. M., Freedman, W. L., Boldt, L., Contreras, C., Folatelli, G., González, S., Hamuy, M., Krzeminski, W., Morrell, N. I., Persson, S. E., Roth, M. R., & Stritzinger, M. D. 2008, *AJ*, 136, 1482

Tan, J. C., Krumholz, M. R., & McKee, C. F. 2006, *ApJ*, 641, L121

Weiss, A., Walter, F., Neininger, N., & Klein, U. 1999, *A&A*, 345, L23

Whitmore, B. C., Brogan, C., Chandar, R., Evans, A., Hibbard, J., Johnson, K., Leroy, A., Privon, G., Remijan, A., & Sheth, K. 2014, *ApJ*, 795, 156

Whitmore, B. C. & Schweizer, F. 1995, *AJ*, 109, 960

Wilson, C., Scoville, N., Madden, S., & Charmandaris, V. 2000, *ApJ*, 542, 120

Zezas, A., Fabbiano, G., Baldi, A., Schweizer, F., King, A. R., Ponman, T. J., & Rots, A. H. 2006, *ApJS*, 166, 211

Zhu, M., Seaquist, E. R., & Kuno, N. 2003, *ApJ*, 588, 243

Formation, evolution, and survival of massive star clusters
Proceedings IAU Symposium No. 316, 2015
C. Charbonnel & A. Nota, eds.

doi:10.1017/S174392131500914X

Cluster Assembly in Hierarchically Collapsing Clouds

Enrique Vázquez-Semadeni[1], Alejandro González-Samaniego[2], Manuel Zamora-Avilés[1] and Pedro Colín[1]

[1]Instituto de Radioastronomía y Astrofísica, UNAM
Antigua Carretera a Pátzcuaro # 8701, Morelia, Michoacán, 58088, México
email: `e.vazquez,m.zamora,p.colin@crya.unam.mx`

[2]Instituto de Astronomía, UNAM,
Circuito Exterior, S/N, México, D.F., 04510, México
email: `ags@astro.unam.mx`

Abstract. We discuss the mechanism of cluster formation in hierarchically collapsing molecular clouds. Recent evidence, both observational and numerical, suggests that molecular clouds (MCs) may be undergoing global, hierarchical gravitational collapse. The "hierarchical" regime consists of small-scale collapses within larger-scale ones. The latter implies that the star formation rate increases systematically during the early stages of evolution, and occurs via filamentary flows onto "hubs" of higher density, mass, and velocity dispersion, and culminates a few Myr after than the small-scale collapses have started to form stars. In turn, the small-scale collapses occur in clumps embedded in the filaments, and are themselves falling into the larger potential well of the still-ongoing large-scale collapse. The stars formed in the early, small-scale collapses share the infall motion of their parent clumps towards the larger potential trough, so that the filaments feed both gaseous and stellar material to the hubs. This leads to the presence of older stars in a region where new protostars are still forming, to a scale-free or fractal structure of the clusters, in which each unit is composed of smaller-scale ones, and to the eventual merging of the subunits, explaining the observed structural features of open clusters.

Keywords. Galaxies: star clusters, Gravitation, Hydrodynamics, ISM: clouds, Stars: formation

1. Introduction

It is presently accepted that most stars form in clusters, although the details of the cluster-formation process, especially the origin of their structural properties, remain a matter of active research (e.g., Lada & Lada 2003). In particular, a number of structural properties of the clusters have been uncovered that still require an adequate theoretical understanding, such as: i) the existence of a mass segregation in the clusters, with the most massive stars lying closer to the cluster's center (Hillenbrand & Hartmann 1998); ii) the distribution of protostellar separations, which appears to have no characteristic scale (Bressert et al. 2010); iii) the likely existence of an age gradient in clusters, with the youngest stars being located in the highest-density regions (Kuhn et al. 2015a).

Numerical simulations have begun to offer some insight about these properties. Kirk et al. (2014) have concluded, from a suite of simulations of self-gravitating, decaying isothermal turbulence, that the most massive stars form in situ at the cluster centers, rather than "sinking" there through dynamical interactions in the cluster itself. However, they gave no physical explanation as to why the most massive stars should form there. More recently, Kuhn et al. (2015b) have suggested, by comparing multi-wavelength observations of stellar clusters with numerical simulations, that clusters form by mergers of "subcluster" structures, although again no explanation of why such mergers should occur is provided. The presence of subunits of somewhat different ages in the clusters has also been pointed out by Rivera-Gálvez *et al.* (2015).

A physical mechanism capable of providing a unifying scenario to these properties is the one advanced by Vázquez-Semadeni *et al.* (2009). These authors proposed that, if molecular clouds (MCs) are assembled by large-scale colliding streams of warm, atomic gas, they rapidly condense into the cold atomic phase, becoming Jeans-unstable and beginning to collapse globally. Moreover, the collision of the streams causes moderately supersonic turbulence (e.g., Koyama & Inutsuka 2002; Heitsch et al. 2005) in the cold gas, which produces a spectrum of density fluctuations, where large-amplitude, small-scale (LASS) fluctuations are superposed on smaller-amplitude, larger-scale (SALS) ones (e.g., Kim & Ryu 2005). Since these density fluctuations are nonlinear, the LASS fluctuations have shorter free-fall times than the SALS ones, therefore completing their collapse earlier. This process is therefore similar to Hoyle's (1953) fragmentation, except that the density fluctuations are of turbulent origin and are therefore nonlinear. In what follows, we will refer to this scenario as "hierarchical gravitational collapse" (HGC).

This scenario also predicts that the star formation rate (SFR) in molecular clouds (MCs) evolves over time, as a consequence of the evolution of the clouds themselves as they go through global gravitational collapse (Zamora-Avilés et al. 2012; Hartmann et al. 2012; Zamora-Avilés & Vázquez-Semadeni 2014).

In this contribution, we describe results of a model on the evolution of the SFR and of cluster assembly in a numerical simulation of MC formation and evolution, emphasizing the role of HGC in the resulting structure of the cluster.

2. A model for the evolution of the SFR in molecular clouds

Several models for the SFR and the core mass function (CMF) have been put forward in the last decade (e.g., Krumholz & McKee 2005; Hennebelle & Chabrier 2011; Padoan & Nordlund 2011), all based on the notion that MCs are globally supported by supersonic turbulence, and with the SFR given by the collapse of local density fluctuations induced by the turbulence.

Operationally, the computation of the SFR in these models is performed by using the probability density function (PDF) of the density field—which is known to have a lognormal form for a non-self-gravitating, isothermal turbulent gas (Vázquez-Semadeni 1994)—to compute the mass fraction at densities above a suitably defined density threshold, and dividing it by a timescale characteristic of its density range. Details of the similarities and differences between these models have been given by Federrath & Klessen (2012). Models of this kind are intrinsically *stationary*, as they assume that the MCs are in approximate virial equilibrium between turbulence and gravity. In these models, the main controlling parameters are those characterizing the turbulence – the Mach number M_s, the *virial parameter* α (Bertoldi &McKee 1992), and perhaps the ratio of compressible to solenoidal energy (Federrath & Klessen 2012).

However, the basic premise of equilibrium in the stationary models is inconsistent with the scenario of HGC, since in the latter, the cloud is assumed to be in global gravitational collapse. A model combining the calculation of small-scale collapse as in the above models with the large-scale collapse of the whole cloud has been presented by Zamora-Avilés *et al.* (2012, hereafter Z+12) and Zamora-Avilés & Vázquez-Semadeni (2014, hereafter ZV14). In what follows we will refer to this as "*the evolutionary model*". In it, in addition to the calculation of the instantaneous SFR as in the models mentioned above, the evolution and collapse of the cloud (with the associated systematic increase of its mean density) are also followed in time. The cloud is assumed to be formed by the collision of warm atomic

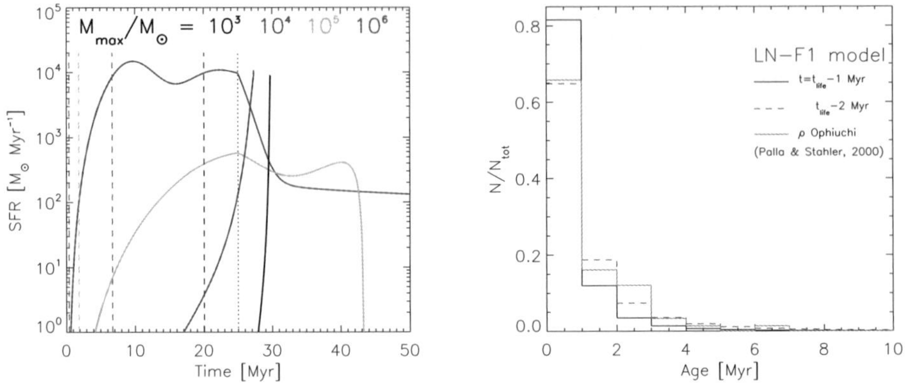

Figure 1. *Left:* Time evolution of the SFR for clouds with $M_{\rm max} = 10^3$, 10^4, 10^5, and 10^6 $M_\odot$ (black, blue, green and red lines respectively). The vertical dashed lines represent the time at which the clouds start to form stars. The vertical black dotted line represents the time at which accretion onto the clouds from the warm medium stops (from ZV14). *Right:* Normalized stellar age histograms for a model cloud at 1 (solid black line) and 2 (dashed blue line) Myr before the cloud is destroyed by the ionizing stellar feedback. Also shown is the corresponding histogram for the embedded cluster in the ρ-Oph cloud (dash-dot-dot magenta line; Palla & Stahler 2000) (from Z+12.).

streams, and to grow in mass by the continued accretion of this material, which undergoes a phase transition to the cold phase as it enters the cloud. The cloud also develops turbulence through a combination of the nonlinear thin-shell instability (Vishniac 1994) and the Kelvin-Helmholz, Rayleigh-Taylor, and thermal instabilities (see, e.g., Heitsch et al. 2005). Because the cloud is collapsing, the instantaneous mass fraction at high densities increases in time, and therefore so does the instantaneous SFR. Knowing the SFR at all times, the total mass in stars can be computed and, assuming a standard IMF, the instantaneous number of massive stars is also computed. These feed back on the cloud through ionizing UV radiation, and the rate of ionization can be computed using standard prescriptions (Franco et al. 1994), eventually destroying the cloud.

Contrary to the stationary equilibrium models, in the evolutionary model the turbulence parameters are not free, but rather they are assumed to have the (fixed) values typical of the cold atomic medium. Thus, the turbulence is assumed to be *moderately* supersonic, with Mach number $M_{\rm s} \sim 3$ (e.g., Heiles & Troland 2003). The only remaining free parameter is the total mass involved in the cloud-formation and evolution process ($M_{\rm max}$). Figure 1 shows, in the *left panel*, the evolution of the SFR for clouds of various masses according to the model. In the *right panel*, this figure shows the histogram of stellar ages in a cluster that evolves according to the evolution of the SFR predicted by the model in a 2000-$M_\odot$ cloud, 1 and 2 Myr before the cloud is destroyed by the ionizing stellar feedback. The figure also shows the normalized stellar age histogram in the ρ-Ophiucus cloud (Palla & Stahler 2000), showing that the model at 2 Myr before destruction matches quite closely the observed histogram. The match is expected to occur some time before destruction because the clusters studied by Palla & Stahler (2000) are still embedded in their parent clouds, suggesting that the destruction of the clouds is not completed yet.

3. A simulation of cluster formation

The results from the model described in the previous section are supplemented by those obtained in a numerical simulation of cloud formation, evolution and destruction,

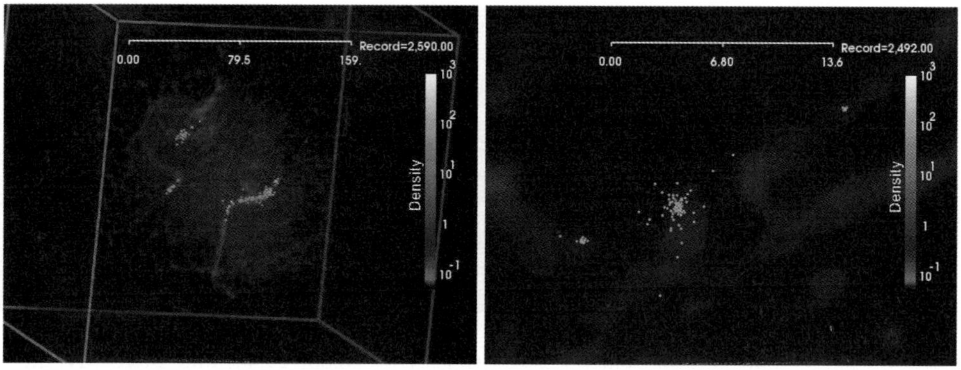

Figure 2. Two views of the numerical simulation. *Left panel:* A global projected view at $t = 25.9$ Myr, showing the three clusters that form within filamentary structures. The box size is 256 pc and the ruler shows a scale of 159 pc. The color bar indicates the local density in cm^{-3}. *Right panel:* A zoom-in around the intermediate-mass cluster at $t = 24.9$ Myr, showing that it is composed of three subunits at this time, which later ($t \sim 30$ Myr) merge to form a single larger cluster. The central object in this image is the cluster whose evolution is shown in Fig. 3, itself having formed from the merger of smaller subunits, and therefore illustrating the self-similar structure of the system. The ruler indicates a scale of 13.6 pc.

presented by Colín et al. (2013, hereafter C+13). This simulation utilized the N-body, AMR-Hydro code ART (Kravtsov *et al.* 1997) with (non-accreting) sink particles, adding a simplified treatment of radiative transfer that allows a first-order estimate for each sink's Strömgren radius. Cells within one Strömgren radius from a sink particle are heated to $T = 10^4$ K. Furthermore, a probabilistic SF scheme was used, which allowed the sink particles to take a power-law mass distribution, with a slope consistent with the Salpeter value. That is, in this simulation, the sink particles correspond to individual stars, with a realistic mass distribution. This allows following the cluster dynamics realistically as well. The box size is 256 pc, and the initial density and temperature in the box were $n = 1$ cm^{-3} and $T = 5000$ K, respectively. Two colliding streams of gas, of radius 64 pc and length 112 pc and with opposite velocities of magnitude 5.9 km s^{-1} each, were set to collide at the $x = 0$ plane in the simulation. A turbulent velocity field with a spectrum peaking at wavenumber $k = 8 \times 2\pi/L_{box}$ (so that the fluctuation size scale is smaller than the radius of the colliding streams) was applied to trigger instabilities in the compressed layer. For more details, we refer the reader to C+13.

In this simulation, three main clusters form, two of which are sufficiently massive to disperse/evaporate the dense gas around them on a timescale of ~ 13 Myr. Stars begin to form at $t \approx 18.9$ Myr, and the first HII regions appear at $t \approx 24.2$ Myr. At $t \approx 37.5$ Myr, the dense gas has been cleared from a radius of ~ 70 pc around the clusters. Figure 2 shows, in the *left panel*, one snapshot of the simulation, at $t = 25.9$ Myr into the evolution, in which the growing HII regions around the two most massive clusters can be seen as faint shells around the clusters, while the filamentary structure of the cloud is still noticeable in general. The clusters themselves have formed along the main filamentary structures in the cloud, and contain 10^3–$10^4 M_\odot$.

A crucial implication of the HGC scenario is that the filaments constitute part of the large-scale gravitational collapse, funneling gas into the cores within them, as observed by Gómez & Vázquez-Semadeni (2014) in a numerical simulation of cloud formation and collapse. Moreover, these authors observed a hierarchy of collapses within the filaments, so that *small clumps, which are sometimes forming stars already, are falling onto larger-scale ones.* This means that the late-stage clusters are assembled from smaller sub-units

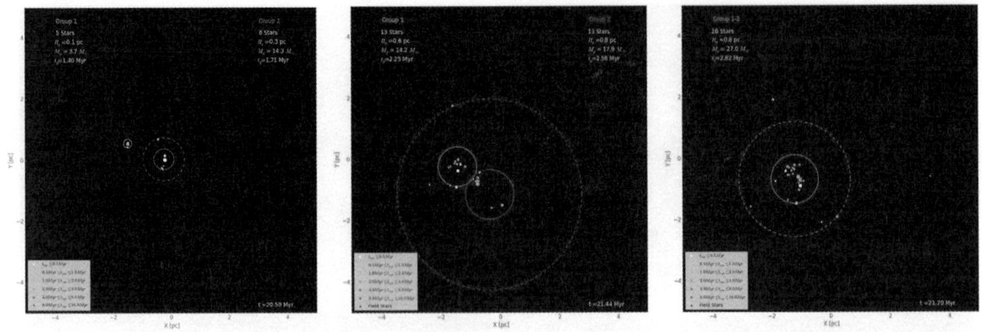

Figure 3. Three snapshots of "cluster 2" at times t = 20.59, 21.44, and 21.70 Myr in the simulation (respectively, left, middle, and right frames), showing the merging of two subunits to form the cluster.

that formed in the filaments in the periphery of the trough of the large-scale gravitational potential well. This is illustrated in the *right panel* of Fig. 2, which shows a zoom-in around the second most massive cluster (hereafter, "cluster 2") of those seen in the left panel, at t = 24.9 Myr. It can be seen that this cluster is actually composed of three groups at this time, which later (at $t \sim 30$ Myr) merge to form a single larger cluster.

Operationally defining a group is a nontrivial task, because the gas clumps continue to form stars over extended periods of time due to the accretion from the filaments onto them. We thus define membership to a group by a sequence of steps. At early times, when the groups are clearly distinct by eye, we apply a "friends-of-friends" algorithm to the stars to define the groups. Subsequently, as new stars are formed, we assign them to the groups whose center of mass is closest to the new star. Also, we define the radius of the group at each time as the distance from its center of mass to the second most distant star. Finally, we say that two groups have merged when the distance between their centers of mass is smaller than the larger of the two radii.

The evolution and assembly of the central group in the *right panel* of Fig. 2 is illustrated in Fig. 3, which shows the projected positions of the sink particles on the (x, y) plane at times t = 20.59, 21.44, and 21.70 Myr into the evolution of the simulation in the *left, middle,* and *right* panels, respectively. It is seen that, at the earlier time, the system consists of two subgroups, which approach each other and finally merge. In general, we observe this mechanism operating at all scales in our simulation, with groups in turn consisting of subgroups, and so on, implying that the clusters must exhibit a self-similar, fractal structure.

4. Discussion

The results described above suggest that the multi-scale, hierarchical gravitational collapse (HGC) of the clouds may be responsible for at least three observed properties of open clusters: 1) that there appears to be an increase of the SFR, so that the age distributions of clusters generally have a tail of older stars (3-7 Myr), while most stars are young (1-2 Myr; Palla & Stahler 2000); 2) that they appear to have a self-similar structure, observationally manifested in an absence of a characteristic scale in the distribution of protostellar separations (Bressert et al. 2010). In our simulation, this is manifested in the hierarchical membership of the stellar units that conform a cluster; 3) that the clusters appear to consist of subunits of slightly different ages that merge to form larger structures (Kuhn et al. 2015b; Rivera-Gálvez *et al.* 2015). These are natural consequences of HGC in globally collapsing clouds. The global collapse produces an increase of the

SFR. The hierarchical nature causes, at large scales, filamentary structures that feed the star-forming clumps for extended periods of time and, at small scales, it produces local star-forming sites that fall into the troughs of the large-scale contracting structures.

References

Bertoldi, F. & McKee, C. F. 1992, *ApJ*, 395, 140

Bressert, E., Bastian, N., Gutermuth, R., *et al.* 2010, *MNRAS*, 409, L54

Colín, P., Vázquez-Semadeni, E., & Gómez, G. C. 2013, *MNRAS*, 435, 1701

Federrath, C. & Klessen, R. S. 2012, *ApJ*, 761, 156

Franco, J., Shore, S. N., & Tenorio-Tagle, G. 1994, *ApJ*, 436, 795

Gao, Y. & Solomon, P. M. 2004, *ApJ*, 606, 271

Gómez, G. C. & Vázquez-Semadeni, E. 2014, *ApJ*, 791, 124

Hartmann, L., Ballesteros-Paredes, J., & Heitsch, F. 2012, *MNRAS*, 420, 1457

Heitsch, F., Burkert, A., Hartmann, L. W., Slyz, A. D., & Devriendt, J. E. G. 2005, *ApJL*, 633, L113

Hennebelle, P. & Chabrier, G. 2011, *ApJL*, 743, L29

Hillenbrand, L. A. & Hartmann, L. W. 1998, *ApJ*, 492, 540

Hoyle, F. 1953, *ApJ*, 118, 513

Kim, J. & Ryu, D. 2005, *ApJL*, 630, L45

Kirk, H., Offner, S. S. R., & Redmond, K. J. 2014, *MNRAS*, 439, 1765

Koyama, H. & Inutsuka, S.-i. 2002, *ApJL*, 564, L97

Kravtsov, A. V., Klypin, A. A., & Khokhlov, A. M. 1997, *ApJS*, 111, 73

Krumholz, M. R. & McKee, C. F. 2005, *ApJ*, 630, 250

Kuhn, M. A., Getman, K. V., & Feigelson, E. D. 2015a, *ApJ*, 802, 60

Kuhn, M. A., Feigelson, E. D., Getman, K. V., *et al.* 2015b, arXiv:1507.05653

Lada, C. J. & Lada, E. A. 2003, *ARA&A*, 41, 57

Palla, F. & Stahler, S. W. 2000, *ApJ*, 540, 255

Rivera-Gálvez, S., Román-Zúñiga, C. G., Jiménez-Bailón, E., *et al.* 2015, arXiv:1510.02835

Vázquez-Semadeni, E. 1994, *ApJ*, 423, 681

Vázquez-Semadeni, E., Gómez, G. C., Jappsen, A.-K., Ballesteros-Paredes, J., & Klessen, R. S. 2009, *ApJ*, 707, 1023

Vishniac, E. T. 1994, *ApJ*, 428, 186

Williams, J. P. & McKee, C. F. 1997, *ApJ*, 476, 166

Zamora-Avilés, M., Vázquez-Semadeni, E., & Colín, P. 2012, *ApJ*, 751, 77

Zamora-Avilés, M. & Vázquez-Semadeni, E. 2014, *ApJ*, 793, 84

Formation, evolution, and survival of massive star clusters
Proceedings IAU Symposium No. 316, 2015
C. Charbonnel & A. Nota, eds.

doi:10.1017/S1743921315010522

Universal and Environmental Influences on Star Clusters

Bradley C. Whitmore[1]

[1]Space Telescope Science Institute, 3700 San Martin Drive, Baltimore, MD 21218, USA
email: whitmore@stsci.edu

Abstract. There is now evidence that some aspects of compact star cluster formation and destruction are quasi-universal in nature, and some aspects depend on environment. But what do we mean by these terms, environmental and universal? Is one the dominant influence? How can things be both universal and environmentally dependent? In this contribution we first provide a brief historical overview, then examine evidence for both universality and environmental dependences, and finish by examining a new approach that both demonstrates the degree to which cluster mass functions are universal (i.e., to a level of roughly 0.2 in the Log over three orders of magnitude when normalizing by the star formation rate), and enables a method for quantifying 2nd-order environmental effects.

1. Introduction - A Brief History and the Size-of-Sample Effect

One of the first major results from the *Hubble Space Telescope*, even while in its aberrated state, was the discovery of Super Star Clusters (SSCs; fainter clusters are often called Young Massive Clusters, YMCs) in merging galaxies (e.g., NGC 1275 - Holtzman *et al.* (1992), NGC 7252 - Whitmore *et al.* (1993), the Antennae - Whitmore *et al.* (1995)). The implication that we could study the formation of young globular clusters in the local universe rather than trying to ascertain how they formed 13 billion years ago caught the interest of a generation of astronomers. An obvious inference was that there was something special about the environment of interacting galaxies that allowed SSCs to form in these systems and no others. This turned out not to be the case.

YMCs were also found in a variety of other types of galaxies, ranging from starburst (NGC 1569 and NGC 1705 -O'Connell *et al.* (1994)), to barred galaxies (NGC 1097 & NGC 6951 -Barth *et al.* (1995)). The finding by Larsen & Richtler (1999) that YMCs could also be found in spiral galaxies, albeit generally with fainter luminosities and lower masses, was another important early discovery. These results suggested that the formation of young massive clusters was more universal than first thought.

The first compilation of cluster luminosity functions for merging galaxies was made by Whitmore (2003). Because the sample was small, a simpler statistic, the luminosity of the brightest cluster versus the Log of the number of clusters in the galaxy brighter than Mv = -9 (i.e., Log N) was also plotted. The expectation was that the distribution would be bi-modal, with the spirals and dwarf galaxies having fainter characteristic luminosities and the mergers having brighter characteristic magnitudes. The surprising result was the finding that a well defined, continuous linear correlation was present. It soon became evident that this relationship could be explained if all the galaxies had the same luminosity function with $dN/dL \propto L^{-2}$ (see Figure 1). It appears that the brightest clusters in merging galaxies are brighter because they come from a larger sample, not because there is necessarily any special physics involved. The larger sample of clusters in these galaxies fills out the distribution more completely, leading to a population of very

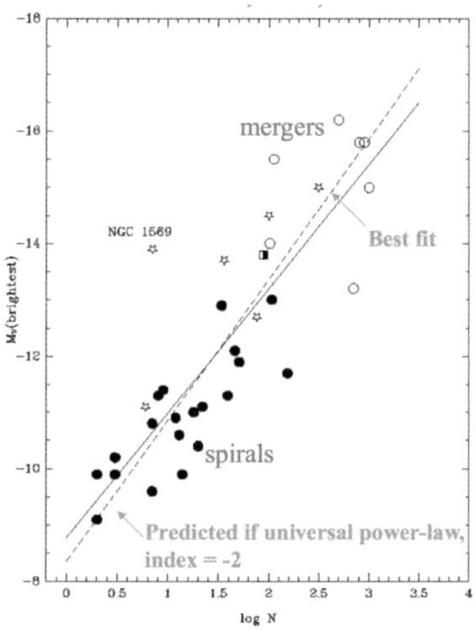

Figure 1. $M_{V\,(brightest)}$ vs. Log N from Whitmore (2003).

bright, massive clusters, along with very large numbers of fainter, less massive clusters. This is generally called the "size-of-sample" effect (De Vaucouleurs (1970)).

A number of other studies found similar results, including Larsen *et al.* (2002), Bastian (2008), Whitmore *et al.* (2014), and Adamo *et al.* (2015). The relationship has now been extended over six orders of magnitudes, from dwarf galaxies (Bastian (2008)) to ULIRGS (Vavilkin (2011)). The use of Log SFR (Star Formation Rate) on the X axis stretches out the diagram by allowing the inclusions of very faint dwarf galaxies, which generally do not have clusters brighter than Mv = -9. The similarity between the versions of the diagram using Log N and using Log SFR, and the tight correlation between Log N and Log SFR (e.g., RMS scatter = 0.11 in Figure 7 in Whitmore *et al.* (2014)) shows that Log SFR is a good stand-in for the sample size, a point that will be important in §5.

Hence, it appears that there is a roughly universal luminosity function for young clusters (see Whitmore *et al.* (2014) for additional discussion). The larger sample (higher SFR) in mergers may simply be due to the fact that conditions for making clusters are globally present in mergers, presumably due to the higher pressure of the ISM due to the collision, but only locally present in the arms of spiral galaxies.

2. What Does Universality Mean and Why Do We Care ?

What do we mean by Universality? In the context used in this paper, it means the fundamental relationships, e.g., the luminosity function, the mass function, and the age distribution, are the same for all star-forming galaxies in all environments, once the size-of-sample effects are taken into account.

Taken to extremes, it is impossible for anything to be perfectly universal, since this represents a limiting case that can never be reached. As soon as there is a deviation, at the 10%, or 1%, or 0.01 % level, perfect universality breaks down. What we are really interested in is what are the *dominant* influences that determine the star cluster demographics. For this reason we use the term "quasi-universal", rather than using the extreme definition.

Why do we care? If the dominant influences are quasi-universal (e.g., deviations are a few tenths over several orders of magnitude), this implies that the primary physical influences are *internal* to the cluster (e.g., gas loss, stellar evolution, stellar

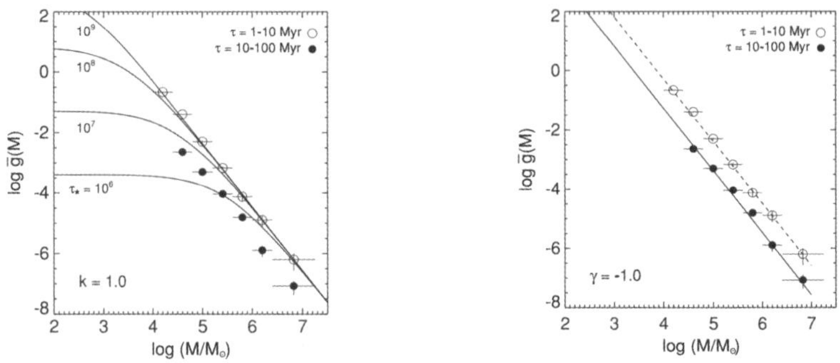

Figure 2. Predicted mass functions from environmental models (left) and the quasi-universal model (right) overlaid on data from the Antennae galaxies. From Fall *et al.* (2009).

ejection, ...). If the dominant influences are environmental, this implies that the primary physical influences are external to the clusters; for example the position in the galaxy (and hence the interstellar pressure, the density of GMCs, the tidal shear), or the morphology of the galaxy itself (e.g., merger, spiral, dwarf).

3. Straight or Curved Lines ?

Papers that have discussed the universal model include Whitmore *et al.* (2004), Whitmore *et al.* (2007), Lamers (2009), Fall *et al.* (2009), and Fall & Chandar (2012).

A very brief, overly simplified, phenomenological description of the difference between the quasi-universal model and the environmentally dependent model is that the former predicts straight lines in the mass function and age distribution diagrams, while the later predicts curved lines.

Figure 2 shows an example using data from the Antennae galaxies in Fall *et al.* (2009). While the data appears to produce nearly linear mass functions, as predicted by the quasi-universal model as shown in the right panel, various environmental models predict curved lines, as shown in the left diagram. If the dissipation time is very long (e.g., 10^9 yrs), the predicted line is nearly linear, but the offset between the two age ranges is still not predicted. Similar results for the age distributions can be found in Figure 11 from Fall *et al.* (2009).

Observational incompleteness also produces curvature, and can mimic environmental effects. A linear relationship, as found in the data, represents a limiting case and is not likely to be the result of some conspiracy between completeness issue, selection effects, or other observational difficulties, all of which typically result in non-linear relations.

Cumulative fits can be useful for showing deviations from linearity. However, they also amplify noise from small number statistics. Several studies by proponents of environmental models employ cumulative fits (e.g., Figs. 9 and 15 from Bastian *et al.* (2012)).

One approach is to show results using both normal binning and cumulative fits, to allow readers to develop a better feel for the degree and reliability of possible deviations. An example where both techniques are shown is Whitmore *et al.* (2014) (e.g., Figures 5 and 9). Note the 1 % arrows in Figure 9 that reminds the reader that these deviations from linearity in a cumulative fit often reflect a very small number of clusters.

4. Recent Evidence of Environmental Effects

Papers that have explored environmental models include Boutloukos, & Lamers (2003), Lamers (2009), Bastian *et al.* (2012), Kruijssen *et al.* (2012), and Adamo *et al.* (2015).

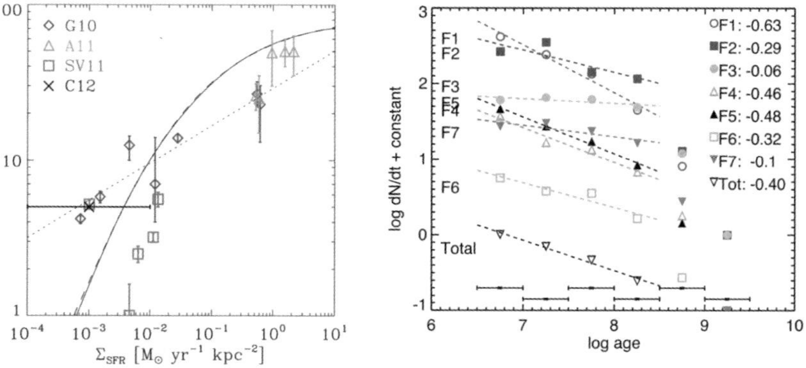

Figure 3. Examples of environmental dependencies. Left - Relationship between the fraction of mass in bound clusters, Γ, vs Specific Star Formation Rate (from Kruijssen *et al.* (2012)). Right - Cluster age distributions for seven fields in M83 (from Silva-Villa *et al.* (2014)).

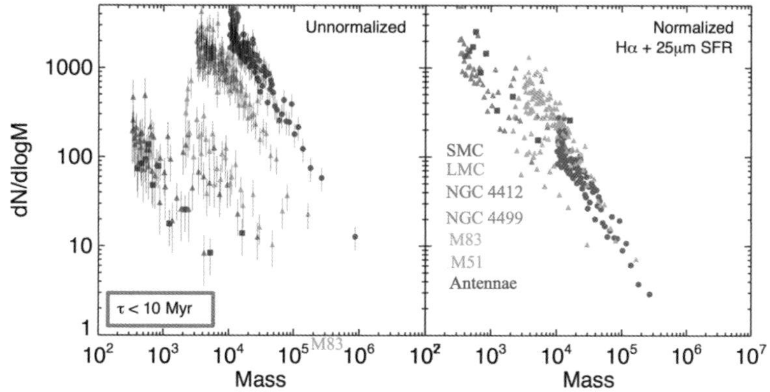

Figure 4. Original (left) and normalized (right) Cluster Mass Functions for clusters with ages less than 10 Myr (from Chandar *et al.* (2015)).

Figure 3 shows two of the most convincing recent examples of environmental dependencies. In the left panel we find both theoretical (dashed lines - Kruijssen *et al.* (2012)) and observational (dotted line and data points from Goddard *et al.* (2010)) indications that the fraction of mass in bound clusters, Γ, depends on the specific Star Formation Rate. While the determination of Γ is rather elaborate, involving a number of measurements (e.g., SFR, field of view used for SFR), assumptions (slope of mass function = -2.0, cluster age range of 7 Myr) and extrapolations (to lower mass clusters in order to determine the total population of clusters), the effect does appear to be real.

In the right panel we find evidence that the slope of the age distribution in the spiral galaxy M83 flattens in the outer parts of the galaxy (Silva-Villa *et al.* (2014)). While there appears to be a relatively clear trend with radius (see paper for details), we note that the mean slope is -0.4. This represents a 60 % disruption rate per decade. We also note that the relationships are all relatively straight in the region of completeness (to $10^{8.5}$ yrs), with downturns due to incompleteness beyond this range.

5. A New Result - CMF/SFR for Seven Galaxies

An underlying implication of the quasi-universal model is that the Star Formation Rate (SFR) and cluster formation rate are directly proportional. If so, it should be possible

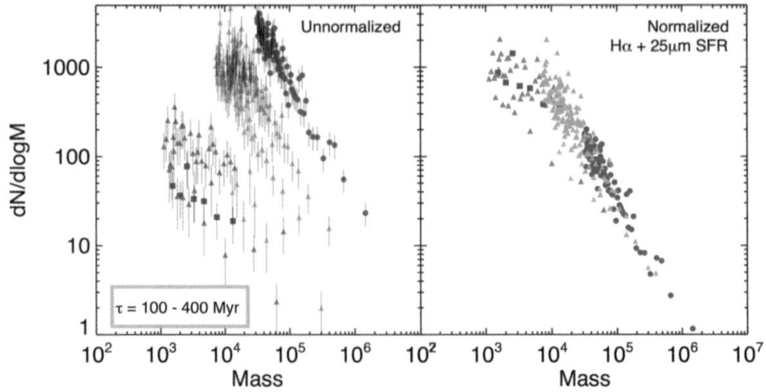

Figure 5. Same as Figure 4 for an older sample of clusters.

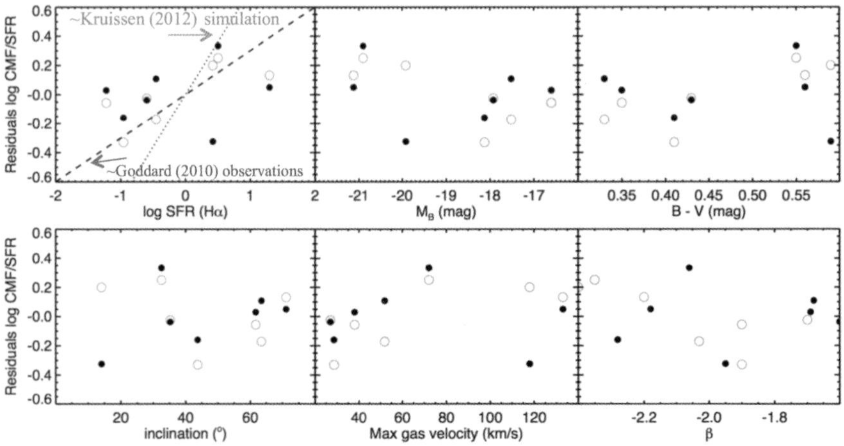

Figure 6. From Chandar *et al.* (2015). See text for discussion.

to use the SFR (e.g., based on extinction corrected Halpha) to scale the Cluster Mass Function (CMF) of all galaxies to fit on the same universal relation. Figure 4 (from Chandar *et al.* (2015) - hereafter CFW) shows that normalizing by SFR collapses the spread from three orders of magnitude to a RMS scatter of a few tenths in the Log.

Somewhat surprisingly, Figure 5 shows that this works just as well for an older population of clusters, where we can be certain that the clusters are 'bound', an issue of some debate when working with clusters less than 10 Myr old.

The scatter in both diagrams is $\approx$0.2 in the Log, roughly the same as expected from the error budget. This suggests that the SFR has been relatively constant over a long period of time, and the disruption rate must also be relatively universal.

Are there any clear 2nd-order correlations? It appears that the dwarf galaxies (e.g., NGC 4412, SMC, LMC) remain slightly low in the two diagrams after normalization, perhaps reflecting a smaller bound fraction of clusters. Figure 6 shows the residuals from the normalized diagrams for a variety of parameters. Focusing on the relationship for Log SFR in the upper left panel, we find that the relationship for the 10 Myr sample (solid points) is not statistically significant. For the older populations the relationship is only significant at the 93 % level (i.e., about a 2 σ result). However, we note that if there is a trend, it is in the same sense as the relationship with Γ from Kruijssen *et al.* (2012), although not as steep, especially with respect to the simulation.

Linear approximations of the Kruijssen *et al.* (2012) simulation, and the observational trend from Goddard *et al.* (2010), are included in Figure 6 (i.e., by assuming the same range in Γ from the faintest to brightest galaxies in both samples). Note that the Kruijssen diagram uses the Specific SFR (hereafter SSFR - i.e., divided by the surface area of the galaxy), while the CFW figure uses Log SFR. This reflects our focus on examining the degree of universality, by using a parameter that is closely related to the sample size (i.e., Log SFR - see discussion in §1) to normalize the CMFs. Kruijssen *et al.* (2012) is primarily interested in the environmental dependence, hence they employ the SSFR rate as the independent variable to remove the size-of-sample (universality) effect.

Both approaches have their merits. As recommended in §3, it is often advantageous to try more than one approach. In this way we can better ascertain to what degree universality can explain the range in different CMFs, and to what degree environmental dependencies are important.

While the results using the CMF/SFR approach look promising, we should keep in mind that the sample is only seven galaxies. The Large Extragalactic Galaxies UV Survey (LEGUS - Calzetti *et al.* (2015)) sample will provide a much larger sample.

References

Adamo, A., Kruijssen, J. M. D., Bastian, N., Silva-Villa, E., & Ryon, J. 2015, MNRAS, 452, 246
Barth, A. J., Ho, L. C., Filippenko, A. V., & Sargent, W. L. 1995, AJ, 110, 1009
Bastian, N. 2008, MNRAS, 390, 759
Bastian, N., Adamo, A., Gieles, M., *et al.* 2012, MNRAS, 419, 2606
Boutloukos, S. G. & Lamers, H. J. G. L. M. 2003, MNRAS, 338, 717
Calzetti, D., Lee, J. C., Sabbi, E., *et al.* 2015, AJ, 149, 51
Chandar, R., Fall, S. M., & Whitmore, B. C. 2015, ApJ, 810, 1
De Vaucouleurs, G. 1970, ApJ, 159, 435
Fall, S. M. & Chandar, R. 2012, ApJ, 752, 96
Fall, S. M., Chandar, R., & Whitmore, B. C., 2009, ApJ, 704, 453
Goddard, Q. E., Bastian, N., & Kennicutt, R. C. 2010, MNRAS, 405, 857
Holtzman, J. A. *et al.* 1992, AJ, 103, 69
Kruijssen, J. M. D. 2012, MNRAS, 426, 3008
Lamers, H. J. G.. L. M.. 2009, Ap&SS, 324,183
Larsen, S. S. 2002, AJ, 124, 1393
Larsen, S. S. & Richtler, T. 1999, A&A, 345, 59
O'Connell, R. W., Gallagher, J. S., & Hunter, D. A. 1994, ApJ, 433, 650
Silva-Villa, E., Adamo, A., Bastian, N., Fouesneau, M., & Zackrisson, E. 2014, MNRAS, 440, 116
Vavilkin, T. 2011, PhD thesis, State University of New York at Stony Brook
Whitmore, B. C. 2003, in A Decade of Hubble Space Telescope Science, ed. Mario Livio, Keith Noll & Massimo Stiavelli (Cambridge, UK: Cambridge Univ. Press)
Whitmore, B. C. 2004, in ASP Conf. Ser. 322, The Formation and Evolution of Massive Young Star Clusters, ed. H. Lamers, L. J. Smith, & A. Nota (San Francisco: ASP), 419
Whitmore, B. C., *et al.* 2014, AJ, 147, 78
Whitmore, B. C., Chandar, R., & Fall, S. M. 2007, AJ, 133, 1067
Whitmore, B. C. & Schweizer, F. 1995, AJ, 109, 960
Whitmore, B. C., Schweizer, F., Leitherer, C., Borne, K., & Robert, C. 1993, AJ, 106, 1354

Formation, evolution, and survival of massive star clusters
Proceedings IAU Symposium No. 316, 2015
C. Charbonnel & A. Nota, eds.

doi:10.1017/S1743921316000478

Formation of the super star cluster RCW 38 triggered by cloud-cloud collision

Yasuo Fukui[1]

[1]Department of Physics, Nagoya University, Furo-cho, Chikusa-ku, Nagoya, Aichi 464-8601, Japan. email: fukui@a.phys.nagoya-u.ac.jp

Abstract. RCW 38 is the youngest super star cluster in the Galaxy and is located at a distance of 1.7 kpc. Molecular observations revealed that the cluster is associated with two molecular clouds having velocity difference of 12 km s^{-1}. We interpret that the two clouds are colliding with each other and the collision triggered the cluster formation. The natal molecular gas still survives within ∼0.5 pc of the central O stars which have an age of 0.1 Myrs as inferred from the collision morphology. We suggest that the high column density of one of the clouds 10^{23} cm^{-2} enabled formation of ∼20 O stars in the cluster center and discuss the implications on massive cluster formation.

Keywords. ISM: clouds — ISM: molecules — ISM: kinematics and dynamics — stars: formation

1. Introduction

It is an important question how massive stellar clusters are formed in the Universe. The globular clusters in the Galactic halo having mass of 10^5 $M_\odot$ are the most extreme of such massive clusters, whereas we are not able to witness their formation which happened more than 10 Byrs ago. An alternative way to approach the question is to make a detailed study of rich young clusters in the Local Group, while they are not as massive as the globular clusters.

The young super star clusters having 10^4 $M_\odot$ may allow us to explore the mechanism of massive cluster formation. There are 13 super star clusters known to date in the Galaxy; they include RCW38 in addition to 12 clusters listed by Portegies Zwart *et al.* (2010). RCW 38 is a rich and young super star cluster having 2000 stars with 30 O stars and is located at 1.7 kpc in the Galaxy (e.g., Wolk *et al.* 2008). RCW 38 whose age is estimated to be ∼0.5 Myrs (Wolk *et al.* 2006) is younger than any of the other 12 clusters. RCW 38 therefore deserves a detailed study of the natal molecular gas which may still hold the initial condition of massive cluster formation.

Recent molecular observations show that the two young super star clusters Westerlund 2 and NGC 3603 are formed by triggering of cloud-cloud collision (Furukawa *et al.* 2009, Ohama *et al.* 2010, Fukui *et al.* 2014). A theoretical study of colliding molecular gas flows (Inoue and Fukui 2013) demonstrates that the collision creates a turbulent interface layer where dense massive cores are formed, leading to high-mass star formation. It is important to explore if the other super star clusters are also formed via cloud-cloud collision, and RCW38 is one of the primary targets in such a follow-up study. We present the main results of the study in this contribution, while a fuller account of this work will be published elsewhere (Fukui *et al.* 2015).

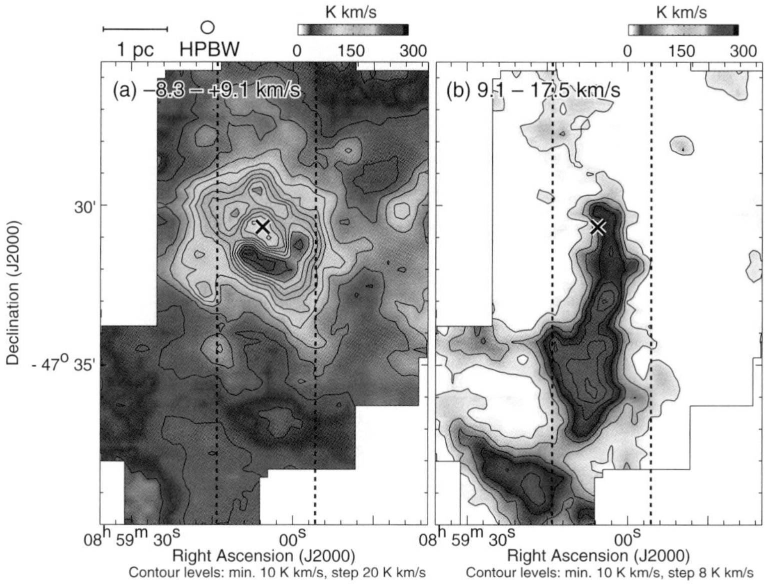

Figure 1. Mopra ^{12}CO J = 1–0 distributions of the two velocity clouds toward RCW 38 are presented. Cross indicates the position of IRS 2. Vertical dashed lines in each panel show the integration range of the declination-velocity diagram shown in Figure 2.

2. Results and interpretation

Observations of the molecular clouds were made in the rotational transition, ^{12}CO J = 1–0, J = 3–2, and ^{13}CO J = 1–0, with the three telescopes NANTEN2, ASTE and Mopra. The Mopra results, obtained through guiding by the NANTEN2 large scale data, are shown in Figure 1. The CO results show two clouds at velocities 2 km s^{-1} and 14 km s^{-1} toward RCW 38. The cloud at 2 km s^{-1} shows a ring-like shape (hereafter "ring cloud") where the cluster is located inside the ring and the other at 14 km s^{-1} is finger-like (hereafter "finger cloud"). We find a high ratio of the J = 3–2 transition to the J = 1–0 transition, $R_{3-2/1-0}$, greater than 0.8 toward the cluster in the two clouds. This high ratio indicates kinetic temperature higher than 30 K according to the LVG calculations and demonstrates that the two clouds are physically associated with the cluster and are radiatively heated up. The two clouds are linked by a bridging feature whose velocity is between the two clouds as shown in the declination-velocity diagram (Figure 2). The location of the central 20 O stars in the cluster shows good spatial correspondence with the top of the finger cloud and the bridging feature (Figure 3). We note similar bridging features are found in two regions in the southern outer skirt of the ring cloud where no star formation is seen (Figure 2). The physical association of the two clouds with the cluster is consistent with the similar distribution of the molecular clouds with the infrared dust features heated by the cluster; the ring cloud shows remarkable correlation with the extend dust features, and the finger cloud corresponds to the infrared ridge IRS 1 (Smith *et al.* 1999) toward the cluster.

Since the total mass of the clouds and the cluster is small, less than $10^5\,M_\odot$, the velocity separation 12 km s^{-1} is too large to be gravitationally bound. Based on these results, we frame a hypothesis that cloud-cloud collision took place recently and triggered formation of the O stars in RCW 38. It has been demonstrated that supersonic collision between molecular clouds forms a compressed interface layer where dense and massive cloud cores are formed (Inoue and Fukui 2013). This explains formation of the O stars and

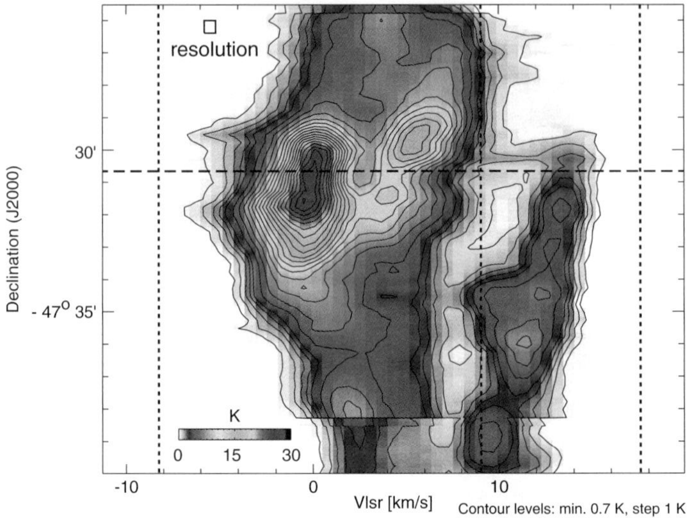

Figure 2. Declination-velocity diagram of the Mopra ^{12}CO $J = 1$–0 emission for the two clouds in RCW 38. Vertical dashed line indicates the direction of IRS 2.

the bridging feature, a typical signature of collision, as resulting from the gas acceleration in the collision (e.g., Haworth *et al.* 2015a, 2015b).

3. Discussion

The timescale of the collision is estimated to be very short $\sim 1 \times 10^5$ yrs as calculated by the ratio of the typical cloud size and the velocity, $\sim 1\,\mathrm{pc}/10\,\mathrm{km\ s^{-1}}$, where we adopt the width of the finger cloud as the cloud size. In the collision scenario, the formation time scale of the O stars is in the same order of magnitude with the collision time scale. The most massive star in RCW 38 is IRS 2 (DeRose *et al.* 2009), and the mass of this O5.5 star is around 40 $M_\odot$. The rest of the member O stars perhaps have $\sim$20 $M_\odot$ for each on average. The mass accretion rate for a 40 $M_\odot$ star is estimated to be $4 \times 10^{-4}\ M_\odot\,\mathrm{yr}^{-1}$ if the star was formed by a constant mass accretion rate in $\sim 1 \times 10^5$ yrs. The molecular mass in the collisional region is estimated to be $\sim$500 $M_\odot$ for typical molecular column density of 10^{23} cm^{-2} of the ring cloud. This suggests very high formation efficiency of the O stars whose total mass is $\sim$400 $M_\odot$ ($\simeq 20 \times 20\,M_\odot$).

RCW 38 is the youngest super star cluster known to date in the Galaxy. The age of the O stars estimated in the present collision interpretation is 0.1 Myrs, which does not contradict the estimate, $\sim$0.5 Myrs, obtained from the stellar photometric data (Wolk *et al.* 2006). RCW 38 is a rare case where the natal molecular gas still survives against the stellar feedback. The other two young super star clusters, Westerlund 2 and NGC 3603, where cloud-cloud collision played a role in triggering cluster formation, have an age of 2–3 Myrs (Furukawa *et al.* 2009, Ohama *et al.* 2010, Fukui *et al.* 2014). In these two clusters the molecular gas is almost fully ionized within a few pc of the cluster, and this does not allow one to see the details of the collisional interaction of the natal molecular gas (Figure 4).

The natal molecular gas in RCW 38 reveals significant details of the O star formation. Most importantly, the size of the O star distribution is similar to that of the collisional interaction, which shows the bridging feature. This suggests that the initial O star distribution is primarily determined by the collision. We also suggest that the O stars

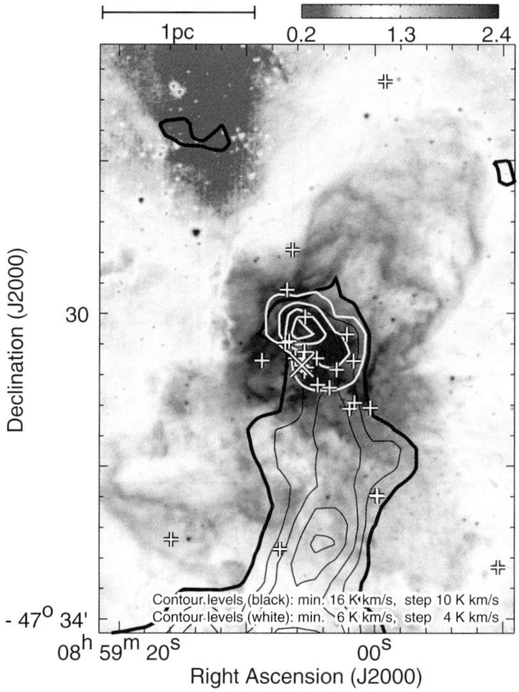

Figure 3. Contour maps of the finger cloud (black contours) and the bridge feature (white contours) are shown superimposed on the Spitzer/IRAC 3.6 μm image (Wolk *et al.* 2008). White crosses indicate IRS 2 (the O 5.5 star) and O star candidates identified by Wolk *et al.* (2006), where the large cross depicts IRS 2.

are formed toward the high molecular column density, 10^{23} cm^{-2}, in the ring cloud. A comparison of RCW 38 with regions of single O star formation in M 20 and RCW 120 indicates that column density is higher in RCW 38 than in M 20 and RCW 120 by an order of magnitude (Torii *et al.* 2011, 2015). This suggests that the high column density 10^{23} cm^{-2} is a necessary condition for multiple O star formation with O star density of 30 (pc)$^{-2}$ as found in RCW 38.

RCW 38 includes 2000 stars, most of which are low mass stars extended over the ring cloud beyond the region of the collisional interaction. These low mass stars are probably formed prior to the collision in the dense ring cloud over a time scale of a few Myrs. The low mass stars are not powerful enough to disperse the ring cloud and were coexistent with the ring cloud until the collision forms the O stars. At present the O stars are violently ionizing the ring cloud.

In the Galaxy, we have 13 super star clusters known to date. Five of them are associated with infrared nebulosity and are young with an age of 0.1 – 3 Myrs. We find that all the five clusters are associated with colliding two clouds including the two clusters, DSB[2003] 179 and Trumpler14 (Kuwahara *et al.* 2016 in preparation, Fukui *et al.* 2016 in preparation).

Magetohydrodynamical numerical simulations of colliding molecular gas demonstrate increase of the mass accretion rate by two orders of magnitude in the shock compressed interface after collision (Inoue and Fukui 2013), providing a theoretical ground for the rapid O star formation. It is possible that low-velocity collision between two molecular clouds causes smaller turbulence in the interface layer and may trigger formation of lower-mass stars by a lower mass accretion rate. Observations of low-mass star formation triggered by cloud-cloud collision are presented and discussed by several authors (Duarte-

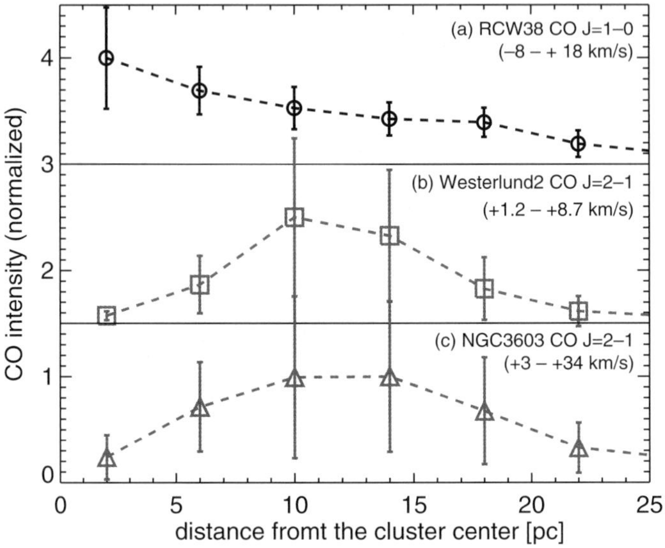

Figure 4. Radial distributions of the CO emission in the three massive star clusters RCW 38 (a), Westerlund 2 (b), and NGC 3603 (c). The CO data are taken from Fukui *et al.* (2015) for RCW 38, Furukawa *et al.* (2009) for Westerlund 2, and Fukui *et al.* (2014) for NGC 3603.

Cabralet al. 2010; Higuchi *et al.* 2010; Nakamura *et al.* 2013). The relative velocity as small as a few km s^{-1} in these works, however, leaves room for non-triggered star formation in a gravitationally bound system.

4. Implications on formation of massive clusters

The two scenarios for high-mass star formation in the literature, the monolithic collapse and competitive accretion, are not yet established by confrontation with observations (e.g., Zinnecker and York 2007, Tan *et al.* 2014). On the other hand, cloud-cloud collision provides another possible scenario for O star formation, and offers directly testable four observational signatures including (1) two clouds associated with young O stars with (2) relative velocity greater than 10 km s^{-1}, as long as the projection effect is not too large to significantly decrease observed velocity difference, (3) the bridging feature between the two velocities, and (4) the nested distribution between the two clouds as expected by numerical simulations of colliding clouds (Habe and Ohta 1992, Anathpindika 2010, and Takahira *et al.* 2014). The present observations of RCW 38 show signatures (1), (2) and (3). The lack of the nested distribution may be ascribed to the extreme youth of RCW38 which does not allow yet formation of a cavity in the extended cloud by the collision. The ionization by the O stars may also tend to destruct the nest distribution.

Although the samples of collision-induced super-star-cluster formation are limited to the five clusters at the moment, Westerlund 2, NGC 3603, RCW 38, DBS[2003] 179 and Tr 14, they are only clusters in the Galaxy where we are able to witness the natal molecular cloud. The rest of the clusters have no associated interstellar medium due to stellar feedback, prohibiting us from observing the possible colliding clouds. Along with the single O star formation in the other regions M 20 and RCW 120, it seems reasonable to consider cloud-cloud collision as one of the formation scenarios of massive cluster including globular clusters. It deserves further intensive efforts to study if cloud-cloud collision is a viable mechanism of massive cluster formation both observationally and theoretically.

5. Conclusions

We have carried out CO J = 1–0 and J = 3–2 observations toward the super star cluster RCW 38 with Mopra, ASTE and NANTEN2 mm/sub-mm telescopes. The main conclusions of the present study are summarized as follows; We observed two molecular clouds at velocities of $2\,\mathrm{km\,s^{-1}}$ and $14\,\mathrm{km\,s^{-1}}$ toward RCW 38. The two clouds, the ring cloud and the finger cloud, are physically associated with the cluster as verified by the high ratio of the J = 3–2 transition to the J = 1–0 transition, $\mathrm{R}_{3-2/1-0}$, toward the cluster. In addition, the two clouds are linked with each other by bridging features in velocity in three places including the direction of the cluster, supporting the physical connection. We present an interpretation that the two clouds collided 0.1 Myrs ago at $\sim$10 $\mathrm{km\,s^{-1}}$ with each other to trigger formation of the O stars in the cluster. The $\sim$20 O star candidates were formed in this timescale with a high mass accretion rate. RCW 38 is the third super star cluster alongside of Westerlund 2 and NGC 3603 where cloud-cloud collisions triggered the O star formation in the cluster, lending further support for an important role of supersonic collision in formation of a super star cluster. RCW 38 is unique because it is the youngest cluster where the initial conditions prior to the O-star formation still hold without significant cloud dispersal by ionization. RCW 38 provides a rare sample to learn how super star clusters are forming in detail.

Acknowledgement

The author acknowledges all the members of the NANTEN team for their invaluable contributions in the present observations. In particular, the author thanks Kauzfumi Torii for his dedicated efforts in the study of RCW 38. This work was financially supported by Grants-in-Aid for Scientific Research (KAKENHI) of the Japanese society for the Promotion of Science (JSPS; grant numbers 15H05694, 15K17607, 24224005, 25287035, and 23540277).

References

Anathpindika, S. V. 2010,*MNRAS*, 405, 1431
Duarte-Cabral, A. *et al.* 2010,*A&A*, 519, A27
Ezawa, H. *et al.* 2004, *SPIE*, 5489, Ground-based Telescopes, 763
Fukui, Y. *et al.* 2014,*ApJ*, 780, 36
Fukui, Y. *et al.* 2015, ArXiv e-prints, arXiv:1504.05391
Furukawa, N. *et al.* 2009, *ApJ*, 696, L115
Habe, A. and Ohta, K. 1992,*PASJ*, 44, 203
Haworth, T. J. *et al.* 2015a, *MNRAS*, 450, 10
Haworth, T. J. *et al.* 2015b, *MNRAS*, 454, 1634
Higuchi *et al.* 2010,*ApJ*, 719, 1813
Inoue, T. and Fukui, Y. 2013,*ApJ*, 774, L31
Nakamura, F. *et al.* 2013,*ApJ*, 791, L23
Ohama, A. *et al.* 2010,*ApJ*, 709, 975
Portegies Zwart, S. F. *et al.* 2010,*ARA&A*, 48, 431
Takahira, K. *et al.* 2014,*ApJ*, 792, 63
Tan, J. C. *et al.* 2014,*Protostars and Planets VI*, 149
Torii, K. *et al.* 2011, *ApJ*, 738, 46
Torii, K. *et al.* 2015, *ApJ*, 806, 7
Wolk *et al.* 2006,*ApJ*, 132, 1100
Wolk *et al.* 2008,*The Embedded Massive Star Forming Region RCW 38*, ed. B. Reipurth, 124
Zinnecker, H. and York, H. W. 2007, *ARA&A*, 45, 481

Formation, evolution, and survival of massive star clusters
Proceedings IAU Symposium No. 316, 2015
C. Charbonnel & A. Nota, eds.

doi:10.1017/S1743921315010868

Inverting the dynamical evolution of globular clusters: clues to their origin

Mark Gieles[1] **and Poul Alexander**[2]

[1]Department of Physics, University of Surrey
Guildford GU2 7XH, United Kingdom
email: m.gieles@surrey.ac.uk

[2]Institute of Astronomy, University of Cambridge
Madingley Road, Cambridge CB3 0HA
email: poul.alexander@gmail.com

Abstract. Scaling relations for globular clusters (GC) differ from the scaling relations for pressure supported (elliptical) galaxies. In this contribution we discuss the relative importance of nature and nurture in the establishment of the scaling between cluster density (or radius), mass and Galactocentric distance for the Milky Way GCs. We show that energy diffusion by stellar encounters (i.e. two-body relaxation) is the dominant mechanism in shaping the bivariate dependence of density on mass and Galactocentric distance for GCs with masses $\lesssim 10^6\,M_\odot$, and it can not be excluded that GCs formed with similar scaling relations as the more massive ultra-compact dwarf galaxies (UCDs). To explore the initial properties that give rise to the distributions of these quantities, we developed a fast cluster evolution model (Evolve Me A Cluster of StarS, EMACSS) and use it in a hierarchical Bayesian framework to fit a parameterised model for the initial properties of Milky Way GCs to the observed present-day properties. The best-fit cluster initial mass function is substantially flatter (power-law with index -0.6 ± 0.2) than what is observed for young massive clusters (YMCs) forming in the nearby Universe (power-law with index -2). This result is driven by the metal-poor GCs, a slightly steeper CIMF is allowed when considering the metal-rich GCs separately ($\alpha \simeq -1.2 \pm 0.4$). If stellar mass loss and two-body relaxation in the Milky Way tidal field are the dominant disruption mechanisms, then GCs formed differently from YMCs.

Keywords. globular clusters: general – galaxies: star clusters – Galaxy: formation – stars: formation

1. Introduction

Globular clusters (GCs) are a common component of almost every galaxy and their properties can be used to put constraints on the formation of their host galaxy. For example, GC ages and metallicities inform us about the assembly history of galaxies (e.g. Searle & Zinn 1978; Leaman *et al.* 2013) and the spatial distribution of GCs within galaxies depends on the epoch of reinoization (Moore *et al.* 2006; Spitler *et al.* 2012). The similarity between structural properties of young halo clusters in the Milky Way and clusters in external satellite galaxies such as the Magellanic Clouds, has been used to distinguish between accreted GCs and those that formed in-situ (e.g. Mackey & Gilmore 2004; Mackey & van den Bergh 2005).

From a combination of data of nearby GCs and extra-galactic stellar systems (e.g. Brodie *et al.* 2011), a break was found in the luminosity-size relation at a luminosity of a few $10^6\,L_\odot$ (e.g. Misgeld & Hilker 2011). Objects above this limit (massive star clusters and ultra-compact dwarf galaxies, UCDs) display a positive correlation between size and luminosity/mass (Kissler-Patig *et al.* 2006), following the extension of the relation of

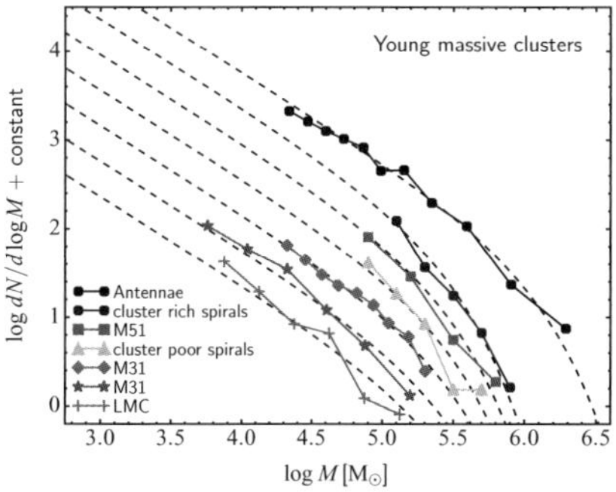

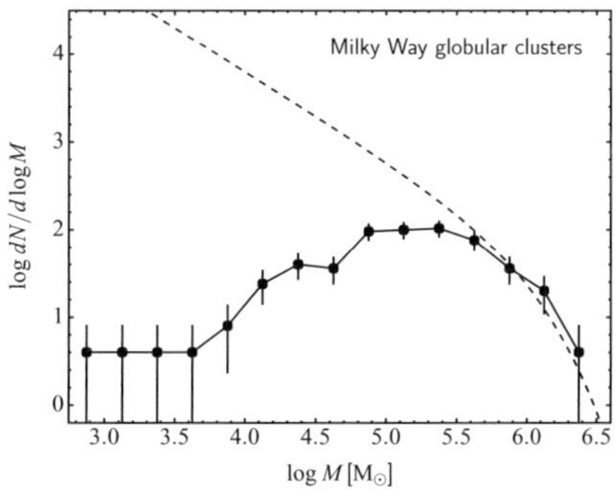

Figure 1. *Left*: Mass distribution of YMCs are well described by $dN/dM \propto M^{-2}$ and an exponential cut-off at $M_* \simeq 10^5 - 10^6\,M_\odot$ (data from Portegies Zwart *et al.* 2010). *Right*: the low-mass end of the Milky Way globular clusters is flat (dN/dM = constant, $dN/d\log M \propto M$, data from Harris 1996).

early type galaxies, whereas the radii of fainter GCs are independent of luminosity/mass, or slightly anti-correlate with luminosity/mass. It is still an open question whether GCs and UCDs share a common formation mechanism, or evolutionary processes are responsible for the difference in structural properties (Hilker *et al.* 2007). The half-mass relaxation timescale ($\tau_{\rm rh}$) of stellar systems with a mass of $10^6\,M_\odot$ and a radius of a few pc, is about a Hubble time and it is therefore plausible that properties of GCs, i.e. systems older than a $\tau_{\rm rh}$, are affected by this process (§ 2).

There are other indications that things change near $10^6\,M_\odot$: the mass-to-light ratio, M/L_V, of GCs is about half that of UCDs (Mieske *et al.* 2008), which may point at the presence of dark matter in UCDs or variations in the stellar initial mass function. For at least one UCD it has been shown that a massive black hole is responsible for the elevated M/L_V (Seth *et al.* 2014). But M/L_V of stellar systems with short relaxation times can be underestimated because of a mass segregation bias, if the assumption is made that light traces mass in deriving the dynamical mass (Shanahan & Gieles 2015; Sollima *et al.* 2015). Finally, if GCs formed embedded in a dark matter halo, it would have been pushed away by the stars as the result of dynamical friction (Baumgardt & Mieske 2008). Considering all of this, it is not clear that the difference in M/L_V between GCs and UCDs points at a genuine difference in nature, or is the result of nurture.

With the discovery of young massive clusters (YMCs) in starburst galaxies with masses and radii similar to old GCs (e.g. Holtzman *et al.* 1992; Whitmore & Schweizer 1995) the idea that GCs are still forming today has gained popularity. However, the mass function of YMCs (power-law with index −2, see Fig. 1) is strikingly different from the Milky Way GC mass function (GCMF, power-law with index 0 below a mass of roughly $10^5\,M_\odot$). The GCMF is very similar across many galaxies (e.g. Harris *et al.* 2014) and this has led to different suggestions for the origin of the GCMF. On one hand, a near universal 'peaked' GCMF could be the result of special conditions in the early Universe causing GCs to form with a preferred mass scale of a few $10^5\,M_\odot$ (Fall & Rees 1985; Bromm & Clarke 2002; Kimm *et al.* 2015). However, low-mass clusters dissolve faster in the Galactic tidal field, hence dynamical evolution also plays a role in shaping the GCMF. Some theoretical studies have claimed that it is possible that GCs formed with the same cluster initial mass function (CIMF) as YMCs and that dynamical evolution can erode the low-mass end of the GCMF by two-body relaxation (Fall & Zhang 2001; Kravtsov & Gnedin 2005; McLaughlin & Fall 2008), but the required escape rate and its dependence on mass and density are inconsistent with results from N-body models

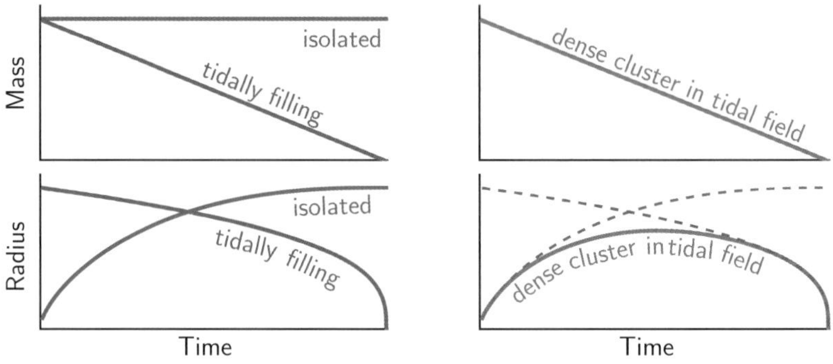

Figure 2. *Left*: Schematic illustration of the evolution of M and $r_{\rm h}$ in the models for isolated clusters (Hénon 1965) and tidally limited clusters (Hénon 1961). *Right*: Evolution of the same parameters in a model that 'stitches' together the two Hénon solutions (Gieles *et al.* 2011).

of clusters dissolving in a tidal field (Vesperini 2001; Gieles & Baumgardt 2008). There may be additional disruptive mechanisms, such as interactions with left-over natal gas, that could be important in shaping the GCMF soon after formation (Kruijssen 2015).

We address the relative contribution of nature and nurture to the properties of GCs, and in particular the GCMF, by using a (inverse) population synthesis approach to include the effect of a Hubble time of dynamical evolution. Because the dynamical evolution of gravitational systems can only be modelled forward in time, we developed a fast modelling techniques that allows us to efficiently explore the parameter space that describe the initial conditions.

In § 2 we present a theoretical framework for GC evolution, with an application to scaling relations of Milky Way GCs. A fast cluster model and results of a population fitting exercise are presented in § 3. Conclusions and a discussion are presented in § 4.

2. Model for cluster evolution and a comparison to Milky Way GCs

2.1. *GC evolutionary tracks*

Michel Hénon introduced two models for the evolution of GCs. His models describe the asymptotic behaviour after many relaxation times, and he showed that then clusters evolve self-similarly. In both models an energy source in the core supplies energy at the correct rate to fuel the two-body relaxation process, that is, the energy source is self-regulating and the evolution is 'balanced'. In the first model (Hénon 1961), the cluster is assumed to be truncated to mimic the effect of a Galactic tidal field, and stars can escape more easily than from an isolated cluster. The cluster loses stars at a constant rate and evolves at a constant density that is set by the tidal density, i.e., $r_{\rm h} \propto (M/\rho_{\rm tid})^{1/3}$. In the second model (Hénon 1965), the cluster evolves in isolation, and the cluster loses no stars, and $t/\tau_{\rm rh}$ becomes constant, such that $r_{\rm h} \propto t^{2/3}$. Both models are highly idealised, because most GCs are neither in isolated, nor completely filling the tidal volume, but rather somewhere in between these two extremes. It is therefore interested to consider how the two models of Hénon connect and consider the two models to describe the early and final phases of GC evolution. The fractional energy increase in Hénon's models is a constant per unit of $\tau_{\rm rh}$ ($\tau_{\rm rh}\dot{E}/E$ =constant) and the constant is approximately the same in the two models (Gieles *et al.* 2011). This latter property allows us to 'stitch' the two solutions together to acquire a single model that evolves the properties of clusters in a tidal field. The cluster is initially following the evolution of the isolated model, and

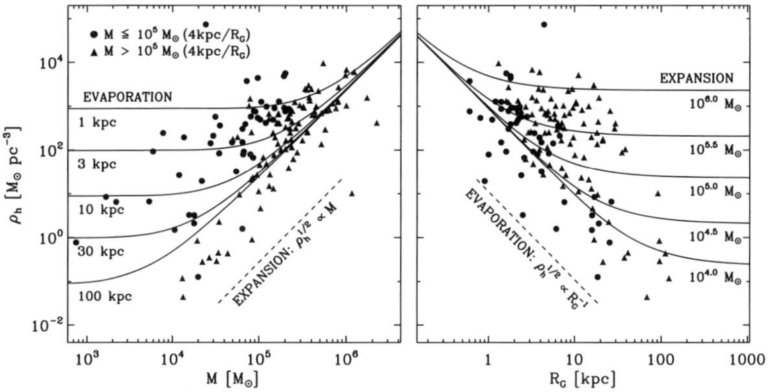

Figure 3. Evolutionary 'isochrones' computed from the model by Gieles *et al.* (2011) overlayed on the properties of Milky Way GCs. *Left:* half-mass density $\rho_{\rm h} = 3M/(8\pi r_{\rm h}^3)$ vs M. The correlation between $\rho_{\rm h} \propto M^2$ is the result of expansion. *Right:* $\rho_{\rm h}$ vs. $R_{\rm G}$. The anti-correlation $\rho_{\rm h} \propto R_{\rm G}^{-2}$ is the result of the tidal limit. From this comparison we see that about half of the GCs is still expanding to the tidal boundary and their properties are insensitive of the tidal field they are in.

while it expands it slowly converges to the constant density solution of the tidally limited model, with the final density set by the strength of the tidal field (see Fig. 2).

2.2. *GC evolutionary 'isochrones'*

With this simple model for the mass and radius evolution we can also compute evolutionary 'isochrones' for GC properties at an age of 12 Gyr in a Milky Way-like Galactic potential. The result is shown in Fig. 3. It shows the density within $r_{\rm h}$ ($\rho_{\rm h}$) as a function of mass M (left) and Galactocentric radius $R_{\rm G}$ (right), together with the values for the Milky Way GCs. It shows that about half of the GCs follows the predicted correlation between $\rho_{\rm h}$ and M of Hénon's isolated cluster (a constant $\tau_{\rm rh}$, or $\rho_{\rm h} \propto M^2$) and the other half of the GCs follows the prediction of Hénon's tidally limited cluster (which for the adopted singular isothermal Galactic halo corresponds to a scaling of $\rho_{\rm h} \propto R_{\rm G}^{-2}$). Because these predicted scaling relations are asymptotic solutions, they are insensitive to the initial mass-radius relation. All we can conclude from this is thus that GCs must have been denser than they are now. It is possible to gain insight in the initial properties by considering the *distributions* of M, $\rho_{\rm h}$ (or $r_{\rm h}$) and $R_{\rm G}$ (i.e. the 'Hess diagrams').

2.3. *GC 'Hess diagrams'*

The simple (analytic) prescription for the combined evolution of $r_{\rm h}$ and M as a function of $R_{\rm G}$ allows us to explore various scenarios for the birth properties of GCs and compare the evolved properties to observations. From data of the ACS Virgo Cluster Survey, it was found that the shape of the $r_{\rm h}$ distribution across galaxies is close to universal (Jordán *et al.* 2005). Motivated by this finding, we use the model to make predictions for the $r_{\rm h}$ distribution of the Milky Way GC population. A Monte Carlo approach was used to simultaneously evolve M and $r_{\rm h}$ for a large number of clusters for different (extreme) assumptions for the CIMF and the initial $r_{\rm h}$ distribution: [A] tidally filling GCs with a -2 power-law CIMF; [B] tidally filling GCs, with a flat CIMF ; [C] tidally under-filling GCs, with a -2 power-law CIMF; and [D] tidally under-filling GCs, with a flat CIMF. In Fig. 4 we show the resulting distributions of $r_{\rm h}$ for the four scenarios, compared to the $r_{\rm h}$ distribution of Milky Way GCs (right). Only in the scenario in which the CIMF was flat at birth and all GCs were tidally under-filling, the narrow shape of the $r_{\rm h}$ distribution can

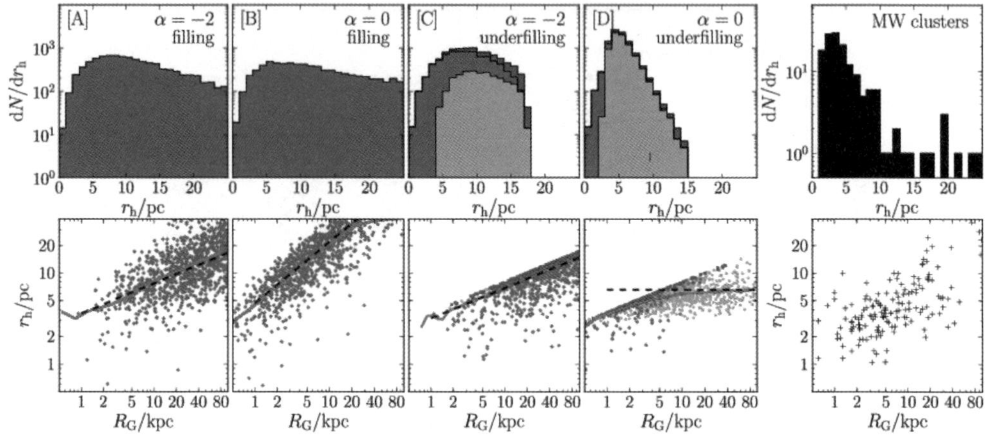

Figure 4. Radius distribution for different assumptions for the CIMF and the initial $r_{\rm h}$ distribution (from Alexander & Gieles 2013) compared to the Milky Way GCs (right panels). Cyan and green marks GCs that are under-filling and filling the Roche radius, respectively, at an age of 12 Gyr and GCs near their maximum $r_{\rm h}$ are shown in purple. A flat CIMF and dense initial conditions results in a near universal $r_{\rm h}$ distribution and a weak correlation between $r_{\rm h}$ and $R_{\rm G}$.

be reproduced. The peak at $\sim 3-4\,$pc is approximately universal, because for about half of the GCs $r_{\rm h}$ is set by expansion (i.e. an internal mechanism), and not by the strength of the tidal field. We see that this scenario also results in the correct scaling between $r_{\rm h}$ and $R_{\rm G}$. From a comparison of model [C] and [D], which have the same initial $r_{\rm h}$ distribution, we see that the resulting $r_{\rm h}$ distribution is sensitive to the CIMF. The shape of the $r_{\rm h}$ distribution, therefore, provides additional constraints on the shape of the CIMF.

3. A fast cluster evolution model (EMACSS) and an hierarchical Bayesian fit to the Milky Way GCs

There are numerous numerical techniques to evolve GC properties, such as solving the Fokker-Planck equations, Monte Carlo modelling and the direct N-body integrations. Each of them has its own draw-backs and advantages (see the supplementary material in Portegies Zwart *et al.* 2010 for a review), but in general it is the case that the more accurate techniques tend to be slower. Alexander & Gieles (2012) developed a method that is comparable to the gas models (e.g. Larson 1970), in which a coupled set of ordinary differential equations (ODEs) for $\dot{N}$ and $\dot{r}_{\rm h}$ is solved, using the assumption that $\tau_{\rm rh}\dot{E}/E$ is constant during the entire evolution (see § 2). In subsequent versions of the model the effect of core evolution (Gieles *et al.* 2014), stellar evolution (Alexander *et al.* 2014) and $R_{\rm G}$ evolution as the result of dynamical friction were included, resulting in a code that can evolve a limited number of GC properties for the entire lifecycle in few thousand ODE integration steps, and the results show good agreement with results from direct N-body simulations†. This fast modelling technique allows us to iteratively explore parameter space for the initial conditions of GCs. An example of an application of EMACSS coupled to a Markov Chain Monte Carlo (MCMC) method to find the posterior distribution of the initial parameters of individual GCs can be found in Pijloo *et al.* (2015).

Thanks to the speed of EMACSS, we are able to get the posteriors of every Milky Way GC. We then take this a step further and use this to establish the parameters of the initial *distributions* of M, $r_{\rm h}$ and $R_{\rm G}$ of the Milky Way GCs, with a hierarchical

† The code Evolve Me A Cluster of StarS (EMACSS) is available from https://github.com/emacss

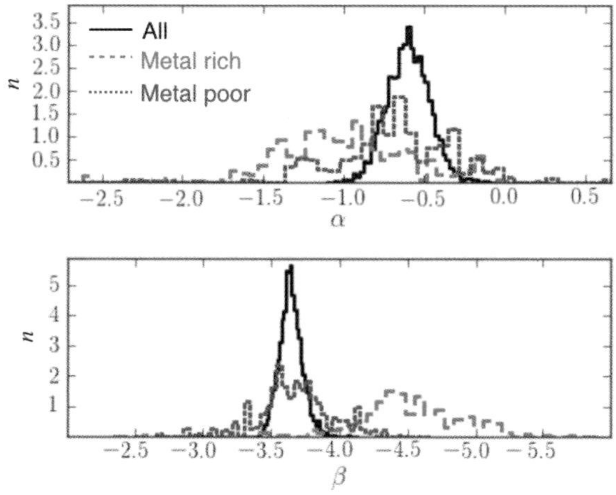

Figure 5. Posterior distribution for the power-law indices α of the CIMF and β of the initial number density in the Galaxy for the Milky Way GC population (from Alexander & Gieles (2106, to be submitted).

Bayesian approach. The technique is similar to the method presented by Hogg *et al.* (2010) to infer the parameters of the eccentricity distribution of planets (or binary stars) from radial velocity measurements. We adopt a Schecher function for the CIMF, with parameters α for the index at low masses and M_* for the exponential truncation and a single power-law $R_{\rm G}^{\beta}$ for the number density distribution of GCs within the Galaxy. Various assumptions for the initial $\rho_{\rm h}$ distribution have been explored and are discussed in Alexander & Gieles (2016, to be submitted). Our five parameters that describe the initial properties are $\Theta^{\rm pop} = (\alpha, M_*, \beta, \mu_\rho, \sigma_\rho)$, where μ_ρ and σ_ρ are the (logarithm of the) mean and the standard deviation of a log-normal for the $\rho_{\rm h}$ distribution. In the first step we determined the posteriors for the initial parameters of each individual GC, $P(\Theta_j^{\rm GC}|X_j^{\rm GC})$, where $\Theta_j^{\rm GC} = (M_{\rm i}, r_{\rm h,i}, R_{\rm G,i})_j$ are the parameters of the individual GC and $X_j^{\rm GC} = (M, r_{\rm h}, R_{\rm G})_j$ are the observed data of that GC. We use EMACSS to map the initial properties into present day properties and in the first iteration we adopt uniform priors for $M_{\rm i}, r_{\rm h,i}$ and $R_{\rm G,i}$. In the second step we use these posteriors as a set of heteroscedastic observations (i.e. $\boldsymbol{X}^{\rm pop} = \{P(\Theta_j^{\rm GC}|X_j^{\rm GC})\}_{j=1}^{N_{\rm GC}})$ to find the posteriors of the GC population: $P(\Theta^{\rm pop}|\boldsymbol{X}^{\rm pop})$. We then repeat the first step and use the posteriors $P(\Theta^{\rm pop}|\boldsymbol{X}^{\rm pop})$ as priors in the fitting of the individual GC parameters $\Theta_j^{\rm GC}$. After three iterations the method converges and we obtain posteriors for $\Theta^{\rm pop}$ and $\Theta_j^{\rm GC}$ for each GC. In Fig. 5 we show the posterior distributions for α and β when considering all GCs, as well as for the metal-poor and metal-rich GCs separately, where we used [Fe/H] $= -1$ as the boundary. The total population is well described by a CIMF with slope of $\alpha \simeq -0.6 \pm 0.2$, with indications for a slightly steeper distribution for the metal-rich GCs ($\alpha \simeq -1.2 \pm 0.4$). The present-day GCMF of metal-rich and metal-poor GCs is very similar, but because of their more centrally concentrated distribution in the Milky Way (see bottom panel), the metal-rich GCs experience on average a stronger tidal field, such that the present-day GCMF can be reconciled with a a slightly steeper CIMF, but within $1-\sigma$ the values of α of the metal-poor and metal-rich GCs are the same.

4. Conclusions and discussion

The scaling relations of GCs with masses $\lesssim 10^6\,M_\odot$ are the result of two-body relaxation driven expansion and evaporation in the tidal field. The radii of most GCs are independent of their initial radii, hence we can not rule out that GCs formed with the same $M - r_{\rm h}$ relation as more massive pressure supported systems, such as UCDs. About half of the GCs is still expanding to fill the Roche volume and their $r_{\rm h}$ are insensitive to the strength of the tidal field. These expanding clusters have similar $\tau_{\rm rh}$ (and hence $r_{\rm h} \propto M^{-1/3}$). The alignment of GCs with lines of constant $\tau_{\rm rh}$ in a $M - r_{\rm h}$ diagram has previously been attributed to the preferential disruption of clusters with short $\tau_{\rm rh}$ (Fall & Rees 1977; Gnedin & Ostriker 1997), but here we show that the GCs that are in the evaporation dominated phase of their evolution (roughly the 2nd half of their life), have adjusted to the tidal density and hence show a relation $\rho_{\rm h} \propto R_{\rm G}^{-2}$ (for a host Galaxy with a flat rotation curve), independent of M. Massive GCs at large $R_{\rm G}$ are still expanding and because this is an internally driven mechanism (i.e. independent of environment), this provides an explanation for the near universal $r_{\rm h}$ distribution of extra-galactic GCs.

Previous studies have shown that it is hard to evolve a -2 power-law CIMF into a peaked GCMF with a universal peak mass, when considering stellar evolution and two-body relaxation in a Milky Way-like tidal field (e.g. Vesperini 2001). We are able to put constraints on the allowed shapes of the CIMF, given the observations, by using the fast cluster evolution code EMACSS to evolve parameterised distributions for the initial properties and fit them to the Milky Way GC data. We find that the low-mass end of the CIMF had a slope of $\alpha \simeq -0.6 \pm 0.2$. A flat CIMF for metal-poor GCs is also inferred from the mass in field stars and GCs in dwarf spheroidals (Larsen *et al.* 2012, 2014). The CIMF of metal-rich GCs in the Milky Way may have been steeper ($\alpha \simeq -1.2 \pm 0.4$).

The models presented here do not include the disruptive effect of interactions with (molecular) gas clouds, which could be important in the early evolution of GCs (Gieles *et al.* 2006; Kruijssen 2015), but several coupled conditions for the initial mass-radius relation and ambient gas density need to be satisfied to establish a near universal GCMF shape. We also only considered a time-independent Milky Way potential. We therefore do not properly account for the effects of the secular growth of the Milky Way and the possibility of GC evolution in a smaller host galaxies that has now been accreted onto the Milky Way, both of which are especially important for the (outer halo) metal-poor GCs. Accounting for the secular growth of the Milky Way results in a higher α (i.e. flatter CIMF), because GCs would have formed further out in a lower-mass Galaxy, and lost less stars than in the models we used to constrain the CIMF (Renaud & Gieles 2015). The effect of the evolution in another host galaxy is harder to quantify. Bianchini *et al.* (2015) showed that the accretion process itself has little effect on the mass, but whether a period of evolution in a dwarf galaxy will result in a higher, or larger α, depends strongly on the dark matter density profile of the progenitor galaxy. In the ΛCDM cosmology, smaller galaxies have higher (dark matter) densities, hence we may naively expect the tides to be stronger in dwarf galaxies, compared to the outer halo of the Milky Way. However, the slope of the density profile is almost as important as the density itself in setting the mass-loss rate. A fraction of the low-mass satellites appears to/is expected to have dark matter 'cores' (i.e. a near constant density) (Read *et al.* 2015). Within constant density cores GCs would hardly lose any stars, because of the compressive tidal field. Because these cored satellites are preferentially destroyed when falling into the Milky Way (Peñarrubia *et al.* 2008), considering the accretion history of GCs within a cosmological context may also result in larger values for α (i.e. flatter CIMF) than reported here. A full consideration of the (tidal) evolution of GCs within the cosmological context will shed light on these issues (Renaud *et al.*, in prep).

References

Alexander, P. E. R. & Gieles, M. 2012, MNRAS, 422, 3415
Alexander, P. E. R. & Gieles, M. 2013, MNRAS, 432, L1
Alexander, P. E. R., Gieles, M., Lamers, H. J. G. L. M., & Baumgardt, H. 2014, MNRAS, 442, 1265
Baumgardt, H. & Mieske, S. 2008, MNRAS, 391, 942
Bianchini, P., Renaud, F., Gieles, M., & Varri, A. L. 2015, MNRAS, 447, L40
Brodie, J. P., Romanowsky, A. J., Strader, J., & Forbes, D. A. 2011, AJ, 142, 199
Bromm, V. & Clarke, C. J. 2002, ApJ, 566, L1
Fall, S. M. & Rees, M. J. 1977, MNRAS, 181, 37P
Fall, S. M. & Rees, M. J. 1985, ApJ, 298, 18
Fall, S. M. & Zhang, Q. 2001, ApJ, 561, 751
Gieles, M., Portegies Zwart, S. F., Baumgardt, H., *et al.* 2006, MNRAS, 371, 793
Gieles, M. & Baumgardt, H. 2008, MNRAS, 389, L28
Gieles, M., Heggie, D. C., & Zhao, H. 2011, MNRAS, 413, 2509
Gieles, M., Alexander, P. E. R., Lamers, H. J. G. L. M., & Baumgardt, H. 2014, MNRAS, 437, 916
Gnedin, O. Y. & Ostriker, J. P. 1997, ApJ, 474, 223
Harris, W. E. 1996, AJ, 112, 1487
Harris, W. E., Morningstar, W., Gnedin, O. Y., *et al.* 2014, ApJ, 797, 128
Jordán, A., Côté, P., Blakeslee, J. P., *et al.* 2005, ApJ, 634, 1002
Hénon, M. 1961, Annales d'Astrophysique, 24, 369
Hénon, M. 1965, Annales d'Astrophysique, 28, 62
Hilker, M., Baumgardt, H., Infante, L., *et al.* 2007, A&A, 463, 119
Hogg, D. W., Myers, A. D., & Bovy, J. 2010, ApJ, 725, 2166
Holtzman, J. A., Faber, S. M., Shaya, E. J., *et al.* 1992, AJ, 103, 691
Kimm, T., Cen, R., Rosdahl, J., & Yi, S. 2015, arXiv:1510.05671
Kissler-Patig, M., Jordán, A., & Bastian, N. 2006, A&A, 448, 1031
Kravtsov, A. V. & Gnedin, O. Y. 2005, ApJ, 623, 650
Kruijssen, J. M. D. 2015, MNRAS, 454, 1658
Larsen, S. S., Strader, J., & Brodie, J. P. 2012, A&A, 544, L14
Larsen, S. S., Brodie, J. P., Forbes, D. A., & Strader, J. 2014, A&A, 565, A98
Leaman, R., VandenBerg, D. A., & Mendel, J. T. 2013, MNRAS, 436, 122
Larson, R. B. 1970, MNRAS, 147, 323
Misgeld, I. & Hilker, M. 2011, MNRAS, 414, 3699
Moore, B., Diemand, J., Madau, P., Zemp, M., & Stadel, J. 2006, MNRAS, 368, 563
Mackey, A. D. & Gilmore, G. F. 2004, MNRAS, 355, 504
Mackey, A. D. & van den Bergh, S. 2005, MNRAS, 360, 631
McLaughlin, D. E. & Fall, S. M. 2008, ApJ, 679, 1272
Mieske, S., Hilker, M., Jordán, A., *et al.* 2008, A&A, 487, 921
Peñarrubia, J., Navarro, J. F., & McConnachie, A. W. 2008, ApJ, 673, 226
Pijloo, J. T., Portegies Zwart, S. F., Alexander, P. E. R., *et al.* 2015, MNRAS, 453, 605
Portegies Zwart, S. F., McMillan, S. L. W., & Gieles, M. 2010, ARA&A, 48, 431
Read, J. I., Agertz, O., & Collins, M. L. M. 2015, arXiv:1508.04143
Renaud, F. & Gieles, M. 2015, MNRAS, 449, 2734
Searle, L. & Zinn, R. 1978, ApJ, 225, 357
Seth, A. C., van den Bosch, R., Mieske, S., *et al.* 2014, Nature, 513, 398
Shanahan, R. L. & Gieles, M. 2015, MNRAS, 448, L94
Sollima, A., Baumgardt, H., Zocchi, A., *et al.* 2015, MNRAS, 451, 2185
Spitler, L. R., Romanowsky, A. J., Diemand, J., *et al.* 2012, MNRAS, 423, 2177
Vesperini, E. 2001, MNRAS, 322, 247
Whitmore, B. C. & Schweizer, F. 1995, AJ, 109, 960

Formation, evolution, and survival of massive star clusters
Proceedings IAU Symposium No. 316, 2015
C. Charbonnel & A. Nota, eds.

doi:10.1017/S1743921315009096

The dynamical importance of binary systems in young massive star clusters

Richard de Grijs[1], Chengyuan Li[1,2] and Aaron M. Geller[3,4]

[1]Kavli Institute for Astronomy & Astrophysics and Department of Astronomy, Peking University, Beijing, China
email: grijs@pku.edu.cn

[2]Purple Mountain Observatory, Chinese Academy of Sciences, Nanjing, China

[3]Center for Interdisciplinary Exploration and Research in Astrophysics (CIERA) and Department of Physics and Astronomy, Northwestern University, Evanston, IL, USA

[4] Department of Astronomy and Astrophysics, University of Chicago, Chicago, IL, USA

Abstract. Characterization of the binary fractions in star clusters is of fundamental importance for many fields in astrophysics. Observations indicate that the majority of stars are found in binary systems, while most stars with masses greater than $0.5M_{\odot}$ are formed in star clusters. In addition, since binaries are on average more massive than single stars, in resolved star clusters these systems are thought to be good tracers of (dynamical) mass segregation. Over time, dynamical evolution through two-body relaxation will cause the most massive objects to migrate to the cluster center, while the relatively lower-mass objects remain in or migrate to orbits at greater radii. This process will globally dominate a cluster's stellar distribution. However, close encounters involving binary systems may disrupt 'soft' binaries. This process will occur more frequently in a cluster's central, dense region than in its periphery, which may mask the effects of mass segregation. Using high resolution *Hubble Space Telescope* observations, combined with sophisticated N-body simulations, we investigate the radial distributions of the main-sequence binary fractions in massive young Large Magellanic Cloud star clusters. We show that binary disruption may play an important role on very short timescales, depending on the environmental conditions in the cluster cores. This may lead to radial binary fractions that initially decline in the cluster centers, which is contrary to the effects expected from dynamical mass segregation.

Keywords. stellar dynamics, methods: statistical, binaries: general, Magellanic Clouds, galaxies: star clusters

1. Introduction

Star cluster physics has a fundamental bearing on many important fields in contemporary astrophysics. Applications range from their use as tracers of the most violent episodes shaping galaxies and their evolution on cosmic timescales to their unique role in helping us determine the shape of the stellar initial mass function and validating theories of stellar evolution. Observational, theoretical and numerical studies have made rapid progress to date, yet most studies focus only on the properties of the populations of single stars, while binary systems remain largely overlooked.

This is perhaps understandable from a practical perspective: characterizing binaries in distant star clusters is difficult and including them in numerical simulations carries a significant computational cost, to the extent that even the most advanced N-body runs often end in unstable configurations and thus lead to crashed programs. In addition, from a theoretical perspective, it is largely unclear how binary systems affect a cluster's color–magnitude diagram (CMD) once we divert our attention from the fairly well-understood main-sequence–main-sequence (MS–MS), MS–red-giant-branch (RGB), and RGB–RGB binaries, i.e., the ubiquitous 'normal,' detached binaries. In the latter context, the formation scenarios of so-called 'blue straggler' systems are now finally reaching a concensus-like state, with most scientists agreeing that these objects have likely originated from either binary mergers or mass transfer between components in binary systems.

At the same time, studies of both the field stellar population in the solar nighborhood and the population properties of star-forming regions and some open clusters imply that the binary fractions at the time of starbirth are non-negligible. Mason *et al.* (1998) derived a binary fraction of ~70% in massive field stars, although the exact numbers depend on the stellar spectral type considered (Raghavan et al. 2010). In a clustered context, Sana & Evans (2011) and Sana *et al.* (2011, 2012a,b) found a binary fraction of around 50% in young open clusters, including in the 30 Doradus region, which ties in very well with the ~55% binary fraction quoted by Kobulnicky *et al.* (2014) for the Cygnus OB2 association. Kouwenhoven *et al.* (2005) studied the Scorpius OB2 association in detail and provided similar binary fractions, notably separated as a function of stellar spectral type. Binary systems are thus apparently ubiquitous in young star-forming regions, including in regions of clustered star formation, and hence one can expect them to have noticeable effects on the dynamics of their host stellar systems.

2. Characterizing binary systems in distant star clusters

Characterizing binary systems in crowded cluster environments is challenging, particularly in distant star clusters. Nevertheless, careful statistical analysis of high-quality CMDs enables quantitative characterization of binary fractions in clusters at distances out to the Magellanic Clouds, 50–60 kpc (e.g., de Grijs *et al.* 2014; de Grijs & Bono 2015), using high-resolution *Hubble Space Telescope* imaging. Unresolved MS–MS binaries captured in a single resolution element will be more luminous than single stars of the same type as the binary's primary component. In fact, unresolved binary systems will attain brightnesses that vary between those defined by the single-star MS and the offset sequence defined by equal-mass (equal-luminosity) MS–MS binary systems. The latter are defined by a binary mass ratio, $q \equiv m_2/m_1$, of unity, where m_1 and m_2 are the binary's primary and secondary components, respectively, and $m_1 \geqslant m_2$. Equal-mass binaries attain luminosities that are twice as bright as those of the equivalent single stars, which thus translates in a binary sequence in a cluster CMD that is offset by 0.75 mag toward brighter magnitudes.† In other words, the presence of a significant fraction of binary systems will broaden a cluster's MS ridge line. If we can accurately characterize that broadening, we will be able to derive the most likely binary fraction.

In practice, characterization of the MS broadening is best done using Monte Carlo simulations (e.g., Hu *et al.* 2010, their Fig. 6; Milone *et al.* 2010, their Fig. 4; and references therein), while also allowing for the unavoidable effects owing to photometric uncertainties and line-of-sight blending. Given that the most important observational effect of the presence of unresolved binary systems in a star cluster's CMD is a spread in the magnitude direction, characterization of the MS broadening is most easily accomplished where the MS is shallowest – and for young clusters, that shallowing of the MS occurs well below the MS turn-off region, so that the effects of rapid stellar rotation do not (yet) dominate.

† Note that while the offset in magnitude is most important, for binaries with $q < 1$ a small offset in color is also implied. The extent of the latter depends on both the actual mass ratio (e.g., Elson *et al.* 1998, their Fig. 3) and the photometric color examined.

Hu *et al.* (2010) were among the first to apply this technique to young massive clusters. For their object of interest, the $\sim 2.8 \times 10^4 M_\odot$ compact, $\sim$ 15–30 Myr-old cluster NGC 1818 in the Large Magellanic Cloud (LMC), they derived a best-fitting MS (F-star) binary fraction of 55% ±10% for $q \geqslant 0.4$. Depending on the mass-ratio distribution governing these binary systems, the cluster's overall binary fraction could be as high as 100% (Hu *et al.* 2010). This study thus established that young star clusters may indeed host large numbers of binary systems. These binaries are expected to result in a significant dynamical signature, thus implying that analysis of N-body simulations of NGC 1818-like clusters without any implementation of binary systems would lead to unreliable physical insights.

3. Radial migration and binary dissolution

While Hu *et al.* (2010) obtained a first, firm constraint on a young star cluster's close-to-primordial binary fraction, from a dynamical perspective determination of the distribution and impact of these systems as a function of clustercentric radius would be more valuable. Intuitively, one might expect binary systems to behave as single objects in an N-body sense, with a somewhat larger cross section than the equivalent single stars. Realistic initial conditions, e.g., a fractal-like stellar distribution determined by the turbulent structure of the interstellar medium (e.g., Goodwin & Whitworth 2004; Goodwin *et al.* 2004a,b), were found to lead to rapid dynamical redistribution of the initial stellar masses – on timescales of $\lesssim$ a few Myr (e.g., Allison *et al.* 2009, 2010) – and hence to a mass-segregated configuration in the inner cluster regions.

Numerical simulations based on realistic initial conditions (i.e., initial substructure and dynamically cool initial configurations) suggest that dynamical mass segregation, at least of the most massive stars, is likely to happen in a crossing time, which is equivalent to the free-fall time defined by the cluster's gravitational potential. Given that NGC 1818 is 15–30 Myr old (or ~5–30 crossing times), one would thus naively expect the cluster to be (dynamically) mass segregated in its core (for confirmation, see de Grijs *et al.* 2002a,b), with an increasing fraction of binary systems at smaller radii.

However, the F-type stars studied by Hu *et al.* (2010), with masses of 1.3–1.6 $M_\odot$, are not expected to have already reached a state close to energy equipartition across the cluster, because for these stellar masses the half-mass relaxation time is much longer than the cluster's age. This process of clusterwide dynamical mass segregation is therefore likely still fully underway. However, contrary to dynamical expectations based on initial conditions adopting Plummer spheres, in de Grijs *et al.* (2013) we first found a hint of an increasing fraction of binary systems in NGC 1818 from the innermost (core) radius out to approximately its half-light radius, which we subsequently confirmed in Li *et al.* (2013b). In light of our naive initial insights, this was surprising. In fact, a previous, preliminary study of the binary population in NGC 1818 (Elson *et al.* 1998) had reached the opposite conclusion for binaries with primary masses between $2M_\odot$ and $5.5M_\odot$ and $q \gtrsim 0.7$. From the core to a clustercentric distance equivalent to about 3 core radii, they found a decrease in the binary fraction from about 35% to 20%. We attributed this discrepant result to blending and the near-vertical morphology of the stellar main sequence within the magnitude range observed by Elson *et al.* (1998). In Geller *et al.* (2015; their Fig. 7) we showed that these results are not mutually exclusive in view of the different stellar mass ranges examined by Elson *et al.* (1998) and Li *et al.* (2013b).

In Li *et al.* (2013b), we used a combination of isochrone fitting and statistical χ^2 minimization to re-investigate the binary fraction as a function of radius in NGC 1818, in essence improving the method applied by de Grijs *et al.* (2013). We also applied the same method to NGC 1805, a similarly aged compact LMC cluster. Our results in Li *et al.* (2013b) exhibit opposite trends as to the binary fractions as a function of radius in the inner regions of both clusters. The binary fraction in NGC 1805 decreases significantly from the cluster's inner core to its periphery. We concluded that while early dynamical mass segregation and the disruption of soft binary systems should be at work in both clusters, time-scale arguments imply that early dynamical mass segregation should be very efficient and, hence, likely dominates the dynamical processes in the core of NGC 1805. Meanwhile, in NGC 1818, the behavior in the core is probably dominated by disruption of 'soft' binary systems, i.e., generally wide binary systems characterized by high kinetic compared to potential energies. We speculate that this may be owing to the higher velocity dispersion in the NGC 1818 core, which creates an environment in which the efficiency of binary disruption is high compared with that of the NGC 1805 core. In other words, by applying the same technique to two young massive clusters in the LMC, we found two opposite radial trends in their binary fractions. This implies that these trends are not driven by the analysis method employed (i.e., our approach is unlikely to introduce a bias one way or the other), but that these trends represent physical reality.

4. Physical implications

Our speculation that the puzzling radial trend in NGC 1818 may have been caused by the disription of soft binaries systems was subsequently confirmed on the basis of N-body simulations (Geller *et al.* 2013). Using a grid of sophisticated N-body simulations, in Geller *et al.* (2013) we showed that the observed surface density profile of NGC 1818 (Mackey & Gilmore 2003) and the radial dependence of the binary frequency (de Grijs *et al.* 2013; Li *et al.* 2013b) can be reproduced simultaneously at the cluster's most likely age using N-body simulations with both initially smooth or substructured and equilibrium or collapsing stellar populations. The radial distribution of the binary frequency in a rich star cluster can transition smoothly over time from a uniform primordial radial distribution, to one that decreases toward the core at early ages, to one that rises toward the core at later ages. Thus both rising and falling radial distributions in binary frequency can arise naturally from the evolution of a binary population within the same rich star cluster as a consequence of both dynamical disruption and mass segregation of the binaries. Indeed, we managed to reproduce the opposite radial trend observed for NGC 1805 in a subsequent N-body study (Geller *et al.* 2015), drawing from the same initial binary distributions in masses and orbital parameters, although with a different initial cluster mass and half-mass radius.

Our recent analysis of the binary fractions in the young massive LMC clusters NGC 1805 and NGC 1818, as a function of radius, thus confirms the dynamical importance of binary systems in such dense environments (see also Schneider *et al.* 2014). This thus elevates previously held convictions that binary dynamics are predominantly important in the context of blue straggler formation to a whole new level. We strongly encourage the community, in particular our colleagues with expertise in N-body simulations, to take up the gauntlet and henceforth fully embrace the opportunities afforded by inclusion of binary systems in their simulations.

5. Binaries as diagnostic tool to trace a cluster's dynamical state

Finally, in Geller *et al.* (2013), we stumbled upon an interesting diagnostic that could potentially indicate a cluster's dynamical state: see their Fig. 2. Our simulations started from a flat distribution of our NGC 1818-like model binary fraction as a function of radius. We found that after one crossing time, the binary fraction decreases toward the cluster core because of the disruption of wide binaries. The higher velocity dispersion in the cluster core compared with its periphery causes binaries in the core to move more rapidly, on average, relative to other stars than their counterparts at larger radii. Therefore, encounters in the core are more energetic and can disrupt tighter binaries, while the higher density results in a higher encounter rate in the core (Leigh & Sills 2011).

Over time, cluster-wide mass segregation begins to dominate the radial dependence of the binary fraction. As a consequence, the theoretical '$r_{\rm min}$' values or 'zones of avoidance' (e.g., Mapelli *et al.* 2004; Ferraro *et al.* 2012) move radially outward. These radii represent the clustercentric distances inside of which the local dynamical friction timescale (Binney & Tremaine 1987) for a binary with a mass equivalent to that of the mean binary mass in the cluster is shorter than the simulated time. In other words, $r_{\rm min}$ predicts the radius inside of which the binaries should experience the effects of dynamical friction and, therefore, migrate toward the cluster center. As a consequence, the binary fraction in the core increases at the expense of that in the cluster's periphery. The observed value of $r_{\rm min}$ is thus a potential indicator of the degree and extent of mass segregation experienced by a given star cluster.

This behavior of the radial distribution of a cluster's binary fraction is reminiscent of that seen for the radial distribution of blue straggler stars in many Milky Way globular clusters (e.g., Ferraro *et al.* 2012) as well as in old massive LMC clusters (Li *et al.* 2013a). Ferraro *et al.* (2012) indeed used the minima in the radial blue straggler frequency as an indication of the host cluster's dynamical state. They also pointed out that blue straggler formation in a cluster's core may have proceeded differently from that in its outer regions. In view of the higher density in the core and the timescales involved in blue straggler formation, core blue stragglers most likely formed through direct stellar collisions, perhaps mediated by binary encounters, while their peripheral counterparts may well have formed less violently and more slowly through mass transfer between both components of their progenitor binary systems. Once again, this underscores the dynamical importance of binary systems on cluster-wide scales, thus reinforcing our message to take the contributions of binary systems seriously in the context of star cluster dynamics.

Acknowledgements

RdG and CL acknowledge financial support from the National Natural Science Foundation of China (grants 11073001 and 11373010); RdG also acknowledges travel support from the IAU to attend this conference. CL is supported by Strategic Priority Research Program 'The Emergence of Cosmological Structures' of the Chinese Academy of Sciences (grant XDB09000000). AMG is funded by a National Science Foundation Astronomy and Astrophysics Postdoctoral Fellowship under Award No. AST-1302765. RdG wishes to thank the organizers of IAU Symposium 316 for inviting him to give a review talk, of which this paper is a summary. He specifically thanks Corinne Charbonnel for her competent organizational leadership of the Scientific Organizing Committee.

References

Allison, R. J., Goodwin, S. P., Parker, R. J., *et al.* 2009, *MNRAS*, 395, 1449

Allison, R. J., Goodwin, S. P., Parker, R. J., Portegies Zwart, S. F., & de Grijs, R. 2010, *MNRAS*, 407, 1098

Binney, J. & Tremaine, S. 1987, *Galactic Dynamics* (Princeton, NJ: Princeton Univ. Press)

de Grijs, R., Johnson, R. A., Gilmore, G. F., & Frayn, C. M. 2002a, *MNRAS*, 331, 228 de Grijs, R., Gilmore, G. F., Johnson, R. A., & Mackey, A. D. 2002b, *MNRAS*, 331, 245

de Grijs, R., Li, C., Zheng, Y., *et al.* 2013, *ApJ*, 765, 4

de Grijs, R., Wicker, J. E., & Bono, G. 2014, *AJ*, 147, 122

de Grijs, R. & Bono, G. 2015, *AJ*, 149, 179

Elson, R. A. W., Sigurdsson, S., Davies, M., Hurley, J., & Gilmore, G. 1998, *MNRAS*, 300, 857

Ferraro, F. R., Lanzoni, B., Dalessandro, E., *et al.* 2012, *Nature*, 492, 393

Geller, A. M., de Grijs, R., Li, C., & Hurley, J. R. 2013, *ApJ*, 779, 30

Geller, A. M., de Grijs, R., Li, C., & Hurley, J. R. 2015, *ApJ*, 805, 11

Goodwin, S. P. & Whitworth, A. P. 2004, *A&A*, 413, 929

Goodwin, S. P., Whitworth, A. P., & Ward-Thompson, D. 2004a, *A&A*, 414, 633

Goodwin, S. P., Whitworth, A. P., & Ward-Thompson, D. 2004b, *A&A*, 423, 169

Hu, Y., Deng, L., de Grijs, R., Liu, Q., & Goodwin, S. P. 2010, *ApJ*, 724, 649

Kobulnicky, H. A., Kiminki, D. C., Lundquist, M. J., *et al.* 2014, *ApJS*, 213, 34

Kouwenhoven, M. B. N., Brown, A. G. A., Zinnecker, H., Kaper, L., & Portegies Zwart, S. F. 2005, *A&A*, 430, 137

Leigh, N. & Sills, A. 2011, *MNRAS*, 410, 2370

Li, C., de Grijs, R., Deng, L., & Liu, X. 2013a, *ApJL*, 770, L7

Li, C., de Grijs, R., & Deng, L. 2013b, *MNRAS*, 436, 1497

Mackey, A. D., & Gilmore, G. F. 2003, *MNRAS*, 338, 85

Mapelli, M., Sigurdsson, S., Colpi, M., *et al.* 2004, *ApJL*, 605, L29

Mason, B. D., Henry, T. J., Hartkopf, W. I., ten Brummelaar, T., & Soderblom, D. R. 1998, *AJ*, 116, 2975

Milone, A. P., Piotto, G., King, I. R., *et al.* 2010, *ApJ*, 709, 1183

Raghavan, D., McAlister, H. A., Henry, T. J., *et al.* 2010, *ApJS*, 190, 1

Sana, H. & Evans, C. J. 2011, *IAU Symp.*, 272, 474

Sana, H., James, G., & Gosset, E. 2011, *MNRAS*, 416, 817

Sana, H., de Mink, S. E., de Koter, A., *et al.* 2012a, *Science*, 337, 444

Sana, H., Dunstall, P. R., Hénault-Brunet, V., *et al.* 2012b, in: *Proc. Scient. Mtg in Honor of Anthony F. J. Moffat*, ASP Conf. Ser. 465, 284

Schneider, F. R. N., Izzard, R. G., de Mink, S. E., *et al.* 2014, *ApJ*, 780, 117

Formation, evolution, and survival of massive star clusters
Proceedings IAU Symposium No. 316, 2015
C. Charbonnel & A. Nota, eds.

doi:10.1017/S1743921315008996

Initial conditions of formation of starburst clusters: constraints from stellar dynamics

Sambaran Banerjee[1,2]

[1]Argelander-Institut für Astronomie
Auf dem Hügel 71, D-53121 Bonn, Germany
email: sambaran@astro.uni-bonn.de

[2]Helmholtz-Instituts für Strahlen- und Kernphysik
Nussallee 14-16, D-53115 Bonn, Germany

Abstract. How starburst clusters form out of molecular clouds is still an open question. In this article, I highlight some of the key constraints in this regard, that one can get from the dynamical evolutionary properties of dense stellar systems. I particularly focus on secular expansion of massive star clusters and hierarchical merging of sub-clusters, and discuss their implications *vis-á-vis* the observed properties of young massive clusters. The analysis suggests that residual gas expulsion is necessary for shaping these clusters as we see them today, irrespective of their monolithic or hierarchical mode of formation.

Keywords. galaxies: star clusters, stellar dynamics, stars: formation, methods: n-body simulations, methods: numerical, ISM: clouds

---

1. Introduction: birth conditions of young massive clusters

Star clusters are found in our Milky Way and in all external galaxies with increasing detail. However, how they form in the first place is still one of the most important challenges of the cosmos. A key question of wide debate is how the exposed clusters' parsec-scale, centrally-pronounced, near-spherical shape, observed at all ages $\gtrsim 1$ Myr, can be explained. This is in direct contrast with the irregular and much more extended (10s of parsecs) structure of molecular clouds, where stars form via gravitational fragmentation of dense substructures. An apparent lack of an age spread among the members of the youngest star clusters (see, e.g., Bastian & Silva-Villa 2013) indicates that these stars form in a "burst" over a short period of time. This, in turn, implies that short-timescale dynamical processes, e.g., violent relaxation (Spitzer 1987), immediately or simultaneously follow the formation of the proto-stars, which shape the newly born cluster.

The unprecedented spatial and spectral resolution in IR and sub-mm wavelengths with *Herschel* and *ALMA* has revealed intricate filamentary networks inside dense regions of molecular clouds (see André *et al.* 2014 for a review). Such observations generally infer that the (projected) widths of the individual filaments and their junctions are as compact as 0.1-0.3 pc. Both theoretical and observational studies suggest that groups of proto-stars preferentially form within these filaments and at their junctions (Schneider *et al.* 2012; André *et al.* 2014).

In Banerjee & Kroupa (2015a), the role of violent relaxation in shaping a young cluster is studied. It is demonstrated that in order to have a single cluster in dynamical equilibrium right from Myr-age, as for massive starburst clusters, the stellar system involved in the mass assembly process must be "near-monolithic". This implies the formation of

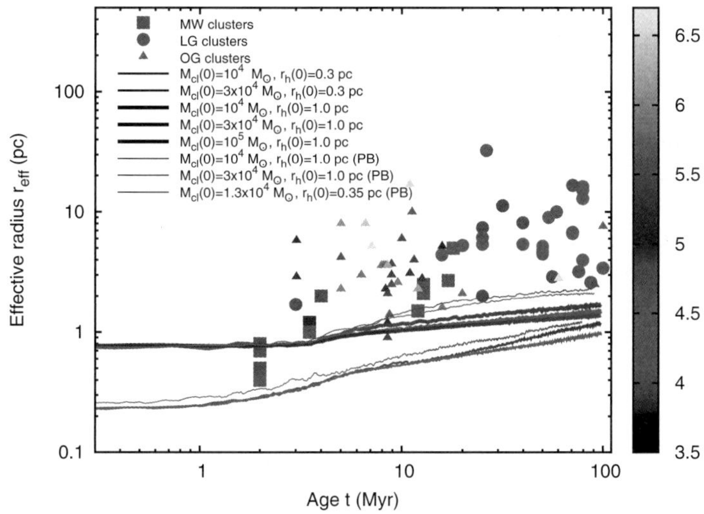

Figure 1. Effective radius, $r_{\rm eff}$, vs. age, t, plot (data from Portegies Zwart *et al.* 2010) for observed young massive clusters (YMCs) in the Milky Way (MW), the Local Group (LG) and external galaxies (OG) which are distinguished by different filled symbols. The symbols are colour-coded according to the clusters' respective photometric mass, $\log_{10}(M_{\rm phot}/M_{\odot})$. Overlaid are the computed curves for the evolution of projected half-mass radius (or effective radius), $r_{\rm eff}(t)$, for model star clusters, without residual gas expulsion ('PB'$\Rightarrow$ primordial binaries included), which are also colour-coded by their instantaneous bound mass.

either a single (monolithic) proto-cluster within a dense molecular clump or of several sub-clusters that merge in $\lesssim 1$ Myr from parsec-scale separation.

In this study, useful constraints are obtained for assembling massive clusters given their birth conditions, as summarized above, and the physics of dynamical evolution. Especially, I shall shed light on one of the most widely debated question regarding massive cluster formation, namely, the relevance of primordial gas blow-out by stellar feedback. Hereafter, the widely accepted definition of young massive clusters (YMCs) is adopted, *i.e.*, star clusters which have a present-day photometric mass exceeding $M_{\rm phot} \gtrsim 10^4 M_{\odot}$ and are younger than $t \lesssim 100$ Myr by stellar age. The youngest subset of them, of $\lesssim 4$ Myr age, are commonly referred to as 'starburst' clusters. These limits are not robust but generally serve as representative values.

2. Secular expansion of star clusters

A possible way in which a star cluster can appear within a molecular-gas filament is through a localized but intense star formation at a privileged location in the filament, say, at a junction of multiple filaments. Such a region is prone to reach a high SFE due to lateral contraction of the filaments and ample gas supply through them. Any such stellar cluster (or sub-cluster) should adapt to the compact cross-section of the filament (see Sec. 1). While formation of proto-stars outside the filaments, *i.e.*, in rarer gas, cannot be ruled out, both observations (e.g., Schneider *et al.* 2012; André *et al.* 2014) and hydrodynamic calculations (e.g., Bate & Bonnell 2004; Girichidis *et al.* 2011) suggest that the majority of the proto-stars must form within the compact dimensions of the dense gas filaments. Now, a cluster with $r_{\rm eff} \lesssim 0.3$ pc is way more compact than the presently observed YMCs (Portegies Zwart *et al.* 2010), *c.f.* Fig. 1. The key question here is whether such a compact star cluster can expand by its own, through its secular evolution, to attain the presently observed sizes.

To reach the observed sizes, such a compact cluster must expand by a factor of $\gtrsim 10$. For a young, massive cluster, several effects contribute to its secular expansion. In the earliest stage (until $t \approx 4.5$ Myr), mass loss due to stellar evolution is the primary driver of the expansion. The most massive stars remain segregated (either primordially or dynamically) at the cluster's center, causing centrally localized mass depletion due to the massive (O-type) stars' strong winds. When these stars undergo supernove, the central mass loss becomes even more severe, causing a higher rate of cluster expansion. This stellar mass loss dominated expansion continues for ≈ 50 Myr. After the stellar mass loss phase, the cluster expansion continues to be driven by dynamical heating due to the centrally segregated black holes (hereafter BHs) (Mackey *et al.* 2008; Banerjee *et al.* 2010; Morscher *et al.* 2013). Finally, young star clusters are observed to contain a substantial fraction ($\gtrsim 70\%$) of tight primordial binaries of massive main-sequence stars (Sana & Evans 2011). These primordial binaries also inject energy to the cluster and contribute to its expansion through binary-single and binary-binary encounters and the associated binary heating and ejections of massive stars (Heggie & Hut 2003).

To compute the secular evolution of model massive clusters in a realistic manner, the direct N-body code `NBODY7` (Aarseth 2012) is used. In addition to computing the individual stars' orbits using the highly accurate fourth-order Hermite scheme and dealing with the diverging gravitational forces, e.g., during close encounters and in hard binaries through regularizations, `NBODY7` includes stellar and binary evolution recipes. The details of the computed models are given in Table. 1. Some of the models contain a primordial binary population, to assess the role of the latter in expanding a young cluster. In these models, a $f_b(0) = 100\%$ primordial binary fraction that follow the "birth orbital period distribution" (Kroupa 1995) is used. For massive stars of $m > 5M_{\odot}$, a much tighter period distribution, given by a (bi-modal) Öpik law (uniform distribution in $\log_{10} P$) for $0.3 < \log_{10}(P/\text{day}) < 3.5$ (Sana & Evans 2011), is used.

The solid curves in Fig. 1 shows the computed evolution of effective radius, $r_{\rm eff}(t)$, for the models in Table 1. Here, the instantaneous $r_{\rm eff}(t)$ is obtained by taking the mean of the projected half-mass radii (50% Lagrange radius integrated over a plane) over three mutually perpendicular planes passing through the cluster's density center. As seen, starting from sizes similar to that of the filamentary substructures in molecular clouds (see Sec. 1), it is practically impossible to attain the observed sizes of YMCs and associations in 100 Myr. Notably, realistic conditions are used in these models including stellar mass loss, retention of $\approx 50\%$ of the BHs and NSs formed via supernovae and a realistic population of tight massive primordial binaries. A few test models also start with more extended size, $r_h(0) \approx 1.0$ pc, which can nearly reach the sizes of the most compact observed YMCs, but still are much more compact than most YMCs. As elaborated in Banerjee & Kroupa (2015c), theoretical uncertainties in stellar and binary evolution and core-collapse supernova, would modify the $r_{\rm eff}(t)$ curve to some extent, but are unlikely to alter the above overall inference. This necessitates additional expansion mechanisms for a newly assembled compact cluster to reach the present-day observed sizes.

3. Non-secular evolution of star clusters: primordial gas expulsion

The calculations in Sec. 2 do not consider any primordial gas present initially in the proto-cluster, *i.e.*, the cluster is taken to be formed with 100% local SFE. A high local SFE has been claimed by several hydrodynamics-based studies (e.g., Bate & Bonnell 2004; Dale *et al.* 2015). Clearly, as shown in Sec. 2, if a cluster is hatched along a

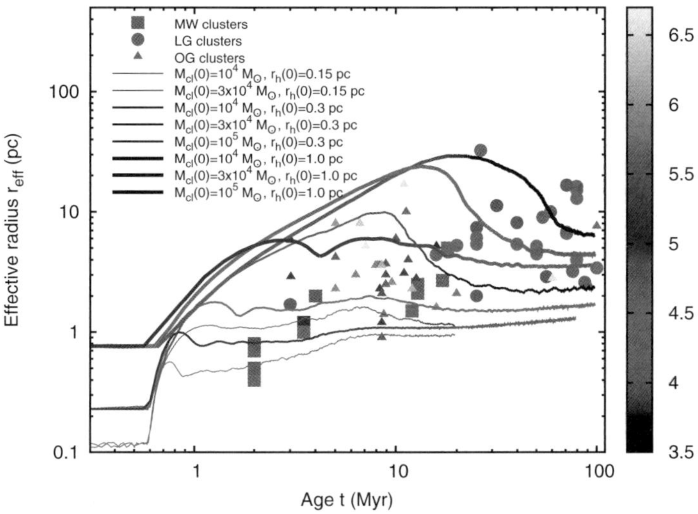

Figure 2. The data points and their colour-code are identical to Fig. 1. The curves show the computed evolution of the effective radius, $r_{\rm eff}(t)$, including a gas dispersal phase with star formation efficiency $\epsilon \approx 33\%$. These curves imply that even if YMCs evolve from filament-like compact sizes, such substantial (and explosive) gas dispersal will expand them to their present observed sizes in the Milky way and in the Local Group. However, to attain the sizes of the most extended Local Group YMCs, one needs to evolve from $r_h(0) \gtrsim 1$ pc half-mass radii, unless such objects are low-mass cluster complexes.

molecular-gas filament or at a filament junction with effectively 100% SFE, it well falls short of observed sizes of YMCs. On the other hand, observations of molecular clouds and embedded clusters suggest that in regions of high star-formation activity, the local SFE typically varies between a few percent to $\approx 30\%$. This is furthermore supported by high-resolution radiation magneto-hydrodynamic (RMHD) simulations of proto-star formation (Machida & Matsumoto 2012; Bate *et al.* 2014). Although these compute-expensive simulations are limited to the spatial scale of a single proto-star, they suggest a maximum feedback-limited $\approx 30\%$ SFE for proto-star formation; see Banerjee & Kroupa (2015b) for more discussion. This implies that the SFE over a gas clump forming a population of proto-stars is also $\lesssim 30\%$. Hence, it would be realistic to consider a proto-cluster (a pre-cluster cloud core) that is initially embedded in primordial gas, the gas being subsequently cleared from the system by the stellar feedbacks (e.g., radiation, outflows).

Fig. 2 shows the $r_{\rm eff}$ evolutions of computed model clusters with $\approx 30\%$ SFE which are compared to the observed $r_{\rm eff}$ values of YMCs and associations (upper and lower panels respectively). Here, the 'gas' is simply a time-varying background potential mimicking the expulsion of ionized Hydrogen with sound speed, as implemented in previous studies (e.g., Kroupa *et al.* 2001; Banerjee & Kroupa 2015a). With such an 'explosive' (*i.e.*, in timescale comparable to the dynamical time of the cluster) gas expulsion of $\approx 70\%$ by mass, the filament-like compact clusters can expand to reach the observed sizes of most of the MW YMCs (*c.f.* Fig. 2). However, most of the LG and OG YMCs are still a few factors larger in size than the computed models with compact initial conditions ($r_h(0) \lesssim 0.3$ pc). It seems likely that these extended YMCs are ensembles of closely-located YMCs ($\sim$ 10s of them), forming low-mass cluster complexes, thus being younger, low-mass versions of "faint-fuzzy" objects (Brüns *et al.* 2009); see Banerjee & Kroupa (2015c) for details.

Table 1. Initial conditions for the computed model clusters without a gas expulsion phase. The initial configurations are Plummer profiles with total mass, $M_{cl}(0)$, and half-mass radius, $r_h(0)$. The corresponding values of the virial velocity dispersion, $V_*(0)$, the crossing time, $\tau_{cr}(0)$, and the two-body relaxation time at half-mass radius, $\tau_h(0)$, are given. The initial clusters are in circular orbits at $R_G \approx 8$ kpc Galactocentric distance.

$M_{cl}(0)/M_\odot$	$r_h(0)$/pc	Primordial binaries	Mass segregation	$V_*(0)$/ km s^{-1}	$\tau_{cr}(0)$/Myr	$\tau_h(0)$/Myr
10^4	0.3	no	no	10.5	0.028	4.65
3×10^4	0.3	no	no	18.1	0.017	8.05
10^4	1.0	no	no	5.8	0.172	28.30
3×10^4	1.0	no	no	10.0	0.100	49.02
10^5	1.0	no	no	18.2	0.055	89.49
1.3×10^4	0.35	yes	yes	11.7	0.030	6.68
10^4	1.0	yes	yes	6.0	0.167	28.30
3×10^4	1.0	yes	yes	10.4	0.096	49.02

4. Cluster formation through hierarchical mergers

So far, monolithic or in-situ formation of star clusters are considered. The substructured and filamentary conditions in molecular clouds (see Sec. 1) make it plausible that YMCs may also arise due to sequential mergers of less massive sub-clusters, which fall in the potential well of the molecular cloud (or clump). In particular, if the sub-clusters are 'decoupled' (e.g., as in Dale *et al.* 2015) from the gas, then the assembly process is independent and proceeds in a timescale longer than the free-fall time of the gas.

If a cluster has to form and evolve from a young age (a few Myr like the youngest Galactic YMCs), the sub-clusters must fall in from sufficiently close separation so that they can merge early enough. This is demonstrated in Banerjee & Kroupa (2015a) for the case of the ≈ 1 Myr old NGC 3603 young cluster (photometric mass $\gtrsim 10^4 M_\odot$), for which the sub-clusters must merge from $\lesssim 2$ pc. This implies that despite star formation is often found to occur over $\gtrsim 10$ pc regions, only a part of the newly formed stellar structure (typically of pc scale) can actually comprise a YMC. There is mounting observational evidence of young stellar and/or proto-stellar sub-clusters packed within pc-scale regions (e.g., Massi *et al.* 2015; Román-Zúniga *et al.* 2015).

Moreover, as shown in Banerjee & Kroupa (2015a), the "prompt" merger of several gas-filament-like compact sub-clusters produces a similarly compact cluster. Therefore, according to the results in Sec. 2, the observed YMC sizes would be unreachable for the newly assembled cluster via their inherent expansion alone. This also holds for the individual sub-clusters. Therefore, as demonstrated in Banerjee & Kroupa (2015a), a post-merger explosive gas expulsion is instrumental in yielding YMCs that are like what we observe. If, on the other hand, the sub-clusters are brewed sufficiently apart that gas blow out happens in them separately prior to their merger, then the likely outcome would be a highly diffuse, massive stellar association with substructures, and it may, as a whole, be super-virial or sub-virial, e.g., Cyg-OB2 (also see Banerjee & Kroupa 2015b).

5. Concluding remarks

The above study shows that if star clusters preferably appear within the overdense filaments of molecular clouds, adapting to their typical dimensions of 0.1-0.3 pc, then their self-driven expansion is generally insufficient for reaching the dimensions of observed YMCs. This is true if either the cluster forms in situ or via mergers of (closely located) sub-clusters, and holds irrespective of the newborn cluster's mass (*c.f.* Fig. 1). Having run out of the other possibilities, feedback-driven rapid expulsion of residual gas from the proto-cluster (see Sec. 3) seems to be indispensable to reach the observed sizes of YMCs. In this line of argument, an important concern is how YMCs gather $\gtrsim 10^4 M_\odot$ within a few Myr, irrespective of any formation channel. Such amount of mass reservoir, over pc scale, is not immediately apparent from typical Galactic star-forming clouds. At the same time, the assembly phase of YMCs have to be short-lived ($\lesssim 1$ Myr), as discussed above, implying that it would be rare to catch an assembling starburst cluster.

The chances of finding a YMC in its assembling phase would be higher in starburst galaxies, where a much larger number of massive clusters are triggered compared to a Milky Way-like galaxy. Indeed, recent ALMA observations of the Antennae galaxy indicate still-forming, deeply-embedded stellar systems of total mass exceeding $10^7 M_\odot$, which are either monolithic or distributed over a few pc (Johnson 2015; Johnson *et al.* 2015). In order to better understand the birth conditions of starburst clusters, improved and more exhaustive observations of starburst galaxies are necessary.

References

Aarseth, S. J. 2012, *MNRAS*, 422, 841

André, P., Di Francesco, J., Ward-Thompson, D., *et al.* 2014, in Beuther, H., Klessen, R., Dullemond, C. and Henning, Th. (Eds.) *Protostars and Planets VI*, University of Arizona Press, Tucson, p.27

Banerjee, S., Baumgardt, H., & Kroupa, P. 2010, *MNRAS*, 402, 371

Banerjee, S. & Kroupa, P. 2015a, *MNRAS*, 447, 728

Banerjee, S. & Kroupa, P. 2015b, in S. W. Stahler eds. "The Birth of Star Clusters", Springer-Verlag, URL: https://astro.uni-bonn.de/sambaran/docs/VYMCformation_Banerjee_KroupaV2.pdf (preprint)

Banerjee, S. & Kroupa, P. 2015c, *A&A*, submitted.

Bastian, N. & Silva-Villa, E. 2013, *MNRAS*, 431, L122.

Bate M. R. & Bonnell, I. A. 2004, in Lamers, H. J. G. L. M., Smith, L. J., Nota A. (Eds.) *The Formation and Evolution of Massive Young Star Clusters*, (ASP Conf. Proc. 322). San Francisco: Astronomical Society of the Pacific, 289

Bate, M. R., Tricco, T. S., & Price, D. J. 2014, *MNRAS*, 437, 77

Brüns, R. C., Kroupa, P., & Fellhauer, M. 2009, *ApJ*, 702, 1268

Dale, J. E., Ercolano, B. & Bonnell, I. A. 2015, *MNRAS*, 451, 5506.

Girichidis, P., Federrath, C., Banerjee, R., & Klessen, R. S. 2011, *MNRAS*, 413, 2741

Heggie, D. C. & Hut, P. 2003, "The Gravitational Millon-Body Problem: A Multidisciplinary Approach to Star Cluster Dynamics". Cambridge University Press, Cambridge, UK

Johnson, K. 2015, IAU General Assembly, Meeting 29, id.2257770

Johnson, K. E., Leroy, A. K., Indebetouw, R., *et al.* 2015, *ApJ*, 806, 35

Kroupa, P. 1995, *MNRAS*, 277, 1507

Kroupa, P., Aarseth, S., & Hurley, J. 2001, *MNRAS*, 321, 699

&Machida, M. N. & Matsumoto, T. 2012, *MNRAS*, 421, 588

Mackey, A. D., Wilkinson, M. I., Davies, M. B., & Gilmore, G. F. 2008, *MNRAS*, 386, 65.

Massi, F., Giannetti, A., & di Carlo, E. 2015, *A&A*, 573, id.A95.

Morscher, M., Umbreit, S.,Farr, W. M., & Rasio, F. A. 2013, *ApJ*, 763, L15

Portegies Zwart, S. F., McMillan, S. L. W., & Gieles, M. 2010, *ARA&A*, 48, 431

Román-Zúniga, C. G., Ybarra, J., Megias, G., *et al.* 2015, *ApJ*, arXiv:1507.00016 (preprint)

Sana, H. & Evans, C. J. 2011, in Neiner, C., Wade, G., Meynet, G. and Peters, G. (Eds.) *Active OB Stars: Structure, Evolution, Mass Loss, and Critical Limits* (IAU Symp. 272). Cambridge Univ. Press, Cambridge, 474

Schneider, N., Csengeri, T., Hennemann, M., *et al.* 2012, *A&A*, 540, L11

Spitzer, L. Jr. 1987, "Dynamical Evolution of Globular Clusters", Princeton University Press.

Formation, evolution, and survival of massive star clusters
Proceedings IAU Symposium No. 316, 2015
C. Charbonnel & A. Nota, eds.

doi:10.1017/S1743921315010674

Effects of Stellar-Mass Black Holes on Massive Star Cluster Evolution

Sourav Chatterjee[1], Meagan Morscher[1], Carl L. Rodriguez[1], Bharat Pattabiraman[1] and Frederic A. Rasio[1]

[1]Center for Interdisciplinary Exploration and Research in Astrophysics
Northwestern University
2145 Sheridan Road, Evanston, IL 60208, USA
email: sourav.chatterjee@northwestern.edu

Abstract. Recent observations have revealed the existence of stellar mass black hole (BH) candidates in some globular clusters (GC) in the Milky Way and in other galaxies. Given that the detection of BHs is challenging, these detections likely indicate the existence of large populations of BHs in these clusters. This is in direct contrast to the past understanding that at most a handful of BHs may remain in old GCs due to quick mass segregation and rapid mutual dynamical ejection. Modern realistic star-by-star numerical simulations suggest that the retention fraction of BHs is typically much higher than what was previously thought. The BH dynamics near the cluster center leads to dynamical formation of new binaries and dynamical ejections, and acts as a persistent and significant energy source for these clusters. We have started exploring effects of BHs on the global evolution and survival of star clusters. We find that the evolution as well as survival of massive star clusters can critically depend on the details of the initial assumptions related to BH formation physics, such as natal kick distribution, and the initial stellar mass function (IMF). In this article we will present our latest results.

Keywords. black hole physics–stellar dynamics–methods: n-body simulations–methods: numerical–globular clusters: general–kinematics and dynamics–galaxies: star clusters

---

1. Introduction

Massive star clusters, such as the GCs in the Milky Way (MW) and other galaxies are expected to form large numbers of BHs simply as a consequence of the IMF and the large initial numbers of stars (N) they are born with. What happens to these BHs after formation, due to a combination of BH formation physics, such as birth-kicks in supernova (SN), and overall dynamical evolution of the host cluster is a long standing unsolved question.

Following the classic work by Spitzer (1969), it was argued that the BHs, being more massive compared to typical cluster stars, will mass segregate on $\lesssim 10^2$ Myr timescales and form a compact subcluster that is dynamically decoupled from the rest of the cluster. Due to the small size, high density, and low (compared to the whole cluster) number of stars in this subcluster, frequent strong scattering interactions would eject most BHs from the cluster on $\sim$ Gyr timescales. As a result the old GCs were expected to contain at most a few BHs at their current old ages (Kulkarni *et al.* 1993, Sigurdsson *et al.* 1993, Portegies Zwart & McMillan 2000, Kalogera *et al.* 2004). This understanding gained popularity because prior to 2012 all observed luminous X-ray sources (XRB) in the MW GCs were accreting neutron stars (NS; e.g., van Zyl *et al.* 2004, Lewin & van der Klis 2006, Altamirano *et al.* 2010, Altamirano *et al.* 2012, Bozzo *et al.* 2011), consistent with no bound accreting BHs in today's GCs.

This classical picture started to change with recent discoveries of candidate BHs in extragalactic as well as MW GCs (Maccarone *et al.* 2007, Irwin *et al.* 2010, Strader *et al.* 2012, Chomiuk *et al.* 2013). Interestingly, the GCs in the MW that contain the detected BH candidates, show large ranges in structural properties, indicating that presence of BHs in today's GCs may not be uncommon in general. Modern star-by-star, multi-physics simulations also find that BH ejection is not nearly as efficient as was previously thought. The primary reason for this difference compared to the earlier understanding comes from the fact that the BH subclusters do not stay decoupled from the rest of the cluster for prolonged periods. The same interactions that dynamically eject BHs from the cluster, also expands the subcluster and the subcluster again dynamically couples with the rest of the cluster (Breen & Heggie 2013, Morscher *et al.* 2015). As a result, the evaporation timescale for the BH subcluster increases dramatically, and even the old GCs may contain hundreds to thousands of BHs today. These simulations also find that only a small fraction of the bound BHs are binaries with one BH and one non-BH components. Thus, only a few such objects may become observable XRBs even when the actual number of bound BHs is high. Factoring in also the typical low duty cycles for these sources (Kalogera *et al.* 2004), detection of just a few BH candidates likely indicates the existence of a much larger number of BHs in these clusters (e.g., Strader *et al.* 2012; Umbreit 2012; Morscher *et al.* 2013, 2015).

In this article, we will focus on the effects of a large number of retained BHs on the global evolution of the host cluster. All past models usually made several initial assumptions related to the details of the BH-formation physics, and the IMF. Several of these so called standard assumptions are actually not well constrained from observations. We will explore the effects of changing some of these assumptions, namely the natal kick distribution for BHs, and the slope of the IMF ($dn/dM \propto M^{-\alpha_1}$) for high-mass stars ($M > 1\,M_\odot$) on the global evolution of the star cluster. In § 2 we describe our numerical setup and assumptions. In § 3 we present our key results. Finally, in § 4 we conclude.

2. Numerical Models

We use a Hénon-type Monte Carlo code (Hénon 1971), `CMC`, developed and rigorously tested by Northwestern University's cluster dynamics group over the past decade (most recently, Fregeau & Rasio 2007; Chatterjee *et al.* 2010, 2013; Umbreit *et al.* 2012; Pattabiraman *et al.* 2013). We follow the initial assumptions of Morscher *et al.* (2015) for our standard model (we call it `S`). This model initially contains $N = 8 \times 10^5$ stars, and overall binary fraction $f_b = 0.05$. The initial positions and velocities of the objects (single stars and center of mass of the binary stars) according to a King profile with $w_0 = 5$ (King 1962, 1965). The initial virial radius $r_v = 2\,\mathrm{pc}$. The cluster is at a galacto-centric distance $r_{\rm GC} = 8\,\mathrm{kpc}$. The primary masses are chosen from an IMF presented in Kroupa (2001) for cluster stars between 0.1 to 100 $M_\odot$. We adopt their middle values for the slopes in all mass ranges for `S`. The secondary masses are chosen assuming a uniform mass-ratio distribution between the primary mass and 0.1 $M_\odot$. Initial binary periods (P) are drawn from a uniform distribution in $\log P$, between 5 times the sum of the stellar radii to the initial local hard-soft boundary. Eccentricities are thermal. We choose metallicity $Z = 0.001$. Single and binary stellar evolution is done using `SSE` and `BSE` (Hurley *et al.* 2000, 2002), updated to implement custom natal kicks for the NSs and BHs. All NSs are assumed to receive natal kicks drawn from a Maxwellian distribution with $\sigma_{\rm NS} = 265\,\mathrm{km\ s^{-1}}$. We adopt a momentum-conserving kick prescription for the BHs, where the kick magnitude is calculated by first drawing from the same Maxwellian as the NSs, but then scaling this velocity by the fallback fraction (Fryer & Kalogera 2001, Belczynski *et al.* 2002).

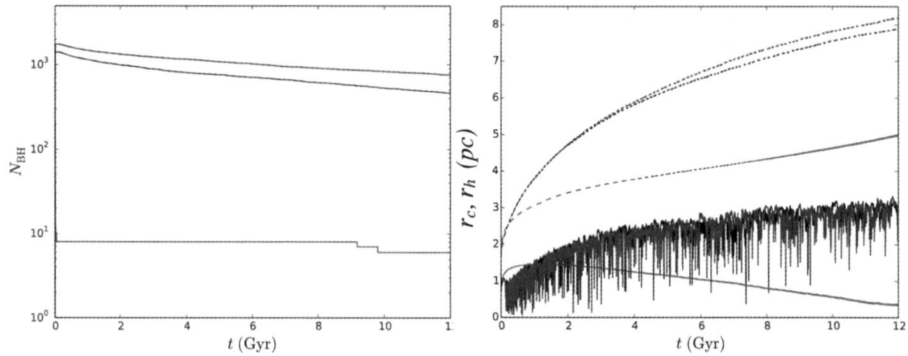

Figure 1. Evolution of the number of bound BHs in the cluster (left), and their structural properties, r_c, and r_h (right), for models `S` (black), `K1` (red), and `K0.1` (blue). The solid and dotted lines on the right panel denote r_c and r_h, respectively, for each model.

For example, if fallback fraction is $x_{FB} = 1$, i.e., no mass is lost during SN, the kick magnitude is zero.

While it is widely accepted that the NSs receive large natal kicks, the distribution of natal kicks for BHs is not well constrained either via observations or via theory. Detailed analysis of kick distributions of individual BH X-ray binaries result in widely varying constraints on kick velocities between these systems (e.g., Brandt *et al.* 1995; Nelemans *et al.* 1999; Willems *et al.* 2005; Gualandris *et al.* 2005; Dhawan *et al.* 2007; Fragos *et al.* 2009; Wong *et al.* 2012, 2014). On the other hand, recently, Repetto *et al.* (2012) suggested that the position offsets of the BH XRBs from the Galactic plane indicates that the natal kicks from BHs can be as high as NSs. Furthermore, they found that there is no clear correlation between the BH mass and its inferred kick magnitude, which would be expected if they all received momentum-conserving natal kicks based on x_{FB}. Modeling of core-collapse SN by Pejcha & Thompson (2015) also suggests that the natal kicks are not related in a simple way to the mass of the BHs. Clearly, the issue of natal kicks for BHs is far from settled. We simulate three additional models by changing our initial assumptions of the natal kick distribution for the BHs. In these models we assume that the natal kicks for the BHs are not mass-dependent. Instead, they receive kicks from a Maxwellian with $\sigma_{BH} = f \times \sigma_{NS}$, where f is a constant scaling fraction. We simulate two models with $f = 1$ and 0.1, and call these models `K1` and `K0.1`, respectively. All other assumptions are exactly the same as in `S`.

The choice of IMF, especially α_1, significantly affect the relative abundance of the high-mass stars, and as a result, the number of BHs a cluster with some given initial N would form. While, the middle values for the slopes presented in Kroupa (2001) are commonly adopted for most simulations, they have large 1σ errorbars, e.g., $\alpha_1 = 2.3 \pm 0.7$. We simulate two other models by changing α_1 to 1.6 and 3, and call these models `I-1.6` and `I-3`, respectively. Our model `S` of course already adopts $\alpha_1 = 2.3$.

3. Results

In our model `S`, $\approx 2 \times 10^3$ BHs are initially formed. The BHs drive repeated core-collapse episodes (Fig. 1). The core re-expands via dynamical binary formation as well as BH ejections due to strong gravitational scattering encounters involving binaries. However, BH ejections are not nearly as efficient as what was previously suggested (Spitzer 1969). Based on the x_{FB}-dependent, momentum-conserving kicks adopted in model `S`, about 400 BHs remain bound to the cluster at integration stopping time $t = 12\,\mathrm{Gyr}$. This result is

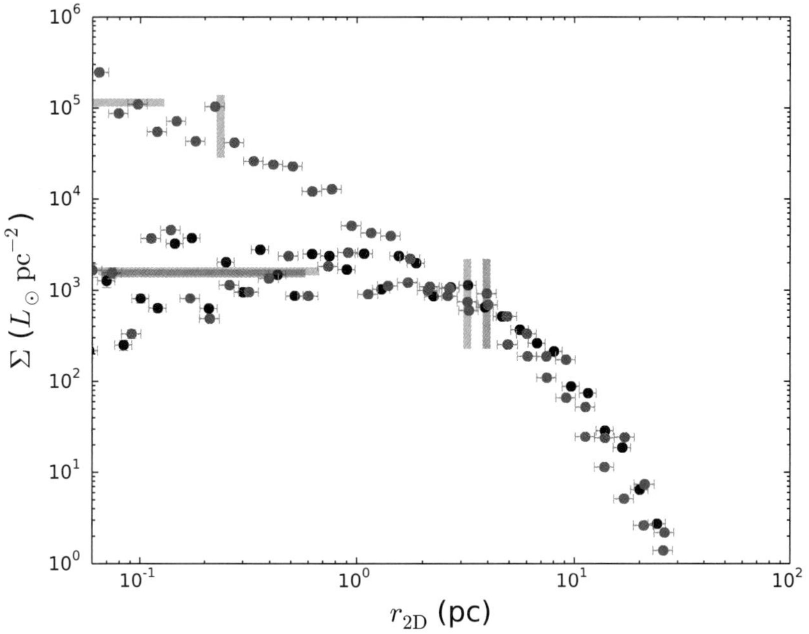

Figure 2. Surface brightness profiles at $t = 12$ Gyr for models S (black), K0.1 (blue), and K1 (red). The horizontal and vertical lines show the central density, and $r_{c,\mathrm{obs}}$ according to the best-fit King profile. The surface brightness profile for K1, where most of the BHs were ejected due to large natal kicks, show a power-law rise of the density, typical of the so-called core-collapsed GCs in the MW. In contrast, the surface brightness profiles indicate rather large and low-density clusters for S, and K0.1, where large numbers of BHs are retained until $t = 12$ Gyr.

in agreement with other recent star cluster models (e.g., Breen & Heggie 2013; Morscher *et al.* 2015). The evolution of our model K0.1 is very similar to S, except, K0.1 retains just shy of 10^3 BHs at $t = 12$ Gyr.

We find that the distribution of natal kicks for the BHs can have a dramatic effect on the overall evolution of the host cluster. If the BHs receive kicks as large as the NSs, as is the case for our model K1, only ≈ 6 BHs are retained in the cluster at late times. Most of the BHs are ejected from the cluster simply because of the kicks received during formation. Dynamical ejections at late times are insignificant. As a result, model K1's overall evolution is dramatically different from S and K0.1. In this case, since not many BHs are retained, the BH-driven core-collapse episodes are unseen. In absence of the BH-driven energy source at the cluster center, r_c starts contracting after about 1 Gyr. In comparison, r_c keeps expanding until integration stopping time of $t = 12$ Gyr due to BH-mediated energy production at the cluster core for models S, and K0.1.

The observable properties of the clusters are also significantly affected due to the difference in assumptions for the distribution of the BHs' natal kicks. For example, the surface brightness profiles (SBP) models S and K0.1 would indicate a rather puffy and low-density cluster, whereas, the SBP for model K0.1 would indicate a high-density core-collapsed cluster (Fig. 2). The only difference between the models is the persistent presence of the additional energy source from the hundreds of retained BHs in models S and K0.1, which is absent in model K1. Models S and K0.1 show central surface luminosity densities ($\Sigma_{c,\mathrm{obs}}$) between 10^3 to $2 \times 10^3\ L_\odot\ \mathrm{pc}^{-2}$. The best-fit King core radii for these models are $r_{c,\mathrm{obs}} \approx 4$, and 3 pc, respectively. In contrast, the surface luminosity density profile for model K1 show a power-law rise, typical of the core-collapse MW GCs

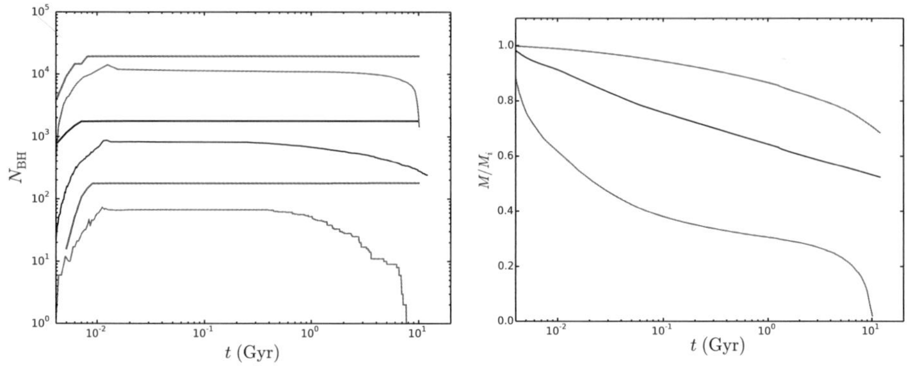

Figure 3. Left: Total number of BHs produced (thick lines) and retained (thin lines) in models `I-1.6` (red), `S` (black), and `I-3` (blue) as a function of cluster age. Right: Evolution of the cluster mass scaled by the initial mass of the cluster for models `I-1.6` (red), `S` (black), and `I-3` (blue). The model with the flat IMF slope, `I-1.6`, forms an order of magnitude higher number of BHs and dissolves within ~ 8 Gyr.

(e.g., Chatterjee *et al.* 2013). For `K0.1`, $\Sigma_{c,\mathrm{obs}} \sim 8 \times 10^4 \, L_\odot \, \mathrm{pc}^{-2}$, and $r_{c,\mathrm{obs}} = 0.2$ pc. The half-light radii for models `S`, `K0.1`, and `K1` are $r_{\mathrm{hl}} = 5.5, 3.4$, and 2.5 pc, respectively.

Similar to the distribution in natal kicks for the BHs, the IMF for high-mass stars also significantly affect the evolution of the clusters. Model `I-1.6`, with an adopted $\alpha_1 = 1.6$ (see §2) forms $\sim 2 \times 10^4$ BHs, almost an order of magnitude more than in model `S`. The high amount of mass loss from stellar winds, compact object formation from the much higher number of high-mass stars, as well as the efficient energy source from the large number of BHs at the cluster's center leads to an eventual dissolution of this cluster after about 8 Gyr (Fig. 3). This is consistent with the predictions of the classic study by Chernoff & Weinberg (1990). We also find that a model with such a flat IMF for high-mass stars would always dissolve, no matter how low the BH natal kicks are. However, significantly different and lower wind mass loss (e..g, Vink *et al.* 2001) may result in clusters that survive longer. In model `I-3`, only about 200 BHs are formed. All of these BHs are ejected from the cluster within 8 Gyr. Devoid of bound BHs, r_c of model `I-3` starts decreasing. At 12 Gyr, for `I-3`, $r_{\mathrm{hl}} = 1.2$ pc, $r_{c,\mathrm{obs}} = 0.15$ pc, and central surface density is $4 \times 10^5 \, L_\odot \, \mathrm{pc}^{-2}$.

4. Discussion

In this article we report results from an ongoing study that investigates the effects of the initial assumptions related to BHs on the overall evolution of the host cluster. Here we present results from two types of variations from a standard model that adopts commonly assumed initial properties. We change the distribution of natal kicks for the BHs and find that starting from clusters that are otherwise identical, variations in this assumption can dramatically alter the evolution of the cluster. For example, if the BHs receive birth kicks as large as the NSs, most BHs are ejected from the cluster and the cluster, at 12 Gyr, appears as a dense core-collapsed GC. On the other hand, if the BHs do not receive large natal kicks, hundreds of BHs may remain bound to the cluster till 12 Gyr, however, in these cases, the resulting cluster appears as puffy, low-density GCs.

Similar to the natal kicks, the power-law index of the IMF for stellar mass $M > 1 \, M_\odot$, α_1, plays a crucial role in determining the fate of a star cluster. We find that adopting $\alpha_1 = 1.6$, which is within the reported 1σ errors presented in Kroupa (2001), results in clusters that dissolve within $\sim 8\,\mathrm{Gyr}$ (model `I-1.6`). For $\alpha_1 = 2.3$ (the middle value of the Kroupa 2001 IMF; `S`), the cluster, at 12 Gyr appears to be puffy, with a large r_c and low $\Sigma_{c,\rm obs}$. For $\alpha_1 = 3$, (1σ upper limit in Kroupa 2001; `I-3`), the cluster with otherwise identical initial properties, attain properties similar to the core-collapsed GCs in the MW.

References

Altamirano, D., Patruno, A., Heinke, C. O., Markwardt, C., *et al.* 2010 *ApJ*, 712, 58
Altamirano, D., Wijnands, R., Heinke, C. O., *et al.*, 2012 *The Astronomer's Telegram*, 4264, 1
Belczynski, K. Kalogera, V., & Bulik, T., 2002 *ApJ*, 572, 407
Bozzo, E., Ferrigno, C., Stevens, J., Belloni, T. M., Rodriguez, J., *et al.*, 2011 *A&A*, 350, 649
Brandt, W. N., Podsiadlowski, P., & Sigurdsson, S., 1995 *MNRAS*, 277, 35
Breen, P. G. & Heggie, D. C., 2013 *MNRAS*, 432, 2779
Chatterjee, S., Fregeau, J. M., Umbreit, S., & Rasio, F. A., 2010 *ApJ*, 719, 915
Chatterjee, S., Umbreit, S., Fregeau, J. M., & Rasio, F. A., 2013 *MNRAS*, 429, 2881
Chernoff, D. F. & Weinberg, M. D., 1990 *ApJ*, 351, 121
Chomiuk, L., Strader, J., Maccarone, T. J., Miller-Jones, J. C. A., *et al.*, 2013 *ApJ*, 777, 69
Dhawan, V., Mirabel, I. F., Ribó, M., & Rodrigues, I., 2007 *ApJ*, 668, 430
Fragos, T., Willems, B., Kalogera, V., Ivanova, N., Rockefeller, G., *et al.*, 2009 *ApJ*, 697, 1057
Fregeau, J. M., & Rasio, F. A., 2007 *ApJ*, 658, 1047
Fryer, C. L., & Kalogera, V., 2001 *ApJ*, 554, 548
Gualandris, A., Colpi, M., Portegies Zwart, S., & Possenti, A., 2005 *ApJ*, 618, 845
Hénon, M. H., 1971 *Ap&SS*, 14, 151
Hurley, J. R., Pols, O. R., & Tout, C. A., 2000 *MNRAS*, 315, 543
Hurley, J. R., Tout, C. A., & Pols, O. R., 2002 *MNRAS*, 329, 897
Irwin, J. A., Brink, T. G., Bregman, J. N., & Roberts, T. P., 2010 *ApJ*, 712, 1
Kalogera, V., King, A. R., & Rasio, F. A. 2004 *ApJ*, 601, 171
King, I., 1962 *AJ*, 67, 471
King, I. R., 1965 *AJ*, 70, 376
Kulkarni, S. R., Hut, P., & McMillan, S. 1993 *Nature*, 364, 421
Lewin, W. H. G. & van der Klis, M. 2006 *Compact Stellar X-ray Sources, Ed. Walter Lewin & Michiel van der Klis*, Cambridge Astrophysics Series, No. 39. Cambridge, UK: Cambridge University Press, ISBN 978-0-521-82659-4
Maccarone, T. J., Kundu, A., Zepf, S. E., & Rhode, K. L., 2007 *Nature*, 445, 183
Morscher, M., Umbreit, S., Farr, W. M., & Rasio, F. A., 2013 *ApJ*, 763, 15
Morscher, M., Pattabiraman, B., Rodriguez, C., Rasio, F. A., & Umbreit, S., 2015 *ApJ*, 800, 9
Nelemans, G., Tauris, T. M., & van den Heuvel, E. P. J., 1999 *A&A*, 352, 87
Pattabiraman, B., Umbreit, S., Liao, W.k., Choudhary, A., *et al.*, 2013 *ApJS*, 204, 15
Pejcha, O., Thompson, T. A., 2015 *ApJ*, 801, 90
Portegies Zwart, S. F. & McMillan, S. L. W. 2000 *ApJ*, 528, 17
Repetto, S., Davies, M. B., & Sigurdsson, S., 2012 *MNRAS*, 425, 2799
Sigurdsson, S. & Hernquist, L. 1993 *Nature*, 364, 423
Spitzer, Jr., L. 1969 *ApJL*, 158, 139
Strader, J., Chomiuk, L., Maccarone, T. J., Miller-Jones, J. C. A., *et al.*, 2012 *Nature*, 490, 71
Umbreit, S., 2012 *Nature*, 490, 46
Umbreit, S., Fregeau, J. M., Chatterjee, S., & Rasio, F. A., 2012 *ApJ*, 750, 31
van Zyl, L., Charles, P. A., Arribas, S., Naylor, T., Mediavilla, E., *et al.* 2004 *MNRAS*, 350, 649
Vink, J. S., de Koter, A., & Lamers, H. J. G. L. M., 2001 *A&A*, 369, 574
Willems, B., Henninger, M., Levin, T., Ivanova, N., Kalogera, V., *et al.*, 2005 *2005*, 625, 324
Wong, T. W., Valsecchi, F., Fragos, T., & Kalogera, V., 2012 *ApJ*, 747, 111
Wong, T. W., Valsecchi, F., Ansari, A., Fragos, T., Glebbeek, E., *et al.*, 2014 *ApJ*, 790, 119

Formation, evolution, and survival of massive star clusters
Proceedings IAU Symposium No. 316, 2015
C. Charbonnel & A. Nota, eds.

doi:10.1017/S1743921315010601

Intermediate-mass black holes in globular clusters: observations and simulations - Update

Nora Lützgendorf[1], Markus Kissler-Patig[2], Karl Gebhardt[3], Holger Baumgardt[4], Diederik Kruijssen[5], Eva Noyola[3], Nadine Neumayer[6], Tim de Zeeuw[7,8], Anja Feldmeier-Krause[7], Edwin van der Helm[8], Inti Pelupessy[8] and Simon Portegies Zwart[8]

[1]ESA, Space Science Department, Keplerlaan 1,
NL-2200 AG Noordwijk, The Netherlands
email: nluetzge@cosmos.esa.int

[2]Gemini Observatory, Northern Operations Center, 670 N. A'ohoku Place,
Hilo, Hawaii, 96720, USA
[3]Department of Astronomy, University of Texas at Austin, Austin,
TX 78712, USA
[4]School of Mathematics and Physics, University of Queensland,
Brisbane, QLD 4072, Australia
[5]Max-Planck Institut für Astrophysik, Karl-Schwarzschild-Straße 1,
D-85748, Garching, Germany
[6]Max-Planck-Institute for Astronomy, Königstuhl 17,
69117, Heidelberg, Germany
[7]European Southern Observatory, Karl-Schwarzschild-Straße 2,
D-85748 Garching, Germany
[8]Leiden Observatory, Leiden University, PO Box 9513,
NL-2300 RA, Leiden, The Netherlands

Abstract. The study of intermediate-mass black holes (IMBHs) is a young and promising field of research. If IMBH exist, they could explain the rapid growth of supermassive black holes by acting as seeds in the early stage of galaxy formation. Formed by runaway collisions of massive stars in young and dense stellar clusters, intermediate-mass black holes could still be present in the centers of globular clusters, today. We measured the inner kinematic profiles with integral-field spectroscopy for 10 Galactic globular cluster and determined masses or upper limits of central black holes. In combination with literature data we further studied the positions of our results on known black-hole scaling relations (such as $M_\bullet - \sigma$) and found a similar but flatter correlation for IMBHs. Applying cluster evolution codes, the change in the slope could be explained with the stellar mass loss occurring in clusters in a tidal field over its life time. Furthermore, we present results from several numerical simulations on the topic of IMBHs and integral field units (IFUs). N-body simulations were used to simulate IFU data cubes. For the specific case of NGC 6388 we simulated two different IFU techniques and found that velocity dispersion measurements from individual velocities are strongly biased towards lower values due to blends of neighbouring stars and background light. In addition, we use the Astrophysical Multipurpose Software Environment (AMUSE) to combine gravitational physics, stellar evolution and hydrodynamics to simulate the accretion of stellar winds onto a black hole. We find that the S-stars need to provide very strong winds in order to explain the accretion rate in the galactic center.

Keywords. stars: kinematics and dynamics, methods: numerical, black hole physics

1. Introduction

Intermediate-mass black holes (IMBHs) provide the missing link between supermassive black holes and stellar-mass black holes. Their masses range from a few hundred solar masses up to $10^5 M_\odot$. Possible formation scenarios are remnants of population III stars (Madau & Rees 2001) or runaway merging in young dense star clusters (e.g. Portegies Zwart *et al.* 2004). Detecting and measuring these black holes would provide important data points to the $M_\bullet - \sigma$ relation at the lower mass end. Furthermore, IMBHs could explain the rapid growth of supermassive black holes at high redshift by acting as seeds in the early universe.

Many attempts have been made to detect IMBHs in globular clusters. These range from spectroscopic and photometric velocity measurements (e.g. Noyola *et al.* 2008) to X-ray and radio observations in order to find signatures of accretion (e.g. Strader *et al.* 2012). The results are disputed and the question whether IMBHs exist in globular clusters is not yet resolved. Each method has its caveats. The low gas content in globular clusters makes the detection of accretion signatures rather difficult. The X-ray and radio signals from low and irregular accretion rates are not yet well understood. On the other hand, kinematic signatures suffer from shot noise caused by a few bright stars or contamination from background light. It is therefore crucial to advance the field of IMBHs in all directions. Simulations on globular clusters and observing techniques provide a valuable tool to verify observing results and to understand the internal processes of globular clusters.

Previous numerical work on IMBHs in globular clusters has been performed by Baumgardt *et al.* (2005) and Noyola & Baumgardt (2011) who found that the surface brightness profiles of clusters hosting an IMBH exhibit weak central cusps, in contrast to core-collapsed clusters with very steep profiles and pre-core collapsed systems with no cusp at all. However, Trenti *et al.* (2010) and Vesperini & Trenti (2010) showed that weak cusp profiles could be confused with that of clusters that are on their way to core-collapse, hence the density profile alone is not a sufficient criterion for a globular clusters hosting an IMBH. Furthermore, it has been shown that IMBHs prevent a cluster from undergoing core collapse and reduce the degree of mass segregation compared with non-IMBH clusters (Baumgardt *et al.* 2004; Gill *et al.* 2008). Trenti *et al.* (2007) predicted that clusters with a high ratio of core radius to half-mass radius are good candidates for hosting an IMBH. This was challenged by Hurley (2007), who showed that the ratios observed for Galactic globular clusters can be explained without the need for an IMBH when treating model data as if they were observational data.

2. Observations

We used integral field spectroscopy provided by the FLAMES (Fiber Large Array Multi Element Spectrograph, Pasquini *et al.* 2002) instrument mounted on UT2 at the Very Large Telescope (VLT) to obtain integrated light spectra from 10 Galactic globular clusters. In addition to the spectroscopic data we used HST images from the archive for each cluster. From these images we obtained the color magnitude diagram, the photometric center and the surface brightness profile.

The velocity dispersion profile was then used to determine the possible existence and mass of an intermediate-mass black hole in the center. For this, the derived density profile of each cluster was used as an input for Jeans models. By applying χ^2 statistics we found the model that fits the data best. Figure 1 shows the fit of the globular cluster NGC 6266 with an IMBH signature. The profile clearly rises and requires a model with a black-hole mass of $\sim 3000\ M_\odot$. This method was applied to all clusters in the sample and for each

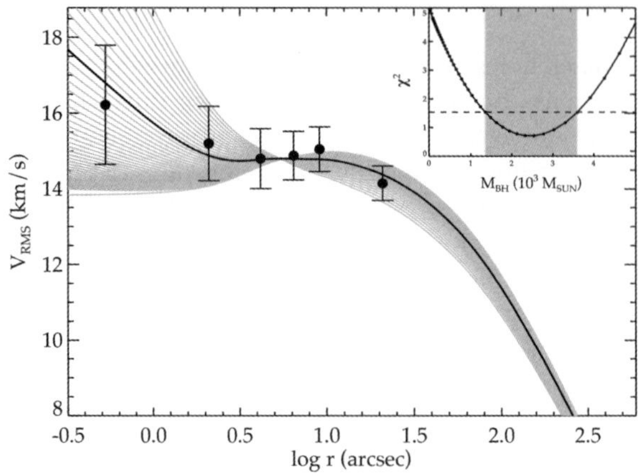

Figure 1. Velocity-dispersion profile of NGC 6266 overplotted by Jeans models with different black-hole masses. The χ^2 values is shown in the upper right and the best fit model indicated by the black sold line.

one we reported an upper limit or a value on the black hole mass (Lützgendorf *et al.* 2011, 2012, 2013b; Feldmeier *et al.* 2013).

Using the data points and upper limits of our sample together with measurements from the literature, we compared the correlation between black-hole mass and host system properties such as velocity dispersion and total mass with those measured for supermassive black holes in galaxies. In Lützgendorf *et al.* (2013c) we analyzed this small and challenging data set by applying survival analysis in combination with Markov-chain Monte Carlo in order to account for uncertainties and upper limits. We found that the major correlations are prominent but shallower than those for supermassive black holes (Figure 2). In a follow up paper (Kruijssen & Lützgendorf 2013) we showed that a possible explanation for the offset and more shallow correlation is the severe mass loss and expansion of globular clusters during their life time in a tidal field which leads to a reduction of the velocity dispersion and total mass of the system (see Figure 2).

3. Simulations

Because of large uncertainties in the measurements and contradicting results from different methods in the field of IMBHs, it is crucial to support the observing strategies, analysis and physics of the observations with sophisticated simulations. The next sections describe the different simulations that were performed to compare observations and verify our observing techniques.

3.1. *IFU simulations*

Integral field spectroscopy has become a popular tool in astronomy over the last years. The ability of simultaneously retrieving spatial and spectral data of an object has significant advantages for dynamical and stellar population studies. However, it is not yet clear what is the effect of this observing method on semi-resolved systems such as galactic globular clusters. For this reason we have developed a tool that allows to simulate IFU observations using star catalogues for different distances and observing conditions (i.e. seeing, Strehl ratio). The program uses a Moffat PSF that is applied to each star and integrated in the grid of the IFU. Using N-body simulations and N-body realizations we have tested the outcome of IFU observations for clusters with different properties, distances, and seeing. The results will be presented in Lützgendorf *et al.* (in prep).

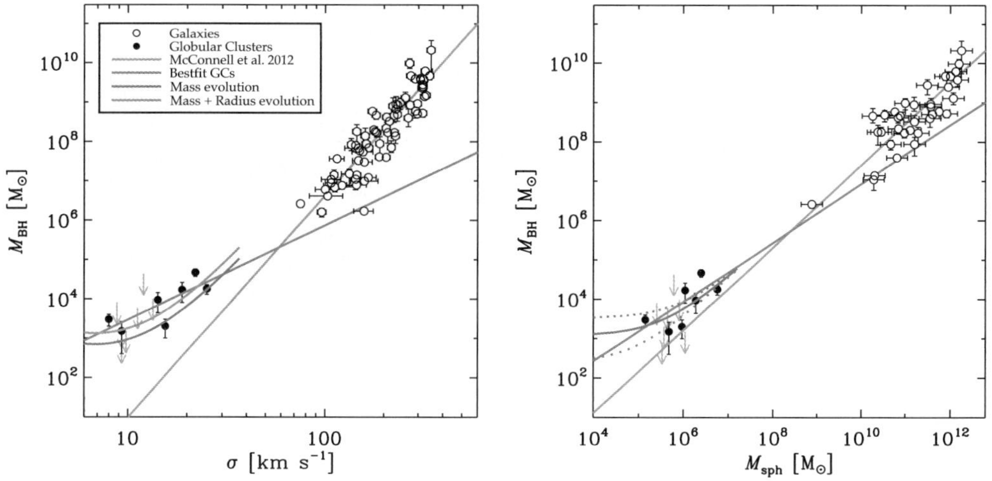

Figure 2. $M_\bullet - \sigma$ and $M_\bullet - M$ relation for supermassive black holes and IMBHs. The clear offset and shallower slope is well reproduced by cluster evolution models considering mass evolution only (magenta line) as well as models that additionally take the radius evolution (i.e. expansion) into account (orange line Kruijssen & Lützgendorf 2013).

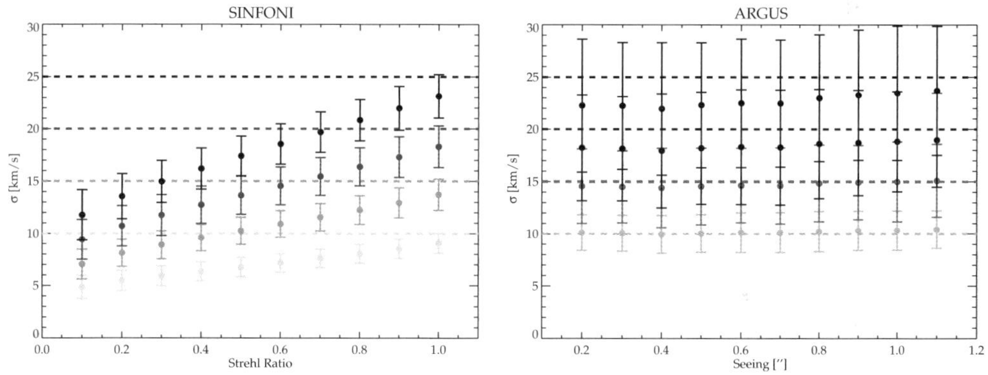

Figure 3. IFU simulations for SINFONI and ARGUS (central bin) with different input velocity dispersions as a function of Strehl ratio (the amount of light contained in the diffraction-limited core of the PSF, with respect to the total flux) and seeing. The input velocity dispersions are shown as dashed lines, the measured velocity dispersions as dots in the corresponding color.

In addition we investigated the specific case of NGC 6388 where two different methods using IFUs brought very different results. While Lützgendorf *et al.* (2011) found a steeply rising velocity dispersion profile and therefore a strong signature for an IMBH in its center, Lanzoni *et al.* (2013) measured a central velocity dispersion by taking individual velocities from adaptive optics supported SINFONI observations that is 40% lower than the value measured with integrated light. We reproduced both observations using the IFU simulation code and find the SINFONI observations being biased towards lower velocity dispersions due to blends of neighboring stars and background light (Lützgendorf *et al.* 2015, see Figure 3).

3.2. *AMUSE simulations*

Recently, many IMBH detections were challenged by contradicting measurements from different observing techniques. Especially the discrepancy between kinematic black-hole measurements and the absence of strong X-ray and radio emission from the centers of

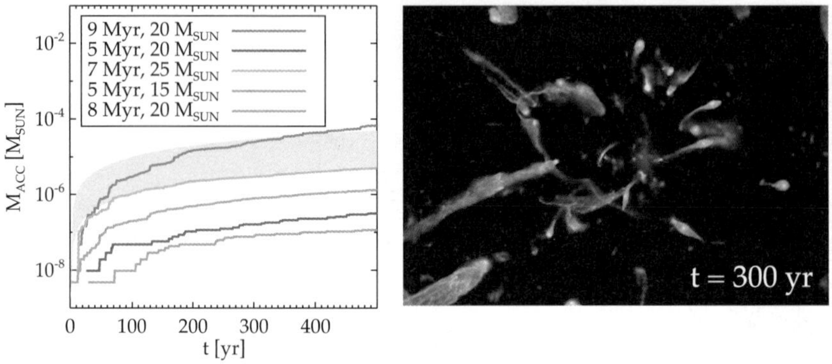

Figure 4. Left: Black-hole accretion rates for different stellar parameters. The gray shaded area marks the observed range. Right: Snapshot of the AMUSE simulation showing all 27 S-Stars and their stellar winds while orbiting the supermassive black hole in the center.

globular clusters remain an unsolved mystery (Strader *et al.* 2012). The main uncertainty when translating X-ray and radio flux measurements to black-hole masses is the amount of gas that is accessible for the black hole to accrete. We set out to investigate the effect of stellar winds on the accretion flow of the black hole. By using the Astrophysical Multipurpose Software Environment (AMUSE, Portegies Zwart *et al.* 2009, 2013; Pelupessy *et al.* 2013) we combined gravitational physics, stellar evolution and hydrodynamics into a single simulation of stars interacting with a black hole in the center of a globular cluster. The first application of this code will be presented in Lützgendorf (submitted to MNRAS) where the accretion rate of the supermassive black hole in the Milky Way from stellar winds of the surrounding S-Stars is studied (Figure 4, right panel).

We find that a) the total accretion is lower when summing up the contributions of individual stars (i.e. when running the simulation for each star separately, measuring the accretion and summing them up) than when including all stars in the simulation, b) cooling is inefficient and the temperatures are high due to wind shocks, and c) the accretion rate is highly dependent on the stellar parameters. In Figure 4 we show the accretion rates for different stellar parameters. The gray area marks the observed accretion rate from various sources in the literature (e.g. Baganoff *et al.* 2003; Bower *et al.* 2003; Nayakshin 2005). In order to reproduce the observed accretion rate of $10^{-8} - 10^{-7}\, M_{\odot}$ year^{-1} the stars have to be in their late evolutionary phase with massive winds, masses of $M = 19 - 25 M_{\odot}$ and ages between 7 and 10 Myr. This, however, is in contrast with observations that have shown that these stars are rather young, main-sequence B-stars. We therefore conclude that the S-stars cannot account for the accretion rate alone.

4. Summary

We present the work done by our group on the field of IMBHs in globular clusters over the past years. Using integral field spectroscopy in combination of high-resolution HST imaging we have investigated the presence of IMBHs in a sample of 10 galactic globular clusters. We found about a third of them show signatures of an IMBH in the center. Using our data points in combination with literature data we put the IMBH mass estimates in context with properties of their host systems and compared this to existing scaling relations. We found an offset, shallower correlation for the main relations ($M_{\bullet} - \sigma$, $M_{\bullet} - L$ and $M_{\bullet} - M$) which can be explained by the mass loss of globular clusters located in a tidal field over its life time. To support our observations we performed N-body simulations on globular clusters with and without a central IMBH. Further simulations aimed on reproducing IFU simulations by creating fake data sets from realistic N-body simulations and realizations with different distances and observing conditions. The simulations show that internal properties are in general well reproduced with IFU observations but that large uncertainties due to shot noise are unavoidable. In the specific case of NGC 6388 we found a strong bias in the velocity dispersion when computing it from individual velocities due to blend effects and background contamination. Finally, we used the Astrophysical Multipurpose Software Environment (AMUSE) to simulate the accretion of stellar winds onto a supermassive black hole to reproduce the S-Star system in the galactic center. We find that in order to explain the accretion rate the stars have to exhibit very massive stellar winds which collides with observational data. We plan to apply the code developed here to globular clusters and intermediate mass black holes in future work.

References

Aarseth, S. J. 1999, PASP, 111, 1333
Baganoff, F. K., Maeda, Y., Morris, M., *et al.* 2003, ApJ, 591, 891
Baumgardt, H., Makino, J., & Ebisuzaki, T. 2004, ApJ, 613, 1143
Baumgardt, H., Makino, J., & Hut, P. 2005, ApJ, 620, 238
Bower, G. C., Wright, M. C. H., Falcke, H., & Backer, D. C. 2003, ApJ, 588, 331
Feldmeier, A., Lützgendorf, N., Neumayer, N., *et al.* 2013, A&A, 554, AA63
Ferrarese, L. & Merritt, D. 2000, ApJ, 539, L9
Gill, M., Trenti, M., Miller, M. C., *et al.* 2008, ApJ, 686, 303
Häring, N. & Rix, H.-W. 2004, ApJ, 604, L89
Hurley, J. R. 2007, MNRAS, 379, 93
Hurley, J. R., Pols, O. R., & Tout, C. A. 2000, MNRAS, 315, 543
Kruijssen, J. M. D. & Lützgendorf, N. 2013, MNRAS, 434, L41
Kustaanheimo, P. & Stiefel, E. 1965, J. Reine Angew. Math., 218, 204
Lanzoni, B., Mucciarelli, A., Origlia, L., *et al.* 2013, ApJ, 769, 107
Lützgendorf, N., Gebhardt, K., Baumgardt, H., *et al.* 2015, A&A, 581, A1
Lützgendorf, N., Baumgardt, H., & Kruijssen, J. M. D. 2013a, A&A, 558, A117
Lützgendorf, N., Kissler-Patig, M., Gebhardt, K., *et al.* 2013b, A&A, 552, A49
Lützgendorf, N., Kissler-Patig, M., Gebhardt, K., *et al.* 2012, A&A, 542, A129
Lützgendorf, N., Kissler-Patig, M., Neumayer, N., *et al.* 2013c, A&A, 555, A26
Lützgendorf, N., Kissler-Patig, M., Noyola, E., *et al.* 2011, A&A, 533, A36+
Madau, P. & Rees, M. J. 2001, ApJ, 551, L27
Nayakshin, S. 2005, A&A, 429, L33
Nitadori, K. & Aarseth, S. J. 2012, MNRAS, 424, 545
Noyola, E. & Baumgardt, H. 2011, ApJ, 743, 52
Noyola, E., Gebhardt, K., & Bergmann, M. 2008, ApJ, 676, 1008
Pasquini, L., Avila, G., Blecha, A., *et al.* 2002, The Messenger, 110, 1
Pelupessy, F. I., van Elteren, A., de Vries, N., *et al.* 2013, A&A, 557, A84
Portegies Zwart, S., McMillan, S., Harfst, S., *et al.* 2009, New Astronomy, 14, 369
Portegies Zwart, S., McMillan, S. L. W., van Elteren *et al.* 2013, Comp. Phys. Comm., 183, 456
Portegies Zwart, S. F., Baumgardt, H., Hut, P. *et al.* 2004, Nature, 428, 724
Strader, J., Chomiuk, L., Maccarone, T. J., *et al.* 2012, ApJ, 750, L27
Trenti, M., Ardi, E., Mineshige, S., & Hut, P. 2007, MNRAS, 374, 857
Trenti, M., Vesperini, E., & Pasquato, M. 2010, ApJ, 708, 1598
Vesperini, E. & Trenti, M. 2010, ApJ, 720, L179

Formation, evolution, and survival of massive star clusters
Proceedings IAU Symposium No. 316, 2015
C. Charbonnel & A. Nota, eds.

doi:10.1017/S1743921315009709

Early dynamical evolution of young substructured clusters

Julien Dorval and Christian Boily

Observatoire Astronomique, Université de Strasbourg, 11 rue de l'Université, 67000 Strasbourg, France
email: `julien.dorval@astro.unistra.fr`
email: `christian.boily@astro.unistra.fr`

Abstract. Stellar clusters form with a high level of substructure, inherited from the molecular cloud and the star formation process. Evidence from observations and simulations also indicate the stars in such young clusters form a subvirial system. The subsequent dynamical evolution can cause important mass loss, ejecting a large part of the birth population in the field. It can also imprint the stellar population and still be inferred from observations of evolved clusters. Nbody simulations allow a better understanding of these early twists and turns, given realistic initial conditions. Nowadays, substructured, clumpy young clusters are usually obtained through pseudo-fractal growth and velocity inheritance. We introduce a new way to create clumpy initial conditions through a "Hubble expansion" which naturally produces self consistent clumps, velocity-wise. In depth analysis of the resulting clumps shows consistency with hydrodynamical simulations of young star clusters. We use these initial conditions to investigate the dynamical evolution of young subvirial clusters. We find the collapse to be soft, with hierarchical merging leading to a high level of mass segregation. The subsequent evolution is less pronounced than the equilibrium achieved from a cold collapse formation scenario.

Keywords. Star cluster, fragmentation, Nbody simulations

1. Introduction

Observed young stellar cluster often display a complex, clumpy structure (Kuhn *et al.* 2014). This is thought to originate from the localized star formation in the turbulent molecular cloud. These substructures are bound to be erased through dynamical evolution and relaxation as older clusters appear much smoother and spherical (Andre *et al.* 2007). This dynamical evolution comes off as a crucial phase in the life of the cluster, setting the longer lived, global structure and ejecting a part of the birth population. A good understanding of this process is of major importance in a study of the life cycle of star clusters. Nbody simulations allow to gain insight into the dynamical processes at play in this problem.

However, a critical aspect of nbody simulations are initial conditions. Older, spherical clusters can be easily reproduced by sampling well known King or Plummer distributions. To generate young substructured clusters constitutes however a real challenge. Several solutions have been adopted in the past to create clumpy initial conditions. For example, one can create several plummer clumps and scatter them in space, effectively creating a clumpy cluster, see McMillan *et al.* (2007). Though simple and straightforward, this approach forces an unjustified internal spatial distribution on clumps. Another widely used method is the fractal growth method from Goodwin *et al.* (2004). Through recursive probabilistic spawning of "children", one can grow a pseudo-fractal star cluster. The outcome of this process is visually very realistic and make convincing clumpy young clusters. However, the velocity distribution, though not absurd, is completely artificial.

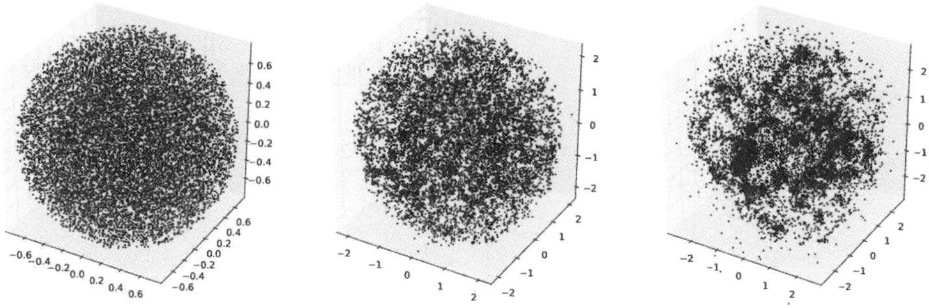

Figure 1. Progressive fragmentation through the Hubble expansion. Axis were rescaled to preserve the model's appearance, even though the radius increases. Left panel shows the initial uniform sphere, middle panel an intermediate step, slightly fragmented with a drastically slowed down expansion, right is the final stage, the expansion has stopped and the fragmentation is fully developed.

When integrated, such a system will undergo immediate relaxation within the clumps, which can be an issue.

We introduce here a new method to create self consistent clumpy initial conditions. We start with an uniform sphere model, in which we spawn the requested number of stars. We then attribute velocities according to a Hubble law $\mathbf{v} = H\mathbf{r}$, H being a "Hubble-like" constant, tuning the rate of expansion. When using the Hénon unit system (Hénon 1973), H must be lower than 1.4 to keep the system bound. We then feed this system to a collisional Nbody integrator, which is NBODY6 in our case. The system expands until it reaches the turning point, at which the expansion stops and the system is about to collapse. We stop the integration at this moment. During the expansion, density fluctuations in the initial uniform sphere grow, which leads to fragmentation. The resulting system is substructured in a fully self-consistent fashion. The clumps † are spontaneous and have self-consistent velocity distributions. When integrated, the system won't be undergoing local initial relaxation like fractal models, as the stars already know of each other's gravitational pull. By construction, the system is subvirial and will collapse when integrated.

2. Clump analysis

To study the resulting substructured system, one first needs to isolate the clumps. To do so we used the Minimum Spanning Tree cut off method, from Maschberger *et al.* (2010). The idea is to first build the three-dimensionnal Minimum Spanning Tree of the system, which is the shortest branching path to link all the points. Then a cut-off length is chosen, and all edges in the tree longer than this length are deleted. The resulting isolated subtrees are considered clumps if they contain more than 12 particles. The cut-off length is the only free parameter of the method. We developped a method to remove this free parameter, which we detail in Dorval *et al.* (2015, MN submitted).

The first aspect of clumps which is of interest to us is the mass segregation. A widely-used method to assess mass segregation is the mass segregation ratio (Allison *et al.* 2009). However, such a method is not adapted to very small systems, of the order of 20 members. We chose instead to implement the radial ranking method from Maschberger *et al.* (2010). For each clump, the distances from stars to the geometrical center of the clumps are computed, then sorted. The positions of the first, second, and third

† In this paper, we use "clump" to mean, loosely, *stellar associations*.

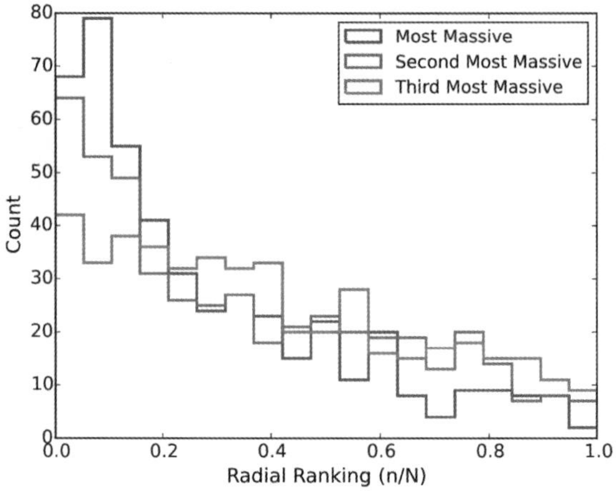

Figure 2. Radial ranking of first, second and third most massive stars for all clumps in a 40,000 particles Hubble fragmented model.

most massive stars in this ranking are converted into a 0 to 1 factor. Plotting the three histograms (first, second and third most massive) once the process is applied to all clumps allows to assess the trend towards mass segregation at a glance. In a non-segregated system, these histograms are flat on average. Figure 1 shows the results for a fragmented Hubble model. The histograms clearly show a mass segregation in the clumps. The results show trends similar to those seen in the analysis of Maschberger *et al.* (2010) applied to stellar cores formed hydrodynamically.

The origin of this primordial segregation was explained by Maschberger *et al.* (2010) through the clump formation process. In their hydrodynamical simulations, the clumps form around the most massive stars, which act as gravitationnal seeds and attract low mass stars. The same phenomenon is at play during the Hubble fragmentation. To confirm this, we turn to the clump stellar content. On figure 2, we show several mass functions in our fragmented model. In green is the global stellar mass function of the cluster, which follows a Salpeter law from 0.35 top 20 $M_{\odot}$. In red is the mass function of all stars registered as part of a clump. Finally, in blue is the theoretical Salpeter distribution for the same number of stars than what was found in clumps, that is to say the expected shape of the red distribution. Clumps are clearly top-heavy, most of the massive stars in the system end up in a clump, resulting in a shallower power-law ($\alpha = 2.15$) for members. This is consistent with them acting as fragmentation seeds.

These two results, mass segregation and top-heavy clumps, confirm that massive stars act as seeds for clump formation. Heavier stars form clumps around them, thus both preferentially belonging to a clump and sitting at the center.

3. Dynamical evolution

Once the fragmented system is obtained, we can let it evolve further to investigate the fate of young substructured and subvirial clusters. We take a 15000 particle system, drawn from a Salpeter stellar mass function from 0.35 to 20 $M_{\odot}$. This gives a mean stellar mass of 1 $M_{\odot}$ and a total mass of $15000 M_{\odot}$. The Nbody calculations to obtain the system were done in Hénon units (Hénon 1973). We pick scales adapted to real systems: for a half-mass radius of 1 pc, the resulting stellar density $\rho = M/(8\pi/3r_h^3) \simeq 10^3 M_{\odot}/pc^3$ is consistent with several observations. We let the system evolve for 40 Henon time units, translating to about 1 Myr. For comparison purposes, we performed similar computations

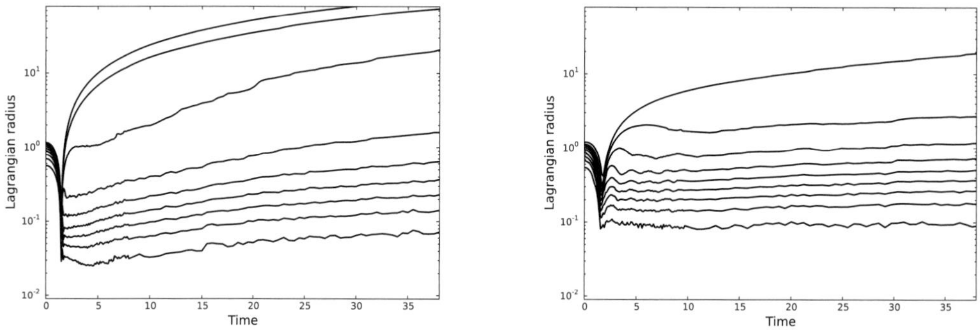

(a) Cold uniform collapse (b) Hubble collapse

Figure 3. Ten percentile lagrangian radii for both types of collapse. Time and radii are in Henon units.

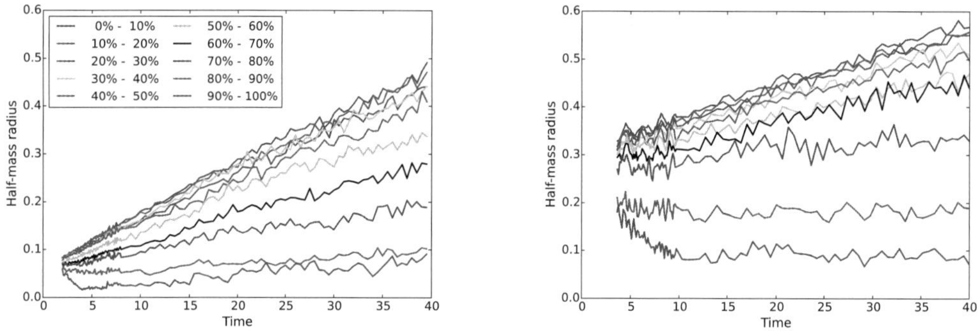

(a) Cold uniform collapse (b) Hubble collapse

Figure 4. Half-mass radii for ten mass-sorted subpopulations. Time and radii are in Henon units.

with a cold uniform sphere model of same mass and size. On figure 3 we show the ten percentile Lagrangian radii for the Hubble fragmented system (3a) and cold uniform (3b). The uniform collapse appears deeper and has a stronger "bounce", as twice as many stars are ejected. From a shallower post-collapse state, the Hubble model displays a slower evolution. The overall structure of the system does not change appreciably from a time $t \approx 5$ units until the end of the run, whereas the cold uniform system is clearly expanding throughout the simulation.

We also wish to investigate the mass segregation in the evolution of our system, as a follow up to the study on the individual clumps. To look at the overall system, we chose another indicator of mass segregation. We take inspiration from Caputo *et al.* (2014) and use their method. The masses are sorted by increasing values, then subdivided into ten equal-mass bins. This means that the first bin contains the least massive stars, and holds the largest number of stars. The number of stars in each bin decreases as we shift to the following bins, since their mean mass increases, and so on until we have binned all the stars. The half-mass radius computed for each bin is then plotted as function of time. In this way the mass segregation un-folds over time : if the stars were not segregated by mass, all half-mass radii would overlap. We show on figure 4 the outcome of the analysis for both uniform and Hubble models. We removed stars that were ejected after collapse, thus no data is available before that. Both uniform and Hubble models develop mass segregation over time, as the different populations separate themselves spatially. However, the Hubble model has a head-start in mass segregation as the different half-mass radii are already well separated immediately after the collapse. This is not surprising as clumps were found to be mass segregated prior to collapse, and inheritance of segregation during mergers is a known result (see McMillan *et al.* 2007). The denser configuration after the collapse in the cold uniform model induce stronger two-body interaction and segregation increases faster than for the Hubble configuration. In the end, after evolving for the same amount of time, the uniform model appears more segregated than the Hubble model.

Whether in regards to the overall spatial distribution or to the mass segregation, the evolution of the Hubble model following collapse is slower than the one of the cold uniform model. The hubble system inherits segregation from its clumps and form a shallower, more stable model, while initially ejecting less stars.

4. Conclusion

We demonstrated a new method to create substructured initial conditions for nbody simulations of young stellar clusters. By applying a Hubble velocity field to an uniform model, we allow continuous fragmentation which stems from pre-existing (Poissonian) density fluctuations. The result is a fully self-consistent, subvirial and fragmented model. By studying clumps individuallys, we showed that they display a primordial mass segregation. This can be explained by a clump formation process in which massive stars act as seeds for substructure growth. This is confirmed by the mass function of clump members, clearly skewed towards massive stars. We find a good agreement between the outcome of our method, which is pure Nbody computations, and the characteristics of more computationnaly intensive hydrodynamical simulations of forming star clusters. When left to relax towards equilibrium, the Hubble-fragmented model undergoes a softer collapse than a cold uniform sphere: the virial equilibrium is reached more quickly and the internal structure evolves much more slowly over time. Having inherited segregation from its clumps, the Hubble model is more segregated than the uniform model after collapse.

The main idea of the Hubble fragmentation model is a new one in the context of stellar clusters and several things remain to be investigated, such as the fate of binaries population in such system, the influence of stellar evolution or the effect of a background galactic potential on the dynamical evolution of the cluster. These will be the object of upcoming contributions. Details on this new method and the analysis can be found in our paper, Dorval *et al.* (2015, MN submitted).

References

Allison, R. J., Goodwin, S. P., Parker, R. J., Portegies Zwart, S. F., de Grijs, R., & Kouwenhoven, M. B. N. 2009, *MNRAS*, 395, 1449

André, P., Belloche, A., Motte, F., & Peretto, N. 2007, *A&A*, 472, 519

Caputo, D. P., de Vries, N., & Portegies Zwart, S. 2014, *MNRAS*, 445, 674

Dorval, J., Boily, C. M., Moraux, & E., Maschberger, T., & Becker, C 2015, *MNRAS*, submitted, MN-15-2591-MJ

Goodwin, S. P., & Whitworth, A. P. 2004, *A&A*, 413, 929

Hénon, M. H. 1973, *A&A*, 24, 229

Kuhn, M. A., Feigelson, E. D., Getman, K. V., Baddeley, A. J., Broos, P. S., Sills, A., Bate, M. R., Povich, M. S., Luhman, K. L., Busk, H. A., Naylor, T., & King, R. R. 2014, *ApJ*, 631, 976

Maschberger, T., Clarke, C. J., Bonnell, I. A., & Kroupa, P. 2010, *MNRAS*, 404, 1061

McMillan, S. L. W., Vesperini, E., & Portegies Zwart, S. F. 2007, *Apj*, 655, L45

Formation, evolution, and survival of massive star clusters
Proceedings IAU Symposium No. 316, 2015
C. Charbonnel & A. Nota, eds.

doi:10.1017/S174392131500900X

Self-shielding clumps in starburst clusters

Jan Palouš[1], Richard Wünsch[1], Soňa Ehlerová[1] and Guillermo Tenorio-Tagle[2]

[1]Astronomical Institute of the CAS,
Boční II 1401, 141 31 Prague 4, Czech Republic
email: palous@ig.cas.cz

[2]Instituto Nacional de Astrofísica Optica y Electronica,
AP 51, 72000 Puebla, Mexico
email: guillermotenoriotagle@gmail.com

Abstract. Young and massive star clusters above a critical mass form thermally unstable clumps reducing locally the temperature and pressure of the hot 10^7 K cluster wind. The matter reinserted by stars, and mass loaded in interactions with pristine gas and from evaporating circumstellar disks, accumulate on clumps that are ionized with photons produced by massive stars. We discuss if they may become self-shielded when they reach the central part of the cluster, or even before it, during their free fall to the cluster center. Here we explore the importance of heating efficiency of stellar winds.

Keywords. Globular clusters: general, Stars: formation, Hydrodynamics

1. Introduction

High-precision photometry with Hubble Space Telescope and stellar spectroscopy with Very Large Telescope discovered multiple stellar generations in globular clusters: enhanced He content in blue main-sequences and anticorrelation of Na - O and other products of hydrogen burning in cluster stars, see Hempel *et al.* (2014), Gratton *et al.* (2012) and Poitto *et al.* (2012). It is assumed that the first stellar generation mixed its nuclear burning products with original gas preparing the medium for formation of further stellar generations. Possible explanations include slow winds of fast rotating massive stars (Krause *et al.* 2013), or winds of AGB stars (D'Ercole *et al.* 2012). We propose the cooling winds of massive stars as the place where the enriched stellar generation forms. Here we discuss when thermally unstable clumps self-shield the UV photons produced by young stars of the cluster to become seeds for secondary star formation.

2. Cooling Winds

In clusters above a critical mass, thermally unstable clumps are created out of hot gas of fast winds of massive stars (Tenorio-Tagle *et al.* 2007, Wünsch *et al.* 2011, Palouš *et al.* 2013, and references therein). They shrink to a smaller volume, and the surrounding hot cluster wind is not able to push them out from the cluster, on the contrary, under the influence of the cluster gravity, they stream down towards the cluster center (see Fig. 1). We derive the minimum time that is needed to accumulate enough mass so that the stream is self-shielded against UV photons from the star cluster. However, when the free fall time of clumps is shorter compared to the self-shielding time, the streams of thermally unstable clumps meet in the cluster center forming central concentration of warm gas. As soon as this central mass concentration is self-shielded, it cools further down to become the seed of a new enriched stellar generation.

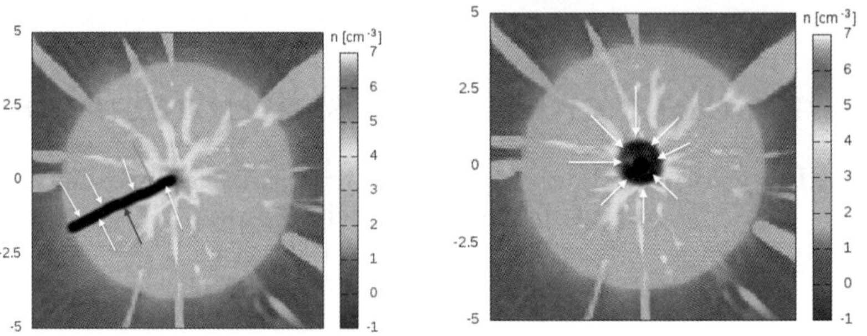

Figure 1. Left: stream of warm gas falling to the cluster center; Right: central condensation of warm gas.

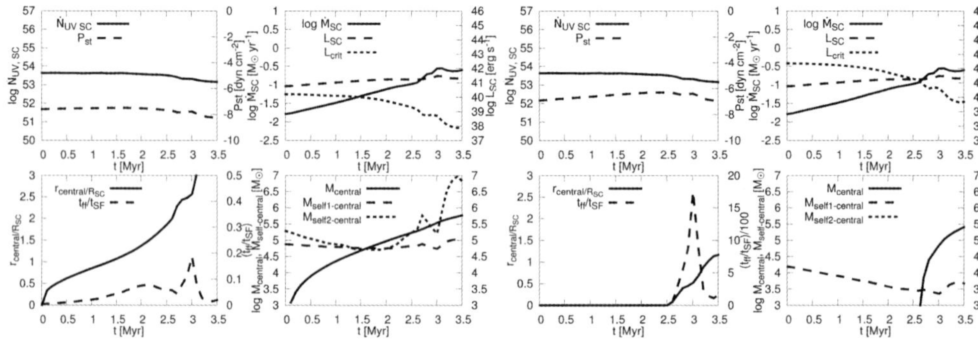

Figure 2. The time evolution of the UV photon flux $\dot{N}_{UV,SWC}$, pressure in the cluster wind P_{st}, total mass flux $\dot{M}_{SC}$ and mechanical energy flux L_{SC} in the cluster wind, radius of the central clump $r_{central}$, self-shielding time of the gas streams t_{SF} and central mass concentration $M_{central}$ during the first 3.5 Myr. Left: 5% heating efficiency; Right 30% heating efficiency.

3. Results

The results of our semi-analytical calculations of the evolution of clusters during its first 3.5 Myr are given in Fig. 2. We assume a first stellar generation with the total mass 10^7 $M_{\odot}$ and a radius 3 pc. In such cluster, its wind is thermally unstable forming after 3.5 Myr clumps with a total mass of a few times 10^5 $M_{\odot}$. An important parameter of the model is the heating efficiency, giving the fraction of the wind mechanical energy that is converted into heat of the hot cluster winds. In the case of a low heating efficiency (~5 %) the free-fall time is shorter compared to the self-shielding time, forming centrally concentrated second stellar generation. On the other hand, when the heating efficiency is high (~30%) the gas streams become self shielded before they reach the cluster center, resulting in a second generation cluster with a radius similar to that of the first stellar generation.

References

D'Ercole, A., D'Antona, F., Carini, R., *et al.*. 2012, *MNRAS*, 423, 1521

Gratton, R. G., Carretta, E., & Bragaglia, A. 2012, *A&ARv*, 20, 50

Hempel, M., Sarajedini, A., Anderson, J., *et al.* 2014, *ApJS*, 211, 1

Krause, M., Charbonnel, C., Decressin, T., Meynet, G., & Prantzos, N. 2013, *A&A*, 552, 121

Palouš, J., Wünsch, R., Martínez-Gonzalez, S, Hueyotl-Zahuantitla, F., Silich, S., & Tenorio-Tagle, G. 2013, *ApJ*, 772, 128

Piotto, G., Milione, A. P., Anderson, J., *et al.* 2012, *ApJ*, 760, 39

Tenorio-Tagle, G., Wünsch, R., Silich, S., & Palouš, J. 2007, *ApJ*, 658, 1196

Wünsch, R., Silich, S., Palouš, J., Tenorio-Tagle, G., & Muñoz-Tuñon, C. 2011, *ApJ*, 740, 75

Formation, evolution, and survival of massive star clusters
Proceedings IAU Symposium No. 316, 2015
C. Charbonnel & A. Nota, eds.

doi:10.1017/S1743921315008893

The UCD Population of the Coma Cluster

Kristin Chiboucas[1], Peter Ferguson[1], R. Brent Tully[2], David Carter[3], Steven Phillipps[4] and Eric Peng[5,6]

[1]Gemini Observatory, Hilo, HI, United States
email: kchibouc@gemini.edu email: petersferguson@gmail.com

[2] Institute for Astronomy, Honolulu, HI, United States
email:tully@ifa.hawaii.edu

[3] Liverpool John Moores University, Liverpool, United Kingdom
email:d.carter@ljmu.ac.uk

[4] University of Bristol, Bristol, United Kingdom
email:s.phillipps@bristol.ac.uk

[5] Peking University, Beijing, China
[6] Kavli Institute for Astronomy and Astrophysics, Beijing, China
email:peng@pku.edu.cn

Abstract. UCDs are super massive star clusters found largely in dense regions but have also been found around individual galaxies and in smaller groups. Their origin is still under debate but consensus is that they formed either during major galaxy mergers as mergers of super massive star clusters, are simply the high mass end of the globular cluster luminosity function and formed in the same way as globular clusters, or that they formed from the threshing of galaxies and are remnant nuclear star clusters, which themselves may have formed from the mergers of globular star clusters within galaxies. We are attempting to disentangle these competing formation scenarios with a large survey of UCDs in the Coma cluster. Using ACS two-passband imaging from the HST/ACS Coma Cluster Treasury Survey, we are using colors and sizes to identify the UCD cluster members. With a large sample within the core region of the Coma cluster, we will use the population size, properties, and spatial distribution, and comparison with the Coma globular cluster and nuclear star cluster populations to discriminate between the threshing and globular cluster scenarios. In particular, previously we have found a possible correlation of UCD colors with host galaxy and a possible excess of UCDs around a non-central giant galaxy with an unusually large globular cluster population, both suggestive of a globular cluster origin. With a larger sample size, we are investigating whether the color correlation with host persists and whether the UCD population is consistent with, or in excess of, the bright end of the GCLF. We present initial results from the survey.

Keywords. galaxies:star clusters - galaxies:dwarf - galaxies: clusters: individual (Coma)

1. Methods and Results

We have searched for UCD candidates in the 25 ACS fields centered on the core of the Coma cluster. Initial selection was based on the SExtractor FWHM, magnitude, half light radius, and $F475W - F814W$ measurements. These measurements for 34 spectroscopically confirmed UCDs in these fields were used to establish criteria for candidate selection. We are currently performing profile fitting on the initial candidates to measure sizes and shapes in order to refine the candidate list. With an accurately known PSF it is possible to measure effective radii for objects down to $\sim 10\%$ of the PSF FWHM. For ACS imaging, we find we can measure sizes as small as ~ 7 pc at the distances of Coma, about the minimum size for UCDs. From the initial selection of candidates, we find an excess of UCDs around 3 prominent giants in the core region (see Figure 1) including

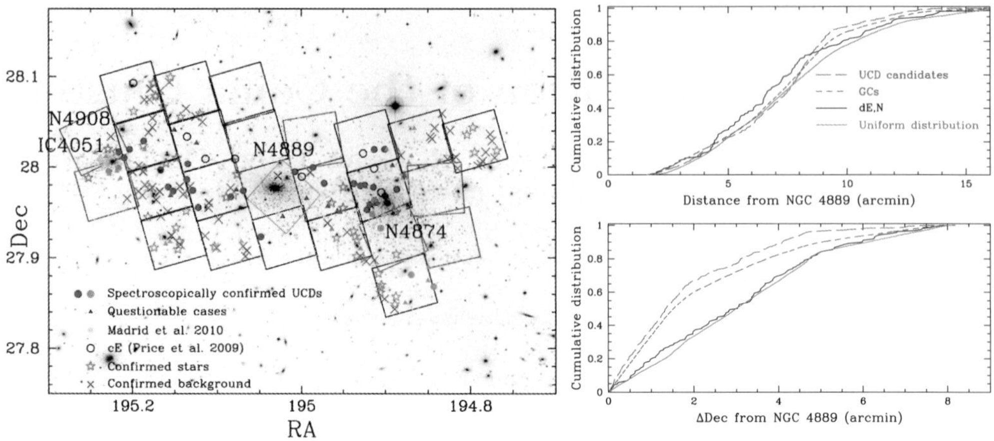

Figure 1. Left: Spatial distribution of UCDs in the Coma cluster. Large filled circles are spectroscopically confirmed. Small green points are UCD candidates chosen based on FWHM, surface brightness, color, and magnitude. Boxes indicate ACS fields from the HST/ACS Coma Cluster Treasury Survey (Carter *et al.* 2008), red boxes are more recent data. Right: Distribution as a function of distance from the most central cD galaxy NGC 4889 including only objects in the original fields. We find that globular clusters (Peng *et al.* 2011) and UCDs share a similar more concentrated distribution in declination with an excess in a band running between NGC 4874 and IC 4051 and around 3 giants in the core region. dEs exhibit a more uniform distribution.

around IC 4051, the same galaxy that hosts an unusually large globular cluster population. Overall, UCD candidates and the Coma cluster globular cluster population share a similar distribution. With properties and distribution more similar to Coma globular clusters than galaxies, we find more evidence for a star cluster origin for these objects. However, other origin scenarios are not yet ruled out.

2. Future work

From size and shape measurements of all candidates, we will determine probable cluster membership for UCDs with sizes > 7pc and establish a more complete sample of UCDs in the core region of the Coma cluster beyond the 34 spectroscopically confirmed members. In order to distinguish between different formation channels, we will investigate the spatial distribution in the cluster in comparison with the globular cluster and dwarf galaxy populations, determine the specific frequency of UCDs and GCs around the 3 major galaxies and analyze whether UCD counts are consistent with the bright tail of the globular cluster luminosity function, establish whether the color affiliations with host found previously are real, and compare the abundance of UCDs and dwarf galaxies in the core region to determine the efficiency of stripping required if UCDs are formed from the threshing of these galaxies.

References

Carter, D., Goudfrooij, P., Mobasher, B., *et al.* 2008 *ApJS*, 176, 424

Madrid, J. P., Graham, A. W., Harris, W. E., Goudfrooij, P., Forbes, D. A., Carter, D., Blakeslee, J. P., Spitler, L. R., Ferguson, H. C. 2010 *ApJ*, 722, 1707

Peng, E. W., Ferguson, H. C., Gudfrooij, P., Hammer, D., Lucey, J. R., Marzke, R. O., Puzia, T. H., Carter, D. 2011 *ApJ*, 730, 23

Price, J., Phillipps, S., Huxor, A., *et al.* 2009, *MNRAS*, 397, 1816

Formation, evolution, and survival of massive star clusters
Proceedings IAU Symposium No. 316, 2015
C. Charbonnel & A. Nota, eds.

doi:10.1017/S174392131501042X

Binary Information from Open Clusters Using SEDS (BINOCS) Project: The Dynamical Evolution of the Binary Populations in Cluster Environments

Peter Frinchaboy[1] **and Benjamin Thompson**[1]

[1]Department of Physics & Astronomy, Texas Christian University,
2800 S. University Ave., Fort Worth, TX, USA 76129
email: p.frinchaboy@tcu.edu, b.a.thompson1@tcu.edu

Abstract. Studying the internal dynamics of stellar clusters is conducted primarily through N-Body simulations. One of the major inputs into N-Body simulations is the binary star frequency and mass distribution, which is currently constrained by relations derived from field binary stars. However to truly understand how clustered environments evolve, binary data from within star clusters is needed including masses. Detailed information on binaries masses, primary and secondary, in star clusters has been limited to date. The primary technique currently available has been radial velocity surveys that are limited in depth. Using previous two-band photometry-based studies that may cover different mass ranges produce potentially discrepant interpretations of the observed binary population. We introduce a new binary detection method, Binary INformation from Open Clusters Using SEDs (BINOCS) that covers the wide mass range needed to improve cluster N-body simulation inputs and comparisons. Using newly-observed multi-wavelength photometric catalogs (0.3 - 8 microns) of the key open clusters with a range of ages, we can show that the BINOCS method determines accurate binary component masses for unresolved cluster binaries through comparison to available RV-based studies. Using this method, we present results on the dynamical evolution of binaries from 0.4 - 2.5 solar masses within five prototypical clusters, spaning 30 Myr to 3.5 Gyr, and how the binary populations evolve as a function of mass.

Keywords. binaries: general, open clusters and associations: general, methods: miscellaneous

---

1. Introduction

Current cluster binary studies are carried out using one of two methods, two-band photometry, and time-baseline radial velocity studies, each of which experience issues which limit their effectiveness in answering the above science questions. However, to deeply understand the binary populations of open clusters, we have created a new method which can determine accurate masses for all members of a cluster within a reasonable amount of telescope time. This new binary detection method is nicknamed BINOCS: BINARY INFORMATION FROM OPEN CLUSTERS USING SEDS. By imaging a star using multiple filters across the spectrum (e.g., $UBVRIJHK_S$[3.6][4.5][5.8][8.0]), one should be able to "rebuild" *spectral energy distribution* (SED) of a star given its parameters: age, metallicity, mass. Similarly, a binary system could not be accurately modelled by a single SED curve, but instead by two SEDs added together. By matching stars to these models, mass can be determined, similar to how temperature could be determined in the idealized black-body case. Since the star is a member of a cluster with known parameters, so age and metallicity are given. By matching stars to models of a library of synthetic single stars SEDs, mass can be determined. Isochrones often come in coarse mass grids. To overcome

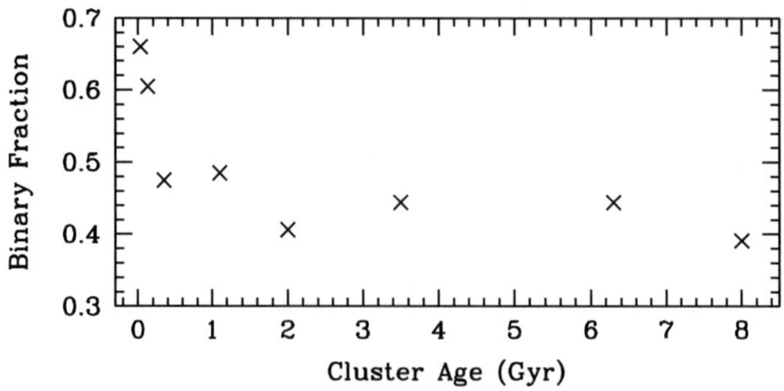

Figure 1. Total binary fraction for the BINCOS clusters sample, which shows a clear rapid decline in binary fraction within the first 200 Myr, and then a slow decline thereafter.

this, stellar parameters and magnitude are cubically interpolated with respect to mass onto a new mass grid, with a spacing of 0.01 $M_\odot$ (Thompson & Frinchaboy, 2016)†.

In this work, we present analysis from nine open clusters (M35, M36, M37, M67, NGC 188, NGC 2158, NGC 2420, NGC 6791. NGC 6819), a sample designed to cover a wide range of ages and metallicites within the open cluster population, as well as provide good ties to other open cluster binaries studies for verification. The method depends on having high-quality photometry covering a wide range of wavelengths.

2. Results: Full Sample BINOCS

Using the sample of clusters from the Table below, we find that there is a significant binary population in open clusters, with 60-70% of stars in binary systems. This significant binary population is quickly disrupted within the first 200 Myr, likely the destruction of wide binaries in the cluster environment, with a much slower rate of disruption thereafter as shown (Fig. 1).

3. Results: Intra-Cluster BINOCS

Binary systems are more massive, on average, than a single star, and should therefore experience mass segregation. This has been observationally confirmed for several globular and open clusters (e.g., Geller & Mathieu 2012, Milone *et al.* 2012). Similar analyses, using the two-band photometric detection method have been conducted on the young (15–30 Myr), massive cluster NGC 1818, located in the Large Magellanic Cloud (LMC), producing conflicting results Elson *et al.* (1998) and De Grijs *et al.* (2013). We leveraged our new BINOCS results (using the cluster M3, with binary and mass info, to show that this discrepancy within single star cluster is due to differing mass showing different dynamical ages (Thompson & Frinchaboy, 2016). This results has now also been seen in N-Body simulations (Geller *et al.* 2015)

Reference

Amari, S., Hoppe, P., Zinner, E., & Lewis R. S. 1995, *Meteoritics*, 30, 490

† Software used in this study can be obtained from: https://github.com/bathompso/BINOCS

Formation, evolution, and survival of massive star clusters
Proceedings IAU Symposium No. 316, 2015
C. Charbonnel & A. Nota, eds.

doi:10.1017/S1743921315009084

Dynamical Evolution of Outer-Halo Globular Clusters

Andreas H. W. Küpper[1] Akram H. Zonoozi[2], Hosein Haghi[2], Nora Lützgendorf[3], Steffen Mieske[4], Matthias Frank[5], Holger Baumgardt[6] and Pavel Kroupa[7]

[1]Department of Astronomy, Columbia University, New York, NY 10027, USA
email: akuepper@astro.columbia.edu

[2]Department of Physics, Institute for Advanced Studies in Basic Sciences, Zanjan, Iran

[3]ESA, Space Science Department, Keplerlaan 1, NL-2200 AG Noordwijk, The Netherlands

[4]European Southern Observatory, Alonso de Cordova 3107, Vitacura, Santiago, Chile

[5]Landessternwarte, Universität Heidelberg, Königsstuhl 12, 69117 Heidelberg, Germany

[6]University of Queensland, School of Mathematics and Physics, Brisbane, QLD 4072, Australia

[7]HISKP, Universität Bonn, Nussallee 14-16, 53115 Bonn, Germany

Abstract. Outer-halo globular clusters show large half-light radii and flat stellar mass functions, depleted in low-mass stars. Using N-body simulations of globular clusters on eccentric orbits within a Milky Way-like potential, we show how a cluster's half-mass radius and its mass function develop over time. The slope of the central mass function flattens proportionally to the amount of mass a cluster has lost, and the half-mass radius grows to a size proportional to the average strength of the tidal field. The main driver of these processes is mass segregation of dark remnants. We conclude that the extended, depleted clusters observed in the Milky Way must have had small half-mass radii in the past, and that they expanded due to the weak tidal field they spend most of their lifetime in. Moreover, their mass functions must have been steeper in the past but flattened significantly as a cause of mass segregation and tidal mass loss.

Keywords. methods: n-body simulations, globular clusters: general, Galaxy: halo

1. Why are outer-halo globular clusters extended?

Detailed photometric and spectroscopic investigations of outer-halo globular clusters (GCs) such as Palomar 4 and Palomar 14 have shown that these clusters are very extended, showing large half-light radii of $\gg 10$ pc, but at the same time show clear signatures of mass segregation within their centers (e.g., Jordi *et al.* 2009; Frank *et al.* 2012, 2014). This is intuitively unexpected, since their present-day two-body relaxation times are larger than a Hubble time. One way of solving this apparent discrepancy is if these GCs were born compact, and the observed mass segregation happened in the early evolution of the clusters.

For the purpose of this investigation, we carried out 16 N-body simulations of the same globular cluster model in a fixed Milky Way-like galaxy potential using the collisional code NBODY6 (Aarseth 2003). The GCs consisted initially of 65536 stars, following a Kroupa (2001) spectrum of stellar masses, and giving the clusters a mass of $M_0 = 26400\,\mathrm{M}_\odot$. The GCs started off compact with an initial half-mass radius of 3 pc. The models were placed on different orbits with a range of apogalactic distances, $r_{apo} = \{16, 32, 64, 128\}$ kpc, and orbital eccentricities, $\epsilon = \{0, 0.25, 0.5, 0.75\}$.

As has been observed in previous studies (e.g., Baumgardt & Makino 2003), the mass of the clusters, M, decreases faster with time for GCs orbiting in a stronger tidal field.

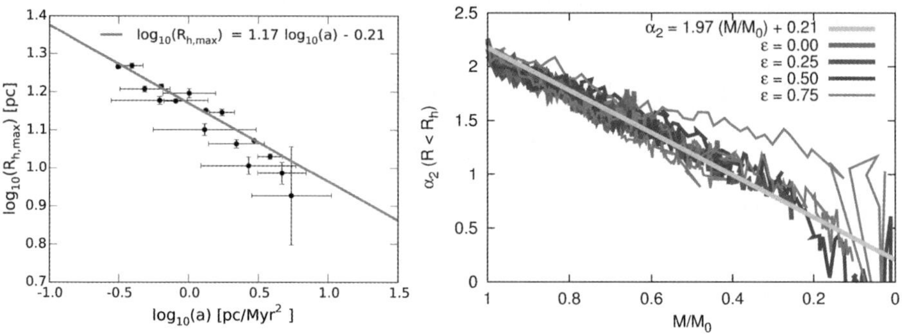

Figure 1. *Left:* the maximum half-mass radius, $R_{h,max}$, a cluster can reach throughout its evolution correlates with the time-averaged acceleration, a, the cluster experiences on its orbit. *Right:* the mass function slope of high-mass stars, α_2, within the half-mass radius depends on the mass the cluster has lost independent of its orbit.

Moreover, the initially compact clusters expand into their Roche lobes due to stellar mass loss and segregation of dark remnants into the cluster core. As a consequence of that, GCs in weaker tidal fields show larger half-mass radii, R_h, throughout their evolution (see, e.g., Madrid *et al.* 2012; Haghi *et al.* 2014). This expansion process can be reversed when the GCs have lost a significant amount of their initial mass. Hence, strongly dissolving clusters shrink after reaching a maximum expansion, $R_{h,max}$. In the left panel of Fig. 1, we show the time-averaged tidal acceleration, a, and its variance of each GC plotted against the maximum half-mass radius. There is a clear correlation between the two quantities. Hence, the half-mass radii of outer-halo clusters reflect the tidal field they live in. Driver of the expansion are mainly dark remnants and high-mass stars segregating into the cluster core, and as such the degree of expansion depends on the initial composition of the cluster (Haghi *et al.* 2014). As a consequence of this segregation, the mass-function slope, α_2, of stars with $m > 0.5\,M_{\odot}$ within the half-mass radius decreases proportionally to mass loss (see, e.g., Vesperini & Heggie 1997). This flattening of mass function is independent of the orbit, as shown in the right panel of Fig. 1. Deviations from this linear trend are only seen for clusters that undergo quick dissolution, and for $M/M_0 < 0.2$. We conclude that the flattened mass functions observed in the inner parts of extended outer-halo globular clusters are characteristic signatures of their expansion histories (cf., Zonoozi *et al.* 2011, 2014).

References

Aarseth, S. J. 2003, *Gravitational N-Body Simulations* (Cambridge University Press)

Baumgardt, H. & Makino, J. 2003, *MNRAS*, 340, 227

Frank, M. J., Hilker, M., Baumgardt, H., Côté, P., Grebel, E. K., Haghi, H., Küpper, A. H. W., & Djorgovski, S. G. 2012, *MNRAS*, 423, 2917

Frank, M. J., Grebel, E. K., Küpper, A. H. W. *MNRAS*, 443, 815

Haghi, H., Hoseini-Rad, M., Zonoozi, A. H., & Küpper, A. H. W. 2014, *MNRAS*, 444, 3699

Jordi, K., *et al.* 2009, *AJ*, 137, 4586

Kroupa, P. 2001, *MNRAS*, 322, 231

Madrid, J. P., Hurley, J. R., & Sippel, A. C. 2012, *ApJ*, 756, 167

Vesperini, E. & Heggie, D. C. 1997, *MNRAS*, 289, 898

Zonoozi, A. H., Küpper, A. H. W., Baumgardt, H., Haghi, H., Kroupa, P., & Hilker, M. 2011, *MNRAS*, 411, 1989

Zonoozi, A. H., Haghi, H., Küpper, A. H. W., Baumgardt, H., Frank, M. J., & Kroupa, P. 2014, *MNRAS*, 440, 3172

Formation, evolution, and survival of massive star clusters
Proceedings IAU Symposium No. 316, 2015
C. Charbonnel & A. Nota, eds.

doi:10.1017/S174392131500887X

GalevNB: from simulations to observations

Xiaoying Pang[1,2], Christoph Olczak[2] and Rainer Spurzem[2]

[1]Shanghai Institute of Technology, 100 Haiquan Road, Fengxian district, Shanghai 201418, P.R. China
email: xypang@bao.ac.cn

[2]National Astronomical Observatories, Chinese Academy of Sciences, 20A Datun Road, Chaoyang District, Beijing 100012, P.R. China

Abstract. We present GalevNB (Galev for N-body simulations), an utility that converts fundamental stellar properties of N-body simulations into observational properties using the GALEV (GAlaxy EVolutionary synthesis models) package, and thus allowing direct comparisons between observations and N-body simulations. It converts the fundamental stellar properties of N-body simulations, i.e., stellar mass, temperature, stellar luminosity and metallicity, into observational magnitudes for a variety of filters of widely used instruments/telescopes (HST, ESO, SDSS, 2MASS), and into spectra that span from far-UV (90 Å) to near-IR (160 μm).

Keywords. N-body, magnitude, spectra

1. Introduction

The output parameters of NBODY6++ (Aarseth 1999) simulations are mostly theoretical values. To make a direct comparison between N-Body simulation data and observations, we combine GALEV (GALaxy EVolutionary synthesis models; Kotulla *et al.* 2009), a flexible algorithm to combine astrophysical colors in many filters and spectra of stars (Lejeune, Cuisinier & Buser 1997, 1998) or sets of stars, with NBODY6++ simulations. In this paper, we present the structure of this new code: GalevNB (Galev for N-Body simulations). Adapting subroutines from GALEV, GalevNB can produce spectra spans the range from far-UV at 90 Å to far IR at 160 μm, with a spectral resolution of 20 Å in the UV-optical and 50-100 Å in the near IR range. Given a list of requested filters in HST, ESO, SDSS, 2MASS etc., GalevNB convolves the spectra with the filter response functions and applies the chosen zero-points (Vegamag, ABmag, and STmag) to yield absolute magnitudes. GalevNB bridges theoretical parameters and their observed values, thus allows us to understand the color and spectra evolution of star clusters, and to determine the initial conditions and parameters of star cluster simulations with a direct comparison to observations.

2. GalevNB Structure and Usage

The main program of GalevNB is GalevNB.f90, which parses single snapshot files (stellar evolution only) generated by NBODY6++ / NBODY6. It uses seven subroutines (startomaginit, specint_initialize, reset_weights, startomag, add_star, spec2mag, spec_output) of GALEV package to convert effective temperature, stellar luminosity, metallicity, and mass into observational magnitudes and spectra. The GalevNB package contains four folders: 1) spectral_templates, in which locate all the spectral template files from the BaSeL library of model atmospheres (Lejeune, Cuisinier & Buser 1997, 1998); 2) standard_filters, contains a large set of filter response functions (FUV, NUV, U, B, V, R, I, J, H, K) that are used as standard reference filters;

Table 1. Column contents for the filter information file: `filterlist.dat`

Column	Content	ID of zero point
1	Filter name	
2	Corresponding path of the filter response function	
3	ID of selected zero point (default value is 1)	
4	Standard zero point in the Vega magnitude system	1
5	Standard zero point in the AB magnitude system	2
6	Standard zero point in the ST magnitude system	3
7	Optional user-defined zero point	4

3) `filter_response_curves`, includes filter response functions from magnitude systems of HST, ESO instruments, 2MASS, SDSS, Johnson, and Cousins in separate subfolders. We also provide a choice of user-specify filter response functions. Information about the entire set of available filters is included in the file `filterlist.dat`. Please aware that `filterlist.dat`, in which the user specify their own choice of magnitude system by uncommenting the line of chosen filter, MUST be presented in the same directory as the `NBODY6++` / `NBODY6` snapshot files. The content of the file, `filterlist.dat`, is presented in Table 1.

To compile `GalevNB`, the user should have C++ and fortran installed. The input file of `GalevNB` should be a sinlge snapshot output from `NBODY6++` / `NBODY6` simulations. In case of a file containing all snapshots (called `sev.83` in `NBODY6++` and `fort.83` in `NBODY6`), we provide the user with a shell script `generate_snapshots.sh` in the folder, `scripts`, for retrieving single snapshot data out of `sev.83` and `fort.83`. The user can select his/her preferred filters (maximum 20) by uncommenting the row of the corresponding filter in `filterlist.dat`, and choose his/her desired magnitude system (Table 1). Magnitudes of individual stars and the whole cluster, and spectra of the cluster or chosen stellar types are produced, respectively.

The code of `GalevNB` is published online†. Users can download it through internet.

Acknowledgements

This work is funded by National Natural Science Foundation of China, No: 11503015. XYP also acknowledge the fund of talents introduction project of Shanghai Institute of Technology, No: 10120K156031-YJ2014-05, and the fund of Shanghai education committee, No: 1021ZK151009027-ZZyy15104.

References

Aarseth, S. J. 1999, *PASP*, 111, 1333

Kotulla, R., Fritze, U., Weilbacher, P., & Anders, P. 2009, *MNRAS*, 396, 462

Lejeune, T., Cuisinier, F., & Buser, R. 1997, *A&A Supplement*, 125, 229

Lejeune, T., Cuisinier, F., & Buser, R. 1998, *A&A Supplement*, 130, 65

† `http://silkroad.bao.ac.cn/repos/galevnb`

Formation, evolution, and survival of massive star clusters
Proceedings IAU Symposium No. 316, 2015
C. Charbonnel & A. Nota, eds.

doi:10.1017/S1743921315008911

Initial Dynamical Evolution of Star Clusters with Tidal Field

So-Myoung Park[1], Simon P. Goodwin[2] and Sungsoo S. Kim[1,3]

[1]School of Space Research, Kyung Hee University,
1732, Deogyeong-daero, Yongin-si 446-701, South Korea
email: smpark12@khu.ac.kr

[2]Dept. of Physics & Astronomy, University of Sheffield,
Sheffield S3 7RH, United Kingdom
email: s.goodwin@sheffield.ac.uk

[3]Dept. of Astronomy & Space Science, Kyung Hee University,
1732, Deogyeong-daero, Yongin-si 446-701, South Korea
email: sungsoo.kim@khu.ac.kr

Abstract. Observations have been suggested that star clusters could form from the rapid collapse and violent relaxation of substructured distributions. We investigate the collapse of fractal stellar distributions in no, weak, and very strong tidal fields. We find that the rapid collapse of substructure into spherical clusters happens quickly with no or a weak tidal field, but very strong tidal fields prevent a cluster forming. However, we also find that dense Plummer spheres are also rapidly destroyed in strong tidal fields. We suggest that this is why the low-mass star clusters cannot survive near the galactic centre which has strong tidal field.

Keywords. Galaxy: open clusters and associations: general, Galaxy: center, Galaxy:kinematics and dynamics, methods: n-body simulations

---

1. Introduction

How star clusters form is one of the important questions in star formation. Observations show that many young stars are in dense clusters (Lada & Lada 2003), even in regions where the tidal field is strong, e.g. the Arches and Quintuplet clusters, near the Galactic centre. Gas-rich star forming regions, however, are often found to be clumpy, filamentary and substructured (Cartwright & Whitworth 2004, Konyves *et al.* 2010). Numerical simulations have shown that star clusters can form from initial fractal distributions with no tidal field (Parker *et al.* 2014 and references therein). But do dense nuclear star clusters like the Arches have to form at high densities, or can they collapse from substructure? Here we present the first results of simulations of how tidal fields can affect the initial evolution of initially clumpy star clusters. We concentrate on the evolution of low-mass (< 1000 $M_\odot$) clusters in this short contribution.

2. Method & Initial Conditions

<u>*Method*</u>. We use the Nbody6 code as modified by Kim (Kim *et al.* 2000). Tidal fields are applied at 30 pc from the Galactic centre with the enclosed mass determined from Launhardt *et al.* (2002).
<u>*Initial conditions*</u>. We use initially fractal (Goodwin & Whitworth 2004) and Plummer (1911) distributions. The number of stars is 999 and the mass of star cluster is about 700 $M_\odot$. Initial mass function follows Macshberger (2013). In a paper to follow we also

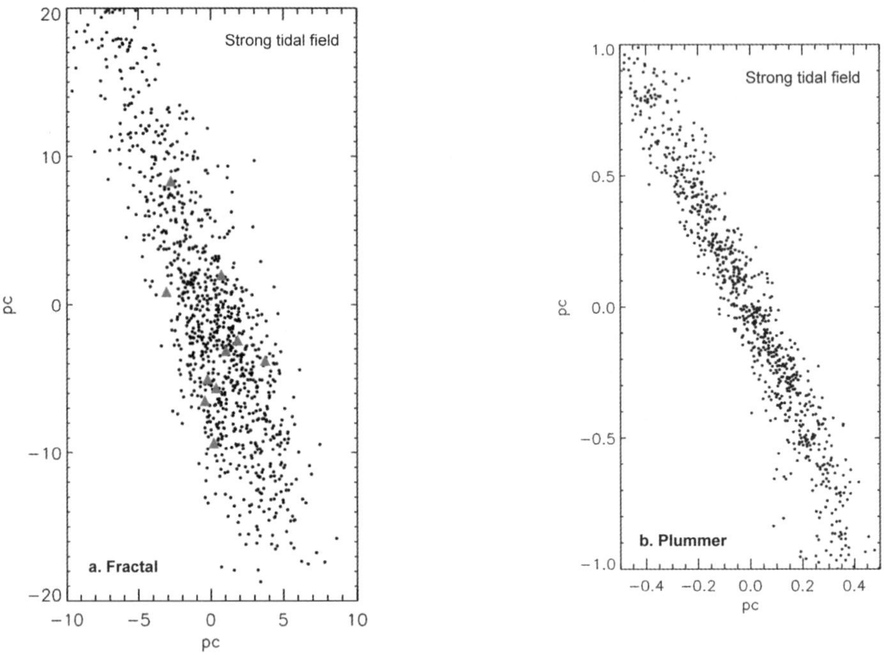

Figure 1. The morphologies of star clusters at 2 Myr, in panel (a) starting with an initially fractal distribution, and in panel (b) starting with a dense Plummer sphere.

examine different strong tidal fields at 100 pc and 200 pc, as well as the effect of weak (Solar radius) tidal fields.

3. Results & Conclusion

We find that a low-mass initially fractal distribution can collapse and form a dense, spherical cluster within 2 Myr with no tidal field (or a weak tidal field). In a strong (30 pc from the Galactic centre) tidal field, however, collapsing clusters are ripped apart within 2 Myr. However, simulations of dense low-mass Plummer spheres in a strong tidal field are also shredded within 2 Myr, despite their initial size being smaller than their tidal radii. Fig.1 shows the state of both the initially fractal (left panel), and Plummer (right panel) distributions after 2 Myr. This means that low-mass star clusters cannot survive near the Galactic Centre, no matter what their initial conditions are.

References

Cartwright, A. & Whitworth, A. P. 2004, *MNRAS*, 348, 589
Goodwin, S. P. & Whitworth, A. P. 2004, *A&A*, 413, 929
Kim, S. S., Figer, Donald F., Lee, Hyung Mok, *et al.* 2000, *ApJ*, 545, 301
Konyves, V., Andre, Ph., Menshchikov, A., *et al.* 2010, *A&A*, 518, L106
Lada, C. J. & Lada, E. A. 1993, *ARA&A*, 41, 57
Launhardt, R., Zylkz, R., & Mezger, P. G. 2002, *A&A*, 384, 112
Maschberger, T. 2013, *MNRAS*, 429, 1725
Parker, R. J., Wright, N. J., Goodwin, S. P., *et al.* 2014, *MNRAS*, 438, 620
Plummer, H. C. 1911, *MNRAS*, 71, 460

Formation, evolution, and survival of massive star clusters
Proceedings IAU Symposium No. 316, 2015
C. Charbonnel & A. Nota, eds.

doi:10.1017/S1743921315008923

From a star cluster ensemble to its formation history

Christine Schulz[1] **and Michael Hilker**[1]
[1]European Southern Observatory (ESO),
Karl-Schwarzschild-Str. 2, 85748 Garching bei München, Germany
email: cschulz@eso.org, mhilker@eso.org

Abstract. The present-day sample of ultra-compact dwarf galaxies (UCDs) and globular clusters (GCs) around NGC 1399 is interpreted to be composed of individual star cluster (SC) populations. It is assumed that such an SC population forms at a constant star-formation rate (SFR), and its mass distribution is described by the embedded cluster mass function (ECMF) up to the upper limit $M_{\rm max}$. The GCs and UCDs probably formed in interactions of the progenitor galaxies during the assembly of the central Fornax galaxy cluster which is why we use them as tracers of those events. After some corrections, the overall GC/UCD mass function is decomposed into separate SC populations, each described by an ECMF. $M_{\rm max}$ of each ECMF is converted to an SFR according to the SFR-$M_{\rm max}$ relation, revealing the SFRs reached during the assembly of galaxies in the central Fornax galaxy cluster.

Keywords. galaxies: clusters: individual (NGC 1399), galaxies: star clusters, globular clusters: general, methods: analytical

1. The background

We assume that present-day GCs and UCDs are accumulated from many individual SC formation events which can be described as follows (Schulz *et al.* 2015, accepted to A&A):

(*a*) Each SC formation epoch has a constant length δt (Table 1) during which the total mass of the SC population, $M_{\rm ECMF}$, is formed at a constant SFR: $M_{\rm ECMF} = {\rm SFR} \cdot \delta t$.

(*b*) The ECMF, $\xi_{\rm ECMF}(M) \propto (M/M_{\rm max})^{-\beta}$, describes the SC mass distribution of each population. $M_{\rm max}$ is the upper mass limit, and $\beta \approx 2.0$ is the index of the ECMF.

2. The observed GC and UCD samples in Fornax

(*a*) Spectroscopic sample ('UCD sample', orange in Fig. 1): A spatially complete sample of massive UCDs and GCs down to $\log(M) \approx 6.7$ within 83 kpc around NGC 1399 whose membership is confirmed by radial velocity measurements.

(*b*) Photometric sample ('GC sample', purple in Fig. 1): A statistically robust sample of GCs from HST/ACS observations (Jordan *et al.* 2007) within 150 kpc around NGC 1399 with masses down to $\log(M) \approx 4.5$, but without complete spatial coverage.

We scale the photometric sample at $\log(M) = 6.7$ to match the spectroscopic sample to obtain an accurate number distribution across the whole mass range. The cumulative number distribution of the combined GC/UCD sample is plotted in red in Fig. 1.

Consistency check I: The combined GC/UCD sample includes 11 615 objects, which agrees very well with the estimate of 11 100 ± 2 400 GCs/UCDs by Gregg *et al.* (2009).

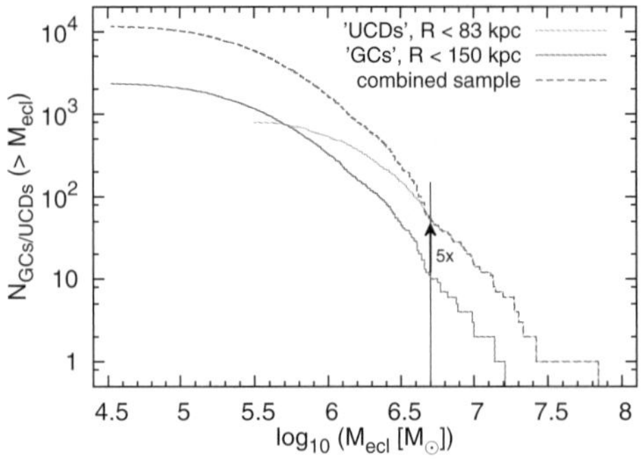

Figure 1. Cumulative 'GC sample' (purple), 'UCD sample' (orange), and the combined GC/UCD sample (red).

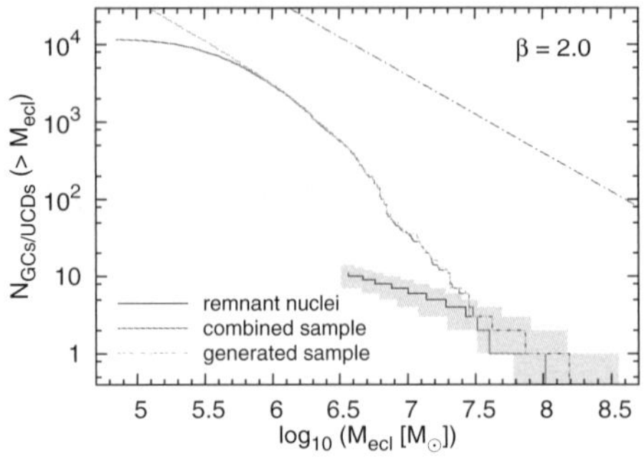

Figure 2. Cumulative combined GC/UCD sample (red) vs. generated sample (green) for $\beta = 2.0$.

3. The analysis

Before the analysis, we have to apply some corrections:

• SCs lose roughly 1/3 of their initial mass due to stellar evolution.

• Remnant nuclei of dwarf galaxies whose envelope was stripped away by the tidal field may contribute to the combined GC/UCD sample at high masses. Their cumulative number estimate by Pfeffer *et al.* (2014) is shown in black in Fig. 2, with the standard deviation in gray.

Considering only objects above the remnant nuclei sample, the combined GC/UCD sample is reproduced iteratively:

(*a*) Find the most massive SC, $M_{\max}$. $M_{\max}$ fully determines the ECMF.

(*b*) Generate all other SCs of the population according to this ECMF.

(*c*) Accumulate all SCs.

Finally, the generated sample (green) and the remnant nuclei sample (black) together match the combined GC/UCD sample (red) as shown for $\beta = 2.0$ in Fig. 2.

Table 1. Results for all β.

β	δt [Myr]	$t_{\rm tot}$ [Gyr]	$M_{\rm tot}$ [$10^{10}\ M_\odot$]
1.5	0.42	1.64	1.39
1.6	0.55	1.59	1.57
1.7	0.77	1.56	1.84
1.8	1.11	1.65	2.34
1.9	1.70	1.74	3.26
2.0	2.80	2.05	5.32
2.1	4.94	2.37	10.05
2.2	9.31	2.62	22.35
2.3	18.57	4.55	62.67
2.4	38.77	8.53	190.51
2.5	83.77	15.25	598.01
2.6	185.82	29.92	2 011.71

4. The result

$M_{\max}$ of each formation epoch is converted to an SFR according to the SFR-$M_{\max}$ relation (Weidner *et al.* 2004). We calculated the total SC formation time, $t_{\rm tot}$, and the total mass of all SC ever formed, $M_{\rm tot}$, as a function of β in Table 1 (Schulz *et al.* in prep.).

Consistency check II: The GCs and UCDs are older than 10 Gyr, so that $t_{\rm tot}$ should not exceed several Gyr. All results for $\beta \lesssim 2.3$ are reasonable.

References

Gregg, M. D., Drinkwater, M. J., Evstigneeva, E., *et al.* 2009, AJ, 137, 498

Jordán, A., Blakeslee, J. P., Côté, P., *et al.* 2007, ApJS, 169, 213

Pfeffer, J., Griffen, B. F., Baumgardt, H., & Hilker, M. 2014, MNRAS, 444, 3670

Schulz, C., Hilker, M., Pflamm-Altenburg, J., & Kroupa, P. in prep.

Schulz, C., Pflamm-Altenburg, J., & Kroupa, P. 2015, arXiv: 1507.00860, accepted for publication in A&A

Weidner, C., Kroupa, P., & Larsen, S. S. 2004, MNRAS, 350, 1503

Formation, evolution, and survival of massive star clusters
Proceedings IAU Symposium No. 316, 2015
C. Charbonnel & A. Nota, eds.

doi:10.1017/S1743921315009011

Bayesian Inference of Kinematics and Mass Segregation of Open Cluster

Z. Shao[1], X. Xie, L. Chen, J. Zhong, J. Hou and C-C. Lin

[1]Shanghai Astronomical Observatory, CAS,
80 Nandan Road, Shanghai 200030, China
email: zyshao@shao.ac.cn

Abstract. Based on the Bayesian Inference (BI) method, the Mixture-Model approach is improved to combine all kinematic data, including the coordinative position($\vec{x}$), proper motion ($\vec{\mu}$) and radial velocity(v), to separate the motion of the cluster from field stars, as well as to determine the intrinsic kinematic status and dynamical effects of the cluster, such as the mass segregation, anisotropy etc.. Meanwhile, the membership probability of individual stars are estimated as by product results. This method has been testified by simulation of toy models and also successfully used for well studied open clusters, such as M67 and NGC188. It is expected to largely help the studies of open clusters while combine the coming GAIA data.

Keywords. open cluster, mass segregation, bayesian inference, mixture model

1. Mixture Model and Bayesian Inference

Cluster members and field stars are assumed to have a mixture distribution in the position, proper motion and radial velocity spaces, $\Phi(\Theta) = \Phi_c(\Theta_c) + \Phi_f(\Theta_f)$, where $\Theta = (\Theta_c, \Theta_f)$ represents a set of parameters describe the distribution shapes of these two components, such as the number of cluster members N_c, the size of the cluster r_c , the average motion of the cluster $(\vec{\mu}_c, v_c)$ and the velocity dispersions of members and field stars (σ_c, σ_f) . The joint likelihood ($\mathcal{L}$) can be written as:

$$\log \mathcal{L} = \sum_i \log \Phi(i|\Theta) = \sum_i \log[N_c \phi_c^{\vec{x}} \phi_c^{\vec{\mu}} \phi_c^{v} + N_f \phi_f^{\vec{x}} \phi_f^{\vec{\mu}} \phi_f^{v}] \tag{1.1}$$

Then the bayesian inference are used to estimate all model parameters(Θ). The prior probability distribution function (PDF) of Θ_f can be obtained from the nearby regions of the cluster, if the data is from a large area survey. The post PDF of all parameters are determined by the Nested Sampling method (Feroz *et al.* (2008)), and then, the most possible value of these parameters, together with their uncertainties are measured from their marginal integrated post PDFs. Bayesian evidence (BE) are used to compare different distribution models, such as, King-model *vs* Gaussian in the coordinative position, whether or not asymmetrical in the proper motion dispersion of cluster members, and if the mass segregation parameters should be involved in. The membership probabilities of individual stars can be regarded as additional parameters that to be functions of the fitting parameters,

$$P_c(i) = \Phi_c(i|\Theta_c)/\Phi(i|\Theta) \tag{1.2}$$

2. Revisit of NGC188

We collected all the coordinative position($\vec{x}$), proper motion ($\vec{\mu}$) and radial velocity(v) data of 8012 stars deep to $V = 22.5$ mag., in the 0.75 deg^2 area of NGC188. Within this

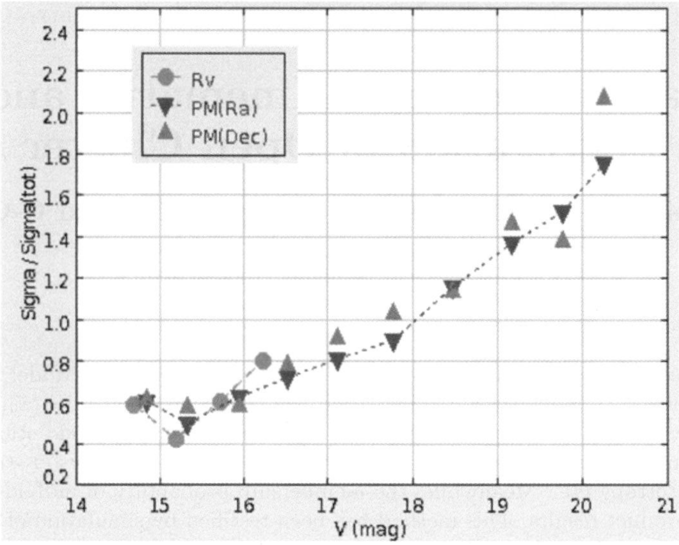

Figure 1. The magnitude dependent velocity dispersions of NGC188 cluster members.

sample, 7771 stars have $\vec{\mu}$ from Platais *et al.* (2003), and 1067 stars have v from Geller *et al.* (2008) or APOGEE.

The main results of NGC188 are: $\sigma_{\vec{\mu}} = 0.48 \pm 0.02$ mas/yr, $\sigma_v = 0.66 \pm 0.03$ km/s and $r_c = 3.12 \pm 0.18$ arcm. It is also found that, having radial velocity data included, the membership determination effectivity (Shao *et al.* (1996)) is largely increased form 0.75 to 0.96. Additionally, according to the BE comparison, the empirical King model is much better than a simple 2D-Gaussian distribution for the surface number density of this cluster. The BE also suggests the mass segregation is significant in either position or velocity spaces. These phenomena can be shown as the magnitude dependences of the velocity dispersion (see Fig.1) and the characteristic size (r_c) of this cluster.

3. Discussion and Further Works

This approach is expected to be able to merge different observational data, especially for those completeness or flux limitations seriously unmatched surveys. The on going work is applying this method to the LAMOST released data which contains a large amount radial velocities cover ~ 200 nearby open cluster regions. The galactic motions of these clusters will be detected by combining archive proper motions. If the coming GAIA data can be used, the accuracy of tangential velocity will be largely improved and the intrinsic kinematics of open cluster can be well investigated, though it is usually less than 1 km/s. There is an additional option that to combine the photometric and kinematic data as well, then the properties and memberships a cluster can be constrained not only in kinematics but also in stellar populations.

References

Feroz, F. & Hobson, M. P. 2008, *MNRAS*, 384, 449
Geller, A. M., Mathieu, R. D., Harris, H. C., & McClure, R. D. 2008, *AJ*, 135, 2264
Platais, I., Kozhurina-Platais, V., Mathieu, R. D., Girard, T. M., *et al.* 2003, *AJ*, 126, 2922
Shao, Z. & Zhao, J. 1996, *Acta Astronomica Sinica*, 37,377 . (*ChAA, 1997, 21, 254*)

Formation, evolution, and survival of massive star clusters
Proceedings IAU Symposium No. 316, 2015
C. Charbonnel & A. Nota, eds.

doi:10.1017/S1743921315009485

Chemical abundances of multiple stellar populations in massive globular clusters

Anna F. Marino

Research School of Astronomy & Astrophysics, Australian National University,
Mt Stromlo Observatory, via Cotter Rd, Weston, ACT 2611, Australia
email: anna.marino@anu.edu.au

Abstract. Multiple stellar populations in the Milky Way globular clusters manifest themselves with a large variety. Although chemical abundance variations in light elements, including He, are ubiquitous, the amount of these variations is different in different globulars. Stellar populations with distinct Fe, C+N+O and slow-neutron capture elements have been now detected in some globular clusters, whose number will likely increase. All these chemical features correspond to specific photometric patterns. I review the chemical+photometric features of the multiple stellar populations in globular clusters and discuss how the interpretation of data is being more and more challenging. Very excitingly, the origin and evolution of globular clusters is being a complex puzzle to compose.

Keywords. stars: abundances, stars: Population II, globular clusters: general, globular clusters: individual (M 4, M 22, NGC 5286, M 2, Omega Centauri)

1. Introduction

The modern picture of globular clusters (GCs) is different from the idea of these objects being a simple stellar population. Thanks to high precision photometry we know that the color-magnitude diagram (CMD) of all investigated Milky Way GCs exhibit multiple photometric sequences along all evolutionary stages, from the main sequence (MS) up to the red giant branch (RGB) and the horizontal branch (HB) (e.g. Milone *et al.* 2012; Piotto *et al.* 2015).

Multiple photometric sequences are the effect of the presence of different stellar populations, with their distinct chemical composition and stellar internal structure (e.g. Marino *et al.* 2008, 2012a). The list of the most investigated chemical elements that have been observed to vary among stars in a given GC includes: He, Li, C, N, O, C+N+O, Na, Mg, Al, Si, K, Fe, and many neutron-capture (n-capture) elements. Not all GCs exhibit the same chemical pattern, and most of them do not show evidence for significant abundance variations in many of these elements. In the same way, each photometric split along different evolutionary stages of the CMD tells us a different story about the nature of the multiple stellar populations hosted in GCs.

I review the chemical abundance variations observed in the Milky Way GCs and the way we identify them photometrically. I will focus on the more massive GCs, that are those exhibiting more complex and puzzling chemical/photometric patterns.

2. Light elements variations

The most common observed star-to-star abundance patterns in Galactic GCs (GGCs) are the C-N and O-Na anticorrelations. Chemical variations in CNONa have been found to be ubiquitous among the GGCs investigated so far (e.g. Carretta *et al.* 2009). On the other hand, the size of these variations appears to change with the mass, as more massive GCs are generally those having the more extended anticorrelations.

Chemical variations in light elements are responsible for many of the multiple sequences observed along the CMDs. As displayed in Fig. 1, carbon, nitrogen and oxygen form strong CN, NH, CH, and OH molecular bands in the ultraviolet region of the spectrum and cause the split RGBs and MSs observed in CMDs constructed by using the UV bands. A clear example of this behaviour is M 4, represented in Fig. 1, which clearly exhibits two RGBs populated by C-O-rich/N-Na-poor and C-O-poor/N-Na-rich stars. Indeed, stars with different content in these elements define different sequences along the CMD when the $(U - B)$ color is used (e.g. Marino *et al.* 2008). The huge effort, through the Legacy GO-13297 (Piotto *et al.* 2015), to analyse the GCs multiple stellar populations in UV bands with *HST* is based on this finding.

Helium variations have been observed within GCs. Multi-wavelength photometry of GCs has been often used to infer the relative helium abundances in GCs. Multiple MSs and RGBs, detected in most clusters (Milone 2015), are interpreted as due to the presence of stellar populations with different He content (e.g. Norris 2004; Piotto *et al.* 2005). A few GCs have highly He-enhanced second populations. Among the GCs with highest He enhancements one of the most studied is NGC2 2808. Photometry suggests the presence of at least five distinct stellar populations that span a wide range of He abundance, up to $Y = 0.40$ (Milone *et al.* 2015). The degree of internal variations in He correlates with the mass of the cluster as shown in Milone (2015).

On the spectroscopic side, the determination of He contents is difficult because IR He lines are subject to chromospheric effects and lines in the optical can only be observed at high temperatures. To date, evidence for the presence of significant He variations using spectroscopic analysis have been provided for NGC 2808 and ω Centauri (Dupree *et al.* 2013; Pasquini *et al.* 2011). For NGC 2808, high He abundances have been inferred for blue HB stars (Marino *et al.* 2014a).

Some GCs also exhibit Mg-Al anticorrelations (e.g. Norris & Da Costa 1995; Yong *et al.* 2003; Sneden *et al.* 2004; Carretta 2015). In many GCs only variations in Al are observed, while Mg does not change significantly.

3. Anomalous GCs

It is a long time that high resolution spectroscopy revealed large internal variations in Fe and in the elements produced via slow n-capture reactions (s-elements) in the most massive GGC ω Centauri (e.g. Norris & Da Costa 1995; Smith *et al.* 2000). More recently we realised that, although this phenomenon is quite rare in Milky Way GCs, it is not confined to ω Centauri. Indeed, a few GCs ("anomalous") have been found to show a genuine internal variation in the bulk metallicity (e.g. Marino *et al.* 2009, Da Costa *et al.* 2009; Ferraro *et al.* 2009; Carretta *et al.* 2011, Yong *et al.* 2014; Marino *et al.* 2015; Johnson *et al.* 2015).

Some of the *anomalous* GCs, specifically NGC 1851, M 22, M 2, NGC 5286 and M 19 exhibit internal variations in s-process elements (Yong & Grundahl 2008; Marino *et al.* 2009, 2011a; Yong *et al.* 2015; Marino *et al.* 2015; Johnson *et al.* 2015); and for NGC 1851 and M 22 significant variations in the total C+N+O have been also detected (Yong *et al.* 2015; Marino *et al.* 2011a). The s-elements and C+N+O abundances are positively correlated to Fe variations; the common light elements variations, typical of *normal* GGCs, are independently present within each population with different Fe/s-elements/C+N+O (Marino *et al.* 2009, 2011a). Both the s-elements and the total C+N+O chemical contents are instead uniform in typical GGCs.

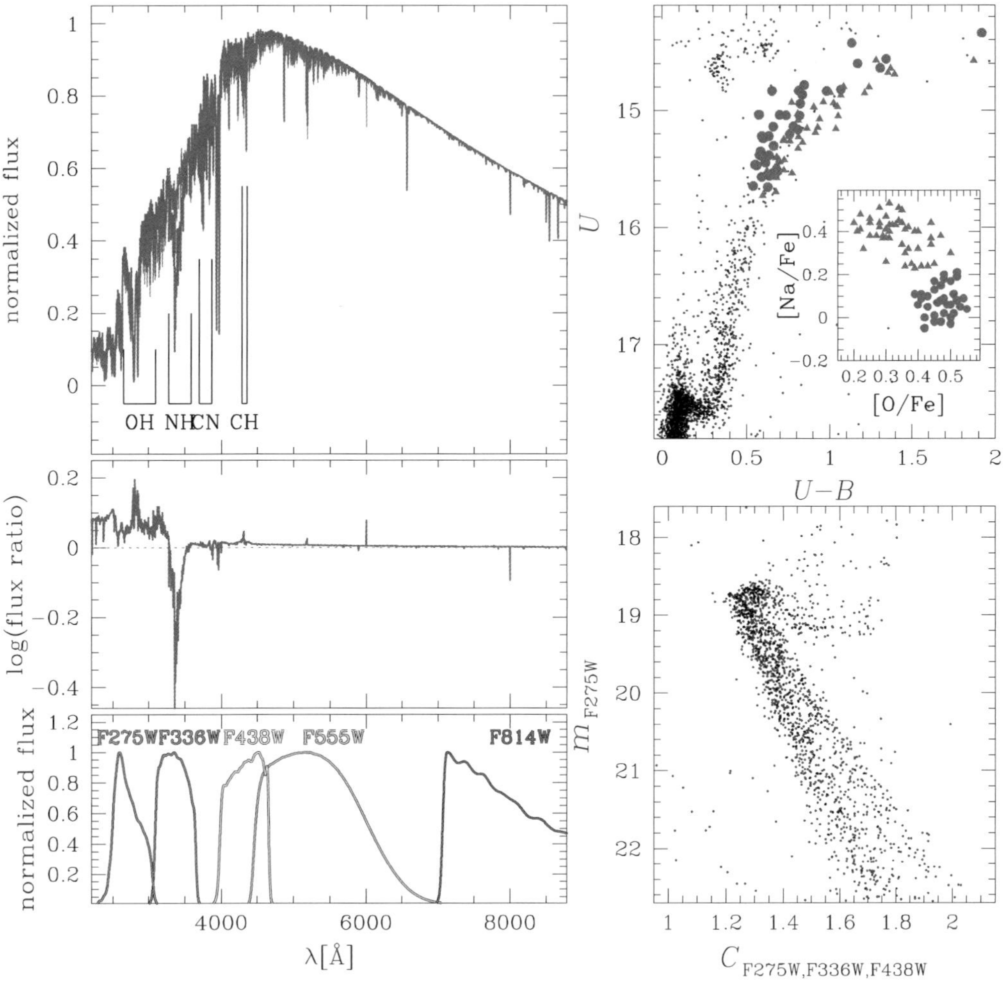

Figure 1. *Right panels:* The U-($U-B$) CMD from ground-based photometry for M 4, with the location of N-Na-poor/C-O-rich and N-Na-rich/C-O-poor stars (Marino *et al.* 2008) (upper panel). The lower panel represents the *HST* m_{F275W}-$c_{F275W,F336W,F438W}$ diagram around the MS of M 4 (Piotto *et al.* 2015). Both the RGB (upper panel) and MS (lower panel) splits in these bands are due to light elements variations. *Left panels:* Synthetic spectra for RGB stars computed with the mean C/N/O abundances of the first (blue) and second (red) population in M 4, with the location of the relevant molecular bands (upper panel). The middle panel displays the difference between the two spectra. The bandpasses of the *HST* filters is shown in the lower panel. The *HST* Survey of multiple populations observes GCs in filters *F275W*, *F336W* and *F438W*.

Photometrically, the most striking feature of *anomalous* GCs can be identified in the SGB region, which is split (Milone *et al.* 2008; Piotto *et al.* 2012). This feature is also peculiar of *anomalous* GCs, since typical GGCs exhibit a single SGB in visual bands. On theoretical background split SGBs are the indication of stellar populations with different C+N+O or different age (Cassisi *et al.* 2008). Spectroscopic+photometric campaigns have confirmed that the SGB splits are produced by stars with different interior structures due to different overall C+N+O and metallicities (Marino *et al.*,2012a; Gratton *et al.* 2012; Marino *et al.* 2014b). The RGB of *anomalous* GCs is also peculiar, as they show well-separated multiple branches in bands where typical GGCs exhibit single sequences, e.g.

in the index c_{BVI} and in the U-$(U - V)$, where the faint and the bright SGB are clearly connected with the red and the blue RGB, respectively (Marino *et al.* 2015).

A summary of some spectroscopic and photometric features of *anomalous* GCs is presented in Fig. 2, where the double SGB and RGB of M 22 due to the C+N+O variations are clearly visible. Chemical abundances for stars distributed on the split SGB has been provided for M 22 and NGC 1851, demonstrating that in both cases the s-rich stars distribute on the *anomalous* faint SGB (see CMDs on the right panels of Fig. 2). Some chemical variations in s-elements for M 22, M 2 and NGC 5286 have been represented in the lower panels of Fig. 2.

The similarity of these objects with ω Centauri is remarkable, not just because of the metallicity variations but also for the chemical pattern is s-elements. Da Costa & Marino (2011) presented a comparison between M 22 and ω Centauri and both the GCs have a very similar rise of the s-elements content as a function of Fe in the common Fe-range ([Fe/H]$\lesssim -1.5$, see right panels in Fig. 3). Additionally both these clusters present similar patterns in the Na-O anticorrelation present among stars with different Fe and C+N+O variations (Marino *et al.* 2011b, 2012b).

3.1. *On the Fe variations of M 22*

Recent claim has been made that the *anomalous* GC M 22 does not host stellar populations with different Fe (Mucciarelli *et al.* 2015). The argument is that, if photometric gravities are used instead of those from the ionisation equilibrium, the difference in Fe II between the two stellar populations with different s/C+N+O content disappears. The Fe I difference instead cannot be removed whatever technique is used to derive the atmospheric parameters. The reason supplied by Mucciarelli *et al.* (2015) for this behaviour was the presence of non-local thermodynamic effects (NLTE) affecting Fe I.

Such an effect, if real, would strongly affect what we know about *anomalous* GCs, including ω Centauri. Indeed, the well-known rise in s-process elements with Fe in ω Centauri lies on the same Fe-range spanned by M 22 stars. Given the similar Fe and chemical patterns of these two GCs, if the common Fe-regime is considered, there is no reason why the same effects should not apply to ω Centauri (see left panels of Fig 3). This would imply that there is no rise of s-elements abundances with Fe also in ω Centauri.

As it will be presented in Marino *et al.* (in prep), however, uncounted NLTE effects on Fe I are not responsible for the appearance of the Fe spread in M 22 (see Lind *et al.* 2012 for discussion of NLTE effects on Fe). This result was already suggested by the variations in CaT lines found in M 22 (Da Costa *et al.* 2009), that should not be affected by the same NLTE effects. On the other hand, the disappearance of the Fe II difference in the Mucciarelli *et al.* analysis is due to the fact that they neglect the C+N+O variations in M 22 discussed in Sect. 3 and get significant systematic offset in gravities, that affect *only* the s-poor population. As shown in Fig. 3, this offset makes the two populations of M 22 lying on the same isochrone, which is unrealistic given their different C+N+O content, as suggested both from spectroscopy and photometry. Such systematic in surface gravity affects the abundances obtained from ionised Fe, but does not affect Fe I significantly.

On the other hand, the main advantage of stellar parameters independent on photometry is that they do not introduce systematic errors that affect one stellar population only. This is suggested by the temperature-gravity distribution obtained from Marino *et al.* (2009, 2011a), which agrees with that expected from two stellar populations with different C+N+O.

Thus, the Fe variations in M 22 and its similarity with ω Centauri are confirmed. There is no support for NLTE affecting this result, but systematics on the atmospheric parameters due to the neglecting of C+N+O variations may introduce spurious results.

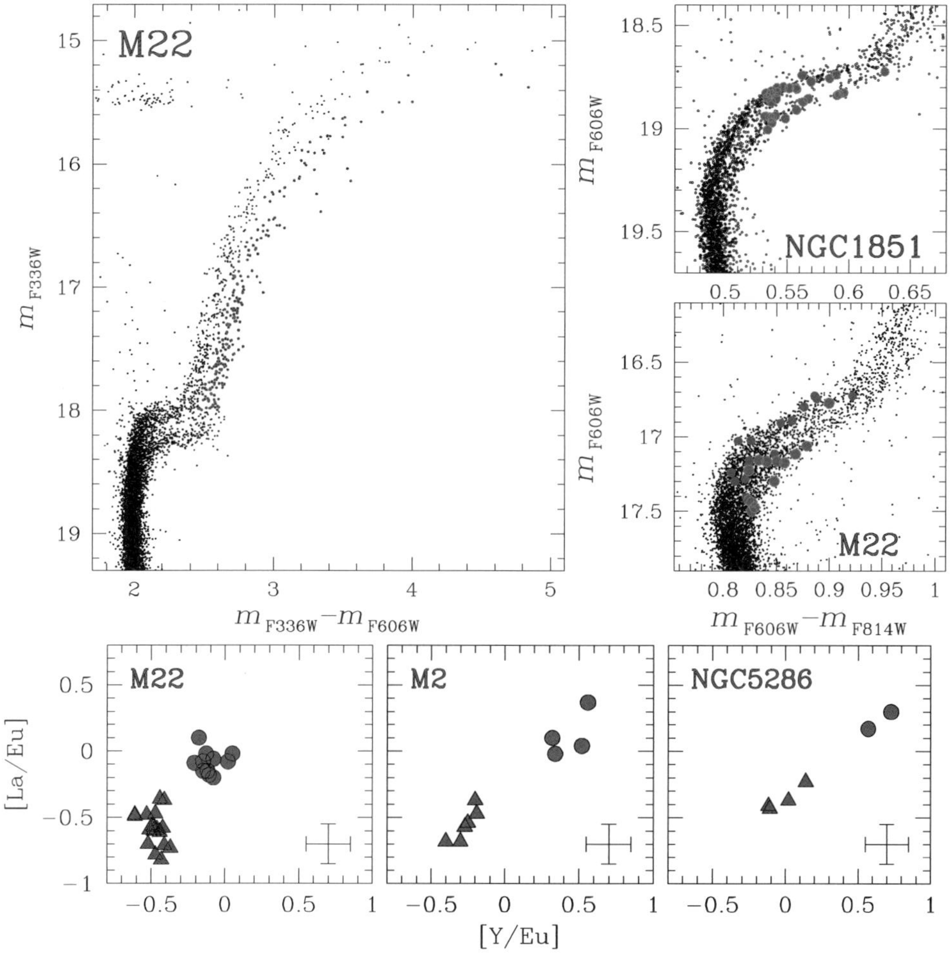

Figure 2. *Upper panels:* CMDs obtained from *HST* photometry for the *anomalous* GCs M 22 and NGC 1851. On the left panel the m_{F336W}-(m_{F336W} $-$ m_{F606W}) for M 22 clearly shows the SGB split. The *anomalous* faint SGB is connected with a redder RGB sequence (red). The right panels display the m_{F606W}-(m_{F606W} $-m_{F814W}$) CMD for NGC 1851 and M 22 around the SGB region. In both clusters *s*-poor and *s*-rich stars, as inferred from spectroscopy (Marino *et al.* 2012a; Marino *et al.* 2014b), have been represented in blue and red, respectively. *Lower panels:* [La/Eu] as a function of [Y/Eu] for the *anomalous* GCs M 22, M 2 and NGC 5286 (Marino *et al.* 2009, 2011a; Yong *et al.* 2009; Marino *et al.* 2015). *s*-poor and *s*-rich stars have been represented with blue triangles and red dots, respectively.

4. The chemical enrichment in GCs

Explaining the whole observational scenario of multiple stellar populations in GCs is difficult. Simply based on the chemical abundances, we may think that *normal* and *anomalous* GCs have experienced a different chemical evolution.

Although we know from recent photometric results that GCs host more than two stellar populations (Milone *et al.* 2015), a "two-populations" scenario can still approximatively explain the light elements chemical patterns observed in *typical* GCs, as sketched in Fig. 4 (left panel). We may assume a self-pollution scenario, e.g. the first generation (1G) is enhanced in O, depleted in Na (just as typical of Galactic halo stars), and, later on, a second generation (2G) forms. If we suppose a similar scenario, 2G stars form from

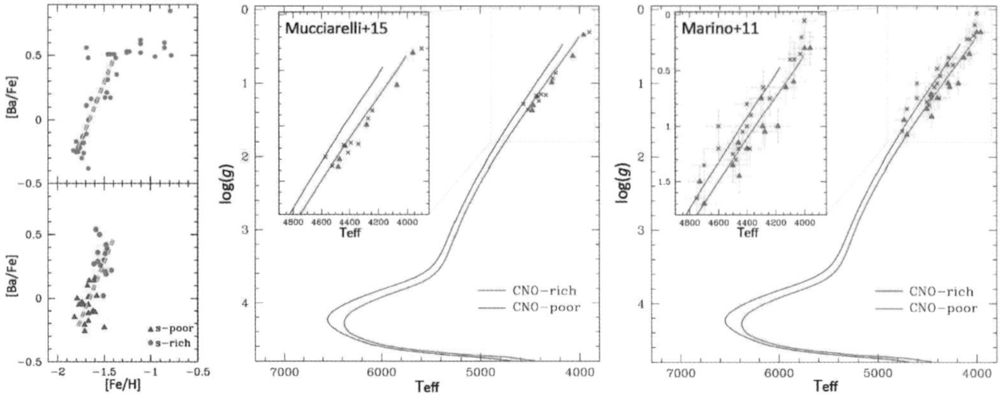

Figure 3. *Left panels*: [Ba/Fe] as a function of [Fe/H] for ω Centauri (Norris & Da Costa 1995) and M 22 (Marino *et al.* 2009, 2011a; shifted by 0.15 dex in [Fe/H]). For M 22 the *s*-poor and *s*-rich stars have been plotted as blue triangles and red dots, respectively. A straight line (green dashed line) tracing the rise of *s*-elements with Fe for stars with [Fe/H]$\lesssim -1.5$ in ω Centauri is super-imposed to the M 22 data. If there are no Fe variations in M 22, they will also disappear in ω Centauri for stars with [Fe/H]$\lesssim -1.5$ dex, removing the well-known *s*-pattern in this GC. *Right panels*: Theoretical isochrones from BaSTI (Pietrinferni *et al.* 2006) for a CNO-poor and CNO-rich population. Superimposed to the isochrones are the atmospheric parameters for M 22 RGBs from Mucciarelli *et al.* (2015) (middle-panel) and Marino *et al.* (2009, 2011a) (right-panel). The atmspheric parameters from Mucciarelli *et al.* (2015) force the stars on a single sequence.

hot-H burning processed material, which has been realised from some kind of polluter. The proposed candidate polluters are fast-rotating massive stars (Decressin *et al.* 2007), asymptotic giant branch stars (D'Ercole *et al.* 2008), or supermassive stars (Denissenkov *et al.* 2015). If we instead do not invoke multiple star-formation episodes, we have to find other mechanisms that can account for the chemical variations. In this context, massive interacting binaries have been proposed to be able of reproduce some of the chemical variations (Bastian *et al.* 2013). None of these scenarios at the moment is accepted, as all of them have serious shortcomings in explaining the observations (see Renzini *et al.* 2015).

If explaining *normal* GCs is complex, *anomalous* GCs present even more challenges. The chemical pattern displayed by these objects is complex and we do not have, at the moment, a solution to interpret the observations in terms of a reasonable chemical enrichment history. As an example, the right panel of Fig. 4 shows the Na and O abundances in M 22. In this GC internal variations in Na and O exist in stars with different Fe/*s*-elements/C+N+O content. This means that, for this object, as well as in the other *anomalous* GCs, we do not know how the intra-cluster chemical enrichment proceeded, e.g. if the first channel to be active was the enrichment in the hot H-burning products (producing light elements patterns) or the enrichment in Fe. Surely, by assuming the self-pollution scenario, polluters of different mass have contributed to the intra-cluster pollution, including low-mass AGBs that produce *s*-elements and increase the C+N+O (e.g. Shingles *et al.* 2014; Straniero *et al.* 2014).

The enrichment in Fe suggests that *anomalous* GCs have been able of retaining Supernovae ejecta. Metallicity variation, which was thought to be an exclusive feature of ω Centauri, proposed to be the remnant of a dwarf galaxy (e.g. Bekki & Freeman 2003), is a more widespread phenomenon in GCs. Thus, it is tempting to speculate that the *anomalous* GCs may be nuclei of disrupted dwarf galaxies, as suggested for ω Centauri.

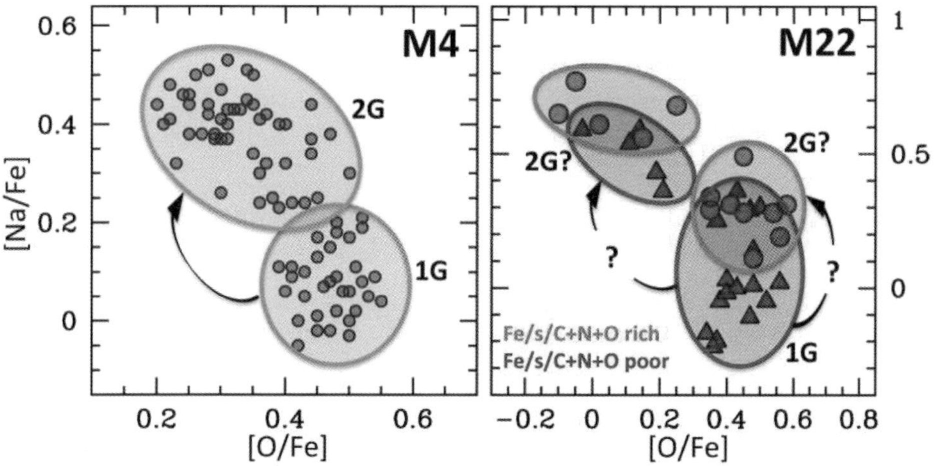

Figure 4. [Na/Fe] versus [O/Fe] for M 4 (Marino *et al.* 2008) and M 22 (Marino *et al.* 2009, 2011a). For M 4 a two-generation model is sufficient with the second generation (2G), being Na-rich and O-poor as it formed from the first generation (1G) polluters. For the *anomalous* GC M 22, the observational scenario is more complex: the Na and O variations are present in both the Fe/*s*-elements/C+N+O-poor (blue) and Fe/*s*-elements/C+N+O rich (red) populations. We do not know yet which is the 2G that formed directly after the 1G, e.g. which enrichment occurred first (in Fe or in hot H-burning products). Similar challenges affect the scenario of massive interacting binaries.

5. Conclusive remarks

Thanks to the large observational material, in the last decade the formation and evolution of GCs has became more difficult to understand. We do not have any model which is able to satisfy all the observational constraints. I conclude by listing some relevant open issues that we should address in the future:

• in general, the origin of the multiple stellar populations in GCs. Will one or more of the proposed scenarios (self-pollution, fast-rotating massive stars, supermassive stars, mass exchange in massive interacting binaries) able to explain the whole observational scenario?

• could the heterogeneity of the multiple population zoo be reconciled with a unique scenario?

• is there some GCs' property, such as mass, determining the length of star formation, and hence the level of chemical enrichment?

• which is the origin of anomalous GCs? Why they exhibit different chemical enrichment with respect to the majority of GGCs? Does their peculiar chemical pattern simply imply a longer star-formation history? Or, could have they originated as nuclei of disrupted galaxies?

Acknowledgements

I acknowledge the organisers of the Symposium for inviting me to review results on the multiple stellar populations in globular clusters. I am grateful to all the collaborators who contribute to the presented results.

References

Bastian, N., Lamers, H. J. G. L. M., de Mink, S. E., *et al.* 2013, *MNRAS*, 436, 2398
Bekki, K. & Freeman, K. C. 2003, *MNRAS*, 346, L11

Carretta, E., Bragaglia, A., Gratton, R. G., *et al.* 2009, *A&A*, 505, 117
Carretta, E., Lucatello, S., Gratton, R. G., Bragaglia, A., & D'Orazi, V. 2011, *A&A*, 533, A69
Carretta, E. 2015, *ApJ*, 810, 148
Cassisi, S., Salaris, M., Pietrinferni, A., *et al.* 2008, *ApJ*, 672, L115
Da Costa, G. S., Held, E. V., Saviane, I., & Gullieuszik, M. 2009, *ApJ*, 705, 1481
Da Costa, G. S. & Marino, A. F. 2011, *PASA*, 28, 28
Decressin, T., Meynet, G., Charbonnel, C., Prantzos, N., & Ekström, S. 2007, *A&A*, 464, 1029
Denissenkov, P. A., VandenBerg, D. A., Hartwick, F. D. A. *et al.* 2015, *MNRAS*, 448, 3314
D'Ercole, A., Vesperini, E., D'Antona, F., McMillan, S. L. W., & Recchi, S. 2008, *MNRAS*, 391, 825
Dupree, A. K. & Avrett, E. H. 2013, *ApJ*, 773, L28
Ferraro, F. R., Beccari, G., Dalessandro, E., *et al.* 2009, *Nature*, 462, 1028
Gratton, R. G., Villanova, S., Lucatello, S., *et al.* 2012, *A&A*, 544, A12
Johnson, C. I., Rich, R. M., Pilachowski, C. A., *et al.* 2015, *AJ*, 150, 63
Lind, K., Bergemann, M., & Asplund, M. 2012, *MNRAS*, 427, 50
Marino, A. F., Villanova, S., Piotto, G., *et al.* 2008, *A&A*, 490, 625
Marino, A. F., Milone, A. P., Piotto, G., *et al.* 2009, *A&A*, 505, 1099
Marino, A. F., Sneden, C., Kraft, R. P., *et al.* 2011a, *A&A*, 532, A8
Marino, A. F., Milone, A. P., Piotto, G., *et al.* 2011b, *ApJ*, 731, 64
Marino, A. F., Milone, A. P., Sneden, C., *et al.* 2012a, *A&A*, 541, A15
Marino, A. F., Milone, A. P., Piotto, G., *et al.* 2012b, *ApJ*, 746, 14
Marino, A. F., Milone, A. P., Przybilla, N., *et al.* 2014, *MNRAS*, 437, 1609
Marino, A. F., Milone, A. P., Yong, D., *et al.* 2014b, *MNRAS*, 442, 3044
Marino, A. F., Milone, A. P., Karakas, A. I., *et al.* 2015, *MNRAS*, 450, 815
Marino, A. F., *et al.* 2015, in prep.
Milone, A. P., Bedin, L. R., Piotto, G., *et al.* 2008, *ApJ*, 673, 241
Milone, A. P., Piotto, G., Bedin, L. R., *et al.* 2012, *ApJ*, 744, 58
Milone, A. P. 2015, *MNRAS*, 446, 1672
Milone, A. P., Marino, A. F., Piotto, G., *et al.* 2015, *ApJ*, 808, 51
Mucciarelli, A., Lapenna, E., Massari, D., *et al.* 2015, *ApJ*, 809, 128
Norris, J. E. & Da Costa, G. S. 1995, *ApJ*, 447, 680
Norris, J. E. 2004, *ApJ*, 612, L25
Pasquini, L., Mauas, P., Käufl, H. U., & Cacciari, C. 2011, *A&A*, 531, A35
Pietrinferni, A., Cassisi, S., Salaris, M., & Castelli, F. 2006, *ApJ*, 642, 797
Piotto, G., Villanova, S., Bedin, L. R., *et al.* 2005, *ApJ*, 621, 777
Piotto, G., Milone, A. P., Anderson, J., *et al.* 2012, *ApJ*, 760, 39
Piotto, G., Milone, A. P., Bedin, L. R., *et al.* 2015, *AJ*, 149, 91
Renzini, A., *et al.* 2015, accepted for publ. in *MNRAS*
Shingles, L. J., Karakas, A. I., Hirschi, R., *et al.* 2014, *ApJ*, 795, 34
Smith, V. V., Suntzeff, N. B., Cunha, K., *et al.* 2000, *AJ*, 119, 1239
Sneden, C., Kraft, R. P., Guhathakurta, P., Peterson, R. C., & Fulbright, J. P. 2004, *AJ*, 127, 2162
Straniero, O., Cristallo, S., & Piersanti, L. 2014, *ApJ*, 785, 77
Yong, D. & Grundahl, F. 2008, *ApJ*, 672, L29
Yong, D., Grundahl, F., Lambert, D. L., Nissen, P. E., & Shetrone, M. D. 2003, *A&A*, 402, 985
Yong, D., Roederer, I. U., Grundahl, F., *et al.* 2014, *MNRAS*, 441, 3396
Yong, D., Grundahl, F., & Norris, J. E. 2015, *MNRAS*, 446, 3319

Formation, evolution, and survival of massive star clusters
Proceedings IAU Symposium No. 316, 2015
C. Charbonnel & A. Nota, eds.

doi:10.1017/S1743921315009278

The *HST* survey of Magellanic-Cloud clusters and of their stellar populations

A. P. Milone

Research School of Astronomy & Astrophysics, Australian National University, Mt Stromlo Observatory, via Cotter Rd, Weston, ACT 2611, Australia.
email: milone@mso.anu.edu.au

Abstract. A large number of intermediate-age (∼1-2-Gyr old) globular clusters (GCs) in the Large and the Small Magellanic Cloud (MC) exhibit either bimodal or extended main-sequence (MS) turn off and dual red clump (RC). Moreover, recent papers have shown that the MS of the young clusters NGC 1844 and NGC 1856 is either broadened or split. These features of the color-magnitude diagram (CMD) are not consistent with a single isochrone and suggest that star clusters in MCs have experienced a prolonged star formation, in close analogy with Milky-Way GCs with multiple stellar populations. As an alternative, stellar rotation or interacting binaries can be responsible of the CMD morphology. In the following I will summarize the observational scenario and provide constraints on the nature of the complex CMD of young and intermediate-age MC clusters from our ongoing photometric survey with the *Hubble Space Telescope* (*HST*).

1. Introduction

It is now widely accepted that the CMD of nearly all the old Galactic GCs (GGCs) is made of multiple sequences that can be identified along the entire diagram, from the bottom of the MS up to the sub-giant branch (SGB), the red-giant branch (RGB) and even the horizontal branch and the asymptotic giant branch (e.g. Milone *et al.* 2012; Piotto *et al.* 2015). The multiple sequences correspond to distinct stellar populations with different helium and light-element abundance and their presence has suggested that GCs have experienced a complex star-formation history.

The discovery of either multimodal or extended MS turn-off (hereafter eMSTO) in several intermediate-age star clusters in both MCs (e.g. Bertelli *et al.* 2003; Mackey & Broby Nielsen 2007; Glatt *et al.* 2008) has suggested that multiple stellar populations could not be a peculiarity of old GGCs. In our survey of MC star clusters, we have shown that the the eMSTO is a common feature among ∼1-2-Gyr old star clusters and has been detected in at least the 70% of the analyzed objects (Milone *et al.* 2009).

2. The CMD of eMSTO clusters

An example of typical CMD of an intermediate-age star cluster is provided in Fig. 1a, where I have plotted $m_{\rm F475W}$ vs. $m_{\rm F475W} - m_{\rm F814W}$ for NGC 411 in the SMC. As mentioned above, the most striking feature is the MSTO, which is widely broadened in color and magnitude. Despite some authors have initially claimed that the eMSTO is due to photometric errors (see e.g. Mucciarelli *et al.* 2007 for the case of NGC 1783), the fact that the other CMD sequences, like the RGB and the lower MS are narrow and well defined, clearly demonstrates that the broadened MSTO is intrinsic and is not due neither to differential reddening nor to photometric errors.

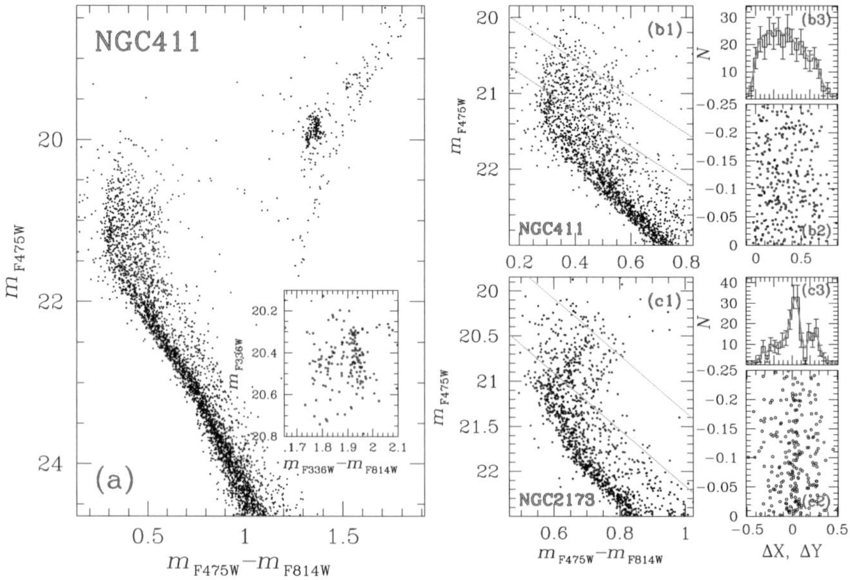

Figure 1. *Panel (a):* $m_{\rm F475W}$ vs. $m_{\rm F475W}-m_{\rm F814W}$ CMD of the intermediate-age GC NGC 411. The inset is a zoom of the $m_{\rm F336W}$ vs. $m_{\rm F336W}-m_{\rm F814W}$ around the RC. Panels b1 and c1 show the $m_{\rm F475W}$ vs. $m_{\rm F475W}-m_{\rm F814W}$ CMD for stars around the MSTO in NGC 411 and NGC 2173, respectively. The verticalized color and magnitude distribution for the MSTO stars between the two gray lines of panels a1 and b1 are plotted in panels a2 and b2, while panels a3 and b3 show the histograms of the ΔX distribution along the direction, X, of the gray lines for these stars. The gray lines overimposed on each histogram are the corresponding kernel-density distributions.

The distribution of stars along the MSTO changes from one cluster to another. Some eMSTO clusters, like NGC 411, exhibit a continuous distribution (see Fig. 1b1-b3). In other clusters there are two main MSTOs, with the faint MSTO hosting approximately 30-40% of the total number of MSTO stars (see Fig. 1c1-c3 for the case of NGC 2173). Another intriguing feature is the RC, which can be either broadened or bimodal (Girardi *et al.* 2009) as shown in the inset of Fig. 1a for NGC 411.

The dual RC has been interpreted as due to two stellar groups: one of them consisting of massive stars, which avoid e^- degeneracy settling in their H-exhausted cores when He ignites. The other group is made of slightly less-massive stars which experience e^- degeneracy before He ignition. This phenomenon is thought to be responsible for the higher brightness of the less-massive stars (Girardi *et al.* 2009, 2013).

2.1. *The RGB*

As discussed in the previous section, most GGCs host two or more stellar populations with different chemical composition. One of them has the same C, N, O, and He abundance as halo-field stars with the same metallicity, while the other population(s) is enhanced in helium and nitrogen and depleted in carbon and oxygen (see Gratton *et al.* 2012; Piotto *et al.* 2015 and references therein). As a consequence of such a specific chemical pattern, most GGCs exhibit either broadened or multiple RGBs, when observed in appropriate CMDs involving the U band (like U vs. $U-B$, Marino *et al.* 2008).

Recent papers, based on spectroscopy and *HST* photometry, have shown that N/He-rich stars of GGCs have larger $m_{\rm F336W}-m_{\rm F438W}$ color than N/He-poor stars with the same luminosity, but smaller $m_{\rm F438W}-m_{\rm F814W}$ color. The $C_{\rm F336W,F438W,F814W}=(m_{\rm F336W}-m_{\rm F438W})-(m_{\rm F438W}-m_{\rm F814W})$ index introduced by Milone *et al.* (2013a) thus maximizes

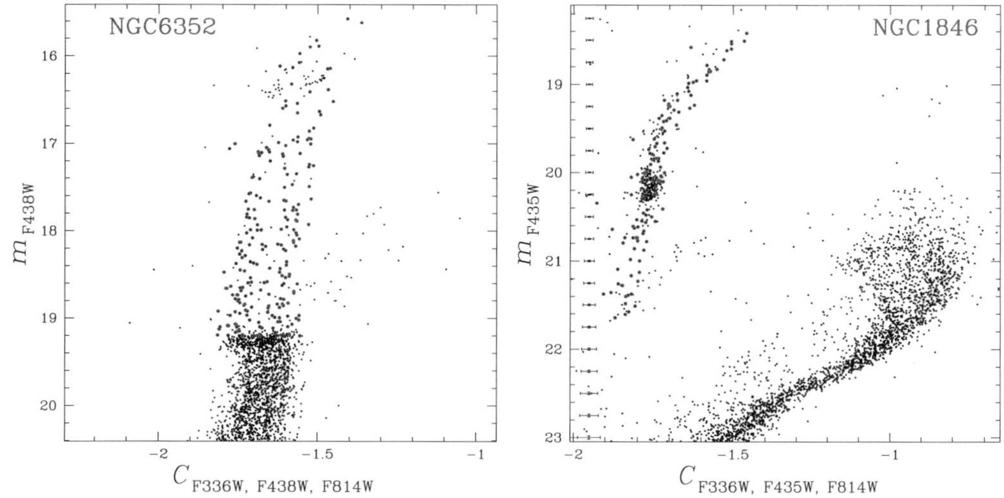

Figure 2. m_{F438W} vs. $C_{F336W,F438W,F814W}$ diagram for the GGC NGC 6352 (left) and m_{F435W} vs. $C_{F336W,F435W,F814W}$ diagram for NGC 1846 in the LMC (right). Red circles mark RGB stars.

the virtues of both $m_{F336W} - m_{F438W}$ and $m_{F438W} - m_{F814W}$, and is a powerful tool to identify multiple MSs and RGBs.

As an example, the left panel of Fig. 2 shows m_{F438W} vs. $C_{F336W,F438W,F814W}$ for NGC 6352, that is a metal-rich GGC ([Fe/H]=−0.64, Harris 1996, 2010 edition) with mass, $\mathcal{M} = 1.2 \times 10^5 \mathcal{M}_\odot$ (Gnedin & Ostriker 1997). NGC 6352 hosts two distinct stellar populations with different He, C, N, Na, and O, that populate the two distinct sequences detected along the entire CMD by Piotto *et al.* (2015) and Nardiello *et al.* (2015). The split RGB of NGC 6352, is clearly visible in the left-panel diagram of Fig. 2 from Piotto *et al.* (2015), where I have marked RGB stars with red circles.

In the right panel of Fig. 2, I show m_{F435W} against $C_{F336W,F435W,F814W}$ for the eMSTO LMC cluster NGC 1846, which has similar mass ($\mathcal{M} = 2.3 \times 10^5 \mathcal{M}_\odot$, Goudfrooij *et al.* 2014) and metallicity ([Fe/H]=−0.4, Milone *et al.* 2009) as NGC 6352. In this case the RGB width in the $C_{F336W,F438W,F814W}$-index, is consistent with the broadening expected from photometric errors, thus demonstrating that star-to-star variations in light-elements and helium, if present, are small and can not be detected from ultraviolet photometry of RGB stars. This finding, corroborates the conclusion by Mucciarelli *et al.* (2014) who have analyzed spectra of eight bright red giants in the eMSTO cluster NGC 1806 and found homogeneous Na and O content.

3. Theoretical interpretation

Several authors have interpreted the eMSTO of intermediate-age MC clusters as due to a prolonged period of star formation with a duration of ~100-500 (see Goudfrooij *et al.* 2014 and references therein). An alternative interpretation has been provided by Bastian & de Mink (2009) who have noticed that rotation significantly affects the stellar structure and the inclination angle between the observer and the star changes the effective temperature. They have thus suggested that a range of stellar rotation in MSTO stars can mimic an eMSTO in a cluster where a simple stellar population exists. Unfortunately, as discussed by Yang *et al.* (2013) the evolution of rotating stars is strongly dependent on the stellar model and the eMSTO can be explained or not depending on the adopted models. As a further interpretation of the CMD of intermediate-age MC clusters, interacting

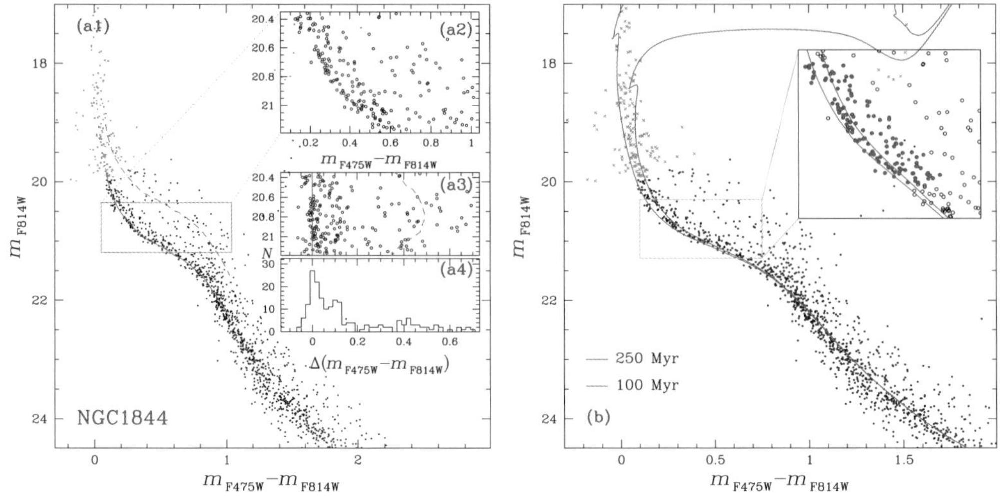

Figure 3. *Left panels:* CMD of the LMC cluster NGC 1844, with saturated stars colored gray. The red continuous and dashed lines mark the MS fiducial line, and the line of equal-mass binaries, respectively (panel a1). A zoom of the CMD region where the MS broadening is more-clearly visible is provided in panel a2. Panel a3 shows the verticalized $m_{\rm F475W}$ vs. $\Delta(m_{\rm F475W} - m_{\rm F814W})$ diagram for the MS stars plotted in panel a2, where $\Delta(m_{\rm F475W} - m_{\rm F814W})$ is obtained by subtracting from the $m_{\rm F475W} - m_{\rm F814W}$ color of each star, the corresponding color of the MS fiducial line at the same F475W magnitude. The histogram distribution of $\Delta(m_{\rm F475W} - m_{\rm F814W})$ is plotted in panel a4. *Right panels:* Reproduction of the same CMD of panel a1 with overimposed BaSTI isochrones with metallicty Z=0.01 and ages of 100 and 250 Myr.

binaries with mass transfer and merged binary systems have been also considered as possible responsible for the eMSTO (Yang *et al.* 2011).

A lot of effort has been done by several groups to discriminate among the different scenarios. Most of their works are based on the comparison between theoretical models and observation of stars along the eMSTO, the RC, and the SGB. I refer to papers by Goudfrooij *et al.* (2015), Li *et al.* (2014), Bastian *et al.* (2015) Niederhofer *et al.* (2015) and references therein for critical discussion about age spread and rotation as responsible of the eMSTO.

4. The young clusters NGC 1844 and NGC 1856

New insight on the eMSTO phenomenon comes from young clusters. In this context the ∼150-Myr old NGC 1844 in the LMC is exemplary. As shown in Fig. 3 the upper MS of this cluster is widely spread in $m_{\rm F475W} - m_{\rm F814W}$ color, with some hints of MS split (Milone *et al.* 2013b). The MS broadening is intrinsic and can not be due neither to field-star contamination, non-interacting binaries, photometric errors, or variation of the foreground reddening. Unfortunately, the interpretation of the CMD of NGC 1844 is still controversial. The comparison of the observations with BaSTI isochrones (Pietrinferni *et al.* 2004) with the same metallicity but different ages reveals that age difference alone can not be responsible for the broadened MS of NGC 1844 (Fig. 3b). Similarly, on the basis of isochrones from the same database, Milone *et al.* (2013b) have excluded that the MS of NGC 1844 is consistent with a range of stellar rotation, while isochrones with different overall C+N+O abundance could provide a better fit with the observed CMD.

Another intriguing GC is the ∼350 Myr-old NGC 1856, which is the only cluster younger than ∼1 Gyr that exhibits an eMSTO (Milone *et al.* 2015; Correnti *et al.* 2015).

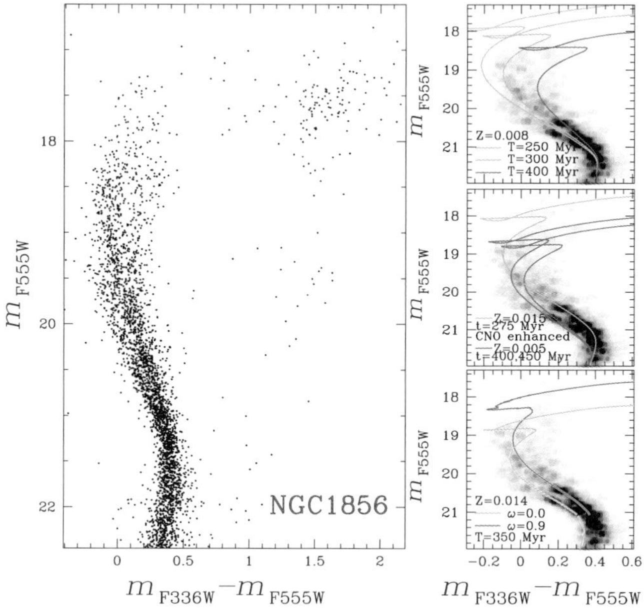

Figure 4. *Left panel.* CMD of the LMC GC NGC 1856 from WFC3/*HST* photometry (Milone *et al.* 2015). *Right panels.* Isochrones are compared with the Hess diagram of NGC 1856. BaSTI isochrones with the same metallicity but different age are plotted in the upper panel, while the middle panel shows BaSTI isochrones with different age, [Fe/H] and C+N+O abundance. The lower panel compares two coeval isochrones from the Geneva database with no rotation (blue line), and with high-rotation (red line, D'Antona *et al.* 2015).

In addition, Milone *et al.* (2015) have revealed that the MS of NGC 1856 is bimodal, with the blue MS hosting about one third of the total number of MS stars. As shown in Fig. 4, the split MS of NGC 1856 is visible only at bright luminosity and the two MS merge together approximately two F555W mag below the MSTO. It is not clear whether the blue MS evolves into the faint MSTO or if it is connected with the bluest part of the MS. Similarly to the MS, the RC of NGC 1856 is bimodal. In contrast, there is no evidence for either split or broadened MS in none of the analyzed intermediate-age GCs.

The interpretation of the CMD of NGC 1856 is still debated. As shown in the upper-left panel of Fig. 4, where I have overimposed isochrones withe the same metallicity on the Hess diagram of NGC 1856, if the cluster stars are chemically homogeneous, the eMSTO is consistent with a prolonged star-formation history of about 150 Myr. In this case, the blue MS would evolve into the brighter MSTO. In contrast, if the two MS merge together and the blue MS is connected with the faint MSTO, the CMD of NGC 1856 is consistent with two stellar populations with different metallicity and age. In this case the blue MS would be enhanced in [Fe/H] and [(C+N+O)/Fe] and younger than the red MS (middle-left panel of Fig. 4). However, while metallicity variations have been detected in some GGCs (see Marino *et al.* 2015 and references therein), as discussed in Sect. 2.1, there is no evidence for star-to-star light-element variations in intermediate-age MC clusters.

As an alternative, according to Geneva stellar models (Georgy *et al.* 2014), the observations of NGC 1856 are consistent with a coeval assembly of stars with a range of rotation. In this context, both the bright MSTO and the red MS of NGC 1856 correspond to a population of rapidly-rotating stars, while the blue-MS and the faint MSTO are made of non-rotating stars (D'Antona *et al.* 2015, see bottom-left panel of Fig. 4).

5. Conclusions

I have used the photometry from our ongoing survey of MC cluster with *HST* to illustrate the main CMD features of intermediate-age and young clusters that are not consistent with a single isochrone. These features include the eMSTO and the dual RC, which are commonly observed in most GCs with age of ∼1-2 Gyr, and the broadened and double MS, recently discovered in the young clusters NGC 1844 and NGC 1856. Moreover, I have shown that the RGB of NGC 1846 shows no evidence for light-elements variations, in contrast with what observed in the majority of old GGCs.

The presence of multiple generations of stars in intermediate and young MC clusters is still controversial. If the eMSTO is due to age difference, these clusters would provide a unique angle to study the multiple-population phenomenon. Indeed they could be similar to the old GGCs just few hundreds Myrs after their formation (e. g. Keller *et al.* 2011). On the contrary, if only one population exists, the eMSTO star clusters would provide unique laboratories to study the physic of rotating stars or interacting binaries.

References

Bastian, N. & de Mink, S. E. 2009, *MNRAS*, 398, L11
Bastian, N. & Niederhofer, F. 2015, *MNRAS*, 448, 1863
Bertelli, G., Nasi, E., Girardi, L., *et al.* 2003, *AJ*, 125, 770
Correnti, M., Goudfrooij, P., Puzia, T. H., & de Mink, S. E. 2015, *MNRAS*, 450, 3054
D'Antona, F., Di Criscienzo, M., Decressin, T., *et al.* 2015, *MNRAS*, 453, 2637
Georgy, C., Granada, A., Ekström, S., *et al.* 2014, *A&A*, 566, A21
Girardi, L., Rubele, S., & Kerber, L. 2009, *MNRAS*, 394, L74
Girardi, L., Goudfrooij, P., Kalirai, J. S., *et al.* 2013, *MNRAS*, 431, 3501
Glatt, K., Grebel, E. K., Sabbi, E., *et al.* 2008, *AJ*, 136, 1703
Gnedin, O. Y. & Ostriker, J. P. 1997, *ApJ*, 474, 223
Goudfrooij, P., Girardi, L., Kozhurina-Platais, V., *et al.* 2014, *ApJ*, 797, 35
Goudfrooij, P., Girardi, L., Rosenfield, P., *et al.* 2015, *MNRAS*, 450, 1693
Gratton, R. G., Carretta, E., & Bragaglia, A. 2012, *A&AR*, 20, 50
Harris, W. E. 1996, *AJ*, 112, 1487
Keller, S. C., Mackey, A. D., & Da Costa, G. S. 2011, *ApJ*, 731, 22
Li, C., de Grijs, R., & Deng, L. 2014, *Nature*, 516, 367
Mackey, A. D. & Broby Nielsen, P. 2007, *MNRAS*, 379, 151
Marino, A. F., Villanova, S., Piotto, G., *et al.* 2008, *A&A*, 490, 625
Marino, A. F., Milone, A. P., Karakas, A. I., *et al.* 2015, *MNRAS*, 450, 815
Milone, A. P., Bedin, L. R., Piotto, G., & Anderson, J. 2009, *A&A*, 497, 755
Milone, A. P., Piotto, G., Bedin, L. R., *et al.* 2012, *ApJ*, 744, 58
Milone, A. P., Marino, A. F., Piotto, G., *et al.* 2013, *ApJ*, 767, 120
Milone, A. P., Bedin, L. R., Cassisi, S., *et al.* 2013, *A&A*, 555, A143
Milone, A. P., Bedin, L. R., Piotto, G., *et al.* 2015, *MNRAS*, 450, 3750
Mucciarelli, A., Origlia, L., & Ferraro, F. R. 2007, *AJ*, 134, 1813
Mucciarelli, A., Dalessandro, E., Ferraro, F. R., Origlia, L., & Lanzoni, B. 2014, *ApJ*, 793, L6
Nardiello, D., Piotto, G., Milone, A. P., *et al.* 2015, *MNRAS*, 451, 312
Niederhofer, F., Georgy, C., Bastian, N., & Ekström, S. 2015, *MNRAS*, 453, 2070
Pietrinferni, A., Cassisi, S., Salaris, M., & Castelli, F. 2004, *ApJ*, 612, 168
Piotto, G., Milone, A. P., Bedin, L. R., *et al.* 2015, *AJ*, 149, 91
Yang, W., Meng, X., Bi, S., *et al.* 2011, *ApJ*, 731, L37
Yang, W., Bi, S., Meng, X., & Liu, Z. 2013, *ApJ*, 776, 112

Formation, evolution, and survival of massive star clusters
Proceedings IAU Symposium No. 316, 2015
C. Charbonnel & A. Nota, eds.

doi:10.1017/S1743921315010613

Formation of globular clusters with multiple populations

T. Decressin

INAF - Osservatorio Astronomico di Roma, via Frascati 33, 00040 Monte Porzio Catone (RM), Italy
email: tdecressin@googlemail.com

Abstract. Spectroscopic and photometric evidences have led to a complete revision of our understanding of globular clusters with the discovery of multiple stellar populations which differ chemically. Whereas some stars have a chemical composition similar to fields stars, others show large star-to-star variations in light elements (Li to Al) while their composition in iron and heavy elements stay constant. This peculiar chemical pattern can be explained by self-pollution of the intracluster gas occurring in the early evolution of clusters. Here the possible impact from a first generation of fast rotating stars to the early evolution of globular clusters is presented. The high rotation velocity will allow the stars to rotate at the break-up velocity and release matter enrich in H-burning which in turn will produce new stars with a chemical composition in agreement with observations. The massive stars have also an important role to clear the cluster from the remaining gas left after the star formation episodes. If the gas expulsion is fast enough, the strong change in the potential well will lead to the loss of stars occupying the outer part of the cluster. As second generation stars are preferentially born in the cluster centre, the ratio of second to first generation stars will increase over time to match the present ratio determined by observations. Considerations on the properties of low-mass stars still present in globular clusters will also be presented.

Keywords. Globular clusters, multiple population, stars: massive stars, stars: low-mass

1. Introduction

Over the last decades our understanding of globular clusters deeply challenged. Indeed with the determination of abundances variations in light elements by spectroscopy (see Gratton *et al.* 2012, for a review) and the broad or multiple main sequences and giant branches by precise photometry (Bedin *et al.* 2004) a new paradigm has risen. Those clusters harbour several population or generations of stars. One has similar abundance pattern as fields stars of similar metallicity while other shows signs of H-burning at high temperature (Prantzos *et al.* 2007) with He increase and anticorrelations between Li-Na, C-N, O-Na and Mg-Al while heavier element stay constant (Gratton *et al.* 2012).

Different scenario have been proposed to explain these features which involved a self-enrichment inside the clusters during its evolution by different stellar polluters: massive AGB stars (Ventura *et al.* 2001, 2013; D'Ercole *et al.* 2010, 2011, 2012), fast rotating massive stars (Maeder & Meynet 2006; Decressin *et al.* 2007b; Krause *et al.* 2013), massive binaries stars (de Mink *et al.* 2009), and super-massive stars (Denissenkov & Hartwick 2014). In the following we will focus on the role of fast rotating massive stars.

2. Effect of massive stars on the evolution of globular clusters

One possible candidate for the self-enrichment in globular clusters are the fast rotating massive stars. If stars more massive than 20-25 $M_{\odot}$ have high enough rotation rate at

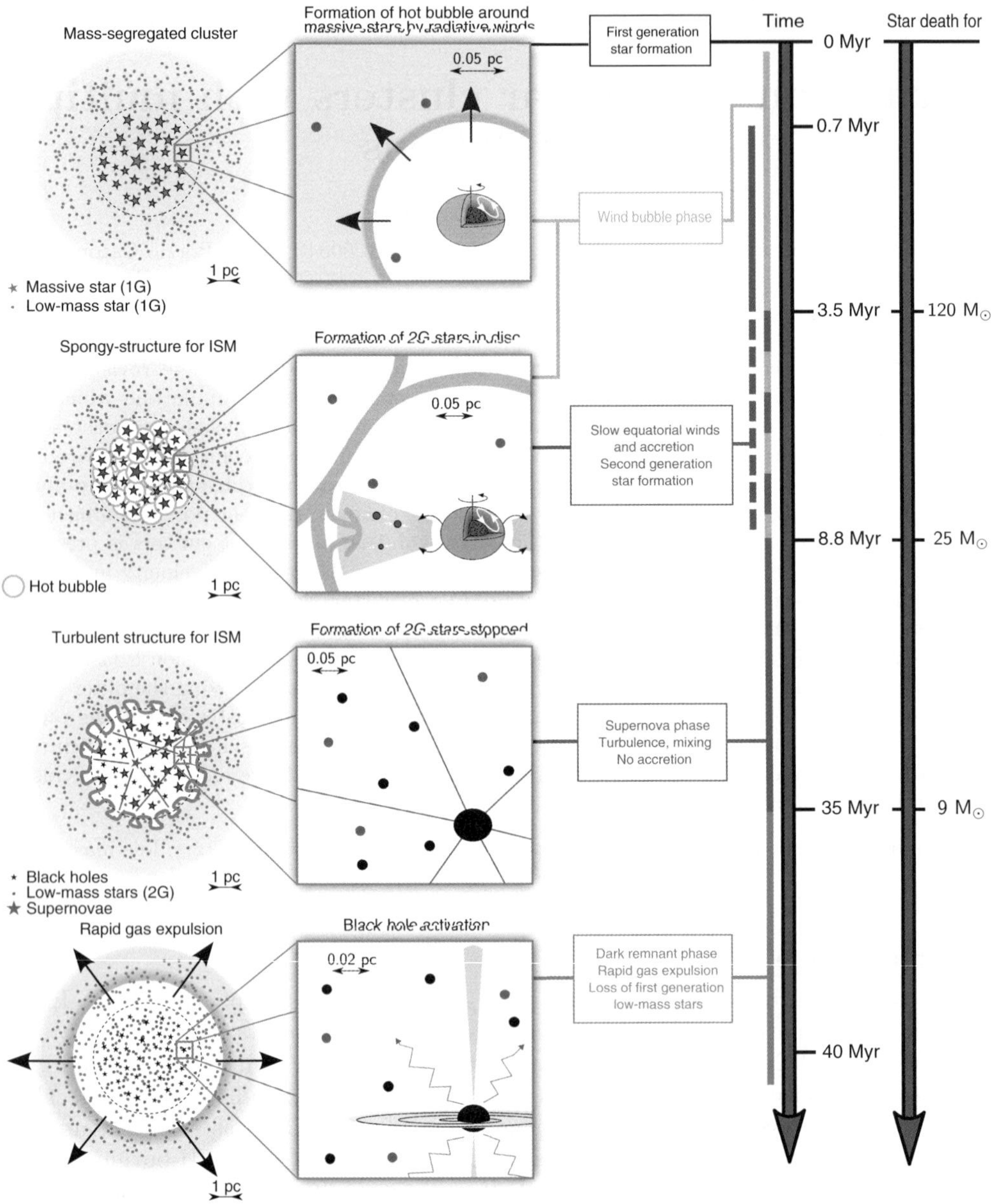

Figure 1. Sketch of the model of first 40 Myr of evolution of a proto-globular cluster populated with fast rotating massive stars leading to two chemically-distinct population of stars as well as the loss of the outer part. A typical timescale is given on the right axis as well as the mass of the most massive star living in the cluster. The massive stars at the centre of the cluster first form hot bubble around them leading to a spongy structure of the ISM. Their fast rotation leads to the creation of a disc around them. The ISM is then been accreted into the disc and new stars will be born with a different chemical composition compared to pristine gas. Then the supernovae phase starting between 3.5 and 8.8 Myr is not enough to inbound the gas due to Rayleigh-Taylor instability but the high turbulence of the gas inhibits further stars formation. When the supernovae cease after about 35 Myr, the ISM can be accreted by dark remnant. Then the activation of dark remnant allow the cluster to efficiently lose its gas left after star formation.

their surface, the rotation-induced mixing will bring matter synthesised in the core to the surface and the centrifugal forces will balance the gravity at the equator so that a slow mechanical wind creates a disk around the stars. This disc will then mix with pristine matter left after the first stars formation to form new stars loaded a H-burning products similar to the observed stars in globular clusters (see Fig. 1 and Decressin *et al.* 2007b).

However this scenario will only produce a limited amount of about 10% of stars with abundance anomalies whereas a much higher fraction is detected in nowadays clusters (about 60% following Carretta *et al.* 2010). One possibility to alleviate this constraint is to assume that the cluster was much more massive at birth (by at least a factor 20-30, see Decressin *et al.* 2007a) and that the cluster lost a large part of its low-mass chemically-normal stars during its evolution. Dynamical studies have shown that two-body relaxation acts in too large timescales to be relevant as shown by Decressin *et al.* (2008) (see also D'Ercole *et al.* 2008).

A more efficient mechanism can be the violent gas expulsion after the stars formation. As the massive stars will seat at the centre of the cluster, the chemically-anomalous stars will also born in the centre. On the contrary the outer part of the cluster will harbour only chemically-normal stars. If the gas expulsion is fast enough (in about less than the cluster crossing time) the potential well of the cluster will be heavily affected and the stars in the outer part of the cluster will be released into the halo of the Galaxy (Decressin *et al.* 2010).

The crucial point to an efficient enough gas expulsion is its short timescale as the physical process responsible for it need to transfer a large amount of energy to the gas rapidly. Supernovae explosion will not be powerful enough to remove efficiently the gas as they lead to the formation of an expanding superbubble that will be destroyed by Rayleigh-Taylor instability before it reaches escape speed. This will lead to a highly turbulent ISM which will inhibit further stellar formation. A more power mechanism is required to expel the gas, which might plausibly be provided by a coherent onset of accretion onto the stellar remnants once the supernovae cease (Krause *et al.* 2012). However as the large fraction of the stars ejected from the cluster will populate the Galactic halo, their number will conflict with observations in Fornax (see Larsen *et al.* 2012).

3. Accretion on pre main sequence stars

One possibility to distinguish between the different proposed scenarios to explain the abundance anomalies in globular clusters is to look at the properties of the resulting low-mass stars. Recently Bastian *et al.* (2013) have proposed an alternative to form chemically-anomalous stars with their scenario of early disk accretion. As the timescale for the evolution of massive stars is shorter than the pre-main sequence (PMS) of low-mass stars, the accretion disk around proto low-mass stars will be replace by matter ejected by massive binaries. In this case only one generation is needed to explain the two population of stars in globular clusters. This model can help to alleviate the mass-budget problem explained previously.

As already noted by D'Antona *et al.* (2014) stars on the PMS will acquire a radiative core and will no longer be fully convective after a few Myr. This poses some additional constraints on the timescale during which the early accretion can take place. As the accreted matter is loaded in H-burning products and has a high helium abundance, its mean molecular weight is heavier than the stellar interior. Such condition is favourable to develop thermohaline instabilities which will mix and homogenise the stellar interior(Ulrich 1972). This mechanism has already been used to explain abundances variations along the

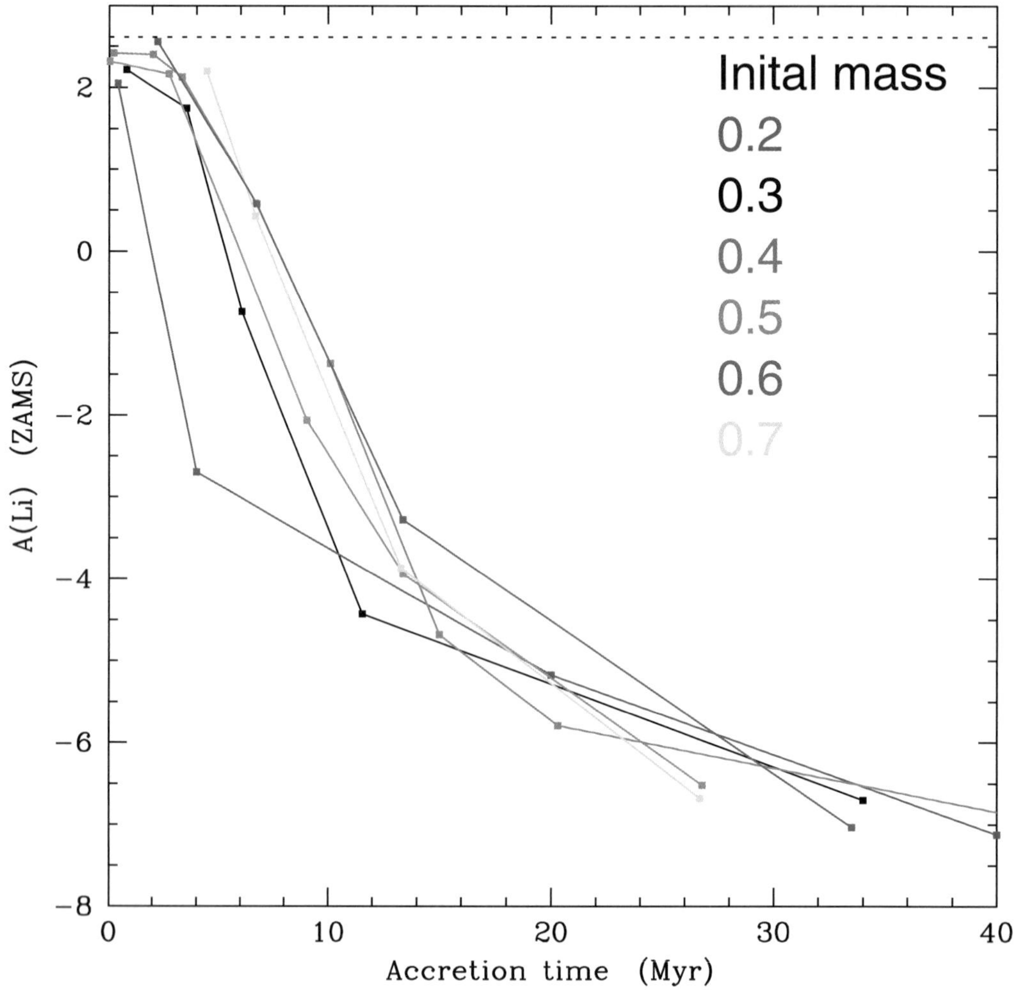

Figure 2. Abundance of Li at the surface of a 0.8 $M_{\odot}$ stars on the ZAMS as function of the timescale of the accretion for a proto-star of 0.2 to 0.7 $M_{\odot}$.

RGB (see e.g., Charbonnel & Lagarde 2010). In PMS stars, fragile element as lithium will be heavily affected by this process as its burning temperature is low enough to be reached below the convective envelope. Thermohaline instabilities will then lead to the complete destruction of lithium inside the star as illustrated in Fig. 2. This is in contradiction with the observations showing that chemically anomalous stars still present lithium on their surface in the main sequence (see e.g., Lind *et al.* 2009; Monaco *et al.* 2012; Dobrovolskas *et al.* 2014).

4. Horizontal branch in ω Cen

One difference between chemically normal and anomalous low-mass stars will also be visible along the horizontal branch (HB). Indeed stars with abundance anomalies have a higher initial He abundance and hence they evolve faster on the main sequence. Assuming similar mass loss on the red giant branch the position on the HB will reflect their mass and composition: lighter stars will become hotter due to their smaller envelope (see e.g. Chantereau *et al.* 2015).

In particular in ω Cen the presence of a large number of stars in the blue hook of the HB (i.e., the hottest in this cluster, see Anderson & van der Marel 2010) has for a long time remained a puzzle. Tailo *et al.* (2015) have shown that this structure can be understood as the progeny of stars which had late helium-core flash ignition followed by deep mixing. However an increase of the core mass up to 0.04 $M_\odot$ is also needed to fully explain their luminosity function. This core-mass increase can be achieved by a dispersion in rotation rates by the progenitors. Indeed, low-mass stars in the dense environment of the cluster's central regions see their PMS accretion disk disrupts at an early stage leaving them with a high rotation at the ZAMS. Following Mengel & Gross (1976) their high rotation rate will lead to higher core mass at the time of the He-flash. Rotation may also account for frequent late-flash-mixing events in massive globular clusters.

5. The extended main-sequence turn-off cluster NGC 1856

Finally it has recently been discovered multiple or extended turn-offs in young clusters in the Magellanic Clouds (see e.g., Mackey *et al.* 2008; Milone *et al.* 2009). Two interpretations of this feature have been proposed: a significant age spread of several 100 Myr in clusters aged 1-2 Gyr (e.g. Milone *et al.* 2009; Girardi *et al.* 2011; Goudfrooij *et al.* 2011; Rubele *et al.* 2013) or a spread in stellar rotation (e.g. Bastian & de Mink 2009; Niederhofer *et al.* 2015).

D'Antona *et al.* (2015) have recently focused on the young cluster NGC 1856, showing a splitting in the upper part of the main sequence observed by Milone *et al.* (2015). With the use of population synthesis programs available from the Geneva stellar models (Georgy *et al.* 2014), this cluster can be seen as the superposition of two main populations having the same age of about 350 Myr. Two third are very rapidly rotating stars while a third are slowly/non-rotating stars. Stars with rapid rotation are common among B and A stars, whereas binary synchronization is a possible process behind the slowly/non-rotating population (see Zahn 1977). One outcome of this explanation is that a large number of the slowly/non-rotating stars should still be in a loose binary system as the cluster is not dense enough to disrupt most of binaries. In such binary systems, Roche lobe overflow will occur when the primary leaves off the main sequence. In this case the primary may not be able to ignite helium in their core and the red clump is only populated by the former fast rotating population which is consistent with the observed cluster.

Acknowledgement

T.D. acknowledges support from the UE Program (FP7/2007/2013) under grant agreement number 267251 of Astronomy Fellowships in Italy (ASTROFit).

References

Anderson, J. & van der Marel, R. P. 2010, *ApJ*, 710, 1032
Bastian, N. & de Mink, S. E. 2009, *MNRAS*, 398, L11
Bastian, N., Lamers, H. J. G. L. M., de Mink, S. E., *et al.* 2013, *MNRAS*, 436, 2398
Bedin, L. R., Piotto, G., Anderson, J., *et al.* 2004, *ApJ*, 605, L125
Carretta, E., Bragaglia, A., Gratton, R. G., *et al.* 2010, *A&A*, 516, A55
Chantereau, W., Charbonnel, C., & Decressin, T. 2015, *A&A*, 578, A117
Charbonnel, C. & Lagarde, N. 2010, *A&A*, 522, A10
D'Antona, F., Di Criscienzo, M., Decressin, T., *et al.* 2015, *MNRAS*, 453, 2637
D'Antona, F., Ventura, P., Decressin, T., Vesperini, E., & D'Ercole, A. 2014, *MNRAS*, 443, 3302

de Mink, S. E., Pols, O. R., Langer, N., & Izzard, R. G. 2009, *A&A*, 507, L1
Decressin, T., Baumgardt, H., Charbonnel, C., & Kroupa, P. 2010, *A&A*, 516, A73
Decressin, T., Baumgardt, H., & Kroupa, P. 2008, *A&A*, 492, 101
Decressin, T., Charbonnel, C., & Meynet, G. 2007a, *A&A*, 475, 859
Decressin, T., Meynet, G., Charbonnel, C., Prantzos, N., & Ekström, S. 2007b, *A&A*, 464, 1029
Denissenkov, P. A. & Hartwick, F. D. A. 2014, *MNRAS*, 437, L21
D'Ercole, A., D'Antona, F., Carini, R., Vesperini, E., & Ventura, P. 2012, *MNRAS*, 423, 1521
D'Ercole, A., D'Antona, F., Ventura, P., Vesperini, E., & McMillan, S. L. W. 2010, *MNRAS*, 407, 854
D'Ercole, A., D'Antona, F., & Vesperini, E. 2011, *MNRAS*, 415, 1304
D'Ercole, A., Vesperini, E., D'Antona, F., McMillan, S. L. W., & Recchi, S. 2008, *MNRAS*, 391, 825
Dobrovolskas, V., Kučinskas, A., Bonifacio, P., *et al.* 2014, *A&A*, 565, A121
Georgy, C., Granada, A., Ekström, S., *et al.* 2014, *A&A*, 566, A21
Girardi, L., Eggenberger, P., & Miglio, A. 2011, *MNRAS*, 412, L103
Goudfrooij, P., Puzia, T. H., Chandar, R., & Kozhurina-Platais, V. 2011, *ApJ*, 737, 4
Gratton, R. G., Carretta, E., & Bragaglia, A. 2012, *A&A Rev.*, 20, 50
Krause, M., Charbonnel, C., Decressin, T., Meynet, G., & Prantzos, N. 2013, *A&A*, 552, A121
Krause, M., Charbonnel, C., Decressin, T., *et al.* 2012, *A&A*, 546, L5
Larsen, S. S., Strader, J., & Brodie, J. P. 2012, *A&A*, 544, L14
Lind, K., Primas, F., Charbonnel, C., Grundahl, F., & Asplund, M. 2009, *A&A*, 503, 545
Mackey, A. D., Broby Nielsen, P., Ferguson, A. M. N., & Richardson, J. C. 2008, *ApJ*, 681, L17
Maeder, A. & Meynet, G. 2006, *A&A*, 448, L37
Mengel, J. G. & Gross, P. G. 1976, *Ap&SS*, 41, 407
Milone, A. P., Bedin, L. R., Piotto, G., & Anderson, J. 2009, *A&A*, 497, 755
Milone, A. P., Bedin, L. R., Piotto, G., *et al.* 2015, *MNRAS*, 450, 3750
Monaco, L., Villanova, S., Bonifacio, P., *et al.* 2012, *A&A*, 539, A157
Niederhofer, F., Georgy, C., Bastian, N., & Ekström, S. 2015, *MNRAS*, 453, 2070
Prantzos, N., Charbonnel, C., & Iliadis, C. 2007, *A&A*, 470, 179
Rubele, S., Girardi, L., Kozhurina-Platais, V., *et al.* 2013, *MNRAS*, 430, 2774
Tailo, M., D'Antona, F., Vesperini, E., *et al.* 2015, *Nature*, 523, 318
Ulrich, R. K. 1972, *ApJ*, 172, 165
Ventura, P., D'Antona, F., Mazzitelli, I., & Gratton, R. 2001, *ApJ*, 550, L65
Ventura, P., Di Criscienzo, M., Carini, R., & D'Antona, F. 2013, *MNRAS*, 431, 3642
Zahn, J.-P. 1977, *A&A*, 57, 383

Formation, evolution, and survival of massive star clusters
Proceedings IAU Symposium No. 316, 2015
C. Charbonnel & A. Nota, eds.

doi:10.1017/S1743921315009138

Contribution of globular clusters to halos

Angela Bragaglia

INAF-Osservatorio Astronomico di Bologna, via Ranzani 1, 40127 Bologna, Italy
email: angela.bragaglia@oabo.inaf.it

Abstract. The contribution of massive star clusters to their hosting halo dramatically depends on their formation mechanism and their early evolution. Massive globular clusters in the Milky Way (and in other galaxies) have been shown to display peculiar chemical patterns (light-elements correlations and anti-correlations) indicative of a complex star formation, confirmed by photometric evidence (spread or split sequences). I use these chemical signatures to try to understand what is the fraction of halo stars originally born in globular clusters.

Keywords. Galaxy, halo, globular clusters

1. Introduction

Globular cluster (GC) systems are found in all kind of galaxies, although the number and frequency of GCs may vary a lot, depending also on the morphological type (e.g., Brodie & Strader 2006). In particular, in the Milky Way (MW) the specific frequency, that is the number of clusters per unit galaxy luminosity, is low, as typical of spirals (see, e.g. the recent work by Harris *et al.* 2013).

The MW hosts about 160 GCs, two-thirds of which reside in the halo (e.g. Harris 1996 and web updates). Possibly this number will increase a lot, since new additions from various surveys such as SDSS, Pan-STARRS, DES, VST ATLAS, etc are nowadays common, see e.g., Koposov *et al.* (2007), Laevens *et al.* (2015), Luque *et al.* (2015), Belokurov *et al.* (2014), even if the objects found are sometimes classified as GCs or ultrafaint dwarfs depending on the study, as is the case of the last example given here. However, these new clusters are all rather small, we do not expect to have missed really massive clusters in the halo.

Clusters lose mass and stars during their evolution (see, e.g., Baumgardt *et al.* 2008). This is due to several effects, first the violent relaxation phase and later two-body encounters and tidal shocks with disk and bulge, on longer timescales. Furthermore, some clusters have a combination of mass and size which put them outside the survival zone (defined in Gnedin & Ostriker 1997), see for instance Jordi & Grebel (2010). The less massive clusters may even be completely dissolved, so that the GCs we see today are the remnants of a larger population and are less massive than at birth, having contributed, and still contributing, stars to the MW halo. One very famous case is Pal 5, with its very extended double tidal tail, discovered by Odenkirchen *et al.* (2001). Many other tails have been found in the halo, some still connected with a cluster (e.g., NGC 288, Grillmair *et al.* 2013 or Pal 14, Sollima *et al.* 2011), but many more without a clearly associated progenitor cluster. For instance, Grillmair (2015) counted 21 such streams and calculated that the original GC number should have been about 450. These streams and tails are direct proofs that clusters contribute(d) to the halo. However, it is not yet clear how large the contribution is.

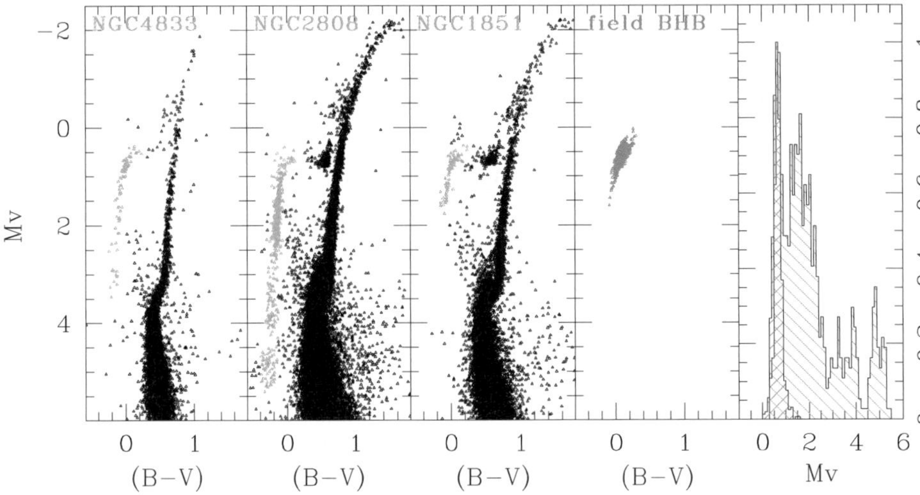

Figure 1. Three CMDs corrected for reddening of GCs (photometry taken from Piotto *et al.* 2002) and CMD for field BHB (Brown *et al.* 2008); in the four panels the BHBs are indicated by coloured points. Right panel: histogram of the absolute magnitudes of field BHBs (in magenta, very peaked, for about 650 stars) and the three GCs (in blue, more extended, for about 630 stars).

2. Field halo and GC stars: is chemical tagging possible?

Globular cluster and field halo stars share in fact many properties. For instance, they have very similar metallicity distributions (see e.g., Fig. 4 in Gratton *et al.* 2012), peaked near [Fe/H]$\simeq -1.5$. They are also indistinguishable when their α-elements are considered (see e.g., Fig. 1 in Carretta 2015), while the same is not not true for other components, like the MW satellites, which generally show lower [α/Fe] ratios.

However, GCs and halo field stars do also show striking differences, for instance, the horizontal branch morphology. The absolute magnitude and the colour of the blue horizontal branch (BHB) stars show very dissimilar distributions. This is shown in Fig. 1, where the colour-magnitude diagrams (CMD) of three GCs, and in particular their BHBs, are compared to field BHBs. The colour and luminosity distributions are clearly not identical; in particular, the GCs seem to possess an additional component, hotter and possibly enriched in helium (e.g., D'Antona & Caloi 2004).

This is most probably connected with the peculiar chemistry shown by GC stars for light elements (C, N, O, Mg, Al, Na). For instance, it has long been known than carbon and nitrogen have large spreads anticorrelated with each other and also show bimodality (e.g., Norris *et al.* 1984, Cohen *et al.* 2002, Pancino *et al.* 2010). Furthermore, GCs show large star-to-star variations in sodium and oxygen (and sometimes in aluminum and magnesium), anticorrelated with each other, which are not present among field stars (see e.g. the reviews in Kraft 1994, Gratton *et al.* 2004, Gratton *et al.* 2012), at least not generally, as we will see next. These (anti)correlations are the signature of hydrogen burning at high temperature; the current understanding is that they are due to a previous generation of stars, more massive than the ones presently alive in GCs, which polluted the material from which (a) second generation(s) of stars formed (as discussed by many in this Symposium, e.g., Charbonnel in the plenary talk, Marino - and Bastian for a distinctive view). These generations will be addressed as first generation (FG) and second generation (SG) in the following.

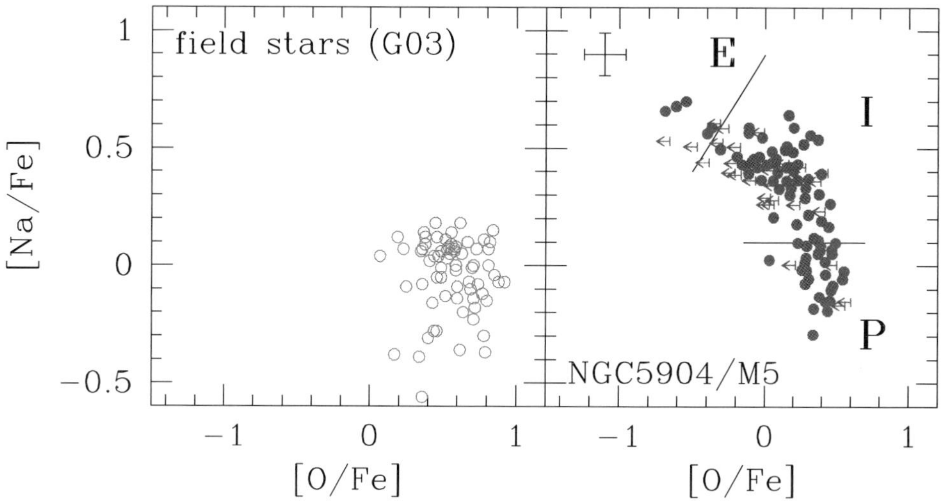

Figure 2. [O/Fe] *vs* [Na/Fe] for field stars (left, data from Gratton *et al.* 2003) and a typical GC, M5 (right, data from Carretta *et al.* 2009a,b). For the GC the separation between FG (here indicated by P, which stands for primordial coposition) and SG stars (indicated by I, E, standing for intermediate and extreme composition) is also presented.

This behaviour is displayed in Fig. 2 for Na and O: the metal-poor field halo stars occupy only a narrow region in the O-Na plane, while the GC stars have an additional large dispersion in both elements. This is shown in Fig. 2 for a single GC, but the same behaviour is seen in practically all the MW GCs massive enough, as demonstrated by many studies summarized in the review papers cited just above, from the small Pal 5 (Smith *et al.* 2002) to the huge ω Cen (e.g. Johnson & Pilachowski 2010, Marino *et al.* 2011). In particular, it is seen in the large, homogeneous survey of red giant stars in 26 MW GCs done with FLAMES@VLT and presented in a series of papers (e.g., Carretta *et al.* 2006 and Bragaglia *et al.* 2015 to cite only the first and the latest). Carretta *et al.* (2010) discussed the "universality" of the Na-O anticorrelation among massive clusters, proposing it as a *defining property of globulars*.

Adopting the mainstream interpretation, in each GC there are two components, one of primordial composition (the FG) and one with modified composition (the SG). Interestingly, the fraction of FG to SG stars in clusters is remarkably constant. Carretta *et al.* (2010) and following papers of the FLAMES series found about one third and two thirds of FG and SG stars, respectively. This has recently been confirmed by Bastian & Lardo (2015), who added a few more clusters from other sources, but also noted that the almost constant fraction of SG to FG stars is not in ageement with models of mass loss from clusters.

As it has been remarked by many studies, this large fraction of SG stars leads to a mass-budget problem. If they formed from the ejecta of FG stars, only a small part of the original stellar cluster mass went into the SG. This seems to imply than the clusters were originally much more massive, even tens of times more massive than today (and/or that the FG had a peculiar initial mass function, as discussed by others in this Symposium. e.g., Charbonnel, Larsen). If clusters were only slightly more massive than today, then the halo contains only a few per cent of stars lost by GCs. But if clusters lost more than 90% of their original stars, as advocated by many models of GC formation in the context of multiple generations, the halo contains instead a significant fraction of stars born in GCs (the exact number depends on the model adopted for FG polluting stars,

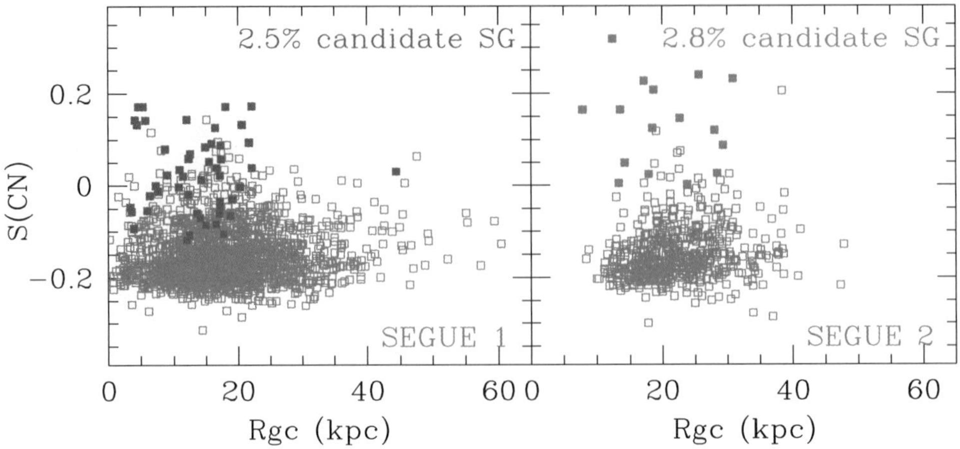

Figure 3. CN enhancement in SEGUE 1 and SEGUE 2 stars, from Martell & Grebel (2010) and Martell *et al.* (2011). Filled, coloured symbols indicate candidate SG-like stars. Note that δS(CN) is used in the original papers, while here S(CN) is plotted.

see e.g., Gratton *et al.* 2012 for a discussion). Especially in the second case, the stars lost were almost all of FG, the ones that have a composition similar to the field halo stars. This means that it is very difficult, if not impossible, to distinguish them using chemical tagging.

3. GC stars: lost and found

On the other hand, it seems rather "easy" to find SG-like stars, given their peculiar chemistry (enhancement in N, Na, and Al, combined with depletion of C, O, and Mg in the same star).

Martell & Grebel (2010) and Martell *et al.* (2011) focused on C and N, measuring the CN and CH band strengths on the low resolution spectra of about 2500 giant stars in the SEGUE 1 and SEGUE 2 samples. They found 49 CN-strong (i.e., SG-like) stars over 1957 and 16 over 561 in the two samples, respectively. Figure 3 is built using data from their papers; in both cases the observed fraction is nearly 3%.

If we consider instead Na and O, we are aware of two different instances. Carretta *et al.* (2010) considered the minimum and maximum Na abundance in their sample of nearly 20 GCs, that are indicative of FG and SG stars respectively, and compared them with the Na abundance of field halo stars of similar metallicity. After discarding 4 stars with high Na abundance because their overabundance was most probably due to a companion AGB star, they were left with 2 stars out of 144, deducing a fraction of 1.4% SG-like stars. Considering the present-day fraction of SG and FG stars, this means a minimum observed GC contribution to the halo of about 3%. A similar fraction was also found by Ramírez *et al.* (2012), who discovered 2 Na-rich and O-poor stars while studying 67 nearby halo stars.

Another possibility for chemical tagging halo stars comes from the Mg-Al anticorrelation. It is not seen in all clusters (see Carretta *et al.* 2009b), but Al can show even huge variations between FG and SG stars, more than 1 dex, so that in principle it is easy to find stars with these extreme chemical properties. Figure 4 (left panel) shows the case for NGC 6752, where Carretta *et al.* (2012) found a very extended distribution in Mg and especially Al. This feature was exploited by Lind *et al.* (2015); they employed the first data release of the large public Gaia-ESO spectroscopic survey (Gilmore *et al.* 2012,

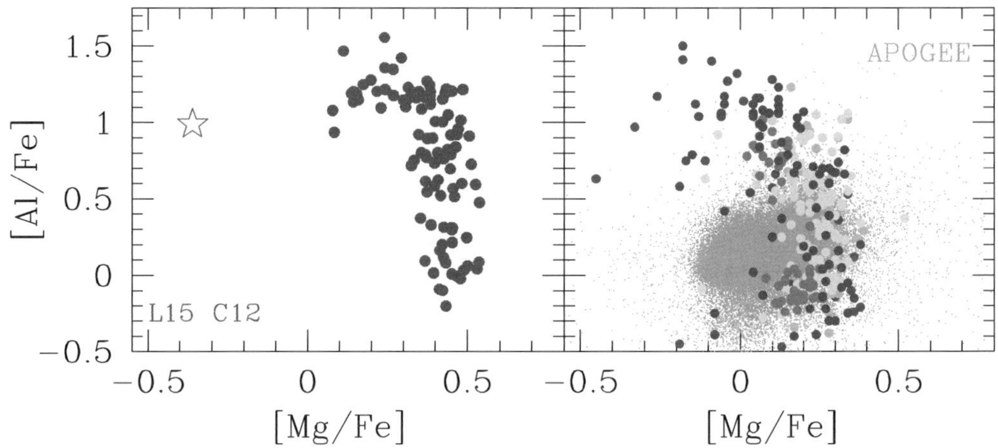

Figure 4. Left panel: [Mg/Fe] *vs* [Al/Fe] for NGC 6752 (blue filled circles, from Carretta *et al.* 2012) and one candidate escapee from a GC (red star, from Lind *et al.* 2015. Right panel: APOGEE data are used here; gray dots are field stars, larger, coloured symbols are stars in five GCs (M2, M3, M5, M13, M15) from Mészáros *et al.* (2015).

Randich *et al.* 2013) to find a good candidate escapee from a GC, shown in Fig. 4 as a star symbol. Unfortunately, the Gaia-ESO setup for the bulk of field MW stars does not cover lines of sodium and oxygen. More candidates will probably be found in more advanced releases and in the final catalogue, so that it will also be possible to do some counts on fractions and calculate the GC contribution to the halo.

The same kind of search could be done also on data from other surveys, for instance APOGEE. The right panel of Fig. 4 shows the Mg and Al abundances taken from the latest data release and the values for a few GCs (Mészáros *et al.* 2015); some of the stars seem to reach the high Al values we are looking for in SG-like stars.

This chemical tagging with connected enhancements and depletions in light elements is being used also to confirm the associations of streams and moving groups with clusters. Examples of a positive and negative identification, respectively, are the Aquarius stream (consistent with being GC debris, Wylie-de Boer *et al.* 2012) and the Kapteyn group (not associated with ω Cen, Navarrete *et al.* 2015).

Finally, I wish to mention one interesting finding that could indicate a connection between halos and GCs. Greene *et al.* (2015), Greene (2015), in their MASSIVE survey of early-type galaxies, found that nitrogen abundance does not decline outwards as iron and other elements do. Apparently the outer regions of these galaxies, which we may associate with the halo, are rich in nitrogen, as SG stars are in GCs; it may be just a concidence, or it may have other explanations, but it is intriguing.

4. Summary

There are theoretical reasons to expect that GCs contribute(d) part (and even most, depending on the proposed scenarios fo multiple stellar populations in GCs) of their stars to the MW field halo. However, using only models of cluster formation and disruption/mass loss, it is not currently assessed how important this contribution is.

On the observational side, there is evidence of disrupting and disrupted clusters in the form of tidal tails, streams, and moving groups. Furthermore, we may use chemical tagging to find stars, presently in the MW halo field, which were originally born in a GC. The halo field stars and the GCs share many chemical properties, but also some striking differences: GC stars show a large star-to-star dispersion in light elements, which is considered proof of the existence of multiple populations in clusters. This is not visible in the bulk of field halo stars.

We may then try to find out halo stars with similar fingerprints, looking in particular to elements like nitrogen, sodium, and aluminum, which show large enhancement in SG GC stars (anticorrelated with carbon, oxygen, and magnesium, respectively). This has been done by a few studies (Carretta *et al.* 2010, Martell *et al.* 2011, Ramirez *et al.* 2012, and Lind *et al.* 2015) and can be done on a larger scale using the large ground-based spectroscopic surveys, like APOGEE and Gaia-ESO.

Using the numbers of observed SG-like stars in the MW halo (about 1.5 to 3%), and bearing in mind that present-day GCs host about 70% SG and 30% FG stars, we may deduce that the minimum contribution of GCs is about 5%. This fraction might increase even a lot if clusters were really much more massive at their birth as advocated by most models trying to explain the multiple populations in GCs. However, the subject is still hotly debated and we need improved models and more observations.

I acknowledge the financial support of IAU and of the Italian MIUR (PRIN "The Chemical and Dynamical Evolution of the Milky Way and Local Group Galaxies", PI F. Matteucci).

References

Bastian, N. & Lardo, C. 2015, *MNRAS*, 453, 357
Baumgardt, H., Kroupa, P., & Parmentier, G. 2008, *MNRAS*, 384, 1231
Belokurov, V., Irwin, M. J., Koposov, S. E., *et al.* 2014, *MNRAS*, 441, 2124
Bragaglia, A., Carretta, E., Sollima, A., *et al.* 2015, *A&A (in press)*, arXiv:1507.07562
Brodie, J. P. & Strader, J. 2006, *ARA&A*, 44, 193
Brown, W. R., Beers, T. C., Wilhelm, R., *et al.*, 2008, *AJ*, 135, 564
Carretta, E., Bragaglia, A., Gratton, R. G., *et al.* 2006, *A&A*, 450, 523
Carretta, E., Bragaglia, A., Gratton, R. G., *et al.* 2009a, *A&A*, 505, 117
Carretta, E., Bragaglia, A., Gratton, R., & Lucatello, S. 2009b, *A&A*, 505, 139
Carretta, E., Bragaglia, A., Gratton, R. G., *et al.* 2010, *A&A*, 516, A55
Carretta, E., Bragaglia, A., Gratton, R. G., Lucatello, S., & D'Orazi, V. 2012, *ApJ (Letters)*, 750, L14
Carretta, E. 2015, *IAUS 317*, IAU General Assembly #29, #2252181, arXiv:1510.00507
Cohen, J. G., Briley, M. M., & Stetson, P. B. 2002, *AJ*, 123, 2525
D'Antona, F. & Caloi, V. 2004, *ApJ*, 611, 871
Gilmore, G., Randich, S., Asplund, M., *et al.* 2012, *The Messenger*, 147, 25
Gnedin, O. Y. & Ostriker, J. P. 1997, *ApJ*, 474, 223
Gratton, R. G., Carretta, E., Desidera, S., *et al.* 2003, *A&A*, 406, 131
Gratton, R., Sneden, C., & Carretta, E. 2004, *ARA&A*, 42, 385
Gratton, R. G., Carretta, E., & Bragaglia, A. 2012, *A&ARv*, 20, 50
Greene, J. 2015, *IAUS 317*, IAU General Assembly #29, #2250981
Greene, J. E., Janish, R., Ma, C.-P., *et al.* 2015, *ApJ*, 807, 11
Grillmair, C. J., Cutri, R., Masci, F. J., *et al.* 2013, *ApJ (Letters)*, 769, L23
Grillmair, C. J. 2015, *IAUS 317*, IAU General Assembly #29, #2255751
Harris, W. E. 1996, *AJ*, 112, 1487
Harris, W. E., Harris, G. L. H., & Alessi, M. 2013, *ApJ*, 772, 82
Johnson, C. I. & Pilachowski, C. A. 2010, *ApJ*, 722, 1373
Jordi, K. & Grebel, E. K. 2010, *A&A*, 522, A71
Koposov, S., de Jong, J. T. A., Belokurov, V., *et al.* 2007, *ApJ*, 669, 337
Kraft, R. P. 1994, *PASP*, 106, 553

Laevens, B. P. M., Martin, N. F., Bernard, E. J., *et al.* 2015, arXiv:1507.07564
Lind, K., Koposov, S. E., Battistini, C., *et al.* 2015, *A&A*, 575, L12
Luque, E., Queiroz, A., Santiago, B., *et al.* 2015, arXiv:1508.02381
Marino, A. F., Milone, A. P., Piotto, G., *et al.* 2011, *ApJ*, 731, 64
Martell, S. L. & Grebel, E. K. 2010, *A&A*, 519, A14
Martell, S. L., Smolinski, J. P., Beers, T. C., & Grebel, E. K. 2011, *A&A*, 534, A136
Mészáros, S., Martell, S. L., Shetrone, M., *et al.* 2015, *AJ*, 149, 153
Navarrete, C., Chanamé, J., Ramírez, I., *et al.* 2015, *ApJ*, 808, 103
Norris, J., Freeman, K. C., & Da Costa, G. S. 1984, *ApJ*, 277, 615
Odenkirchen, M., Grebel, E. K., Rockosi, C. M., *et al.* 2001, *ApJ (Letters)*, 548, L165
Pancino, E., Rejkuba, M., Zoccali, M., & Carrera, R. 2010, *A&A*, 524, A44
Piotto, G., King, I. R., Djorgovski, S. G, *et al.*, 2002, *A&A*, 391, 945
Ramírez, I., Meléndez, J., & Chanamé, J. 2012, *ApJ*, 757, 164
Randich, S. & Gilmore, G., Gaia-ESO Consortium 2013, *The Messenger*, 154, 47
Smith, G. H., Sneden, C., & Kraft, R. P. 2002, *AJ*, 123, 1502
Sollima, A., Martínez-Delgado, D., Valls-Gabaud, D., & Peñarrubia, J. 2011, *ApJ*, 726, 47
Wylie-de Boer, E., Freeman, K., Williams, M., *et al.* 2012, *ApJ*, 755, 35

Formation, evolution, and survival of massive star clusters
Proceedings IAU Symposium No. 316, 2015
C. Charbonnel & A. Nota, eds.

doi:10.1017/S1743921315009023

Bimodal regime in young massive clusters leading to subsequent stellar generations

Richard Wünsch[1], Jan Palouš[1], Guillermo Tenorio-Tagle[2], Casiana Muñoz-Tuñón[3] and Soňa Ehlerová[1]

[1] Astronomical Institute of the Czech Academy of Sciences, v.v.i.
Boční II 1401, 141 31 Prague, Czech Republic
email: richard@wunsch.cz

[2] Instituto Nacional de Astrofísica Optica y Electrónica
AP 51, 72000 Puebla, México

[3] Instituto de Astrofísica de Canarias
38200 La Laguna, Tenerife, Spain

Abstract. Massive stars in young massive clusters insert tremendous amounts of mass and energy into their surroundings in the form of stellar winds and supernova ejecta. Mutual shock-shock collisions lead to formation of hot gas, filling the volume of the cluster. The pressure of this gas then drives a powerful cluster wind. However, it has been shown that if the cluster is massive and dense enough, it can evolve in the so–called bimodal regime, in which the hot gas inside the cluster becomes thermally unstable and forms dense clumps which are trapped inside the cluster by its gravity. We will review works on the bimodal regime and discuss the implications for the formation of subsequent stellar generations. The mass accumulates inside the cluster and as soon as a high enough column density is reached, the interior of the clumps becomes self-shielded against the ionising radiation of stars and the clumps collapse and form new stars. The second stellar generation will be enriched by products of stellar evolution from the first generation, and will be concentrated near the cluster center.

Keywords. globular clusters: general, galaxies: starburst, galaxies: star clusters: general, galaxies: star formation, stars: winds, outflows, hydrodynamics, radiative transfer

1. Introduction

It has been found by photometric observations that globular clusters contain two or more populations of stars differing by age and/or the chemical composition (see e.g. Pancino *et al.* 2000; Bedin *et al.* 2004; Piotto *et al.* 2007, 2015, and references therein). Furthermore, spectroscopic observations found anticorrelations between abundances of Na and O and other pairs of elements (Carretta *et al.* 2009; Mucciarelli *et al.* 2012) suggesting that a certain fraction of stars in globular clusters contain products of hydrogen burning at high temperatures. Several hypotheses were suggested to explain the above observations (Decressin *et al.* 2007; D'Ercole *et al.* 2008; Bastian *et al.* 2013; de Mink *et al.* 2009). Here we propose a cooling winds scenario in which the second stellar generation is formed out of fast stellar winds enriched by products of hydrogen burning in massive stars, that cool down inside the cluster evolving in the so–called bimodal regime.

Young massive clusters with masses $10^5 - 10^7$ M$_\odot$ and ages $< 10^7$ yr include a high number of massive stars ($\sim 2 \times 10^4$ per 10^6 M$_\odot$ of the cluster stellar mass assuming the standard IMF) concentrated in a small volume of several pc in radius (see e.g. Portegies Zwart *et al.* 2010; Larsen 2010). These stars insert through stellar winds large amounts of gas moving with velocities several thousands km s^{-1} into their surroundings. As the winds collide with each other, their kinetic energy is thermalised and the gas is heated

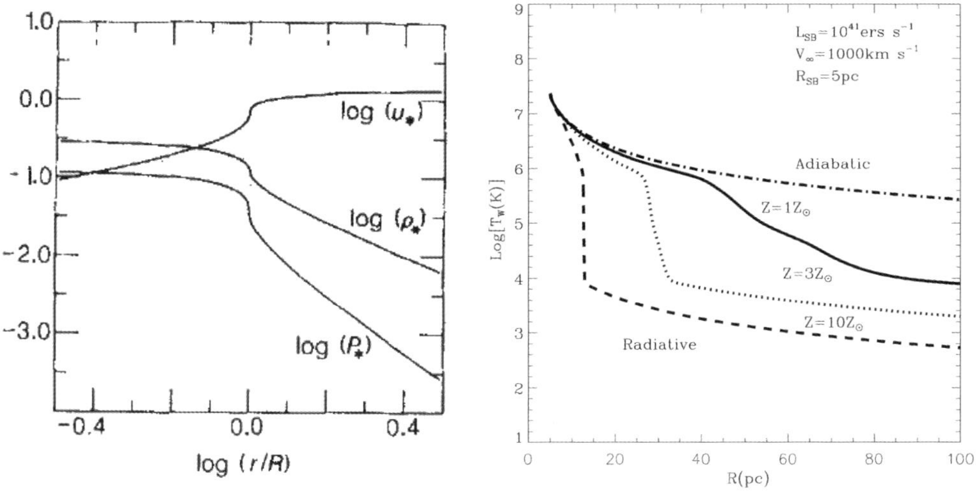

Figure 1. Left: Cluster wind solution by Chevalier & Clegg (1985, Fig. 1, reprinted by permission from Nature Publishing Group), it shows radial profiles of the wind density ($\rho_\star$), pressure ($P_\star$) and velocity ($u_\star$) in logarithmic scale. Right: Radial profile of the wind temperature, comparison of the adiabatic solution to the radiative solution by Silich *et al.* (2003, Fig. 1, reprinted by permission from AAS) for three different metallicities.

to temperatures $\sim 10^7$ K. The high pressure of this hot gas then drives the star cluster wind.

Winds of young massive star clusters were studied analytically by Chevalier & Clegg (1985) who derived a stationary solution of spherically symmetric hydrodynamic equations. They assumed that sources of mass and energy are distributed uniformly in a sphere with a given radius and they neglected radiative cooling of the hot gas. A remarkable property of their solution is that the wind velocity reaches the sound speed always exactly at the cluster border. Radial profiles of the basic hydrodynamic quantities given by their solution are shown on the left panel of Fig. 1.

The Chevalier&Clegg wind solution was later tested and further explored by e.g. Cantó *et al.* (2000); Raga *et al.* (2001) and many other authors. An important field of research is an interaction of the cluster with the ambient gas, in particular with the remnant of the parent molecular cloud. This subject is not covered here and we refer e.g. to Tenorio-Tagle *et al.* (2006); Harper-Clark & Murray (2009); Krause *et al.* (2012, 2013); Rogers & Pittard (2013); Herrera & Boulanger (2015) and references therein. Another interesting problem is the interaction of cluster winds with each other. It was studied by Tenorio-Tagle *et al.* (2007a) who show that it can lead to the formation of the super galactic wind as observed for instance in M82 galaxy.

The influence of the radiative cooling on the star cluster wind was studied by Silich *et al.* (2003) who included cooling into the spherically symmetric wind solution similar to the one by Chevalier & Clegg (1985). They show that the wind rapidly cools at some distance from the cluster, which depend on the cluster parameters, in particular on its mass and the wind metallicity (see Fig. 1). The radiative solution was compared to X-ray observations of young clusters by (Silich *et al.* 2004), for instance it was shown that it is in good agreement with measured X-ray fluxes of the nuclear cluster NGC 4303 (Jiménez-Bailón *et al.* 2003).

It has also been found that if the cluster is massive and compact enough, *i.e.* if its wind mechanical luminosity $L_{\rm crit}$ exceeds a certain value, no stationary wind solution exists

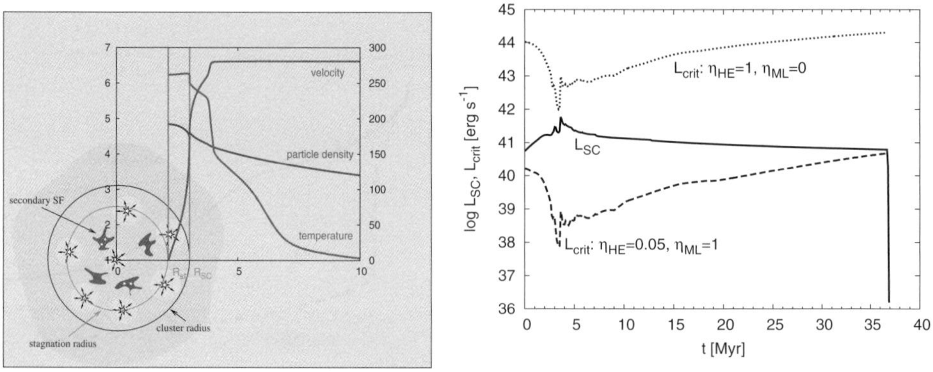

Figure 2. Left: schematic view of the cluster evolving in the bimodal regime. Individual stars are surrounded by free wind regions, however, most of the cluster volume is filled by the hot gas heated up the wind reverse shock. The hot gas is thermally unstable and forms dense warm clumps in some regions inside the cluster. The graph shows the wind density, temperature and velocity in the outer part of the cluster where the stationary solution exists. Right: time evolution of the wind mechanical luminosity (for cluster with mass 10^7 $M_\odot$, solid) compared to the critical luminosity calculated for heating efficiency 1 and no mass loading (dotted) and heating efficiency 5% and mass loading 1 (dashed).

(Silich *et al.* 2003). This can be easily understood from basic scaling relations. The energy available for driving the wind is directly proportional to the total energy of stellar winds which is directly proportional to the cluster mass. On the other hand, energy losses due to cooling are proportional to the square of the wind density, and therefore to the square of the cluster mass. As a result, the energy losses will always dominate if the cluster is massive and compact enough. This led Tenorio-Tagle *et al.* (2005) to suggest that the star formation feedback in massive clusters takes an extreme positive form leading to a high star formation efficiency.

2. Bimodal regime and secondary star formation

Tenorio-Tagle *et al.* (2007b) studied winds of clusters with mechanical luminosities higher than $L_{\rm crit}$. They found a new, so–called bimodal regime, for the wind solution, in which the volume of the cluster is divided into two regions (see Fig. 2 left). In the outer region, the stationary solution still exists with the wind velocity starting from zero at the stagnation radius, $R_{\rm st}$, and reaching the sound speed at the cluster border $R_{\rm SC}$. In the inner region below $R_{\rm st}$, random parcels of hot gas become thermally unstable, cool down and get compressed by the ambient hot gas into dense clumps. These dense clumps may stay warm, maintained at temperature $T \sim 10^4$ K by the stellar ionising radiation, however, if their column densities become large enough, they can self-shield themselves (see Section 3), cool to lower temperatures and feed the secondary star formation inside the cluster. The existence of the bimodal regime was confirmed by 2D hydrodynamic simulations by Wünsch *et al.* (2008) where also the estimates of the mass accumulated inside the cluster were provided. This prediction of the mass accumulation and the secondary star formation is particularly interesting in context of formation of globular clusters and their observed multiple stellar populations.

There is an observational evidence that the temperature of the hot gas inside young massive clusters is lower than several times 10^7 K which correspond to the thermalisation of all the kinetic energy of individual stellar winds. It led to the introduction of the so–called heating efficiency, $\eta_{\rm HE}$, denoting a fraction of the stellar wind kinetic energy that is converted into the hot gas internal energy (Silich *et al.* 2007). For instance, hydrogen recombination lines of super star clusters in the Antennae galaxies exhibit moderately supersonic widths (Gilbert & Graham 2007), larger than individual velocities of stars, but smaller than typical velocities of stellar winds. They were interpreted in terms of the bimodal wind solution by Tenorio-Tagle *et al.* (2010) who suggest that the line profiles are consistent with $\eta_{\rm HE} \lesssim 0.2$. Silich *et al.* (2009) found a similarly small values $\eta_{\rm HE} \lesssim 0.1$ by measuring and analysing sizes of HII regions coinciding with super star clusters in M82 galaxy. On the other hand, Strickland & Heckman (2009) found relatively high values of the heating efficiency, $\eta_{\rm HE} = 0.3 - 1.0$ by comparing X-ray observation of M82 with a set of 1D and 2D hydrodynamic models. Another effect that can decrease the temperature of the hot gas is mass loading of the wind with the primordial gas. Strickland & Heckman (2009) estimate the mass loading factor, $\eta_{\rm ML}$, having a moderate values between 1 and 2.8. Additionally, there is evidence coming from observations of the Li abundance (Decressin *et al.* 2007), that stars in subsequent generations in globular clusters contain approximately 30% of the primordial gas. In summary, the heating efficiency and the mass loading are important parameters of the wind solution, however, they are not yet very well constrained.

Evolution of the star cluster wind for the first 40 Myr was computed by Wünsch *et al.* (2011) by combining output from the stellar population synthesis code Starburst99 (Leitherer *et al.* 1999) with a semi-analytic code calculating the spherically symmetric wind solution. Fig. 2 (right) shows the evolution of the wind mechanical luminosity for a cluster with mass $M_{\rm SC} = 10^7$ $M_\odot$and radius $R_{\rm SC} = 3$ pc compared to the evolution of the the critical luminosity $L_{\rm crit}$ given for two combinations of the heating efficiency and the mass loading. It can be seen that the most conservative and rather unrealistic values $\eta_{\rm HE} = 1$ and $\eta_{\rm ML} = 0$ represent a marginal case for which the wind is not bimodal for all the period shown. On the other hand, a more realistic case with $\eta_{\rm HE} = 0.05$ and $\eta_{\rm ML}$ gives a solution in the bimodal regime (and hence mass accumulation) for the whole period of the existence of massive stars.

3. Influence of stellar ionising radiation

If the cluster evolves in the bimodal regime, a certain fraction of the gas reinserted within the cluster by massive star in a form of their winds gets thermally unstable, cools down and accumulates inside the cluster. However, the stellar ionising radiation can still maintain the gas warm and ionised if the column density of dense warm clumps does not exceed the value necessary for self-shielding. The minimum mass, $m_{\rm self}$, of a spherical clump which is able to self-shield against the ionising radiation was estimated analytically by Palouš *et al.* (2014):

$$m_{\rm self} = \dot{N}_{\rm UV,SC} \frac{\mu m_{\rm H}}{\alpha_\star} \frac{kT_{\rm ion}}{P_{\rm hot}} \tag{3.1}$$

where $\dot{N}_{\rm UV,SC}$ is the total production rate of ionising photons, $T_{\rm ion}$ is the temperature of the warm ionised gas, $P_{\rm hot}$ is the pressure in the hot gas, $\alpha_\star$ is the recombination coefficient to the second and higher levels, μ is the average molecular weight of particles in the hot gas, $m_{\rm H}$ is the hydrogen nuclei mass and k is the Boltzmann constant.

If the gravity of the cluster is taken into account, dense clumps start to sink into the cluster centre and evolve in more or less steady streams, as shown by the hydrodynamic simulations (see below). The gas streams collide in the centre forming a massive central clump. If self-shielding occurs, it may happen either in both streams and the central clump, or only in the central clump. It has interesting implications for the second stellar

Figure 3. A frame from the 3D radiation-hydrodynamic simulation of a cluster in the bimodal regime at age 1 Myr. Cluster parameters are given in Tab. 1. Individual panels show: logarithm of the gas volume density at plane $z = 0$ (top left), logarithm of the gas temperature (top right) at $z = 0$, logarithm of the gas column density integrated in z-direction (bottom left), and the gas radial velocity (bottom right). In the last panel in the color version of this figure, the red/green color shows the outward/inward velocity and the cyan line marks the points where the flow changes from subsonic to supersonic. (see online edition for the colour version)

generation, because in the latter case, it should be very compact, because the stars are formed out of gas with a very small velocity dispersion. To decide which of the two cases takes place, time $t_{\rm SS}$ needed for self-shielding of the stream can be estimated and compared to the free fall time $t_{\rm ff}$ of the dense clump into the cluster centre. The self-shielding time is

$$t_{\rm SS} = \frac{4\pi^2}{9} N_{\rm UV}^2 R_{\rm SC}^{-1} R_{\rm st}^{-2} (1+\eta_{\rm ML})^{-1} \dot{M}_{\rm SC}^{-1} \mu m_{\rm H} \alpha_*^{-2} \left(\frac{kT_{\rm ion}}{P_{\rm hot}} \right)^3 . \qquad (3.2)$$

In order to test the above model and to explore it in more detail, we run 3D radiation hydrodynamic simulations of the cluster wind in the bimodal regime. It is based on the publically available hydrodynamic code Flash Fryxell *et al.* (2000), regions that are kept warm by the ionising radiation are determined by our code TreeRay, which works in this setup in a similar way as the TreeCol algorithm (Clark *et al.* 2012). The numerical model includes radiative cooling, the background gravitational field of the cluster and self-gravity of the gas. The mass and energy are inserted smoothly in a spherical volume with a radial distribution given by the Schuster profile (Palouš *et al.* 2013). More details will be given in the forthcoming publication (Wünsch *et al.*, 2015).

quantity	symbol	value
cluster half mass radius	$R_{\rm h}$	2.67 pc
stellar mass of the 1st generation	$M_{\rm SC}$	10^7 $M_\odot$
mass inserted by stellar winds	$M_{\rm sw}$	4×10^5 $M_\odot$
heating efficiency	$\eta_{\rm HE}$	0.05
mass loading	$\eta_{\rm ML}$	1
mass loaded	$M_{\rm ML}$	4×10^5 $M_\odot$
mass of the 2nd stellar generation at 3.5 Myr	$M_{\rm 2G}$	7×10^5 $M_\odot$
mass of the warm gas remaining inside cluster at 3.5 Myr	$M_{\rm warm}$	1.4×10^4 $M_\odot$

Table 1. Parameters of the star cluster model for which the presented RHD simulation was carried out. The last two lines show simulation results: mass of the 2nd stellar generation, *i.e.* mass of all sink particles and the mass remaining in the simulation in a form of warm gas.

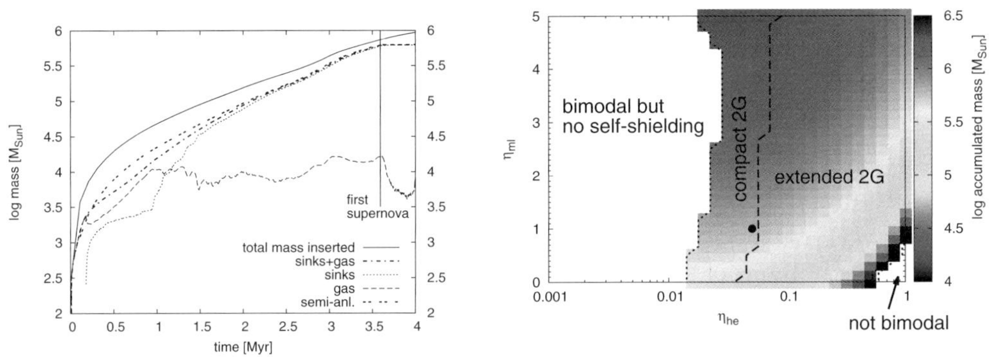

Figure 4. Left: evolution of the total mass inserted into the simulation (solid) and its fraction that stays in the simulation (dash-dotted). The latter can be split into mass of sink particles (thin dotted) and the mass in the cold/warm clumps (thin dashed). Prediction of the accumulated mass by the semi-analytic code is shown by the double dashed line. Right: mass of the second stellar generation calculated by the semi-analytic code as a function of the heating efficiency, $\eta_{\rm HE}$, and mass loading, $\eta_{\rm ML}$. The second stellar generation is predicted to be extended if the thermally unstable gas is able to self-shield against ionising radiation before it falls into the centre, and compact if it is not. The black circle marks parameters used in the presented RHD simulation. (see online edition for the colour version)

Fig. 3 shows a snapshot from the simulation with parameters given in Tab. 1 at time 1 Myr. It can be seen that thermally unstable gas evolves into dense streams that flow into the cluster centre forming there a massive clump. The streams are fully ionised and stay at $T = 10^4$ K, however, the central clump cools to much lower temperatures in its central part. There, the gas becomes gravitationally unstable and forms sink particles. Due to relatively poor spatial resolution due to missing proper treatment of thermal and chemical processes taking place at low temperature, only a few unrealistically massive sink particles are formed. Therefore, the model is able to predict only the total mass of the second stellar generation, $M_{\rm 2G}$, and the approximate position where the second generation stars are formed. The latter is interesting in terms of mass loss of the first generation stars, because if the second stellar generation is highly concentrated, preferential removal of the first generation stars will be easier independently of the removal mechanism.

Evolution of the mass accumulated in the simulation is shown in Fig. 4 (left) where it is compared to the total mass inserted into the simulation. The accumulated mass is divided into mass in sink particles and mass in warm clumps/streams, and it is also compared to the calculation by the semi-analytic code. It can be seen that the agreement between the complex RHD simulation and the semi-analytic code is very good. The simulation also includes a short period after 3.5 Myr when supernovae start to explode. As a result, mass of sink particles stops to grow, *i.e.* star formation is stopped, and the amount of the warm gas decreases by almost one order of magnitude, *i.e.* most of the warm gas is removed.

Finally, we explore how properties of the second stellar generation depend on the heating efficiency $\eta_{\rm HE}$ and mass loading $\eta_{\rm ML}$. Fig. 4 (right) shows whether the second generation stars are formed and if they are, what is their total mass and whether they are formed only in the central clump (concentrated 2G) or also in the infalling streams (extended 2G). The mass of the first generation is 10^7 M$_\odot$and the metallicity of the stellar winds is assumed to be solar. The $\eta_{\rm HE} - \eta_{\rm ML}$ plane includes four qualitatively different regions defined mainly by the heating efficiency. If $\eta_{\rm HE}$ is very low, the wind solution is bimodal, however, the accumulated gas is unable to self-shield against ionising radiation due to its relative low density (given by a low hot gas pressure). For $0.02 \lesssim \eta_{\rm HE} \lesssim 0.07$, self-shielding is possible only in the central clump and a concentrated second stellar generation is formed. If $\eta_{\rm HE} \gtrsim 0.07$, self-shielding is possible also in infalling streams and the extended second generation is formed. In the small region around $\eta_{\rm HE} = 1$ and $\eta_{\rm ML} = 0$, the wind solution is not bimodal, and no mass accumulation occurs.

4. Summary

We have reviewed properties of the so–called bimodal star cluster wind solution which are interesting in terms of secondary star formation in young massive clusters. We have shown that if the cluster is massive enough, the thermal instability in the hot gas inside the cluster is inevitable. The exact mass limit depends on the cluster parameters, in particular, on the poorly constrained heating efficiency $\eta_{\rm HE}$ and mass loading $\eta_{\rm ML}$. However, with the most conservative values $\eta_{\rm HE} = 1$ and $\eta_{\rm ML} = 0$, the thermal instability and the bimodal regime should always occur for clusters with masses above 10^7 M$_\odot$.

We suggest the cooling winds model as the possible explanation of the multiple stellar generations in massive star clusters. Numerical simulations show that clusters evolving in the bimodal regime accumulate inside them mass reinserted by massive stars in a form of stellar winds. With realistic cluster parameters (particularly $\eta_{\rm HE}$ being not extremely low), the accumulated gas is able to self-shield against the ionising radiation of massive stars and it highly probably leads to secondary star formation. The heating efficiency, $\eta_{\rm HE}$, also determines where the accumulated gas becomes self-shielding and by that whether the second stellar generation is concentrated (low $\eta_{\rm HE}$) or extended throughout the whole cluster (high $\eta_{\rm HE}$). This is interesting in terms of the so–called mass budget problem, because all models explaining the subsequent stellar populations observed in globular clusters as stars formed out of winds and outflows of the first generation stars have to assume that a substantial fraction of the first generation is removed from the cluster. If the second generation is concentrated near the cluster centre, the preferential removal of the first generation is more probable.

Acknowledgements

We acknowledge support by the Czech Science Foundation grant 209/15/06012 and by the institutional project RVO:67985815 of the Astronomical Institute, Academy of Sciences of the Czech Republic.

References

Bastian, N., Lamers, H. J. G. L. M., de Mink, S. E., *et al.* 2013, *MNRAS*, 436, 2398
Bedin, L. R., Piotto, G., Anderson, J., *et al.* 2004, *ApJ Let*, 605, L125
Cantó, J., Raga, A. C., & Rodríguez, L. F. 2000, *ApJ*, 536, 896
Carretta, E., Bragaglia, A., Gratton, R., & Lucatello, S. 2009, *A& A*, 505, 139
Chevalier, R. A. & Clegg, A. W. 1985, *Nature*, 317, 44
Clark, P. C., Glover, S. C. O., & Klessen, R. S. 2012, *MNRAS*, 420, 745
de Mink, S. E., Pols, O. R., Langer, N., & Izzard, R. G. 2009, *A& A*, 507, L1
Decressin, T., Meynet, G., Charbonnel, C., Prantzos, N., & Ekström, S. 2007, *A& A*, 464, 1029
D'Ercole, A., Vesperini, E., D'Antona, F., McMillan, S. L. W., & Recchi, S. 2008, *MNRAS*, 391, 825
Fryxell, B., Olson, K., Ricker, P., *et al.* 2000, *ApJS*, 131, 273
Gilbert, A. M. & Graham, J. R. 2007, *ApJ*, 668, 168
Harper-Clark, E. & Murray, N. 2009, *ApJ*, 693, 1696
Herrera, C. N. & Boulanger, F. 2015, IAU General Assembly, 22, 52184
Jiménez-Bailón, E., Santos-Lleó, M., Mas-Hesse, J. M., *et al.* 2003, *ApJ*, 593, 127
Krause, M., Charbonnel, C., Decressin, T., Meynet, G., & Prantzos, N. 2013, *A& A*, 552, A121
Krause, M., Charbonnel, C., Decressin, T., *et al.* 2012, *A& A*, 546, L5
Larsen, S. S. 2010, Royal Society of London Philosophical Transactions Series A, 368, 867
Leitherer, C., Schaerer, D., Goldader, J. D., *et al.* 1999, *ApJS*, 123, 3
Mucciarelli, A., Bellazzini, M., Ibata, R., *et al.* 2012, *MNRAS*, 426, 2889
Palouš, J., Wünsch, R., Martínez-González, S., *et al.* 2013, *ApJ*, 772, 128
Palouš, J., Wünsch, R., & Tenorio-Tagle, G. 2014, *ApJ*, 792, 105
Pancino, E., Ferraro, F. R., Bellazzini, M., Piotto, G., & Zoccali, M. 2000, *ApJ Let*, 534, L83
Piotto, G., Bedin, L. R., Anderson, J., *et al.* 2007, *ApJ Let*, 661, L53
Piotto, G., Milone, A. P., Bedin, L. R., *et al.* 2015, *AJ*, 149, 91
Portegies Zwart, S. F., McMillan, S. L. W., & Gieles, M. 2010, *ARA& A*, 48, 431
Raga, A. C., Velázquez, P. F., Cantó, J., Masciadri, E., & Rodríguez, L. F. 2001, *ApJ Let*, 559, L33
Rogers, H. & Pittard, J. M. 2013, *MNRAS*, 431, 1337
Silich, S., Tenorio-Tagle, G., & Muñoz-Tuñón, C. 2003, *ApJ*, 590, 791
Silich, S., Tenorio-Tagle, G., & Muñoz-Tuñón, C. 2007, *ApJ*, 669, 952
Silich, S., Tenorio-Tagle, G., & Rodríguez-González, A. 2004, *ApJ*, 610, 226
Silich, S., Tenorio-Tagle, G., Torres-Campos, A., *et al.* 2009, *ApJ*, 700, 931
Strickland, D. K. & Heckman, T. M. 2009, *ApJ*, 697, 2030
Tenorio-Tagle, G., Muñoz-Tuñón, C., Pérez, E., Silich, S., & Telles, E. 2006, *ApJ*, 643, 186
Tenorio-Tagle, G., Silich, S., & Muñoz-Tuñón, C. 2007a, *New Astronomy Reviews*, 51, 125
Tenorio-Tagle, G., Silich, S., Rodríguez-González, A., & Muñoz-Tuñón, C. 2005, *ApJ Let*, 628, L13
Tenorio-Tagle, G., Wünsch, R., Silich, S., Muñoz-Tuñón, C., & Palouš, J. 2010, *ApJ*, 708, 1621
Tenorio-Tagle, G., Wünsch, R., Silich, S., & Palouš, J. 2007b, *ApJ*, 658, 1196
Wünsch, R., Silich, S., Palouš, J., Tenorio-Tagle, G., & Muñoz-Tuñón, C. 2011, *ApJ*, 740, 75
Wünsch, R., Tenorio-Tagle, G., Palouš, J., & Silich, S. 2008, *ApJ*, 683, 683

Formation, evolution, and survival of massive star clusters
Proceedings IAU Symposium No. 316, 2015
C. Charbonnel & A. Nota, eds.

doi:10.1017/S1743921315009692

A critical assessment of models for the origin of multiple populations in globular clusters

Nate Bastian

Astrophysics Research Institute, Liverpool John Moores University,
146 Brownlow Hill, Liverpool L3 5RF, UK
email: N.J.Bastian@ljmu.ac.uk

Abstract. A number of scenarios have been put forward to explain the origin of the chemical anomalies (and resulting complex colour-magnitude diagrams) observed in globular clusters (GCs), namely the AGB, Fast Rotating Massive Star, Very Massive Star, and Early Disc Accretion scenarios. We compare the predictions of these scenarios with a range of observations (including young massive clusters (YMCs), chemical patterns, and GC population properties) and find that all models are inconsistent with observations. In particular, YMCs do not show evidence for multiple epochs of star-formation and appear to be gas free by an age of ~ 3 Myr. Also, the chemical patterns displayed in GCs vary from one to the next in such a way that cannot be reproduced by standard nucleosynthetic yields. Finally, we show that the "mass budget problem" for the scenarios cannot be solved by invoking heavy cluster mass loss (i.e. that clusters were 10-100 times more massive at birth) as this solution makes basic predictions about the GC population that are inconsistent with observations. We conclude that none of the proposed scenarios can explain the multiple population phenomenon, hence alternative theories are needed.

Keywords. globular clusters, multiple populations, young massive clusters

---

1. Introduction

Multiple populations are now established as a near-ubiquitous property of globular clusters in the Galaxy (e.g., Gratton *et al.* 2012) as well as within nearby galaxies where their GCs can be resolved into individual stars (Mucciarelli *et al.* 2011; Larsen *et al.* 2012). Abundance differences between stars in light elements such as He, Na, C, N, O and Al (within an individual cluster) can be found spectroscopically, and their influence on the lifetimes, bolometric corrections and colours of the stars results in complex colour-magnitude diagrams of the clusters. This can be seen as split or spread main sequences, turn-offs, red giant branches and complex horizontal branch morphologies, depending on the filter combinations used (e.g, Sbordone *et al.* 2011).

Stars that are enriched in He, Na, N & Al while being depleted in C and O are often referred to as "2nd generation" (2G) stars, while the stars that have the abundance patterns observed in halo field stars of the same metallicity are often referred to as "1st generation" (1G) stars. This is somewhat misleading, as this is due to the interpretation of the patterns as due to multiple generations of star formation within the cluster, which is now looking as an unlikely explanation for the anomalies. One point to note is that while age differences between the populations is possible (up to a few hundred Myr - e.g., Nardiello *et al.* 2015), age differences are not the reason for the complex CMDs.

A number of scenarios have been put forward in order to explain the origin of the multiple populations within GCs. Most of the scenarios invoke multiple generations of star-formation within globular clusters, where a 2nd generation of stars is formed from material processed through first generation stars (i.e., the stellar ejecta of massive stars or

AGB stars). The models differ as to where the processed material comes from in order to form the 2G. The main candidates are Fast Rotating Massive Stars (FRMS), AGB stars, and Very Mass stars ($> 10^4\ M_{\odot}$). The FRMS and AGB scenarios have been developed in some detail and examples of the scenarios can be found in Decressin *et al.* (2007); D'Ercole *et al.* (2008); Conroy (2012); and Krause *et al.* (2013).

An additional constraint, not discussed in the present work, is that most GCs show small or absent spreads in Fe and other heavy elements. Hence, any successful scenario will need to explain this lack of variation, in particular, how SNe ejecta did not pollute any of the gas if multiple generations of star-formation is invoked (c.f., Renzini 2013).

These scenarios differ in many of their details but make some of the same predictions. In the present work we will compare three of their basic predictions to observations. 1) The models do not invoke any special physics that only operates in the early Universe, so they should be happening in young massive clusters (YMCs) forming today. 2) Since they use the yields of evolved stars, there should be clear chemical patterns (modulo dilution with 'primordial' gas) observed in GC stars (i.e. relations between Na enhancement, O depletion and He spreads). 3) They all suffer from a "mass budget problem" where there is simply not enough material being processed through 1G stars to form the observed number of 2G stars. In order to solve this, these models all suggest that GCs were much more massive at birth (by factors of 10-100 times their present day masses), and invoke heavy mass loss (through gas expulsion or galactic tides) to explain the current fractions of enriched stars. This heavy mass loss should leave a clear imprint on the GC cluster population.

Alternatively, Bastian *et al.* (2013a) suggested a scenario that only entails a single generation of stars to form, and that low-mass stars with protoplanetary discs sweep up material shed from interacting binary or rapidly rotating massive stars. This is referred to as the Early Disc Accretion (EDA) scenario and like the above scenarios, it predicts clear chemical patterns for GC stars today.

In the following sections we compare the models to a variety of observations. No model is found that can satisfy the observational constraints.

2. Constraints from Young Massive Clusters

Of course GC formation at high redshift may have been entirely different than YMC formation happening locally. However, none of the models put forward so far invoke any special physics applicable only to the early Universe, so observations of YMCs can be used to directly test the proposed models. There have been a number of recent works on this subject which we summarise here:

- **A lack of ongoing star-formation within YMCs:** Bastian *et al.* (2013b) studied $\sim$ 130 Galactic and extragalactic YMCs with ages between 10-1000 Myr and masses between $10^4 - 10^8\ M_{\odot}$ looking for evidence of ongoing star-formation within them based on emssion lines. No evidence for secondary star-formation events where found, which is in conflict with predictions of the AGB scenario.
- **The star-formation history (SFH) of a YMC:** Cabrera-Ziri *et al.* (2014) used a high S/N optical spectrum of NGC 34-cluster 1 to estimate its SFH. The cluster was found to be best fit by a single age populations (age$\sim$ 100 Myr; mass$\sim 10^7\ M_{\odot}$). This is in conflict with predictions of the AGB scenario.
- **CMDs of YMCs:** Larsen *et al.* (2011) and Niederhofer *et al.* (2015a) studied resolved CMDs of YMCs (with masses between $10^4 - 10^6\ M_{\odot}$ and ages between a few Myr and 1 Gyr) in comparison with stellar isochrones (essentially working out the SFH of

the clusters). While age spreads could not be ruled out for all clusters (see also Niederhofer *et al.* 2015b) none were consistent with the age spreads expected in the AGB scenario.

• **Young gas free clusters:** Hollyhead *et al.* (2015) and Bastian *et al.* (2014) studied young (< 10 Myr) massive clusters and found that they were gas free with ~ 3 Myr of their formation. This is in conflict with the FRMS scenario which requires left-over gas from the formation of the 1G stars to remain in the cluster for $7 - 20$ Myr.

• **A lack of gas in older clusters** Longmore (2015) showed that in the AGB scenario (as detailed in D'Ercole *et al.* 2008) clusters with ages between 30-100 Myr should have large amounts of gas within them, which will result in heavy extinction within the cluster, even out to 3-5 pc from the cluster centre. This is in basic disagreement with observations of YMCs which often find low extinction for clusters in this age range.

• **ALMA observations of YMCs:** Similarly, Cabrera-Ziri *et al.* (2015) used ALMA to search for gas in and around YMCs in the Antennae galaxies (ages$\sim 50 - 200$ Myr, mass$\sim 10^6$ $M_\odot$) which is predicted to remain in or near the clusters. No such gas was found, down to limits of $< 10\%$ of the stellar mass, in conflict with the AGB model.

Observations of massive stellar clusters forming today are inconsistent with the AGB and FRMS scenarios. At the very least the proposed scenarios would not be generally applicable to all massive clusters, but instead would need some additional criteria that is only in operation in the early Universe. At worst, the basic idea of multiple generations of star-formation within massive clusters is not valid.

3. Constraints from Chemical Abundance Patterns

All of the proposed models use the yields of certain stars (AGB, FRMS, VMS, or interacting binary stars), together with the abundance patterns observed in the 'primordial' population (i.e. those stars with abundance patterns similar to the field) to make predictions for the expected trends in certain elements. For example, by diluting the yields of AGB stars with gas with the abundance patterns of the primordial stars, one expects an anti-correlation between Na and O (whereas the yields themselves predict a correlation). This also applies to other elements, so we can test, for example, if the predicted spreads in Na and O are consistent with the observed spreads in He. This is a key constraint, as basic nucleosynthesis and stellar evolution predict clear correlations between these three elements. See Bastian *et al.* (2015) for a further description of the tests and expectations.

In Fig. 1 we show such dilution models for the AGB scenario, using the yields of AGB stars from Ventura *et al.* (2013). These represent some of the most optimistic yields for AGB stars if they are to be the origin of the anomalies seen in GCs †. The solid lines show the yields for the AGB stars of different masses, and the dashed lines show the dilution model for each mass. The dash-dotted lines show regions of constant He. So, for a given position in the Na-O plane, one should be able to explicitly predict the He abundance for that star (relative to the "primordial" stars). By looking at the most extreme stars in the Na-O plane, the maximum He spread for that cluster can be estimated. The filled (red) circles show data for stars within a given GC. Finally, for each GC, the allowed range in the Na-O plane for cluster stars, for the measured He spread within the cluster, is shown as a filled region.

It is clear that all the GCs have significant numbers of stars outside the allowed range. In fact, most of the GCs have similar spreads in Na and O, while having extremely

† Compare these yields with the AGB yields of Doherty *et al.* 2014, who conclude that AGB stars cannot be the origin of the chemical anomalies in GCs.

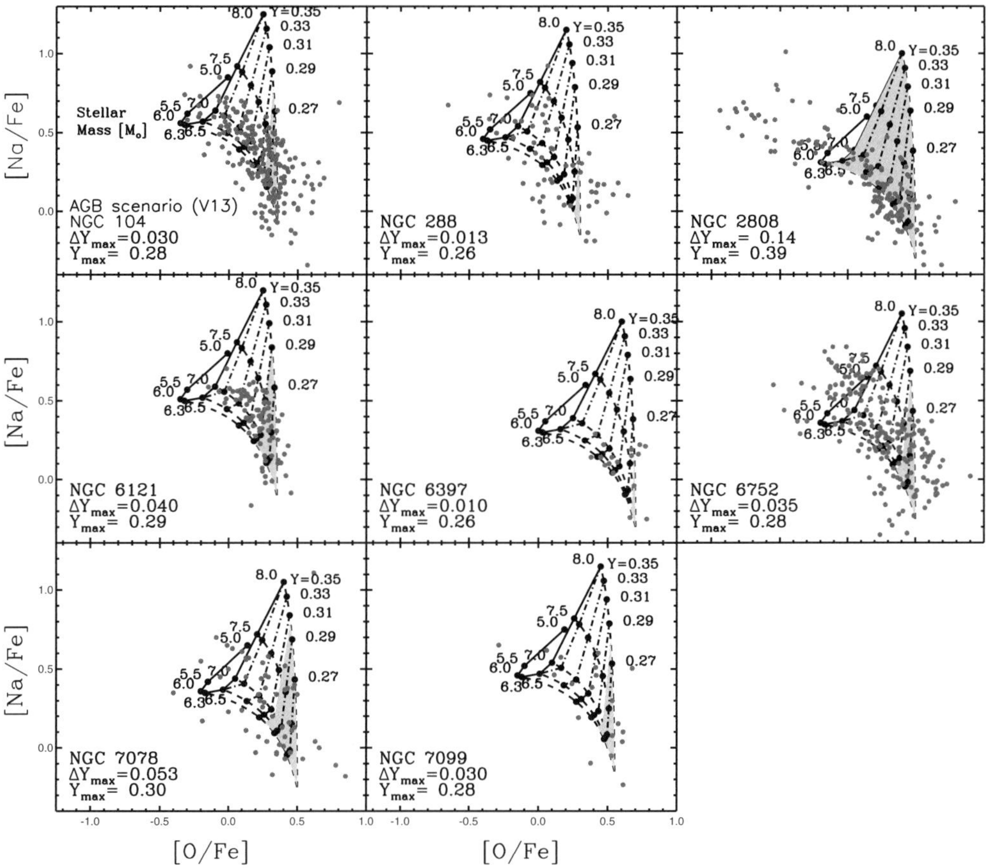

Figure 1. Dilution models for the AGB scenario using the yields of Ventura *et al.* (2013). Solid lines show the yields, while dashed lines show the dilution model adopting the yields for each stellar mass. Lines of constant He enrichment are shown as dash-dotted lines. For a given He spread (ΔY) the allowed range of Na-O values for these yields is shown as a shaded region. Note that all clusters violate this constraint. The same applies to all AGB yields in the literature as well as the yields of rotating massive stars, interacting massive binaries or very massive stars. Taken from Bastian *et al.* (2015).

different He contents. Hence, these specific yields are not consistent with the observations. We have also compared the observations (not shown) with the yields of FRMSs (Decressin *et al.* 2007), a number of other available AGB models, Very Massive Stars (Denissenkov & Hartwick 2014), and massive interacting binaries (de Mink *et al.* 2009). All of these sources are at odds with observations, with the general problem that they all over-predict the amount of He spread for a given Na-O spread. **Hence, none of the yields put forward in the literature so far are consistent with the observations.**

Additionally, the yields of AGBs and FRMS are expected to be metallicity dependent, so that trends with the type of Na-O spreads observed should vary as a function of metallicity. To date, no such variations have been found (e.g., Bastian *et al.* 2015).

However, the problem is in fact much worse than the yields being wrong. Bastian *et al.* (2015) also made empirical yields, by taking the observed Na-O spread of a given cluster and associating this with the maximum allowed spread in He. By comparing the "empirical yield" from each cluster to others, it is clear that all clusters are different, i.e.

if a hypothetical source could explain the trends in one cluster, say 47 Tuc, they would not be able to fit any other cluster, for example, NGC 2808.

Hence, a large range in stochasticity is required in the yields, if nucleosynthesis is the source of the enriched material. However, none of the suggested sources (AGBs, FRMS, etc) will be able to produce the required amount stochasticity.

The need for large amounts of stochasticity in the source of the enriched material effectively rules out all the proposed sources (e.g., AGB, FRMS, interacting massive binaries, VMS, etc).

4. Constraints from Globular Cluster Population Properties

Next we consider constraints on the models from the properties of the GC population in general. As discussed in the introduction, scenarios that invoke the formation of a second generation of stars from the ashes (i.e. nuclear processed material) of a first generation all suffer from a severe mass budget problem. This is because AGB or massive stars contribute very little to the overall mass of the cluster (due to the stellar IMF), so they can only process a rather limited amount of material. On the contrary, enriched stars are observed to make up $> 50\%$ of stars within GCs (whereas the expected fraction would be $1 - 10\%$ depending on the source and efficiency of the process). In order to solve this basic mass budget problem, it has been suggested that GCs were much more massive at birth and have preferentially lost most of the 1G stars within them.

In order to do this, the scenarios have invoked cluster expansion due to stellar mass loss in a mass segregated cluster. As the clusters expand, their outer regions will be lost due to galactic tides (as the 2G stars are expected to be more centrally concentrated), changing the enriched fraction ($\mathrm{F_{enrich}} = N_{\mathrm{enriched}}/N_{\mathrm{total}}$) from the initial value of ~ 0.05 to the observed value of > 0.5. This can only work if the galactic tidal field is strong enough, and due to this, scenarios that invoked this mechanism have put the young GCs at small Galactocentric radii (~ 4 kpc) in order to maximise the loss of 1G stars.

A clear prediction of this kind of GC dissolution is that clusters that experience weaker tidal fields (e.g., clusters at larger Galactocentric radii) should have a lower fraction of $\mathrm{F_{enrich}}$, tending towards the initial value. Bastian & Lardo (2015) have searched for such a relation using 34 GCs with abundance measurements (or HST imaging that can separate the different populations within GCs with certain filter combinations) and the results are shown in Fig. 2. The data show no correlation between $\mathrm{F_{enrich}}$ and Galactocentric distance, while the expected relation in such a heavy mass loss scenario is shown as a solid line. In fact, the data are consistent with a constant $\mathrm{F_{enrich}} = 68 \pm 7\%$. The predictions clearly inconsistent with this, suggesting that this mass-loss mechanism cannot be responsible for the heavy mass loss required in such multiple-generation scenarios.

Khalaj & Baumgardt (2015) have studied another potential mass loss mechanism, the rapid expulsion of gas from young GCs, which also causes the clusters to expand. This type of mass loss is also expected to result in a strong relation between $\mathrm{F_{enrich}}$ and Galactocentric distance, but also a strong relation with GC mass. This is tested in the middle panel of Fig. 2, and is again incompatible with the data, in agreement with the conclusions of Khalaj & Baumgardt (2015). Finally, the in right panel we show the observed relation between $\mathrm{F_{enrich}}$ and GC metallicity, and again there is no correlation. This suggests that the location and age of the GC does not affect the fraction of enriched stars.

Finally, Kruijssen (2015) have shown that because the tidal environment experienced by a GC is dependent on the galaxy host of which it forms, which in turn is related to the metallicity of the host galaxy (due to the mass-metallicity relation), $\mathrm{F_{enrich}}$ would be

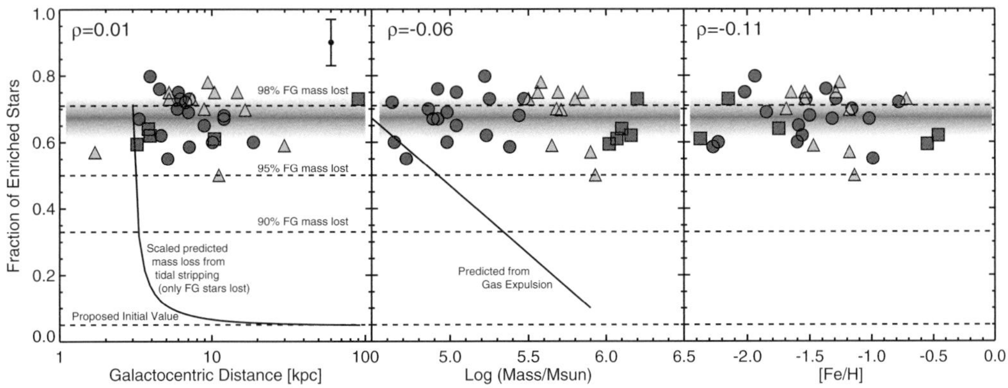

Figure 2. The fraction of enriched stars (F_{enrich}) as a function of Galactocentric distance, GC mass and metallicity in a sample of 34 Galactic GCs. The lines in the left and centre panels denote predictions of mass loss mechanisms that have been invoked to make GCs lose significant amounts of stellar mass in order to solve the "mass budget problem". The constancy of F_{enrich} is inconsistent with any of the scenarios suggested so far. Taken from Bastian & Lardo (2015).

expected to be low for low-metallicity GCs and increase for high-metallicity GCs. This is clearly not compatible with the data, as shown in the right panel of Fig. 2. The author concludes that GCs were, on average, only a factor of $2-3\times$ more massive at birth than they are presently.

Taken together, this suggests that if GCs did undergo strong mass loss, they must all have lost between 95 and 99% of their initial masses (assuming only 1G stars are lost) independent of the mass, Galactocentric distance, mass, age, location at birth or subsequent migration. No mass loss mechanism proposed to date can match these criteria. Hence, we conclude that GCs did not in fact undergo heavy mass loss and that the observed fraction of enriched stars is representative of their initial value. This effectively rules out standard nucleosynthesis as the origin of the chemical anomalies.

5. Other constraints on the Mass Budget

Larsen *et al.* (2012; 2014) have investigated the GC properties in a sample of three dwarf galaxies, namely the Fornax, WLM and IKN dwarf galaxies. These galaxies have very high specific frequencies (the fraction of light in GCs relative to the field), especially at low metallicities. The authors found that below [Fe/H]$= -2$, GCs make up between 10-30% of the total stellar mass of the galaxy. This is incompatible with the idea of heavy cluster mass loss, as the stars lost from the GCs would share the same metallicity as the GCs and make up a substantial fraction of the field. Fornax, in particular, is a very isolated galaxy and is not expected to have undergone any strong interactions which may have resulted in the loss of stars from the galaxy. In any case, it is very improbably to strip field stars from a galaxy without stripping GCs as well.

Schiavon *et al.* (2015) have found, based on the APOGEE survey of stars in the Galactic bulge, a relatively high percentage of "enriched" stars in the field, relative to primordial stars. Under the extreme assumption that the all the field stars are from dissolved GCs, the authors find a lower limit to F_{enrich} of 0.07, in tension with model predictions. Under more realistic assumptions, this value increases substantially.

	YMCs	Abundances	Variety/stochasticity	Mass budget	He spread correlation with GC mass	Constant $F_{enriched}$	Lack of trends with metallicity
AGB	✗	✗	✗	✗	✗	✗	✗
FRMS	✗	✗	✗	✗	✗	✗	✗
VMS	✓	✗	✗	✗	✗	✗	✗
EDA	✓	✗	✗	✗	✗	✗	✗

Figure 3. A summary of the results presented in the current work. A green check means that a given constraint does not directly violate a prediction/requirement of a given scenario, while a red cross means that the observational constraints directly contradict a prediction/requirement of a scenario.

The observations by Larsen *et al.* and Schiavon *et al.* directly contradict a basic (and necessary) premise of scenarios that invoke multiple generations of star-formation within GCs, that heavy loss of 1G stars is required to solve the "mass budget problem".

Finally, it has been found that integrated [N/Fe] of GCs correlates with GC mass in M31 (Schiavon *et al.* 2013; Colucci *et al.* 2014; and likely in the Galaxy as well) and that the spread of He (ΔY) within Galactic GCs also is proportional to mass (Milone 2015). This is inconsistent with self-enrichment scenarios, as the amount of N or He produced should be constant per unit stellar mass (as the IMFs are fully sampled). In order to accommodate the observations, low mass clusters must have lost substantial fractions of the total amount of material processed through 1G stars. This further exacerbates the mass budget problem by a factor of 3 or more (Bastian *et al.* 2015).

6. Summary

The constraints presented here effectively rule out all the main scenarios for the formation of multiple populations within GCs. First, we have shown that observations of young massive clusters are incompatible with predictions from the FRMS and AGB scenarios, so at the very least an additional component is needed in these models to make them only operate in the early Universe, and at worst this shows that the basic idea behind multiple generations of stars forming in the same cluster (outside nuclear clusters) is not valid.

Secondly, none of the proposed nucleosynthetic sources can produce the observed chemical patterns in GC stars. They cannot 1) produce the observed variations from GC to GC (i.e., a large amount of stochasticity is required that none of the sources can provide) and 2) all sources predict a strong correlation between the extent of the Na-O anti-correlation and the He spread within the GC, in contrast with observations.

Finally, the observed fraction of enriched stars within GCs is remarkably constant, $F_{\rm enrich} = 68 \pm 7\%$, independent of a cluster's mass, galactocentric distance, metallicity, location of birth or subsequent migration. This is incompatible with basic predictions of a scenario where GCs preferentially lost a large fraction of their initial mass (i.e., lost large amounts of FG stars). Hence, the mass budget problem cannot be solved by invoking a scenario where GCs lost large amounts of mass, confirming previous results from studies of GCs and field stars in dwarf galaxies.

The chemical patterns observed in GCs, when combined with limits on the mass budget of GCs, rule out standard nucleosynthesis the process responsible for the observed abundance anomalies. Hence, with the exclusion of all current models, new scenarios are desperately needed.

We conclude with a visual summary of the constraints discussed in the present work which is shown in Fig. 3.

References

Bastian, N., Lamers, H. J. G. L. M., de Mink, S. E., *et al.* 2013a, *MNRAS*, 436, 2398
Bastian, N., Cabrera-Ziri, I., Davies, B., & Larsen, S. S. 2013b, *MNRAS*, 436, 2852
Bastian, N., Hollyhead, K., & Cabrera-Ziri, I. 2014, *MNRAS*, 445, 378
Bastian, N., Cabrera-Ziri, I., & Salaris, M. 2015, *MNRAS*, 449, 3333
Bastian, N. & Lardo, C. 2015, *MNRAS*, 453, 357
Cabrera-Ziri, I., Bastian, N., Davies, B., *et al.* 2014, *MNRAS*, 441, 2754
Cabrera-Ziri, I., Bastian, N., Longmore, S. N., *et al.* 2015, *MNRAS*, 448, 2224
Colucci, J. E., Bernstein, R. A., & Cohen, J. G. 2014, *ApJ*, 797, 116
Conroy, C. 2012, *ApJ*, 758, 21
D'Ercole, A., Vesperini, E., D'Antona, F., McMillan, S. L. W., & Recchi, S. 2008, *MNRAS*, 391, 825
Decressin, T., Meynet, G., Charbonnel, C., Prantzos, N., & Ekström, S. 2007, *A&A*, 464, 1029
Denissenkov, P. A. & Hartwick, F. D. A. 2014, *MNRAS*, 437, L21
de Mink, S. E., Pols, O. R., Langer, N., & Izzard, R. G. 2009, *A&A*, 507, L1
Doherty, C. L., Gil-Pons, P., Lau, H. H. B., *et al.* 2014, *MNRAS*, 441, 582
Gratton, R. G., Carretta, E., & Bragaglia, A. 2012, A&ARv, 20, 50
Hollyhead, K., Bastian, N., Adamo, A., *et al.* 2015, *MNRAS*, 449, 1106
Khalaj, P. & Baumgardt, H. 2015, *MNRAS*, 452, 924
Krause, M., Charbonnel, C., Decressin, T., Meynet, G., & Prantzos, N. 2013, *A&A*, 552, A121
Kruijssen, J. M. D. 2015, *MNRAS*, in press (arXiv:1509.02163)
Larsen, S. S., de Mink, S. E., Eldridge, J. J., *et al.* 2011, *A&A*, 532, A147
Larsen, S. S., Strader, J., & Brodie, J. P. 2012, *A&A*, 544, L14
Larsen, S. S., Brodie, J. P., Grundahl, F., & Strader, J. 2014, *ApJ*, 797, 15
Longmore, S. N. 2015, *MNRAS*, 448, L62
Milone, A. P. 2015, *MNRAS*, 446, 1672
Mucciarelli, A., Origlia, L., Ferraro, F. R., & Pancino, E. 2009, *ApJL*, 695, L134
Nardiello, D., Piotto, G., Milone, A. P., *et al.* 2015, *MNRAS*, 451, 312
Niederhofer, F., Hilker, M., Bastian, N., & Silva-Villa, E. 2015a, *A&A*, 575, A62
Niederhofer, F., Georgy, C., Bastian, N., & Ekström, S. 2015b, *MNRAS*, 453, 2070
Sbordone, L., Salaris, M., Weiss, A., & Cassisi, S. 2011, *A&A*, 534, A9
Schiavon, R. P., Caldwell, N., Conroy, C., *et al.* 2013, *ApJ*, 776, L7
Schiavon, R. P., *et al.* 2015, *MNRAS* (submitted)
Renzini, A. 2013, *Mem. S. A. It.*, 84, 162
Ventura, P., Di Criscienzo, M., Carini, R., & D'Antona, F. 2013, *MNRAS*, 431, 3642

Formation, evolution, and survival of massive star clusters
Proceedings IAU Symposium No. 316, 2015
C. Charbonnel & A. Nota, eds.

doi:10.1017/S1743921315010650

Multiple populations in globular clusters: constraints from kinematics and dynamics

Vincent Hénault-Brunet
Department of Physics, University of Surrey,
GU2 7XH, Guildford, UK
email: v.henault-brunet@surrey.ac.uk

Abstract. We discuss constraints on the formation of multiple populations in globular clusters (GCs) imposed by their present-day kinematics (velocity dispersion and anisotropy) and spatial distribution. We argue that the observational evidence collected so far in the outer parts of clusters is generally consistent with an enriched population forming more centrally concentrated compared to the primordial population, in agreement with all the scenarios proposed to date (in some cases by design), but not sufficient to favour a particular scenario. We highlight that the differential rotation of subpopulations is a signature that may provide crucial new constraints and allow us to distinguish between various scenarios. Finally, we discuss the spatial distribution of subpopulations in the central regions of GCs and speculate that mass segregation between subpopulations may be due to a difference in their binary fraction.

Keywords. star clusters, globular clusters: general, stars: kinematics, Galaxy: kinematics and dynamics

1. Introduction

It is convenient to split the problem of multiple populations in globular clusters (GCs) into two challenges or questions. First, what is producing the chemical abundance anomalies giving rise to observable features such as the Na-O anti-correlation and multiple sequences in the colour-magnitude diagrams of GCs (e.g. Gratton *et al.* 2012, Piotto *et al.* 2012)? But another important challenge is to understand how this polluted material makes its way into a large fraction of the cluster stars (the observed enriched fraction is $\sim 68\%$ today; see e.g. Bastian & Lardo 2015). Whether it is through multiple generations of star formation (e.g. D'Ercole *et al.* 2008), via accretion onto low-mass pre-main sequence stars born as part of the same burst of star formation (Bastian *et al.* 2013), or via a yet unknown alternative process (as suggested by Bastian *et al.* 2015), the mechanism by which enriched material is incorporated into stars will dictate the initial configuration of the enriched (i.e. polluted) stars compared to the primordial (i.e. normal) stars. This will set the initial conditions for the long-term dynamical evolution of the cluster and its multiple populations, which is the focus of this contribution.

Because GCs are collisional systems, their long-term dynamical evolution leads to two phenomena that can affect the present-day kinematics and spatial distribution of subpopulations: phase-space mixing and mass segregation. Due to two-body encounters, populations of stars with different initial phase-space distributions will exchange energy and angular momentum and gradually mix (Decressin *et al.* 2008, Vesperini *et al.* 2013, Hénault-Brunet *et al.* 2015). The trend towards kinetic energy equipartition also leads massive stars and remnants to slow down and sink towards the centre of the cluster, effectively pushing the lower-mass stars outwards. Because the two-body relaxation timescale (on which these processes become important) is inversely proportional to the local density (e.g. Spitzer 1987), the importance of these processes will vary significantly between different regions of the cluster. In the outer regions, where the relaxation timescale is longer, there will be less mixing and we may expect to still see imprints of the initial conditions and formation process. In the inner regions, the shorter relaxation timescale leads to more mixing, potentially erasing any initial differences between subpopulations, and mass segregation can also proceed more efficiently and introduce differences in the spatial distribution and kinematics of different mass species. In the following sections, we review constraints on the formation of multiple populations imposed by their present-day kinematics and spatial distribution.

2. Spatial distribution & kinematics of subpopulations in the outer regions of GCs

N-body simulations of clusters with two subpopulations starting spatially segregated and with comparable numbers of stars have shown that complete spatial mixing is expected when about 60 – 70% of the initial mass of the cluster has been lost due to two-body relaxation (Vesperini *et al.* 2013). This also applies to the distinct kinematics of different subpopulations, which will be erased on a similar timescale (e.g. Decressin *et al.* 2008, Hénault-Brunet *et al.* 2015). Given their current mass and galactocentric radius, a large fraction of Galactic GC are expected to still be expanding within their tidal radius and to have lost less than 50% of their initial mass by relaxation-driven evaporation (Gieles *et al.* 2011), so some memory of the initial configuration of subpopulations should be preserved in the outer parts of many GCs. And indeed, when looking at the spatial distribution of subpopulations in the outer regions of clusters (from around the half-mass radius - r_h - and beyond), a more centrally concentrated enriched population has been found in many clusters (e.g. Sollima *et al.* 2007, Bellini *et al.* 2009, Lardo *et al.* 2011, Milone *et al.* 2012), suggesting that enriched stars formed more centrally concentrated. That said, evidence for a fully mixed cluster has also been reported (NGC 6362; Dalessandro *et al.* 2014) and full mixing may be more common than previously thought (Vanderbeke *et al.* 2015; although see their caveats about using Horizontal Branch stars as tracers).

Kinematic differences have been found in some cases in the outer regions of clusters, with the (Na) enriched population having a lower velocity dispersion (Bellazzini *et al.* 2012, Kučinskas *et al.* 2014). The more centrally concentrated population is generally expected to be dynamically cooler (see e.g. Bekki 2011; Mastrobuono-Battisti & Perets 2013; Hénault-Brunet *et al.* 2015), regardless of its initial configuration (flattened disc-like or spherical subsystem), so again this is consistent with an enriched population that started more centrally concentrated. Using Hubble Space Telescope (HST) proper motions of stars near $2\,r_h$ in 47 Tuc, Richer *et al.* (2013) found that the presumably He-enriched stars (bluer main-sequence stars) have a more radially anisotropic velocity distribution. Bellini *et al.* (2015) observed a similar behaviour for stars between ~ 1.5 and $2\ r_h$ from HST proper motions in NGC 2808, with stars coinciding with the presumably He-enchanced populations in the middle and blue main sequences showing a smaller tangential velocity dispersion than the populations corresponding to primordial stars, while no significant differences were found in the radial-velocity dispersion. We showed in Hénault-Brunet *et al.* (2015) that a more radially anisotropic velocity distribution for the polluted stars is also predicted if this population started more centrally concentrated and then diffused outward, and it is by no means a unique signature of any of the scenarios proposed. For example, whether gas collected into the core of the cluster to form a new generation of polluted stars (possibly adopting a flattened initial distribution if the cluster had some net rotation; e.g. D'Ercole *et al.* 2008, Bekki 2010), or if the

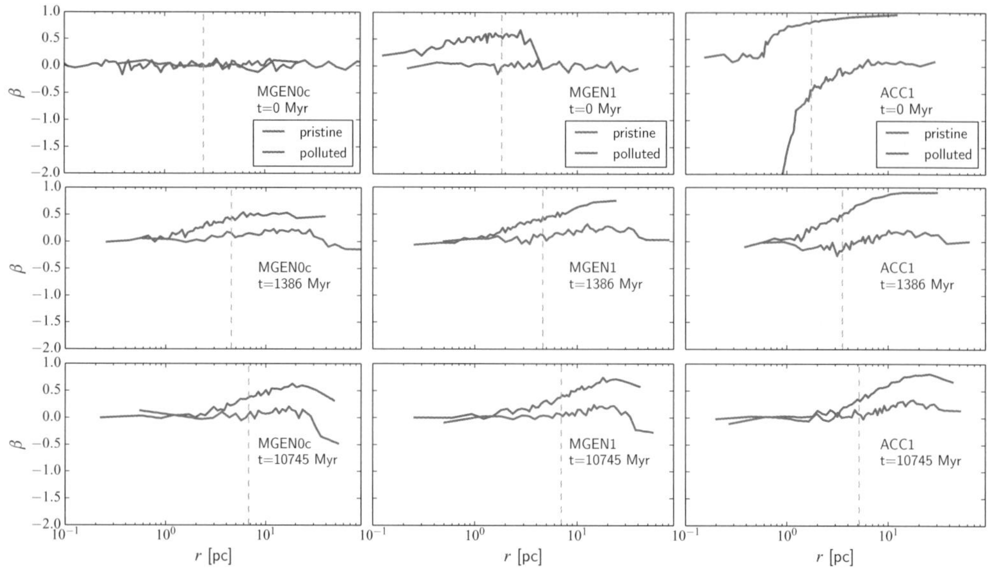

Figure 1. Velocity anisotropy (β) as a function of radius (in 3D) for the polluted (in blue) and pristine (i.e. primordial composition - in red) stars as a function of time, for three different initial configurations: a spherical and isotropic subsystem of polluted stars embedded in a more extended spherical cluster (model MGEN0c - left), a flattened rotating subsystem of polluted stars embedded in a non-rotating spherical cluster (model MGEN1 - centre), a rotating cluster where the polluted stars are flagged as the ones with an orbit crossing the cluster core at $t = 0$ (model ACC1 - right). The half-mass radius is indicated by grey dashed lines. For more details on these simulations see Hénault-Brunet *et al.* (2015), from which this figure is adapted.

polluted stars are the ones initially crossing the core of the cluster and preferentially on radial orbits (e.g. as in the early disc accretion scenario of Bastian *et al.* 2013), any remaining present-day velocity anisotropy signature is expected to be very similar (see Fig. 1).

3. Differential rotation as a potential way to distinguish between formation scenarios

All the observational evidence collected so far in the outer parts of GCs points to an enriched population forming more centrally concentrated, but this does not help to discard or favour any particular formation mechanism since all current scenarios either predict such a trend or were designed to match these observational constraints. To make progress, we need to identify unique imprints that would allow to distinguish between various scenarios. For example, Mastrobuono-Battisti & Perets (2013) proposed to look for morphological features such as a larger flattening of the enriched population, which is a possible outcome of a scenario with multiple generations of stars (although this may prove challenging to measure). We recently suggested the differential rotation of subpopulations as a promising kinematic signature (Hénault-Brunet *et al.* 2015). Given some net angular momentum initially, models for which a second generation forms from gas that collects in a cooling flow into the core of the cluster predict an initially larger rotational amplitude for the polluted stars compared to the primordial stars (a consequence of angular momentum conservation). Hypothetically removing $> 90\%$ of first-generation stars (in order to alleviate the "mass budget" problem plaguing scenarios with multiple

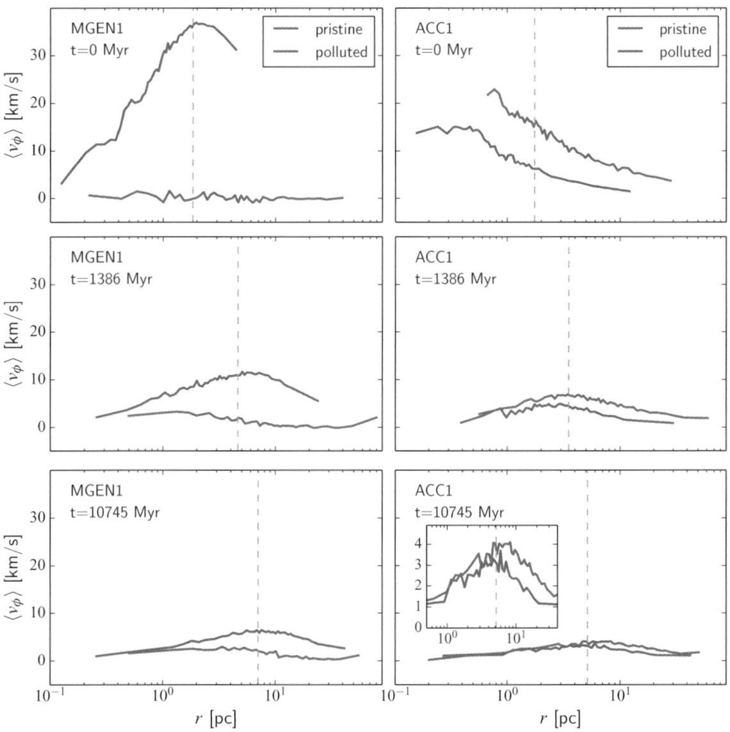

Figure 2. Mean azimuthal velocity as a function of radius - i.e. rotation curve - (in 3D) for the polluted (blue) and primordial (red) stars as a function of time, for two of the models also considered in Fig. 1: a flattened rotating subsystem of polluted stars embedded in a non-rotating spherical cluster (model MGEN1 - left), a rotating cluster where the polluted stars are flagged as the ones with an orbit crossing the cluster core at $t = 0$ (model ACC1 - right). Adapted from Hénault-Brunet *et al.* (2015).

generations of star formation) would also remove a large fraction of the angular momentum of the more extended primordial population (see Hénault-Brunet *et al.* 2015). The opposite trend is expected from a scenario in which the polluted stars are the ones initially crossing the core of the cluster (e.g. Bastian *et al.* 2013) and preferentially on radial (low-angular momentum) orbits. In this case, the rotational amplitude of the polluted stars would be lower initially (assuming the cluster has some net rotation). This is illustrated in Fig. 2, where we also show that a differential rotation between the polluted and primordial stars of the order of 1 km s^{-1} or larger can remain in the outer parts of a cluster after a Hubble time of dynamical evolution (the simulations shown are scaled to match the properties of 47 Tuc). Marginal evidence for differential rotation between subpopulations has been reported in a few clusters (e.g. NGC 6441 and NGC 6388; Bellazzini *et al.* 2012), with polluted stars displaying a smaller rotational amplitude, but these are based on relatively small datasets and will need to be confirmed with more observations.

4. Mass segregation and the spatial distribution of subpopulations in the centre of GCs

Larsen *et al.* (2015) found a surprising result when investigating the spatial distributions of subpopulations in M15 using HST/WFC3 photometry of red giants. Unlike

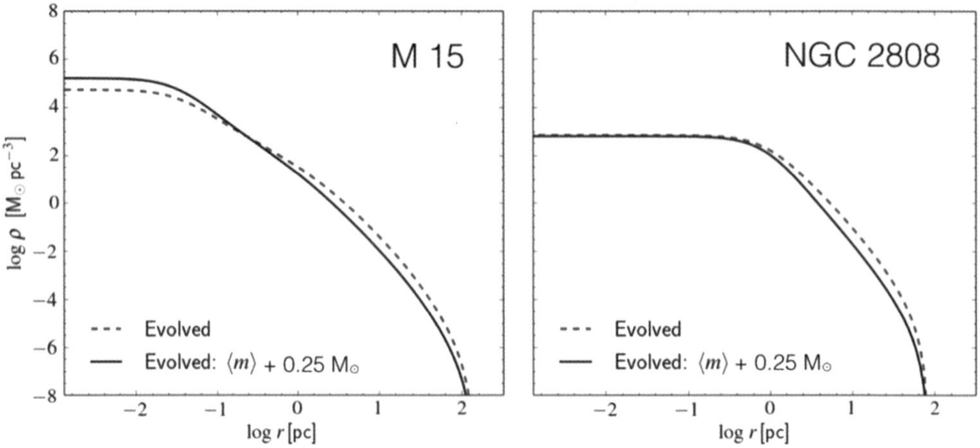

Figure 3. Predicted mass density profile of two tracer populations of evolved stars with a mean stellar mass differing by 0.25 $M_\odot$ in GCs M15 and NGC 2808.

what is typically observed in the outer regions of clusters (including M15), they found that the giants with primordial composition are more centrally concentrated in the inner parts of the cluster (within the half-light radius). They considered mass segregation as a possible culprit for the observed trend, but showed with N-body simulations that a significant mean mass difference would be needed, with primordial giants required to be ~ 0.25 $M_\odot$ more massive than their enriched counterparts. If interpreted as a difference in helium content, this mean mass difference would imply extreme He enhancement ($Y \gtrsim 0.40$), which is clearly incompatible with the colour-magnitude diagram of M15. Larsen *et al.* therefore concluded that the radial trends seen in M15 presumably reflect initial conditions, but we consider below another possibility.

A larger binary fraction has been inferred for primordial stars compared to enriched stars ($\sim 5-12\%$ vs. $\sim 1\%$; D'Orazi *et al.* 2010, Lucatello *et al.* 2015) around the half-mass radii of GCs, in keeping with the idea that the enriched stars were born in a denser environment where (loose) binaries are more efficiently destroyed (e.g. Hong *et al.* 2015). In the core of clusters, the binary fraction can increase to be as high as $40-50\%$ (e.g. Hurley *et al.* 2007), so this may account for the large mean mass difference needed to explain the results of Larsen *et al.* (2015). Whether or not this can work in detail however remains to be shown, and it will depend on the complex interplay between many factors: the initial binary properties and spatial distribution of subpopulations, the rate of creation and disruption of binaries, the amount of dynamical evolution, the structure of the cluster. Studies of the spatial distribution of subpopulations in the central regions of other GCs should also help to clarify this picture, but we should keep in mind that the perceived effect of mass segregation on the distribution of different mass species will vary from cluster to cluster. For example, this effect would be stronger in more concentrated clusters like M15 in comparison to clusters with a very large core over half-light radius like NGC 2808. We illustrate this in Fig. 3 by showing, for M15 and NGC 2808, the mass density profile (in 3D) predicted for two tracer populations of evolved stars with a mean stellar mass differing by 0.25 $M_\odot$, when assuming a fully mass segregated cluster. The underlying potential was constrained by fitting multi-mass models from Gieles & Zocchi (2015) to the surface brightness and velocity dispersion profiles of these clusters in order to model their mass distribution.

5. Summary

We emphasised that observations of the kinematics and spatial distribution of multiple populations around the half-mass radius of GCs and beyond are consistent with the enriched population forming more centrally concentrated. In contrast, the trend seen in the inner regions of M15 appears to clash with this interpretation and remains difficult to explain by mass segregation. We speculated that a larger binary fraction for the primordial population in the centre of M15 may reconcile these findings. We finally suggest possible future avenues to make progress in understanding the formation of multiple populations in GCs through their long-term kinematics and dynamics: (1) constrain any differential rotation of subpopulations, (2) quantify their central spatial distributions in a larger sample of GCs to compare with the puzzling results in M15, and (3) measure the respective binary fractions of subpopulation in the central regions of clusters.

Acknowledgements

I would like to thank Mark Gieles and Alice Zocchi for useful comments during the preparation of this contribution. I also thank the IAU for a travel grant.

References

Bastian, N., Lamers, H. J. G. L. M., de Mink, S. E., *et al.* 2013, *MNRAS*, 436, 2398
Bastian, N., Cabrera-Ziri, I., & Salaris, M. 2015, *MNRAS*, 449, 3333
Bastian, N. & Lardo, C. 2015, *MNRAS*, 453, 357
Bekki, K. 2010, *ApJL*, 724, L99
Bekki, K. 2011, *MNRAS*, 412, 2241
Bellazzini, M., Bragaglia, A., Carretta, E., *et al.* 2012, *A&A*, 538, A18
Bellini, A., Piotto, G., Bedin, L. R., *et al.* 2009, *A&A*, 507, 1393
Dalessandro, E., Massari, D., Bellazzini, M., *et al.* 2014, *ApJL*, 791, L4
Decressin, T., Baumgardt, H., & Kroupa, P. 2008, *A&A*, 492, 101
D'Ercole, A., Vesperini, E., D'Antona, F., *et al.* 2008, *MNRAS*, 391, 825
D'Orazi, V., Gratton, R., Lucatello, S., *et al.* 2010, *ApJL*, 719, L213
Gieles, M., Heggie, D. C., & Zhao, H. 2011, *MNRAS*, 413, 2509
Gieles, M. & Zocchi, A. 2015, *MNRAS*, 454, 576
Gratton, R. G., Carretta, E., & Bragaglia, A. 2012, *A&ARv*, 20, 50
Hénault-Brunet, V., Gieles, M., Agertz, O., & Read, J. I. 2015, *MNRAS*, 450, 1164
Hong, J., Vesperini, E., Sollima, A., *et al.* 2015, *MNRAS*, 449, 629
Hurley, J. R., Aarseth, S. J., & Shara, M. M. 2007, *ApJ*, 665, 707
Kučinskas, A., Dobrovolskas, V., & Bonifacio, P. 2014, *A&A*, 568, L4
Lardo, C., Bellazzini, M., Pancino, E., *et al.* 2011, *A&A*, 525, A114
Larsen, S. S., Baumgardt, H., Bastian, N., *et al.* 2015, *ApJ*, 804, 71
Lucatello, S., Sollima, A., Gratton, R., *et al.* 2015, arXiv:1509.05014
Mastrobuono-Battisti, A. & Perets, H. B. 2013, *ApJ*, 779, 85
Milone, A. P., Piotto, G., Bedin, L. R., *et al.* 2012, *ApJ*, 744, 58
Piotto, G., Milone, A. P., Anderson, J., *et al.* 2012, *ApJ*, 760, 39
Richer, H. B., Heyl, J., Anderson, J., *et al.* 2013, *ApJL*, 771, L15
Sollima, A., Ferraro, F. R., Bellazzini, M., *et al.* 2007, *ApJ*, 654, 915
Spitzer, L. 1987, Princeton University Press, 191 p.
Vanderbeke, J., De Propris, R., De Rijcke, S., *et al.* 2015, *MNRAS*, 451, 275
Vesperini, E., McMillan, S. L. W., D'Antona, F., & D'Ercole, A. 2013, *MNRAS*, 429, 1913

Formation, evolution, and survival of massive star clusters
Proceedings IAU Symposium No. 316, 2015
C. Charbonnel & A. Nota, eds.

doi:10.1017/S1743921315009308

Binary interactions and multiple stellar populations in globular clusters

Dengkai Jiang[1,2], Zhanwen Han[1,2] and Lifang Li[1,2]

[1]Yunnan Observatories, Chinese Academy of Sciences, P.O. Box 110, Kunming, Yunnan Province, 650011, China

[2]Key Laboratory for the Structure and Evolution of Celestial Objects, Chinese Academy of Sciences, Kunming, 650011, China
email: dengkai@ynao.ac.cn

Abstract. Globular clusters (GCs) have multiple stellar populations, which show star-to-star abundance variations and multiple sequences (or spreads) in the Hertzsprung-Russell diagrams. It is explained by multiple generations of star-formation in GCs. However, the observed evidence of ongoing star-formation was not found within any clusters. Here we present a binary interactions scenario for the formation of multiple stellar populations in GCs, where GC stars were born in a single burst of star formation, but some of them are members of binary systems. Binary interactions can produce peculiar stars, e.g. the merged stars and the accretor stars. They are more massive than normal single stars in the same evolutionary stage, and they are rapidly rotating stars at the moment of their formation. Rotationally induced mixing can cause the variations of their surface chemical composition. This results in the single-generation GCs showing abundance anomalies.

Keywords. globular clusters: general, binary: general, stars: evolution, stars: rotation

1. Introduction

Globular clusters (GCs) are the oldest objects in our Galaxy, and they are gravitationally bound systems of thousand to about 10^6 stars. In the traditional view, all stars in the same GCs are commonly accepted to born at the same time and out of the same material, and then have the same age and the same metallicity. As a result, GCs have been considered as the best examples of single, simple stellar population for several decades, and the ideal laboratory for testing and calibrating the theory of stellar evolution and the dynamical evolution of star clusters.

However, recent spectroscopic and photometric observational results suggest that the parent GCs are more complex than a single, simple stellar population, and may present multiple stellar populations (Gratton, Sneden & Carretta 2004). Many GCs have an anomalous pattern in their chemical abundances, although all stars in a single GC share the same Fe abundance (see Gratton, Sneden & Carretta 2004, Piotto 2009, and references therein). Star-to-star variations are found in light-element abundances, rich in N, Na and Al, but poor in C, O and Mg, which are the typical results of the CNO, NeNa and MgAl cycles of hydrogen burning. There abundances of C and N, Na an O, or Mg and Al are anti-correlated with each other. However, Na-O and Al-Mg anticorrelations are almost absent among field stars (Gratton, Sneden & Carretta 2004). On the other hand, some GCs show rather than a single evolutionary sequence in the Hertzsprung-Russell (HR) diagrams (Piotto 2009).

Among the evolved stars of some GCs, the presence of star-to-star chemical variations has been known for several decades, and it is considered as the results of peculiar evolution of individual stars. But recently the extensive high-resolution spectroscopic

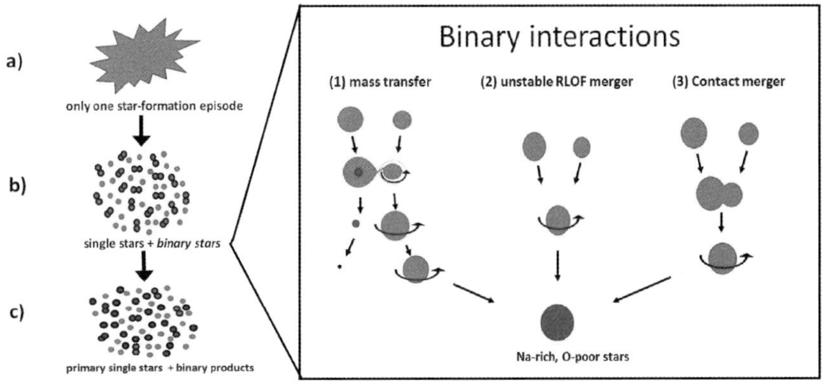

Figure 1. Schematic view for the formation of multiple stellar populations in GCs from binary interactions. a) all stars in a GC come from a single star-formation process; b) some stars in GC are born as close binary components while others are single stars; c) today, there are primary single stars and binary products (anomalous stars) in GC. Three main binary interactions are included in our model: (1) mass transfer; (2) unstable RLOF merger; (3) contact merger.

surveys in many GCs show that main-sequence and turn-off stars also have these variations (Gratton *et al.* 2001), which are considered as the evidence of primordial origin. Therefore, multiple stellar populations in GCs are explained to be formed from multiple star-formation processes, where the subsequent stellar generation (second generation stars) forms from gas chemically polluted by the ejecta from the previous generation (first generation stars) (Gratton, Sneden & Carretta 2004). However, many aspects of this self-pollution scenario are still unclear and currently debated (Gratton, Carretta, & Bragaglia 2012, and references therein). The nature of the first-generation polluter is uncertain, and the main candidates are asymptotic giant branch stars (D'Ercole *et al.* 2008), fast rotating massive stars (Decressin *et al.* 2007) and massive stars in interacting binary systems (de Mink *et al.* 2009). This scenario also does not agree with the observations of young massive clusters that appear to be well represented by a single burst of star formation (Kudryavtseva *et al.* 2012; Bastian & Silva-Villa 2013). Due to these shortcomings, Bastian *et al.* (2013) recently present a model for the formation of multiple stellar populations in GCs that does not require multiple star-formation episodes, where the low-mass stars accrete the enriched ejecta of high-mass stars through disc accretion.

It should be noted that some stars in GCs are members of binaries and many peculiar objects in GCs are closely related to binary interactions, e.g. W UMa contact binaries, blue stragglers and millisecond pulsars. Binary stars have peculiar evolution relative to single stars due to binary interactions. They can produce stars more massive than normal single stars in the same evolutionary stage by mass transfer and merger. These binary products may be hot enough for CNO cycle and NeNa cycle. Furthermore, they will be rapid rotating because binary interactions can convert orbital angular momentum into their spin angular momentum (de Mink *et al.* 2013). Rotationally induced mixing can bring part of the nuclear-processed products from the core to the surface (Decressin *et al.* 2007). Therefore, these binary products, e.g. the merged stars and the accretor stars, may have different surface abundances, effective temperatures, and luminosities.

2. Binary interactions scenario

In our scenario (Fig. 1), GC stars were still born in a single burst of star formation, but some of them are members of binary systems. The anomalous stars observed in GCs

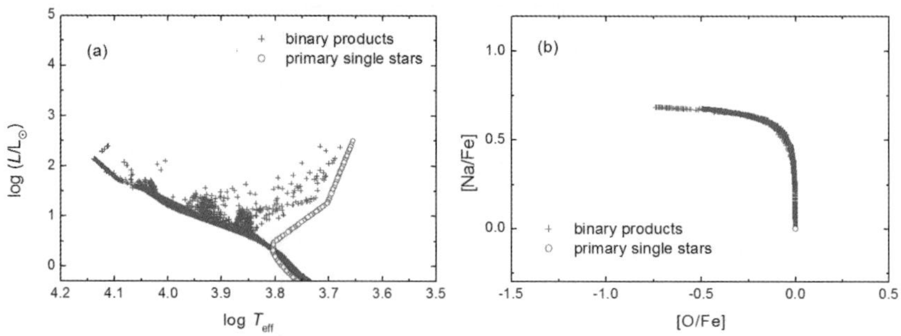

Figure 2. Distributions of the binary population at 10 Gyr. (a) the HR diagram. (b) the surface abundances (Na-O) plane. Blue pluses and red circles are binary products and primary single stars, respectively.

today are merged stars and the accretor stars produced by binary interactions. These binary products are rapidly rotating and more massive than the normal stars. They have different effective temperatures, luminosities, and surface abundances from single stars in the same evolutionary stage due to their evolution. Thus, the stellar population with binaries (the binary population) includes two sub-populations: (1) primary single stars with normal abundance (the initial abundance of GCs); (2) merged stars and accretor stars (produced by binary interactions) that have abnormal surface abundances through their own evolution.

There are three binary evolutionary channels for the formation of rapidly rotating stars: (1) a binary evolves into Roche lobe overflow (RLOF) and experiences stable mass transfer where the secondary obtains the mass and angular momentum transferred from the primary and subsequently becomes a rapidly rotating star (mass transfer channel for the accretor stars), (2) a binary evolves into RLOF and experiences unstable mass transfer while both components are still MS stars then merges into a rapidly rotating star (unstable RLOF merger channel for the merged stars), and (3) a binary evolves into contact and then merges into a rapidly rotating star (contact merger channel for the merged stars).

Using binary population synthesis, we simulate the distributions of the binary population at 10 Gyr(Fig. 2). The binary products show a different distribution from single stars in the HR diagrams, so the binary population show multiple/extended main sequences, turn-off, sub-giant branch and red-giant branch. It should be noted that most of these binary products will be blue stragglers or blue straggler progeny. On the other hand, these binary products also have anomalous surface abundances. They exhibit a large spread in sodium and oxygen abundances, while single stars do not show significant variations. The oxygen-depleted binary products have higher sodium abundances, e.g. the Na-O anticorrelation. Therefore, due to the present of binary products, the binary population has anomalous abundances and distribution in HR diagrams.

3. Conclusion

We suggested that multiple stellar populations in GCs may be formed from binary interactions. The anomalous stars are the binary products, include merged stars and accretor stars. This scenario does not require multiple star formation in GCs. Our results show that the binary population can reproduce the Na-O anticorrelation and multiple sequences in the HR diagrams. Therefore, binary interactions seem to be a possible scenario and we will further research in this direction.

References

Bastian, N. & Silva-Villa, E. 2013, *MNRAS*, 431, L122
Bastian, N., Lamers,H., de Mink, S., Longmore, S., Goodwin, S., & Gieles, M. 2013, *MNRAS*, 436, 2398
D'Ercole, A., Vesperini, E., D'Antona, F., McMillan, S. L. W., & Recchi, S. 2008, *MNRAS*, 391, 825
Decressin, T., Meynet, G., Charbonnel, C., Prantzos, N., & Ekström, S. 2007, *A&A*, 464, 1029
de Mink, S. E., Pols, O. R., Langer, N., & Izzard, R. G. 2009, *A&A*, 507, L1
de Mink, S. E., Langer, N., Izzard, R. G., Sana, H., & de Koter, A. 2013, *ApJ*, 764, 166
Kudryavtseva, N. *et al.* 2012, *ApJ*, 750, L44
Gratton, R., *et al.* 2001, *A&A*, 369, 98
Gratton, R., Sneden, C., & Carretta, E. 2004, *ARAA*, 42, 385
Gratton, R., Carretta, E., & Bragaglia, A. 2012, *A&AR*, 20, 50
Piotto, G. 2009, *IAUS*, 258, 233

Formation, evolution, and survival of massive star clusters
Proceedings IAU Symposium No. 316, 2015
C. Charbonnel & A. Nota, eds.

doi:10.1017/S1743921315008881

Theoretical evolutionary framework for low-mass stars in massive star clusters

Santi Cassisi

INAF – Astronomical Observatory of Teramo,
Via M. Maggini, Teramo, 64100, Italy
email: cassisi@oa-teramo.inaf.it

Abstract. Recent spectroscopic and photometric surveys of Galactic Globular Clusters have shown that these stellar systems host distinct sub-populations of stars characterised by peculiar chemical patterns. In the following we wish to address the issue of how these specific chemical patterns affect both the structural and evolutionary properties of stars as well as their spectral energy distribution. The implications of these effects on the photometric appearance of multiple stellar populations in different photometric planes are also briefly discussed.

Keywords. stars: Population II, stars: low-mass, stars: interiors, stars: evolution

1. Introduction

Extensive spectroscopical surveys of Galactic GCs have shown the existence of well-defined and peculiar chemical patterns among the stars belonging to the same cluster such as the existence of light-elements (anti-) correlations - the most famous being the Na-O anti-correlation - as discussed in the reviews by Gratton *et al.*(2004) and Gratton, Carretta & Bragaglia(2012). In the same time, very accurate photometric investigations performed by means of the HST have revealed the existence in the Color-Magnitude Diagram (CMD) of various GCs of multiple sequences such as distinct Main Sequence (MS) and Sub-Giant Branch (SGB) loci, and multiple Red Giant Branch (RGB) loci. Actually, the features observed in the CMD change significantly from a cluster to another one, and their properties strongly depend on the adopted photometric systems (see Piotto 2010 and Piotto *et al.* 2015 and references therein).

The commonly accepted scenario is that, in any GC, a second (and in some cases also more) generation(s) of stars can form from the ejecta of intermediate-mass and/or massive stars belonging to the first stellar population formed during the early phase of the cluster evolution, whose initial chemical composition is modified by high-temperature proton captures. Although some amount of dilution between pristine (unpolluted) matter and nuclearly processed matter seems to be unavoidable in order to explain many observational evidence, the second generation stars would be formed via matter characterised by light-element (anti-)correlations and He enhancement.

The ability to trace both spectroscopically and photometrically the various sub-populations hosted by each individual GC, allows now the identification of both the primordial stellar component (the first generation, FG) and the second generation (SG) stars. The existence of a strong correlation between the spectroscopic signatures of the distinct sub-populations and their distribution along the multiple CMD sequences, has suggested that the peculiar chemical patterns of SG stars have to affect both the evolutionary properties of these stars as well as their spectral energy distribution (SED).

The investigations performed so far in order to understand the physical reasons for the occurrence of these multiple evolutionary sequences in the CMD of GCs have provided sound evidence of the fact that the distinct sequences can be interpreted as due to 'quantised' He abundances such as in the case of the massive GCs ω Cen and NGC 2808, distinct CNO abundance pattens as for M 22 and of NGC1851, and the presence of light-elements anti-correlations.

It is evident that in order to properly trace the various sub-populations in a given GCs, the availability of an evolutionary theoretical framework properly - and self-consistently - accounting for the observed peculiar chemical patterns is mandatory. In this last decade several theoretical investigations have been performed in order to investigate the effects of the chemical patterns characteristic of the multiple population phenomenon on both the evolutionary and structural properties of stars (Salaris *et al.* 2006, Cassisi *et al.* 2008,Pietrinferni *et al.* 2009,Vandenberg *et al.* 2012) and model atmospheres and, hence on the corresponding color - $T_{\rm eff}$ transformations and bolometric corrections (Sbordone *et al.* 2011, Cassisi *et al.* 2013).

2. On the impact of the peculiar chemical patterns of SG stars on stellar evolution

He-enhancement: Helium is one of the most important chemical species in the context of stellar evolution, because any change of its abundance hugely affects the structural and evolutionary properties of stars. In more detail, a change of the He abundance affects low-temperature, radiative opacities, for an increase of He causes a reduction of the opacity. This is due to the fact that when increasing the helium abundance at fixed metallicity, the H abundance has to decrease, and given that H is a major opacity source via the $^{-}$H ion, this causes a global reduction of the opacity. This effect explains why He-enhanced stellar models have hotter $T_{\rm eff}$ values in comparison with normal He abundance stars.

The opacity reduction due to He-enhancement contributes also to make brighter the stars during the core H-burning stage, although the larger contribution to the change in the stellar surface luminosity is due to the change in the mean molecular weight associated to the He abundance increase. In fact, the H-burning efficiency is strongly dependent on the value of the mean molecular weight μ: $L_{\rm H} \propto \mu^7$. When He increases the mean molecular weight (at fixed metallicity) has to increase and this translates in a larger H-burning efficiency. Given that $\Delta L_{\rm H}/L_{\rm H} = 7\Delta\mu/\mu$, a change $\Delta Y = 0.10$ – lower than the maximum He enhancements expected for the bluest MS stars in GCs ω Cen (King *et al.* 2012) and NGC 2808 (Piotto *et al.* 2007, Milone *et al.* 2012) – causes a $\sim 50\%$ variation of the H-burning efficiency compared to normal-He stars.

The combined effect of radiative opacity decrease and H-burning efficiency increase causes He-rich stellar models to be brighter and hotter during the MS stage. As a consequence, their core H-burning lifetime (t_H) is significantly reduced: for a $0.8M_\odot$, t_H decreases by $\sim 48\%$ when increasing the He content from the primordial value Y=0.245 to 0.40.

Figure 1 shows isochrones computed for the same [Fe/H] value and age, but distinct values for the initial He content. There are some interesting features disclosed by this plot: *i)* the He-rich MS runs parallel in the luminosity interval from the MS Turn-off (TO) and the location of stars with mass $\sim 0.5M_\odot$; *ii)* the MS effective temperature is sensitive to the He increase ($\Delta T_{\rm eff}/\Delta Y \approx 2.3 \times 10^3$K); *iii)* for any given age, the SGB is not affected by a He change; *v)* the $T_{\rm eff}$ values of RGB stellar models are also affected by an He increase, although to a smaller extent than the MS locus. The predicted behaviour of the MS as a function of the initial He abundance justifies the interpretation of the MS splitting observed in the optical CMDs of some GCs, like ω Cen and NGC 2808 (see also the following discussion on this issue).

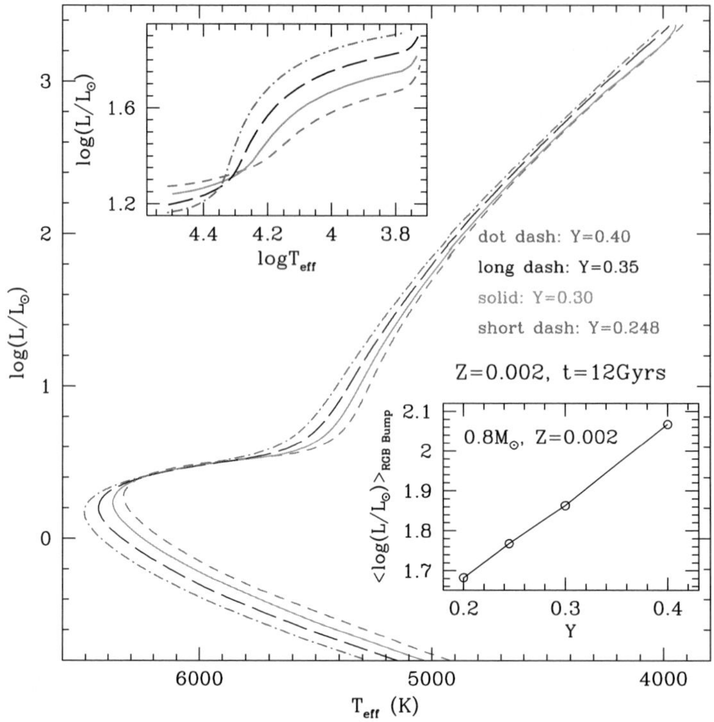

Figure 1. 12 Gyr-old, Z=0.002, isochrones computed for various assumptions about the initial He abundance. The lower inset shows the trend of the average RGB bump brightness as a function of the initial He abundance for a 0.8M$_\odot$ model, while the upper inset shows the location in the H-R diagram of ZAHB loci corresponding to various initial He abundances but for the same assumption about the mass of the RGB progenitor, $M_{RGB} = 0.8M_\odot$, namely.

It is also worth noting that, at fixed age, the mass at the MS TO significantly decreases when increasing the initial He content: for a 12Gyr old isochrone it is equal to $0.806M_\odot$ for Y=0.245 and $0.610M_\odot$ for Y=0.40. As it will be discussed in the following, this occurrence has an important implication during the core He-burning stage.

When considering the RGB evolution, there are two important features that are affected by a change of the initial He abundance: the RGB bump and the brightness of the RGB tip (TRGB). Model computations predict: *a)* an increase of the bump brightness (see Fig. 1), and *b)* a smaller luminosity excursion during the RGB bump stage. The first effect is due to the lower envelope opacity of He-rich stars, that causes the discontinuity of the H abundance left over by the *first dredge-up* to be located in more external layers. As a consequence, the H-burning shell encounters the discontinuity at later times, hence at a brighter luminosity. The second effect is caused by the fact that in He-rich stars, the jump of the H abundance at the discontinuity is smaller than in normal He stars. As a consequence the surface stellar luminosity is less affected when the H-burning shell crosses the discontinuity. From an observational point of view, the impact of an He enhancement on the RGB bump brightness in GCs has been discussed by Bragaglia *et al.*(2010) .

The TRGB brightness is significantly affected by an He-enhancement. For a given total stellar mass increase of the initial He content decreases the TRGB brightness. This is due to the fact that the interiors of He-rich stars are hotter at the end of the core H-burning stage, and when they reach the RGB stage they develop a significantly lower level of electron degeneracy in their He core, as well as they are less affected by neutrino energy losses. At the same time, as a consequence of the larger H-burning efficiency, the He core mass grows at a faster rate. Both effects make 'easier' to achieve the thermal conditions required by the 3α reaction ignition, and He ignition is attained with a smaller He core mass (for more details see Cassisi & Salaris 2013). Due to the existence of a *He core mass - luminosity* relation for RGB stars, the TRGB brightness decreases in He-rich, low-mass giants. Unfortunately, due to the low number of stars populating the brighter portion of the RGB and the still unsettled issue of the GC distance scale, it is almost impossible to verify observationally this prediction.

We also note that, as a consequence of the reduction of the value of t_H for He-rich stars, the mass of stars at the TRGB is expected to be significantly smaller in He-enhanced stellar populations, when age is kept constant (and if RGB mass loss does not have a strong dependence on He). This occurrence explains why He-enhanced stellar populations are characterised by a, 'on average', bluer Horizontal Branch (HB) morphology with respect to a FG stellar population.

When moving to the core He-burning stage, one notices that Zero Age HB (ZAHB) brightness is a strong function of the initial He content. When Y increases, the ZAHB becomes brighter for T_{eff} lower that $\sim$ 20000K, and fainter at higher T_{eff} (see Fig. 1). This behaviour is the consequence of both the decrease of the He-core mass at the TRGB, and of the increased efficiency of the shell H-burning in He-rich stars. In ZAHB objects cooler than $\sim$ 20000 K, the second effect dominates over the first one and the stars appear brighter, whereas in the hottest portion of the ZAHB – due to the tiny envelope mass – the H-burning shell is not efficient enough, and the decrease of the He core mass is the dominating effect. This has the important implication that the slope of the ZAHB in the H-R diagram (as well as in the various observational planes) is strongly dependent on the (spread in the) initial He abundances of the various sub-populations hosted within a given GC. This theoretical prediction can provide a direct explanation for the existence of a tilted HB in GCs like NGC6388, NGC6441 and NGC1851.

He-enhancement has an additional important implication for the global HB morphology. Due to the combined effect of the lower RGB evolving mass and more extended blue loops that characterise the off-ZAHB evolution of He-rich stars, for a given average efficiency of the mass loss along the RGB the predicted HB location of He-enhanced models will be on average hotter, i.e. bluer, than that of He-normal stars. This helps explaining the existence the very blue HB morphology – as well as the presence of extended HB blue tail – in GCs hosting He-enhanced sub-populations. Therefore, it appears evident that the presence of He-enhanced stellar populations in Galactic GCs can contribute to solve the longstanding 'second parameter problem', i.e. the presence of different HB morphology among clusters with the same metallicity. This issue has been recently discussed by Milone *et al.*(2014).

Light-element (anti-)correlations: light element abundance changes can affect the stellar properties via the effects induced on both the radiative opacity evaluations and – at least in the case of C, N and O, i.e. the CNO-cycle catalysts – the H-burning efficiency. Many investigations have been devoted to this topic as those by Salaris *et al.*(2006), Cassisi *et al.*(2008), Pietrinferni *et al.*(2009), Ventura *et al.*(2009), Vandenberg *et al.*(2012) and the main results can be summarised as follows:

- even in case of extreme light-element anti-correlations, as long as the sum of CNO elements is kept constant, the evolution in the H-R diagram is unchanged compared to standard $\alpha-$enhanced stellar models;
- if the CNO sum is enhanced, the morphology of the evolutionary tracks in the H-R diagram is modified. For a given iron content and stellar mass, the SGB appears fainter

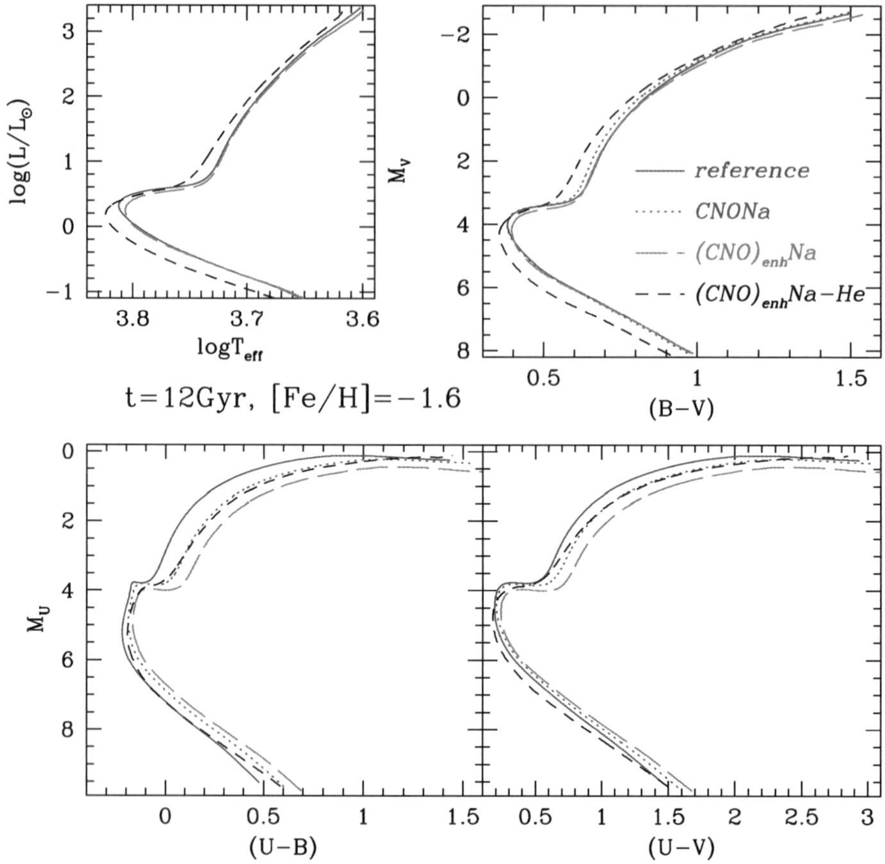

Figure 2. Theoretical isochrones for the same age and metallicity (see label) but various assumptions about the light-element distribution and He-enhancement: *reference*: canonical $\alpha-$enhanced mixture, Y=0.248; *CNONa*: light-element (anti-)correlation but the same CNO sum and He abundance of reference mixture; $(CNO)_{enh}Na$: as before but the CNO sum is enhanced by a factor ~ 2; $(CNO)_{enh}Na - He$: as before but now the initial He abundance is equal to Y=0.40.

in comparison with standard $\alpha-$enhanced models. This behaviour is mainly due to the fact that the efficiency of the CNO-cycle increases when the CNO sum is enhanced. The $T_{\rm eff}$ scale for both MS and RGB CNO-enhanced stellar models is marginally affected (by less than 20 K);

• when computing isochrones, unless the CNO sum is increased with respect the reference $\alpha-$enhanced mixture, the effect of the light-element (anti-)correlations on the H-R diagram location is negligible. When the C+N+O sum is increased, a separation appears along the SGB. When enhancing by a factor of $\sim$ 2 the CNO sum, a CNO-enhanced isochrone is almost perfectly mimicked by a 'canonical' $\alpha-$enhanced, $\sim (1.5 - 2)$ Gyr older, isochrone.

3. On the impact of SG chemical patterns on the SED

As shown in the upper, left panel of fig. 2, in the theoretical H-R plane, the only possibility to observe a separation between isochrones with a standard $\alpha-$enhanced isochrone and those with the chemical composition of SG stars, is to assume a huge He enhancement (that affect MS and RGB) and/or a significant CNO-enhancement (that affect the SGB sequence). Needless to say that this theoretical evidence is in contrast with observations, that reveal the presence of splitting/broadening of the photometric sequences (in particular, the RGB one) also in those GCs that do not show (spectroscopical) evidence of a significant CNO enhancement and/or huge He enhancement. The chemical peculiarities of second generation stars must therefore have an effect also on the spectral energy distribution.

A detailed analysis of the impact of the peculiar chemical patterns of SG stars on their SED has been performed by Sbordone *et al.*(2011), Cassisi *et al.*(2013), and Dotter *et al.*(2015). Figure 3 summarises the results of their investigation displaying the comparison between synthetic spectra for a specific SG composition computed for selected MS, TO and RGB models covering the wavelength range from 200 to 1000 nm, and theoretical spectra for the same stars but with a FG chemical composition. Some prominent absorption features (mostly molecular bands), whose strength changes significantly between FG and SG compositions are labelled, and the transmission curves for the Johnson-Cousins UBVI, and the Strömgren uvby filters are also displayed.

Starting with the RGB fluxes, the SG mixture shows much stronger NH and CN absorption bands in the U, B, I passbands compared to the FG one. This is due to the higher N abundance, despite the fact that the C abundance is lower, because the N abundance acts as bottleneck in forming CN molecules. The G-band – a CH feature within the B passband – is stronger in the FG mixture, because of the higher C abundance. The increased opacity in the blue part of the SG spectrum increases the continuum flux redwards of $\sim$450 nm.

When moving to the spectrum of a MS TO star, the much higher temperature prevents the formation of CN and CH molecules, as well as most of the effect on the G-band. Only the very strong NH band around 340 nm is still visible. The cool MS star displays again features similar to the RGB one, with a much stronger OH absorption at the blue edge of the U filter in the FG flux distribution. The NaD appears much stronger in the SG spectrum, for the Na abundance is increased due to the O-Na anti-correlation.

On the other hand, it has been found that the impact of Mg-Al abundance variations at fixed [Fe/H] and helium abundance is negligible on stellar models and isochrones (from the main sequence to the tip of the red giant branch) and bolometric corrections, when compared to the effect of C-N and O-Na variations.

These results show the crucial role played by CN and NH molecules (whose abundances in SG stars are very different from those in FG stars) in modifying the stellar spectrum at wavelengths shorter than $\sim$ 400 nm. This has the important implication that only magnitudes corresponding to photometric filters bluer than the standard Johnson B filter are affected by the peculiar chemical patterns of multiple populations; as a consequence, the photometric appearance of these sub-populations depends on the adopted photometric systems. How theoretical isochrones suitable for SG chemical compositions compare with isochrones appropriate for standard -(FG) chemical abundances in the CMD in various photometric systems can be summarised as follows:

- *Visible to near-infrared CMDs* (e.g. BVI or Strömgren vy) are generally unaffected by the presence of SG stars, unless this stellar component has a different initial He content (see the upper, right panel in fig. 2). In this case multiple sequences appear along the MS. This behaviour explains very nicely the multiple MS in the optical CMD of NGC 2808 discussed by Piotto *et al.*(2007). A variation of the CNO sum also affects optical CMDs by altering the location of TO and SGB. This occurrence provides a sound interpretation for the double SGB of NGC 1851 in optical filters (see Cassisi *et al.* 2008), as well as for the double SGB observed in NGC 6656 (Marino *et al.* 2012). Both He and CNO affect

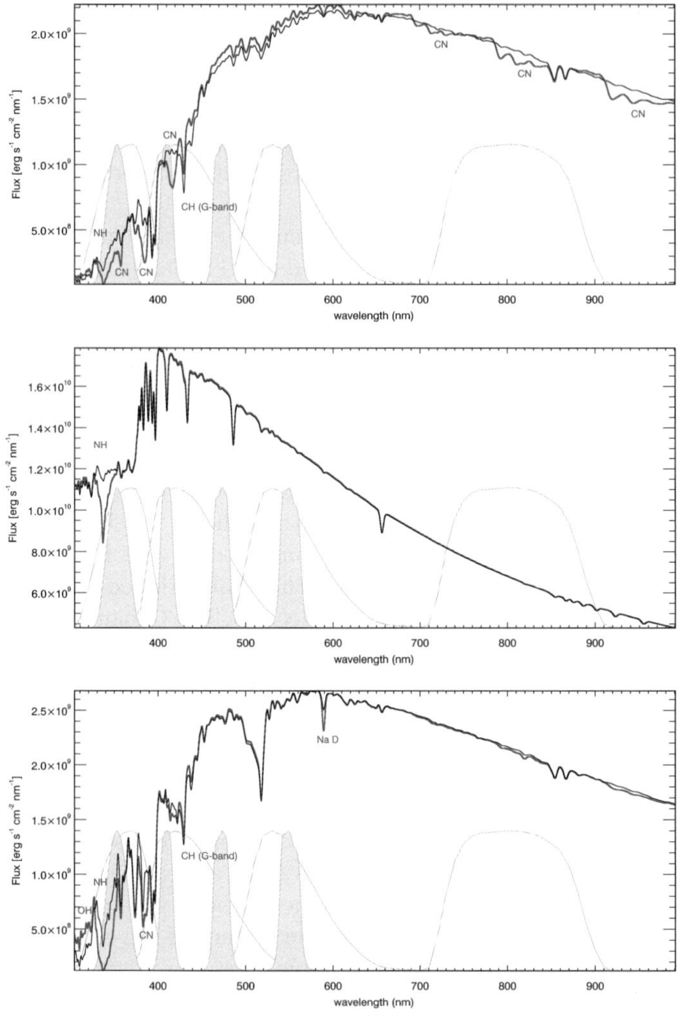

Figure 3. Synthetic spectra for a *reference* mixture (thin black line) and $(CNO)_{enh}Na$ mixture (red thick line) for selected stellar models. The top panel displays the spectra for a RGB model with $T_{\rm eff} = 4476$ K, $\log(g) = 4.1.2$, the middle panel panel the flux for a MS TO model with $T_{\rm eff} = 6490$ K, $\log(g) = 4.22$, while the bottom panel corresponds to a MS model with $T_{\rm eff} = 4621$ K, $\log(g) = 4.47$. The transmission curves for the Johnson-Cousins UBVI filters (thin, black lines) and for the Strömgren uvby filters (grey-shaded regions) are also displayed (adapted from Cassisi & Salaris 2013).

optical CMDs because they hugely affect the stellar evolutionary properties, and hence modify the isochrone location in the H-R diagram.

• *Optical-UV CMDs* (e.g. UBV, UVIS@WFC3, and Strömgren uy) are hugely affected by the presence of SG stars. In fact, C-N-O-Na anti-correlations affect the bolometric correction scale for photometric passbands bluer than ~ 400 nm, so that even at constant CNO sum and He abundance, SG and FG stars will be distributed amongst several separate sequences. The quantitative effect will depend on the filter combination employed and the exact over- and under-abundances with respect to the baseline FG composition. Variations of He and CNO sum add an additional level of separation in the CMDs, as a consequence of their effect on the underlying isochrones as shown in the bottom panels of fig. 2.

Acknowledgements

SC warmly thanks the organizers for inviting him to this interesting and pleasant conference. Financial support from PRIN INAF 2014 (PI: S. Cassisi) and PRIN MIUR 2010-2011, project "The Chemical and Dynamical Evolution of the Milky Way and Local Group Galaxies", prot. 2010LY5N2T " is acknowledged.

References

Bragaglia, A., Carretta, E., Gratton, R., D'Orazi, V., Cassisi, S., & Lucatello, S. 2010, *A&A*, 519, 60

Cassisi, S., Salaris, M., Pietrinferni, A., Piotto, G., Milone, A. P., Bedin, L. R., & Anderson, J. 2008, *ApJ* (Letters), 672, L115

Cassisi, S., Mucciarelli, A., Pietrinferni, A., Salaris, M., & Ferguson, J. 2013, *A&A*, 554, A19

Cassisi, S. & Salaris, M. 2013, *Old stellar populations: how to study the fossil record of galaxy formation*, Wiley-VCH

Dotter, A., Ferguson, J. W., Conroy, C., *et al.* 2015, *MNRAS,* 446, 1641

Gratton, R., Sneden, C., & Carretta, E. 2004, *ARAA*, 42, 385

Gratton, R., Carretta, E., & Bragaglia, A. 2012, *A&ARv*, 20, 50

King, I. R., Bedin, L. R., Cassisi, S., *et al.* 2012, *AJ*, 144, 5

Marino, A. F., Milone, A. P., Sneden, C., *et al.* 2012, *A&A*, 541, A15

Milone, A. P., Piotto, G., Bedin, L. R., *et al.* 2012, *A&A*, 537, 77

Milone, A. P., Marino, A. F., Dotter, A., *et al.* 2014, *ApJ*, 785, 21

Pietrinferni, A., Cassisi, S., Salaris, M., Percival, S., & Ferguson, J. W. 2009, *ApJ*, 697, 275

Piotto, G. Bedin, L. R., Anderson, J., *et al.* 2007, *ApJ* (Letters), 661, L53

Piotto, G. 2010, *Publications of the Korean Astronomical Society*, 25, 91

Piotto, G., Milone, A. P., Bedin, L. R. *et al.* 2015, *AJ*, 149, 91

Salaris, M., Achim, W., Ferguson, J. W., & Fusilier, D. J. 2006, *ApJ*, 645, 1131

Sbordone, L., Salaris, M., Weiss, A., & Cassisi, S. 2011, *A&A*, 534, 9

VandenBerg, D. A., Bergbusch, P. A., Dotter, A., Ferguson, J. W., Michaud, G., Richer, J., & Proffitt, C. R. 2012, *ApJ*, 755, 15

Ventura, P., Caloi, V., D'Antona, F., Ferguson, J., Milone, A. P., & Piotto, G. P. 2009, *MNRAS, 399, 934*

Formation, evolution, and survival of massive star clusters
Proceedings IAU Symposium No. 316, 2015
C. Charbonnel & A. Nota, eds.

doi:10.1017/S1743921315010546

Globular clusters and the evolution of their multiple stellar populations

W. Chantereau[1], C. Charbonnel[1,2] and G. Meynet[1]

[1]Department of Astronomy, University of Geneva,
Chemin des Maillettes 51, CH-1290 Versoix, Switzerland
email: william.chantereau@unige.ch

[2]RAP, UMR 5277 CNRS and Université de Toulouse,
14 Av. E. Belin, F-31400 Toulouse, France

Abstract. Our knowledge of the formation and early evolution of globular clusters (GCs) has been totally shaken with the discovery of the peculiar chemical properties of their long-lived host stars. Therefore, the interpretation of the observed Colour Magnitude Diagrams (CMD) and of the properties of the GC stellar populations requires the use of new stellar models computed with relevant chemical compositions. In this paper we use the grid of evolution models for low-mass stars computed by Chantereau *et al.* (2015) with the initial compositions of second-generation stars as predicted by the fast rotating massive stars scenario to build synthesis models of GCs. We discuss the implications of the assumed initial chemical distribution on 13 Gyr isochrones. We build population synthesis models to predict the fraction of stars born with various helium abundances in present day globular clusters (assuming an age of 13 Gyr). With the current assumptions, 61 % of stars on the main sequence are predicted to be born with a helium abundance in mass fraction, Y_{ini}, smaller than 0.3 and only 11 % have a Y_{ini} larger than 0.4. Along the horizontal branch, the fraction of stars with Y_{ini} inferior to 0.3 is similar to that obtained along the main sequence band (63 %), while the fraction of very He-enriched stars is significantly decreased (only 3 % with Y_{ini} larger than 0.38).

Keywords. globular clusters: general, Hertzsprung-Russell diagram, stars: evolution, stars: low-mass, stars: abundances

1. Introduction

It is now well accepted that GCs host at least two populations of low-mass stars, one presenting the chemical patterns typical of the protoglobular cluster gas and of halo stars of similar metallicity (first population, 1P), while the other one displays peculiar abundances (second population, 2P). This overall picture has been set up thanks to numerous spectroscopic surveys (see e.g. Gratton *et al.* 2001; Carretta *et al.* 2009a, 2009b), and the famous O-Na anti-correlation has been adopted as the main characteristic of a *bona-fide* GC (Carretta *et al.* 2010). In addition, with the advent of high resolution photometry, this paradigm has been strongly supported by the discovery of multiple discrete structures in CMDs of individual GCs (see e.g. Bedin *et al.* 2004; Piotto *et al.* 2007; Piotto 2009; Milone *et al.* 2013). Different scenarios implying different polluters have been developed to explain the nucleosynthetic sources of ashes of H-burning at high temperature needed to explain the chemical composition of 2P stars. The main polluters advocated in the literature are fast rotating massive stars (FRMS, Decressin *et al.* 2007; Krause *et al.* 2013), asymptotic giant branch stars (AGB, Ventura *et al.* 2001, 2013), massive binary stars (de Mink *et al.* 2009), supermassive stars (Denissenkov & Hartwick 2014), and also low-mass pre-main sequence stars coupled with interactive binaries/and or rotating massive stars (Bastian *et al.* 2013).

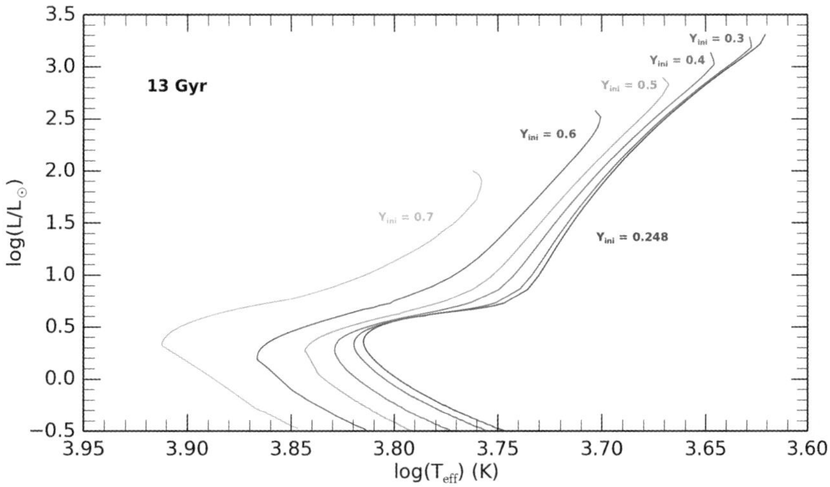

Figure 1. Theoretical isochrones at 13 Gyr for different initial helium composition between 0.248 and 0.7 (mass fraction).

Despite the numerous differences between the scenarios developed to explain the typical chemical patterns of GCs, it is widely accepted by the community that 2P stars have formed with a higher helium content than their 1P counterparts. It is then mandatory to take into account the impact of various helium content on the evolution of stars in order to provide a fully consistent scenario. Each of the above scenarios predict different initial helium content in the 2P stars. The various initial He distributions associated will then evolve into different distributions as a result of the evolution of stars and give therefore various outcomes for the helium distribution in the present day GCs. This is the point that we study in the present paper. More specifically, we start from the initial abundances predicted by the FRMS scenario and we predict the helium distribution that one should observe in the cluster NGC 6752. Although helium abundances remain very difficult to measure directly by spectroscopy in cool stars, there exist indirect ways to constrain the present day helium distribution in GCs (see e.g. Salaris *et al.* 2004).

2. General framework

In the FRMS scenario, 2P stars form in the immediate vicinity of very massive and fast rotating stars (into the mass range 25-120 $M_\odot$ as investigated by Decressin *et al.* 2007a,b; Krause *et al.* 2013) from their H-burning ashes diluted at various degrees with the surrounding intra-cluster pristine gas. Therefore 2P stars are mainly C-, O-, and Mg-depleted while they are He-, N-, Na-, and Al-enriched, compared to their 1P counterparts that display a chemical composition typical of the pristine intra-cluster material and of the surrounding halo field stars. At the metallicity investigated in this study ([Fe/H] = -1.75), 2P stars are expected to be born with $Y_{\rm ini}$ ranging between 0.248 (as 1P) and 0.8.

3. Isochrones

Stellar models enriched in helium have a higher luminosity and effective temperature ($T_{\rm eff}$), thus evolve faster (see Chantereau *et al.* (2015), hereafter paper I, for more details). At a given age, He-enriched isochrones are thus located in the HRD at a higher $T_{\rm eff}$, and are populated by lower masses than the reference isochrone ($Y_{\rm ini} = 0.248$, see fig. 1).

More quantitatively, the effective temperature of stars with a solar luminosity are hotter by 1,400 K when their initial helium abundance is 0.7 instead of 0.248. At this luminosity, the 2P stars with $Y_{ini} = 0.7$ display a mass lower by $\sim 0.4\ M_{\odot}$ than their 1P counterparts. These differences are much smaller for models moderately enriched in helium, still at solar luminosity, we expect 'only' an increase of less than ~ 100 K between the reference isochrone and the $Y_{ini} = 0.3$ one, which is similar to the error in T_{eff}. Along the RGB, we expect the same behaviour for the He-rich models as for the previous phase, i.e. mainly an increase of the T_{eff} with the helium enhancement. This time the dispersion is smaller than the one at solar luminosity, slightly above 700 K for the $Y_{ini} = 0.7$ case in the middle of the RGB ($\log(L/L_{\odot}) = 1.5$). Again, for lower helium enrichments (e.g. $Y_{ini} = 0.4$), the increase in T_{eff} is only ~ 100 K. Finally, as expected (see paper I and Chantereau *et al.*, in prep.), the RGB-tip is fainter for He-enriched models, therefore the maximum helium enrichment and the temperature spread decreases as we go to higher luminosities.

4. Synthetic globular clusters

4.1. *Initial conditions and basic assumptions*

The construction of a synthetic GC is based on the grid of models (see paper I) assuming a initial mass function (IMF) and an initial distribution for the initial helium mass fraction of the stars populating the cluster. We use a Paresce & De Marchi (2000) IMF for masses below $0.85\ M_{\odot}$, and a Salpeter IMF (Salpeter 1955) with $\alpha = 2.35$ for higher masses. The distribution of the initial helium distribution is build to retrieve the sodium distribution of the bright RGB stars of NGC 6752 (Carretta, 2013), a GC which metallicity is close to the one of our grid ([Fe/H] = -1.56, Carretta *et al.* 2009a). Indeed from the sodium abundance we can infer the helium content thanks to the helium-sodium correlation provided by the FRMS scenario (see fig. 1 in the paper I). Then with Monte-Carlo simulations of 300'000 stars (0.3 - 1.0 $M_{\odot}$), by iteration on the Y_{ini} distribution we reproduce this sodium distribution observed on the RGB at 13 Gyr (13.4±1.1 Gyr for NGC 6752, Gratton *et al.* 2003).

The initial helium mass fraction distribution inferred this way (see fig. 2) is composed of 49 % of stars displaying an initial helium enrichment above 0.3 and 21 % above 0.4†. After 13 Gyr, from the initial sample of 300'000 stars, only 200'000 stars (M = 0.30 - 0.81 $M_{\odot}$) are still alive.

4.2. *Helium distribution along the HRD at 13 Gyr*

In this section we present the distribution of stars according to their helium content at birth (Y_{ini}) and after 13 Gyr of evolution. This is shown in fig. 3 where we present the relative number of stars as a function of their initial helium content, for different regions of the HRD.

Along the MS (here $-0.5 \leqslant \log\left(\frac{L}{L_{\odot}}\right) \leqslant 0.3$; top panel in fig. 3), the highest initial helium content that we obtain is 0.724, which implies in principle a broadening of the MS band in the HRD by about 1,900 K. However in this domain in luminosity, only 28 % of the main sequence stars were born with Y_{ini} between 0.3 and 0.4, 11 % with Y_{ini} higher than 0.4, and 2 % with Y_{ini} higher than 0.6. Therefore, the photometric impact of the

† We used a grid of standard stellar models for low-mass stars (0.3-1.0 $M_{\odot}$) at low metallicity ([Fe/H] = -1.75), with an alpha-enhancement of +0.3 and initial mass fraction of helium between 0.248 and 0.8 (paper I)

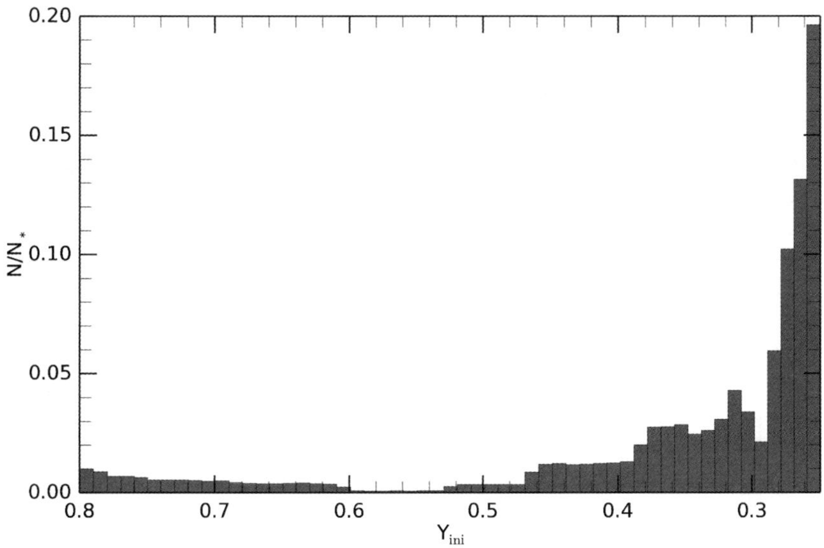

Figure 2. Initial distribution of the helium mass fraction for our sample of 300'000 stars with assumptions described in § 4.1.

high helium tail ($Y_{ini} > 0.4$) would be extremely moderate.

On the bright part of the RGB ($\log\left(\frac{L}{L_{\odot}}\right) \geqslant 1.2$; third panel in fig. 3, upper RGB), the maximum Y_{ini} found is nearly the same as on the MS (0.729), and the proportion of stars born with a Y_{ini} above 0.3 is 38 %, similar to the one found for the MS. In this luminosity domain, only 10 % of the stars were born with an initial helium content above 0.4, and 2 % with Y_{ini} higher than 0.6.

The RGB-tip being fainter for stars with higher initial helium content (see fig 1), as we go to higher luminosities, the maximum Y_{ini} and the fraction of present He-rich stars drops quickly, therefore the distribution of helium on the RGB is very sensitive to the bin of luminosity chosen.

At 13 Gyr all the stars with an initial helium content above 0.41 evolve directly towards the cooling white dwarf curve without undergoing the He-flash and end their life as He-WD. Therefore we do not expect to see stars born with a higher helium content on the horizontal branch (HB) and *a fortiori* on the AGB (see Charbonnel *et al.* 2013). A star with a $Y_{ini} = 0.41$ have also the lower mass on this phase (0.473 $M_{\odot}$) while the 1P stars on this phase have a current mass of 0.6 $M_{\odot}$. We also expect to still find 37 % of stars born with a helium content above 0.3 Y_{ini} but only 3 % with a helium value between 0.38 and 0.41.

Surprisingly, we find that the maximum Y_{ini} for stars on the HB is similar to the maximum He enrichment for 2P stars in the AGB scenario (~ 0.38, e.g. Doherty *et al.* 2014). Therefore we see that the determination of He for HB stars can not help discriminating between the two scenarios.

Concerning the AGB phase, the maximum Y_{ini} value found is even lower due to stars that ignite helium later on the HB (hot and late-hot flashers, see e.g. Miller Bertolami *et al.* 2008), and that miss the AGB phase, the so-called AGB-manqué stars (see e.g.

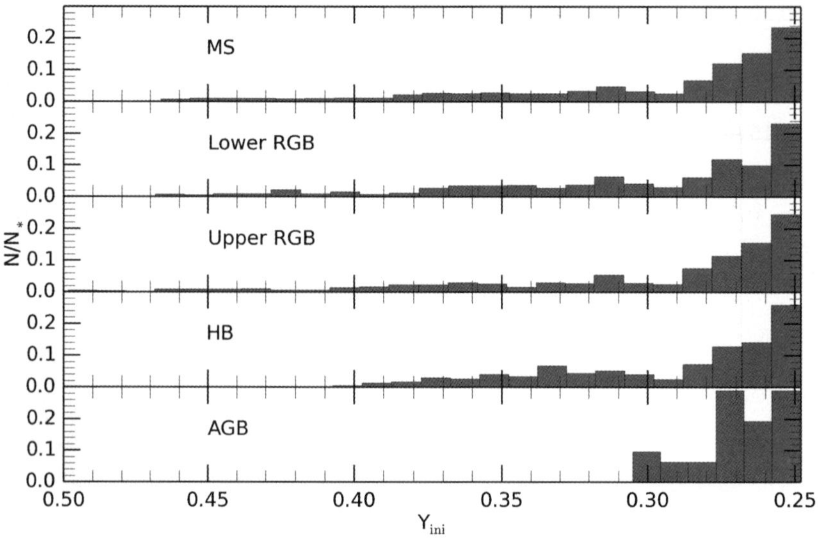

Figure 3. Y_{ini} distribution along a GC at 13 Gyr. The MS corresponds to a bin of luminosity between $\log\left(\frac{L}{L_\odot}\right)$ = -0.5 and 0.3, The RGB phase is separated in two parts, the lower one displays the distribution for a bin of luminosity between $\log\left(\frac{L}{L_\odot}\right)$ = 0.8 and 1.2 while the upper part shows all the brighter stars.

Charbonnel *et al.* 2013, Cassisi *et al.* 2014). At 13 Gyr, the most helium enriched model displays an initial helium content of 0.305 and a mass of 0.536 $M_\odot$ whereas its 1P counterpart has a mass of 0.598 $M_\odot$. The maximal Y_{ini} value is very sensitive to the initial helium mass fraction distribution coupled to the IMF because of the very low number of stars at this phase (31 out of the 200'000 here).

5. Conclusion

Starting with initial helium abundances for 2P stars as given by the FRMS models and using standard assumptions for the IMF, we present predictions for the helium distribution in a typical 13 Gyr old GC with [Fe/H] = -1.75.

At birth, 49 % of the stars have a helium content above 0.3, among which 21 % have Y_{ini} higher than 0.4. At 13 Gyr, due to stellar evolution, the stars born with $Y_{ini} \geqslant 0.3$ correspond to 39 % of the stars present in the GC (sum over all the phases) among which only 10 % born with a $Y_{ini} \geqslant 0.4$.

The low fraction of stars with very high helium enrichments (or their total absence) shows that it would be complicated (or impossible) to use the maximum helium enrichment as a strong discriminating parameter between the different scenarios trying to explain this multiple populations phenomenon (for this typical age and metallicity).

Acknowledgements

We warmly thank Eugenio Carretta for having kindly provided us his valuable data on NGC 6752. We acknowledge support from the Swiss National Science Foundation (FNS). We thank the International Space Science Institute (ISSI, Bern, CH) for welcoming the activities of ISSI Team 271 "Massive Star Clusters across the Hubble Time" (2013 - 2015; team leader C.C.). C.W. acknowledges financial support from the "Société Académique de Genève" and the "Fondation Ernst et Lucie Schmidheiny".

References

Bastian, N., Lamers, H. J. G. L. M., de Mink, S. E., Longmore, S. N., Goodwin, S. P., & Gieles, M. 2013, *MNRAS*, 436, 2398

Bedin, L. R., Piotto, G., Anderson, J., Cassisi, S., King, I. R, Momany, Y., & Carraro, G. 2004, *ApJ*, 605, L125

Carretta, E., Bragaglia, A., Gratton, R. G., Lucatello, S., Catanzaro, G., Leone, F., Bellazzini, M., Claudi, R., D'Orazi, V., Momany, Y., *et al.* 2009a, *A&A*, 505, 117

Carretta, E., Bragaglia, A., Gratton, R., D'Orazi, V., & Lucatello, S. 2009b, *A&A*, 508, 695C

Carretta, E., Bragaglia, A., Gratton, R. G., Recio-Blanco, A., Lucatello, S., D'Orazi, V., & Cassisi, S. 2010, *A&A*, 516, A55

Carretta, E. 2013, *A&A*, 557, A128

Cassisi, S., Salaris, M., Pietrinferni, A., Vink, J. S., & Monelli, M. 2014, *A&A*, 571, 81C

Chantereau, W., Charbonnel, C., & Decressin, T. 2015, *A&A*, 578, A117

Charbonnel, C., Chantereau, W., Decressin, T., Meynet, G., & Schaerer, D. 2013, *A&A*, 557, L17

Decressin, T., Charbonnel, C., & Meynet, G. 2007a, *A&A*, 475, 859

Decressin, T., Meynet, G., Charbonnel, C., Prantzos, N., & Ekström, S. 2007b, *A&A*, 464, 1029

de Mink, S. E., Pols, O. R., Langer, N., & Izzard, R. G. 2009, *A&A*, 507, L1

Denissenkov, P. A. & Hartwick, F. D. A. 2014, *MNRAS*, 437, L21

Doherty, C. L., Gil-Pons, P., Lau, H. H. B., Lattanzio, J. C., Siess, L., & Campbell, S. 2014, *MNRAS*, 441, 582

Gratton, R. G., Bonifacio, P., Bragaglia, A., Carretta, E., Castellani, V., Centurion, M., Chieffi, A., Claudi, R., Clementini, G., D'Antona, F., *et al.* 2001, *A&A*, 369, 87

Gratton, R. G., Bragaglia, A., Carretta, E., Clementini, G., Desidera, S., Grundahl, F., & Lucatello, S. 2003 *A&A*, 408, 529

Krause, M., Charbonnel, C., Decressin, T., Meynet, G., & Prantzos, N. 2013, *A&A*, 552, A121

Miller Bertolami, M. M., Althaus, L. G., Unglaub, K., & Weiss, A. 2008, **A&A** 491, 253

Milone, A. P., Marino, A. F., Piotto, G., Bedin, L. R., Anderson, J., Aparicio, A., Bellini, A., Cassisi, S., D'Antona, F., Grundahl, F., Monelli, M., & Yong, D. 2013, *ApJ*, 767, 120

Paresce, F. & De Marchi, G. 2000, *ApJ*, 534, 870

Piotto, G., Bedin, L. R., Anderson, J., King, I. R., Cassisi, S., Milone, A. P., Villanova, S., Pietrinferni, A., & Renzini, A. 2007, *ApJ*, 661, L53

Piotto, G. 2009, *IAUS*, 258, 233P

Salaris, M., Riello, M., Cassisi, S., & Piotto, G. 2004, *A&A*, 420, 911S

Salpeter, E. E. 1955, *ApJ*, 121, 161

Ventura, P., D'Antona, F., Mazzitelli, I., & Gratton, R. 2001, *ApJ*, 550, L65

Ventura, P., Di Criscienzo, M., Carini, R., & D'Antona, F. 2013, *MNRAS*, 431, 3642

Formation, evolution, and survival of massive star clusters
Proceedings IAU Symposium No. 316, 2015
C. Charbonnel & A. Nota, eds.

doi:10.1017/S1743921315008960

Accretion onto Protoplanetary Discs: Implications for Globular Cluster Evolution

T. P. G. Wijnen[1,2], O. R. Pols[1], F. I. Pelupessy[2] and S. Portegies Zwart[2]

[1]Department of Astrophysics/IMAPP, Radboud University Nijmegen,
P.O. Box 9010, 6500 GL Nijmegen, The Netherlands
email: thomas.wijnen@astro.ru.nl

[2]Leiden Observatory, Leiden University,
PO Box 9513, 2300 RA Leiden, The Netherlands

Abstract. In the past decade, observational evidence that Globular Clusters (GCs) harbour multiple stellar populations has grown steadily. These observations are hard to reconcile with the classical picture of star formation in GCs, which approximates them as a single generation of stars. Bastian *et al.* recently suggested an evolutionary scenario in which a second, chemically distinct, population is formed by the accretion of chemically enriched material onto the protoplanetary disc of low-mass stars in the initial GC population. Using assumptions that represent the (dynamical) conditions in a typical GC, we investigate whether a low-mass star surrounded by a protoplanetary disc can accrete sufficient enriched material to account for the observed abundances in 'second generation' stars. We compare the outcome of two different smoothed particle hydrodynamics codes and focus on the lifetime and stability of the disc and on the gas accretion rate onto both the star and the disc.

Keywords. accretion, accretion discs, protoplanetary discs, planetary systems: formation, stars: formation, globular clusters: general

1. Context

The idea of the early disc accretion scenario is that the massive stars reside in binaries in the cluster core and expel enriched material through binary interactions. The low-mass, pre-main sequence stars of the same generation then sweep up the gas into their protoplanetary disc as they move through the cluster core (Bastian *et al.* 2013, B13). To explain the most extreme abundance spreads observed in globular clusters, they have to accrete an amount of enriched material of the order of their own mass (as inferred from, e.g., the main sequence of NGC2808, Piotto *et al.* 2007) within the expected disc lifetime of 10^6 -10^7 years. We test this scenario by simulating a protoplanetary disc embedded in a laminar flow of gas with 2 smoothed particle hydrodynamics (SPH) codes, Fi (Pelupessy *et al.* 2004) and Gadget2 (Springel 2005), within the AMUSE framework (Pelupessy *et al.* 2013). We test against different implementations of the artificial viscosity in each SPH code.

2. Results & Implications

Our comparison shows that the rate at which the protoplanetary disc can sweep up interstellar material (ISM loading rate, the green symbols in Fig. 1a) is consistent for both codes and all resolutions. It is a factor 2 lower than the rate expected from geometrical arguments (and used in B13):

$$\dot{M}_{\rm ISM} = \rho_{\rm ISM} v_{\rm ISM} \pi R_{\rm disc}^2,$$

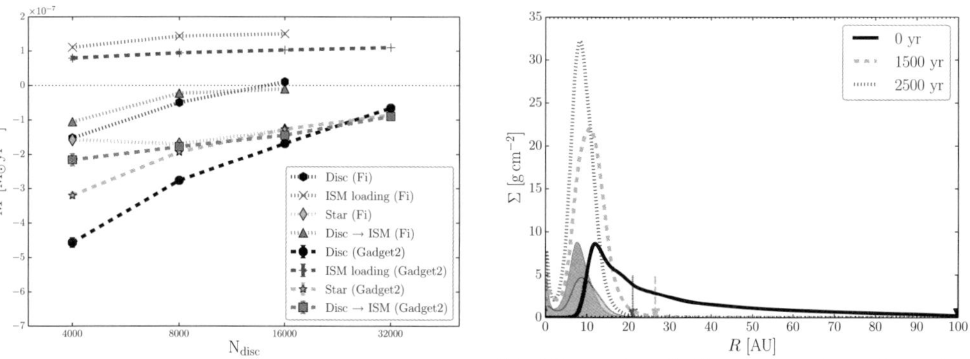

(a) Average mass loss and gain rates as a function of the number of disc particles, N_{disc}, for both SPH-codes Fi (dotted) and Gadget2 (dashed). The change of the disc mass (black) consists of 3 components: mass is lost due to accretion onto the star (purple) and stripping at the outer edge (brown), and gained from sweeping up ISM by the disc and star (green).

(b) The azimuthally averaged surface density profile for a simulation with Gadget2 and 16.000 disc particles at the start (solid black), after 1500 year (dashed purple) and 2500 year (dotted green). The contribution of swept-up ISM at each moment is plotted in the corresponding transparent colour, while the coloured arrow indicates the disc radius at that moment.

Figure 1.

where $\rho_{\rm ISM}$ and $v_{\rm ISM}$ are the density and velocity of the ISM with respect to the protoplanetary disc and $R_{\rm disc}$ the radius of the disc. This means that the effective surface area of the disc is smaller than its actual surface area. We can explain this using the surface density profile of the discs at different times (Fig. 1b). The swept up ISM (shaded regions) is only located at smaller radii and never extends to the outer radius of the disc (coloured arrows), indicating that the ISM flows around the disc at the outer edge instead of being entrained by it.

The outer parts of the disc are continuously being stripped (brown symbols in Fig. 1a) by the ram pressure of the flow. The disc therefore not only loses mass (black symbols in Fig. 1a), but also angular momentum. In our simulations, the disc loses both its mass and angular momentum on a time scale of $10^3 - 10^4$ years, which would make the early disc accretion scenario infeasible. However, our convergence study suggests that this is a numerical rather than physical effect and that it can be attributed to (the treatment of) the artificial viscosity in the disc. This work will be presented extensively in Wijnen *et al.* 2015.

References

Bastian N., Lamers H. J. G. L. M., de Mink S. E., Longmore S. N., Goodwin S. P., & Gieles M. *MNRAS*, 436, 2398

Pelupessy F. I., van der Werf P. P., & Icke V. 2004, *A&A*, 422, 55

Pelupessy F. I., van Elteren A., de Vries N., McMillan S. L. W., Drost N., & Portegies Zwart S. F. 2013, *A&A*, 557, 84

Piotto G., Bedin L. R., Anderson J., King I. R., Cassisi S., Milone A. P., Villanova S., Pietrinferni A., & Renzini A. 2007, *ApJ*, 661, L53

Springel V. 2005, *MNRAS*, 364, 1105

Wijnen T. P. G.., Pols O. R., Pelupessy F. I., & Portegies Zwart S. 2015, *submitted to MNRAS*

Formation, evolution, and survival of massive star clusters
Proceedings IAU Symposium No. 316, 2015
C. Charbonnel & A. Nota, eds.

doi:10.1017/S174392131500890X

Globular Clusters within Dark Matter Halos : Case Studies of 47 Tuc, NGC 1851 and M 15

Joowon Lee[1], Jihye Shin[2] and Sungsoo S. Kim[1,3]

[1]School of Space Research, Kyung Hee University,
Yongin, Gyeonggi , 446-701, Korea
email: joowon.lee@khu.ac.kr

[2]Kavli Institute for Astronomy and Astrophysics, Peking University,
Yi He Yuan Lu 5, Hai Dian District, Beijing 100871, China

[3]Department of Astronomy and Space Science, Kyung Hee University,
Yongin, Gyeonggi , 446-701, Korea

Abstract. Globular clusters (GCs) are known to have a very small amount of or no dark matter (DM). Even if GCs are formed in individual DM halos, they must have lost the majority of the DM through dynamical processes such as mass segregation or tidal stripping. Using Fokker-Planck (FP) calculations, we investigate the dynamical evolution of three Galactic GCs with an assumption that they were formed in mini DM halos. We trace the amount of DM of 47 Tuc, NGC 1851, and M 15, which are a 'disk/bulge' cluster, an 'old halo' cluster, and a 'young halo' cluster, respectively. We find that these three GCs must have initially had insignificant amounts of DM, less than 10 percent of the initial stellar mass of each cluster.

Keywords. stellar dynamics, methods: numerical, galaxy: globular clusters

---

1. GC formation within DM halo

Since Peebles & Dicke (1968) first proposed a primordial scenario of globular clusters (GCs) within individual dark matter (DM) halos, the existence of DM halo in GC formation has been debated. However, it was found that many Galactic GCs have tidal tails. Thus, numerical simulations showed that these tidal tails cannot be formed if DM halos envelop the GCs (Moore 1996). Meanwhile, several numerical simulations showed that even DM which once embedded the GCs can be tidally stripped away by Galactic tidal field or removed by mass segregation (Bromm & Clarke 2002; Mashchenko & Sills 2005; Baumgardt & Mieske 2008).

So far, numerical studies on the initial amount of DM in two GCs have been done. NGC 2419 could still have DM halo because it is less affected by Galactic tidal field due to the large Galactocentric radius. However, previous studies suggested that NGC 2419 unlikely had significant amount of DM (Baumgardt *et al.* 2009; Conroy *et al.* 2011). NGC 6397 is one of the typical Galactic GCs, and it might suffer significant mass segregation and tidal stripping, more than NGC 2419. However, the amount of the maximum initial DM of NGC 6397 is less than the initial stellar mass (Shin *et al.* 2013).

Although previous numerical studies found that two GCs did not show any evidence of significant amount of DM, further studies are still needed to see if there are any GC that have been formed with a significant amount of DM. Therefore we select three Galactic GCs (47 Tuc, NGC 1851 and M 15) which have different properties and origins. Then we find their initial amount of DM by comparing the results of Fokker-Planck (FP) calculation to the observational data (surface brightness and velocity dispersion).

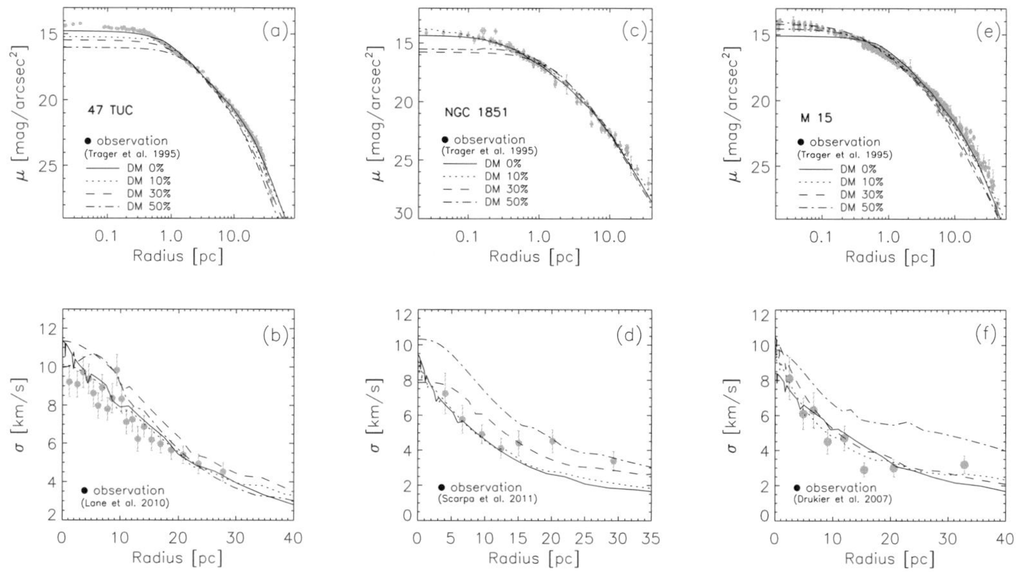

Figure 1. Surface brightness and velocity dispersion profiles of 47 Tuc, NGC 1851 and M 15. Grey dots are observational data (references are in each panel) and black lines are results of FP calculations which have different value of M_{DM}/M_{star}.

2. Initial amount of DM

We compare our FP results with the present-day observational data. We try hundreds of models which have different sets of initial conditions and find the best-fit model whose $M_{DM}/M_{star} = 0$ (no DM), 1, 3, and 5. Figure 1 shows the surface brightness (upper panel) and velocity dispersion profiles (lower panel) of three clusters. We find that 47 Tuc, NGC 1851 and M 15 have the best-fit model when M_{DM}/M_{star}=0 or 0.1 . When M_{DM}/M_{star} increases, the surface brightness profiles of models underestimate the central value while the velocity dispersion profiles overestimate the overall value.

47 Tuc and NGC 1851 are bulge/disk and old halo clusters which are believed to be native GCs. Otherwise, M 15 is a young halo cluster which might be accreted into the Galaxy. Therefore, if M 15 has lost the majority of DM in early phase of accretion, M 15 possibly had more DM mass initially.

We conclude that the three Galactic GCs, 47 Tuc, NGC 1851 and M 15 did not contain significant amount of DM regardless of their origins. Our result is similar to that of NGC 2419 and NGC 6397.

References

Baumgardt, H. & Mieske, S. 2008, *MNRAS*, 391, 942
Baumgardt, H., Côté, P., & Hilker, M., *et al.* 2009, *MNRAS*, 396, 2051
Bromm, V. & Clarke, C. J. 2002, *ApJ*, 566, L1
Conroy, C., Loeb, A., & Spergel, D. N. 2011, *ApJ*, 741, 72
Mashchenko, S. & Sills, A. 2005, *ApJ*, 619, 258
Moore, B. 1996, *ApJ*, 461, L13
Peebles, P. J. E. & Dicke, R. H. 1968, *AJ*, 154, 891
Shin, J., Kim, S. S., & Lee, Y.-W. 2013, *JKAS*, 46,173

Formation, evolution, and survival of massive star clusters
Proceedings IAU Symposium No. 316, 2015
C. Charbonnel & A. Nota, eds.

doi:10.1017/S1743921315009758

The ODD Old, Super-Metal-Rich Open Cluster, NGC 6791

Ann Merchant Boesgaard[1,3], Michael G. Lum[1,3] and Constantine P. Deliyannis[2,3]

[1]Institute for Astronomy, University of Hawai'i at Manoa, 2680 Woodlawn Drive, Honolulu, HI 96822

[2]Department of Astronomy, Indiana University 727 East 3rd Street, Swain Hall West 319, Bloomington, IN 47405-7105

[3]Visiting Astronomer, W. M. Keck Observatory jointly operated by the California Institute of Technology and the University of California.

Abstract. We report on the composition of turn-off stars in the intriguing open cluster, NGC 6791, which is old, but super-metal-rich using Keck/HIRES spectra. We find [Fe/H] = +0.30 ±0.02[O/Fe]$_n$ −0.06 ±0.02,[Mg/Fe], [Si/Fe], [Ca/Fe], and [Ti/Fe] near solar and the two Fe-peak elements, Cr and Ni, are consistent with Fe.

1. Introduction

NGC 6791 is a fascinating, unique cluster. For an open cluster it is very massive at ∼4000 $M_{\odot}$ (King *et al.* (2005)). Its age is ∼8.3 Gyr (Brogaard *et al.* (2012)). The metallicity is at least twice that of the sun (e.g., Boesgaard *et al.* (2009)). NGC 6791 has a heliocentric distance of ∼4 kpc (King *et al.* (2005)). It has a Galactic latitude of +11° which makes it 1 kpc above the Galactic plane. Most open clusters are near or in the Galactic plane. Bedin *et al.* (2006) found the absolute proper motion of the cluster. They determine orbital parameters and find that it has a *boxy* orbit with high eccentricity (ϵ ∼0.5. Its perigalactic distance is about 3 kpc and its apogalactic distance is about 10 kpc. The orbital period is ∼130 Myr and it has crossed the Galactic plane several times. It has remained as an intact cluster due to its high mass and density. It may have originated in the inner regions of the Galaxy (Bedin *et al.* (2006)). Chemical evolution models are able to produce [Fe/H] $\leqslant$ +0.30 for galactocentric distances of <4-5 kpc even at early ages. Following its formation in the inner part of the Milky Way, NGC 6791 may have been ejected into its current high eccentricity orbit by a massive feature, such as a bar (Jilkova *et al.* (2012)).

2. Observations and Abundances

Previous studies of [Fe/H] in NGC 6791 have been mostly low and medium resolution, primarily of K and M giants. We chose to observe faint (V ∼17.4) main-sequence turn-off at high spectral resolution (∼ 46,000) with the Keck 10 m telescope and its upgraded HIRES spectrograph. We used B and V values are from Stetson *et al.* (2003), the differential reddening corrections of Brogaard *et al.* (2012) and the calibrations of Casagrande *et al.* (2010) to determine temperatures. We also found temperatures spectroscopically from the agreement of Fe lines with a range of excitation potentials. The mean difference between $T_{\rm eff}$s determined from the colors and those from excitation potentials is −6 K ±53 K. The details can be found in Boesgaard *et al.* (2015). We used model atmospheres

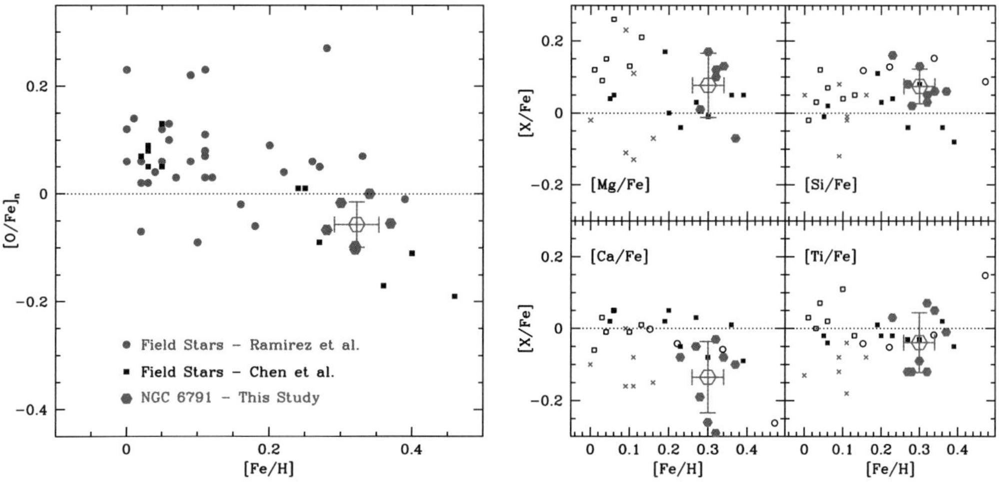

Figure 1. Left: Our results for [O/Fe] as corrected for nLTE effects are the (red) hexagons and are compared with those from (Ramirez *et al.* (2013) and Chen *et al.* (2003). The stars selected from Ramirez are shown as filled (blue) circles and are metal-rich ([Fe/H] >0.00) and old (age >7.7 Gyr). The old (8–10 Gyr) metal-rich ([Fe/H] > 0.01) field stars from Chen *et al.* (2003) are filled squares (black). The turn-off stars in NGC 6791 have values of $[O/Fe]_n$ that are lower than solar $[O/Fe]_n$. The trend of decreasing [O/Fe] with increasing [Fe/H] can be seen. Right: Abundances of the four alpha-elements relative to Fe shown as filled (red) hexagons. The cluster mean with error bars are shown as open hexagons. The comparison field stars are all old (>8 Gyr) and metal rich. The filled squares are from Chen *et al.* (2003), the open squares from Edvardsson *et al.* (1993), the open circles from Feltzing & Gonzales (2001), the crosses (blue) are from Reddy *et al.* (2006) as selected by age and [Fe/H] from (Ramirez *et al.* (2013).

of Kurucz (1993). We measured the equivalent widths for Fe I and Fe II lines which were unblended and unsaturated. The lines are unsaturated when $\log W/\lambda \leqslant -4.82$. We typically measured 37-46 Fe I lines and 5-8 Fe II lines. The values of [Fe I/H] and [Fe II/H] agreed very well and the mean for the eight turn-off stars is [Fe/H] = 0.30 ±0.02

The O I triplet lines at 7774 Å are well-separated. In order to make a differential analysis with respect to the Sun, we used a 10 s exposure of the Moon that we had taken with Keck/HIRES in 2003 January 11 UT. The measured S/N in the region of the O I triplet is 350. We measured the equivalent widths for the 3 lines in the lunar and stellar spectra and derived the LTE abundance for each line. Takeda (2003) has determined the nLTE corrections for each O I line for an array of temperatures, log g values, and Fe abundances. We applied the appropriate corrections for each line by interpolating between Takeda's tables for our parameters. In Figure 1a (left) we show a comparison of our $[O/Fe]_n$ versus[Fe/H] with results for field stars. The main field star sample is from Ramirez *et al.* (2013), selected to have [Fe/H] greater than 0.0 and ages older than 7.7 Gyr. Those 34 stars are from their total sample of 835 nearby FGK stars. Also shown in the figure are the [O/Fe] and [Fe/H] from the sample of old, metal-rich stars of Chen *et al.* (2003). These field stars are between 8–10 Gyr and have [Fe/H] values > 0.00. The values for [O/Fe]n in NGC 6791 turn-off stars fit the pattern of decreasing [O/Fe] with increasing [Fe/H]. There is no sign of an intrinsic spread in $[O/Fe]_n$ in our six turn-off stars, which is consistent with a single population. We determined abundances of the alpha-element from 2-3 lines of Mg I, 7-9 Si I lines, 7-9 Ca I lines and 6-8 Ti I lines. The cluster mean values are [Mg/Fe] = +0.08 ±0.04, [Si/Fe] = +0.07 ±0.02, [Ca/Fe] = −0.13 ±0.04, and [Ti/Fe] = −0.04 ±0.03. Figure 1b shows a plot of each of the four alpha-elements relative to Fe in each star as compared to the old and metal-rich field stars. The comparison stars are from Chen *et al.* (2003), Edvardsson *et al.* (1993), Feltzing & Gonzales (2001), and Reddy *et al.* (2006). We were able to measure 5-6 unsaturated, unblended lines of Cr I and 11-16 unsaturated, unblended lines of Ni I. Both Fe-peak elements are similar to Fe in our turn-off stars. The comparison with old, metal-rich field stars hints that the NGC 6791 stars might possibly be slightly enriched in Cr and Ni.

References

Bedin, L. R., Piotto, G., Carraro, G., King, I. R., & Anderson, J. 2006, *A&A*, 460, L27
Boesgaard, A. M., Jensen, E. C., & Deliyannis, C. P. 2009, *ApJ*, 633, 398
Boesgaard, A. M., Lum, M. G., & Deliyannis, C. P. 2015, *ApJ*, 599, 202
Brogaard, K., VandenBerg, D. A., Bruntt, H. *et al.* 2012, *AAP*, 543, A106
Casagrande, L., Ramìrez, I., Melèndez, J., Bessell, M. & Asplund, M. 2010 *AAP*, 512, 54
Chen, Y. Q., Zhao, G., Nissen, P. E., Bai, G., & Qui, H. M. 2003, *ApJ*, 591, 925
Edvardsson, B., Andersen, J., Gustafsson, B. *et al.* 1993, *A&A*, 275, 101
Feltzing, S. & Gonzales, G. 2001, *A&A*, 367, 253
Jilkova, L., Carraro, G., Jungwiert, B., & Minchev, I. 2012, *AAP*, 541, 64
King, I. R., Bedin, L. R., Piotto, G., Cassisi, S., & Anderson, J. 2005, *AJ*, 130, 626
Kurucz, R. L. 1993, CD-ROM 13 (Cambridge: Smithsonian Astrophys. Obs.)
Ramirez, I., Allende Prieto, C., & Lambert, D. L. 2013, *ApJ*, 764, 78
Reddy, B., Lambert, D. L., & Allende Prieto, C. 2006, *MNRAS*, 367, 1329
Stetson, P. B., Bruntt, H., & Grundahl, F. 2003, *PASP*, 115, 413
Takeda, Y. 2003, *A&A*, 402, 343

Formation, evolution, and survival of massive star clusters
Proceedings IAU Symposium No. 316, 2015
C. Charbonnel & A. Nota, eds.

doi:10.1017/S1743921316011790

The Composition of the Bulge Globular Cluster NGC 6273

C. A. Pilachowski[1], C. I. Johnson[2], R. M. Rich[3], N. Caldwell[2], M. Mateo[4], J. I. Bailey, III[4] and J. D. Crane[5]

[1]Indiana University Bloomington
Department of Astronomy SW319
727 E 3rd St.
Bloomington, IN 47405 USA
email: cpilacho@indiana.edu
[2] Harvard-Smithsonian Center for Astrophysics
60 Garden St. MS-15
Cambridge, MA 02138 USA
email: cjohnson@cfa.harvard.edu
[3]University of California at Los Angeles
[4]University of Michigan at Ann Arbor
[5]The Observatories of the Carnegie Institution of Washington

Abstract. Observations of red giants in the Bulge globular cluster NGC 6273 with the Michigan/Magellan Fiber System (M2FS) mounted on the Nasmuth-East port of the Magellan-Clay 6.5-m telescope at the Las Campanas Observatory reveal a spread in metallicity. Members have been confirmed with radial velocity. NGC 6273 has at least two populations separated by 0.2-0.3 dex in [Fe/H]. The sodium and aluminum abundances are correlated while the magnesium and aluminum abundances are anti-correlated. The cluster also shows a rise in the abundance of the s-process element lanthanum with [Fe/H] similar to other massive clusters. The cluster contains a possible third population depleted in most elements by 0.3 dex.

Keywords. globular clusters: individual (NGC 6273), stars: abundances

1. Introduction

The Bulge globular cluster NGC 6273 (M19) lies only 1.7 kiloparsecs from the Galactic Center (Harris 1996, 2010 edition), and is among the most massive and luminous globular clusters in the Milky Way. The cluster resembles ω Centauri in many ways, including its overall metallicity ([Fe/H]$\approx$–1.75), its high ellipticity ($e = 0.28$), and its multi-modal horizontal branch.

As part of a study of clusters in the Galactic Bulge, we obtained high resolution spectra of giant stars in NGC 6273 to examine its composition in more detail and specifically to determine if the cluster's stars exhibit a distribution of metallicities similar to ω Cen. Spectra were obtained using the Michigan/Magellan Fiber System (M2FS) on the Magellan-Clay 6.5-m telescope at the Las Campanas Observatory.

2. Results

Analysis of the spectra reveals a range of [Fe/H] abundance among the giants of NGC 6273. The cluster contains an extended metallicity distribution similar to that of ω Cen, as shown in Figure 1 from Johnson *et al.* (2015b). We also include metallicity distribution functions for four other clusters with extended metallicity distributions, including ω Cen (Johnson & Pilachowski 2010), M22 (Marino *et al.* 2011), M2 (Yong *et al.* 2014), and

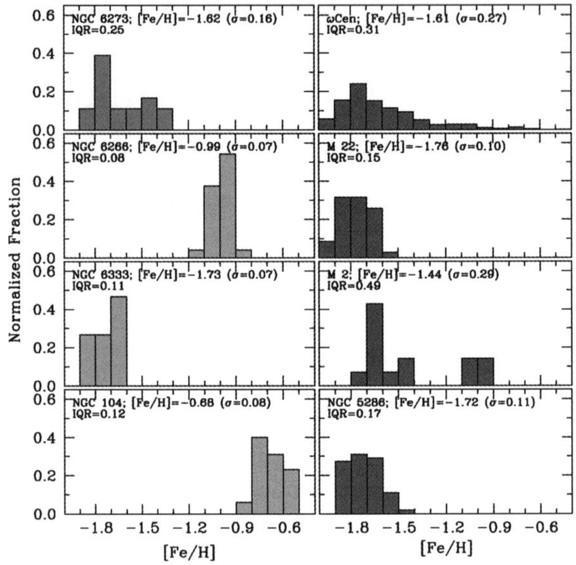

Figure 1. Metallicity distribution function for NGC 6273. Also shown are distribution functions for four globular clusters with a range of observed iron abundances, and for three clusters with no observed spread in [Fe/H]. See text for references. This figure is reproduced from Johnson *et al.* (2015b).

NGC 5286 (Marino *et al.* 2015). We also include three clusters with essentially uniform iron abundances: 47 Tuc (Johnson *et al.* 2015a), and NGC 6266 and NGC 6333, which were also observed with M2FS.

NGC 6273 shows an extended metallicity distribution function with at least two populations of stars. The width of the metallicity distribution function is highly reminiscent of what is seen in ω Cen. NGC 6273 appears to contain at least two populations of stars, a metal-poor population centered at [Fe/H]=–1.75, and a higher metallicity population centered at [Fe/H]=–1.50. We also found one anomalous star at [Fe/H]=–1.3 that may indicate the presence of a third population. The anomalous star's composition is also unusual, showing [m/Fe] deficiencies of roughly –0.3 dex for most elements.

Element to iron abundance ratios in NGC 6273 also share similarities with ω Cen, including a pattern of decreasing dispersion in [Na/Fe] with increasing metallicity, a bimodal signature in [Al/Fe], and an increase in [La/Fe] with metallicity. The increase in the [La/Eu] ratio suggests the increase in [La/Fe] is due to s-process enrichment.

References

Dotter, A., Chaboyer, B., Jevremović, D. *et al.* 2008, *ApJS*, 178, 89
Harris, W. E. 1996 (2010 edition), *AJ*, 112, 1487
Johnson, C. I. & Pilachowski, C. A. 2010, *ApJ*, 722, 1371
Johnson, C. I., McDonald, I., Pilachowski, C. A., *et al.* 2015a, *AJ*, 149, 71
Johnson, C. I., Rich, R. M., Pilachowski, C. A., *et al.* 2015b, *AJ*, 150:63
Marino, A. F., Sneden, C., Kraft, R. P., *et al.* 2011, *A&A*, 532, A8
Marino, A. F., Milone, A. P., KaraKas, A. I, *et al.* 2015, *MNRAS*, 450, 815
Yong, D., Roderer, I., Grundahl, F. *et al.* 2014, *MNRAS*, 441, 3396

Formation, evolution, and survival of massive star clusters
Proceedings IAU Symposium No. 316, 2015
C. Charbonnel & A. Nota, eds.

doi:10.1017/S1743921315009369

The birthrates of SNe Ia in globular clusters

Bo Wang[1,2] **and Dongdong Liu**[1,2]

[1]Yunnan Observatories, Chinese Academy of Sciences, Kunming 650216, China (email: wangbo@ynao.ac.cn)

[2]Key Laboratory for the Structure and Evolution of Celestial Objects, Chinese Academy of Sciences, Kunming 650216, China

Abstract. Globular clusters has been proposed as testbeds for type Ia supernovae (SNe Ia). In this work, using a detailed binary population synthesis approach, we studied the birthrates of SNe Ia from various progenitor models in globular clusters, including the single-degenerate model, the double-degenerate model and the Sub-Chandrasekhar model. Here, a single starburst with a total mass of $10^6 M_\odot$ is assumed.

Keywords. binaries: close – supernovae: general – white dwarfs

1. Introduction

Type Ia supernovae (SNe Ia) have high values in astrophysics, especially in Cosmology and the chemical evolution of galaxies (e.g., Howell 2011). They are thought to be thermonuclear explosions of accreting carbon–oxygen white dwarfs (CO WDs) in close binaries, however, the nature of the mass donor is still uncertain (see Nomoto *et al.* 1997).

Several progenitor models have been proposed to produce SNe Ia, that is, the single-degenerate (SD), double-degenerate (DD) and Sub-Chandrasekhar (Sub-Ch) mass models (e.g., Wang & Han 2012; Maoz *et al.* 2014). In the SD model, a CO WD can accrete H- or He-rich matter from a non-degenerate star to increase its mass to approach the Chandrasekhar mass, and then form an SN Ia, in which the donor star could be a main-sequence star (WD+MS channel), a red giant (WD+RG channel), or a He star (WD+He star channel). In the DD model, SNe Ia arise from the merging of two CO WDs. The closeness of two WDs is due to common envelope evolution, which then enables gravitational wave radiation to drive orbital inspiral to merger. In the Sub-Ch model, the explosion of a CO WD is triggered by the detonation of a substantial surface layer of accreted He.

The delay times of SNe Ia are defined as the time interval between the star formation and SN explosion. The various progenitor models of SNe Ia can be examined by comparing the delay time distributions expected from a progenitor model with that of observations. Globular clusters has been proposed as testbeds for SNe Ia (e.g., Pfahl *et al.* 2009). In this work, we will give the delay time distributions of SNe Ia for various progenitor models in globular clusters using a binary population synthesis method.

2. Model and Results

In order to give the delay times of SNe Ia from various models, we first employed the Eggelton stellar evolution code (Eggleton 1973) and adopted the prescription of Hachisu *et al.* (1999) for the accretion efficiency of a CO WD, and carried out detailed binary evolution calculations for close CO WD binaries, and mapped out the initial parameters in the orbital period-secondary mass plane for various WD masses which lead to an SN

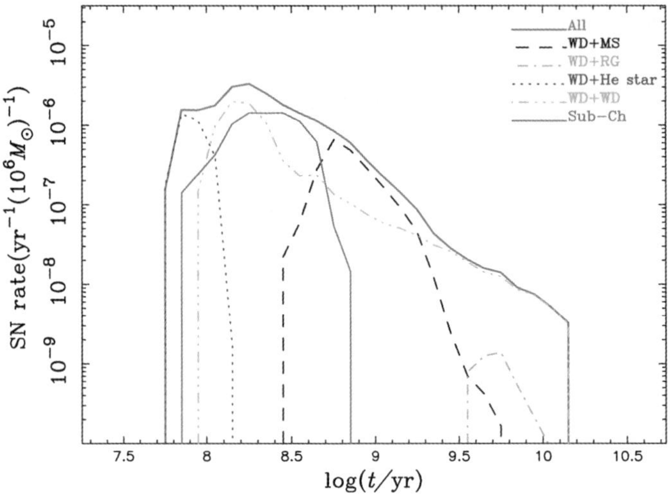

Figure 1. Delay time distributions of SNe Ia for different progenitor models of SNe Ia in the globular clusters, including the WD+MS, WD+RG, WD+He star, WD+WD and Sub-Ch models, in which a single starburst with a total mass of $10^6 M_\odot$ is assumed.

Ia. We then implemented these results into a binary population synthesis study, which is to follow the evolution of millions of primordial binaries, to obtain the birthrates of SNe Ia. We performed a series of Monte Carlo simulations in the binary population synthesis study. In each simulation, we followed the binary evolution via the rapid binary evolution code developed by Hurley *et al.* (2002).

In Figure 1, we present the delay time distributions of SNe Ia for different progenitor models of SNe Ia in the globular clusters. Here, a single starburst with a total mass of $10^6 M_\odot$ is assumed. From this figure, we can see that the WD+He star and Sub-Ch models produce SNe Ia with short delay times, while the WD+MS model contributes to SNe Ia with intermediate and long delay times. In addition, the WD+RG model contributes to long delay time SNe Ia. Furthermore, the DD model may produce both short time delay SNe Ia and long time delay ones. For details of the calculations, we refer the reader to Wang *et al.* (2010a) for the WD+MS model, to Wang *et al.* (2009a,b) for the WD+He star model, Wang & Han (2010) for the WD+RG model, Wang *et al.* (2010b) for the WD+WD model, Wang *et al.* (2013) for the Sub-Ch model.

References

Eggleton, P. P. 1973, *MNRAS*, 163, 279
Hachisu, I., Kato, M., Nomoto, K., & Umeda, H. 1999, *ApJ*, 519, 314
Howell, D. A. 2011, *Nature Communications*, 2, 350
Hurley, J. R., Tout, C. A., & Pols, O. R. 2002, *MNRAS*, 329, 897
Maoz, D., Mannucci, F., & Nelemans, G. 2014, *ARA&A*, 52, 107
Nomoto, K., Iwamoto, K., & Kishimoto, N. 1997, *Science*, 276, 1378
Pfahl, E., Scannapieco, E., & Bildsten, L. 2009, *ApJ*, 695, L111
Wang, B. & Han, Z. 2012, *New Astron. Rev.*, 56, 122
Wang, B., Justham, S., & Han, Z. 2013, *A&A*, 559, A94
Wang, B., Li, X.-D., & Han, Z. 2010a, *MNRAS*, 401, 2729
Wang, B., Liu, Z., Han, Y., *et al.*, 2010b, *Sci. China Ser. G*, 53, 586
Wang, B., Meng, X., Chen, X., & Han, Z. 2009a, *MNRAS*, 395, 847
Wang, B., Chen, X., Meng, X., & Han, Z. 2009b, *ApJ*, 701, 1540

Formation, evolution, and survival of massive star clusters
Proceedings IAU Symposium No. 316, 2015
C. Charbonnel & A. Nota, eds.

doi:10.1017/S1743921315006870

Evidences of tidal distortion and mass loss from the old open cluster NGC 6791

Giovanni Carraro[1,2] and Emanuele Dalessandro[3]
[1]ESO, Alondo de Cordova 3107, 19001, Santiago de Chile, Chile
[2]Dept. of Physics and Astronomy, University of Padova, Vicolo Osservatorio 3, 35122, Padova, Italy
email: gcarraro@eso.org
[3]Dept. of Physics and Astronomy, University of Bologna, Via Ranzani 1, 40127, Bologna, Italy
email: emanuele.dalessandr2@unibo.it

Abstract. We present the first evidence of clear signatures of tidal distortions in the density distribution of the fascinating open cluster NGC 6791. We find that the 2D density map shows a clear elongation and an irregular distribution starting from $\sim 300''$ from the cluster center and two tails extending in opposite directions beyond the tidal radius. These features are aligned to both the absolute proper motion and to the Galactic centre directions. Accordingly we find that both the surface brightness and star count density profiles reveal a departure from a King model starting from $\sim 600''$. These observational evidences suggest that NGC 6791 is currently undergoing mass-loss likely due to gravitational shocking and interactions with the tidal field of the Milky Way. We derive the expected mass-loss due to stellar evolution and tidal interactions and we estimate the initial cluster mass to be $M_{ini} = (1.5 - 4.0) \times 10^5 M_{\odot}$.

Keywords. Open clusters: general - Open clusters: individual: NGC 6791

1. Context and results

NGC 6791 is one of the most massive open cluster (Carraro 2014) in the MW, possibly harbouring more than a stellar population (Geisler *et al.* 2012, Bragaglia *et al.* 2014).We used deep images obtained with the wide field imager MegaCam mounted at the Canada-France-Hawaii Telescope (CFHT) to cover a $2^o \times 2^o$ area around the cluster (Dalessandro *et al.* 2015). The upper left panel in Fig 1 shows the obtained (g', $g' - r'$) CMDs of the innermost region of NGC 6791 (left) and the external *control field* (right) including stars located at a distance $r \geqslant 3000''$ from the cluster center. Large scale 2D colour-coded surface density map of NGC6791 obtained by using the optical matched filter technique is shown in the upper right panel of Fig. 1. The contour levels span from 3σ to 40σ with irregular steps. The solid arrow represents the direction of the absolute proper motion while the dashed ones mark the direction of the Galactic Center and that perpendicular to the Galactic plane (Z=0).The observed star count density profile is shown in the upper left panel of Fig. 1 (open grey squares). The dashed line represents the density value of the background as obtained in the *control field*. The black filled dots are densities obtained after background subtraction. The best single-mass King model is also over-plotted to the observations (solid line). For $r \geqslant 600''$ the density profile clearly deviates from the King model following a power-law with exponent $\alpha \sim -1.7$.By using a simple analytic approach (Lamers *et al.* 2005), we estimated the mass likely lost by NGC 6791 during its evolution because of the effect of both stellar evolution and dynamical interactions. On this basis we estimated the cluster initial mass as a function of the dissolution time parameter t_0. In particular the red curve (lower right panel in Fig. 1) shows the dependence for the current cluster mass (5000 $M_{\odot}$) and its actual age of 8 Gyr.

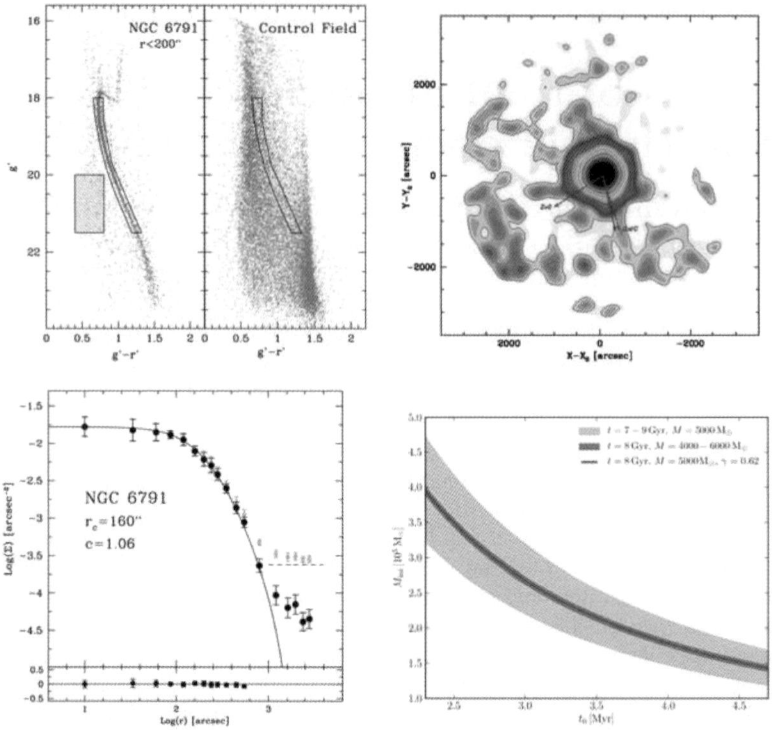

Figure 1. **Upper left**: CMD of NGC 6791 and nearby field. **Upper right**: star counts . **Lower left**: King profile fitting . **Lower right**: Mass at birth estimate.

2. Conclusions

NGC 6791 shows clear evidence of tidal features in its star distribution in the form of irregular but evident elongations and tidal tails. These features are present also in the star density and surface brightness profile and they represent clear indication of recent mass-loss. By using the simple recipes we derived the initial mass of NGC6791 to be $M_{ini} = (1.5 - 4) \times 10^5 M_{\odot}$, i.e. several tens larger than its present day mass. This finding would qualitatively explain why the cluster could have survived for such a long time contrary to the expectations of current estimates of the destruction rate of Galactic open clusters.

References

Bragaglia, A., Sneden, C., Carretta, E., Gratton, R., Lucatello, S., Bernath, P., Brooke, J. S. A., & Ram, R. S. *ApJ*, 796, 68

Carraro, G. 1995, *ASPC*, 482, 245

Dalessandro, E., Miocchi, P., Carraro, G., Jilkova, L., & Moitinho, A. 2015, *MNRAS*, 449, 1811

Geisler, D., Villanova, S., Carraro, G., Pilachowski, C., Cummings, J., Johnson, C. I., & Bresolin, F. 2012, *ApJ*, 756, 40

Lamers, H., Gieles, M., Bastian, N., Baumgartd, H., Kharchenko, N., & Portegies Zwart, S. 2005, *A&A*, 441, 17

Formation, evolution, and survival of massive star clusters
Proceedings IAU Symposium No. 316, 2015
C. Charbonnel & A. Nota, eds.

doi:10.1017/S1743921315006845

Ages and chemical compositions of massive clusters in NGC147 and M31

Margarita Sharina[1] and Vladislav Shimansky[2]

[1]Special Astrophysical Observatory, Russian Academy of Sciences, Nizhnii Arkhyz, 369167 Russia

[2] Kazan (Volga region) Federal university, Kazan, 420008 Russia
email: sme@sao.ru, Slava.Shimansky@kpfu.ru

Abstract. We present estimates of ages, [Fe/H], helium content (Y) and abundances of C, N, Mg, Ca, and several other elements for the following globular clusters (GCs): GC7 in NGC147, and Mayall II, Mackey 1 and Mackey 6 in M31. Medium-resolution integrated-light spectra of the GCs were conducted with the 6m telescope. To derive the ages and abundances for the GCs we carried out their population synthesis using model stellar atmospheres, the Padova YZVAR isochrones and the Chabrier mass function. We compare the results with the corresponding data obtained using the same method for several massive Galactic GCs. We show that the differences in the light-element abundances between GCs with similar ages and metallicities may reach 0.5-0.6 dex. The corresponding differences for other elements are usually 2-3 times smaller. We suggest that at least partially the detected differences may be due to light-element abundance variations in the atmospheres of high-luminosity red giant branch stars as a consequence of the transportation of the produced elements to the surface layers.

Keywords. galaxies: individual (M31, NGC147) - galaxies: star clusters: abundances

---

Old massive GCs in different galaxies show similar structural properties and photometric colors (e.g. Veljanoski *et al.* /, 2013). It is not known whether the processes of their formation were also similar and whether element abundance variations exist in most massive of them as it is observed in our Galaxy (Gratton *et al.* /, 2012).

We carry out population synthesis of medium-resolution (FWHM$\sim 5\mathring{A}$) integrated light spectra of massive GCs near M31 using model stellar atmospheres and estimate abundances of chemical elements. The observational instrumental setup and methods of the spectroscopic reduction are described by Makarov *et al.* (2015). Our method of chemical element abundance analysis and determination of evolutionary parameters using stellar atmosphere models was described by Sharina *et al.* (2014, 2013, hereafter: S14, 13) and Khamidullina *et al.* (2014a,b, hereafter: Kh14a, b). In these papers the method was tested using medium-resolution integrated light spectra of Galactic GCs. In particular, it was shown by S13 that it is possible to derive simultaneously age and helium abundance using medium-resolution integrated light spectra of GCs, because of the influence of these two parameters on the hydrogen absorption line profiles.

Table 1. Age [Gyr], metallicity [dex] and helium abundance [dex] for the studied GCs.

Object	Log t (ours)	[Fe/H] (ours)	Y (ours)	Log t (Lit.)	[Fe/H] (Lit.)
GC7-N 147	10.1±0.1	-1.70±0.1	0.30±0.03	9.9±0.12[7]	-1.5±0.2[7], -1.8±0.3[8]
Mayall-II	10.15±0.05	-1.00±0.05	0.26±0.02	10.176[6], 10.26±0.4[4]	-0.95±0.09[6], -1.08±0.09[3], -0.9[2]
MGC 1	10.0±0.05	-2.20±0.1	0.30±0.03	9.9±0.1[4]	-2.14[5], -2.15[2]
MGC 6	10.0±0.1	-1.85±0.1	0.30±0.03	10.3±0.1[4],10.134[1]	-2.07±0.18[1], -2.14[5], -1.8[2]

Notes: 1: Fan *et al.* (2012); 2: Federici *et al.* (2012); 3:Huchra *et al.* (1991); 4: Ma (2012); 5: Mackey *et al.* (2007); 6: Meylan *et al.* (2001); 7: Sharina & Davoust (2009); 8: Veljanoski *et al.* (2013)

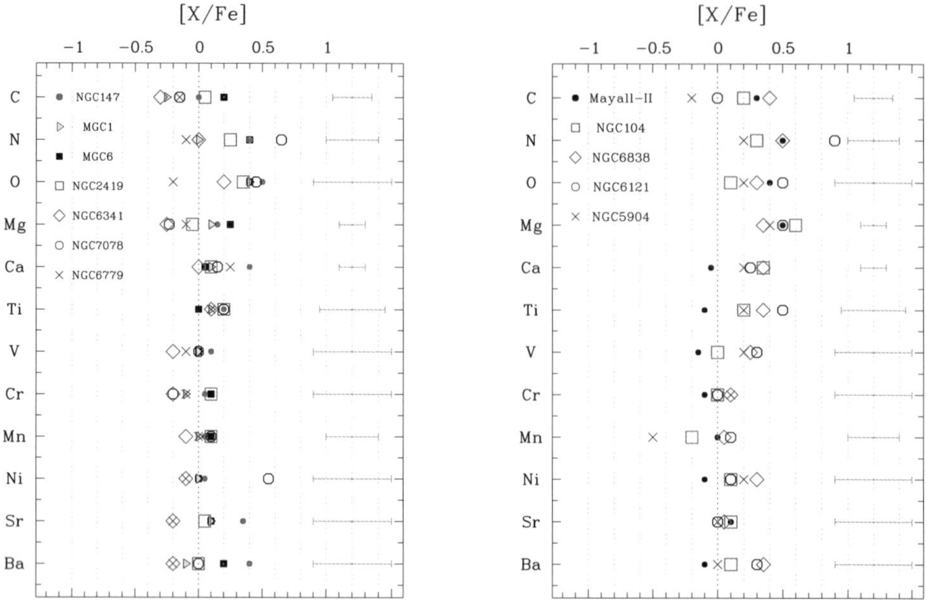

Figure 1. Comparison of chemical abundances derived using our method for GC7-NGC147, MGC1, MGC6, Mayall-II and for several Galactic GCs. Left: [Fe/H] ~ -2.0 dex; right: ~ -1.0 dex. Errors of abundance determination are shown in the right parts of the two plots.

Our results of age, metallicity and helium abundance determination and data from the literature are listed in Table 1. Comparison of chemical abundances for the four extragalactic GCs with the corresponding data for Galactic GCs of similar metallicity derived using the same method (S14, 13; Kh14a, b) is shown in Fig. 1. It is seen that the differences in the light-element abundances between individual GCs in the low-metallicity group (left panel of Fig. 1) and in the high-metallicity group (right panel) are on average larger than the corresponding differences in the abundances of other elements.

Acknowledgements. This presentation was supported by the Russian Scientific Foundation grant 14-12-00965.

References

Fan, Zh., Huang, Y.-F., Li, J.-Z., Zhou, Xu, Ma, Jun, & Zhao, Y.-H. 2012, *Research in Astronomy and Astrophysics*, 12, 829

Federici, L., Cacciari, C., Bellazzini, M., Fusi, Pecci F., Galleti, S., & Perina, S 2012, *A&A*, 544, 155

Gratton, R. G., Carretta, E., & Bragaglia A. 2012 *A&AR*, 20, 50

Huchra, J. P., Brodie, J. P., & Kent, S. M. 1991, *ApJ*, 370, 495

Khamidullina, D., Sharina, M., Shimansky, V., & Davoust, E. 2014a, *Astrophysical Bulletin*, 69, 409

Khamidullina, D., Sharina, M., Shimansky, V., & Davoust, E. 2014b, *Baltik Astronomy*, 23, 260

Ma, Jun 2012, *Research in Astronomy and Astrophysics*, 12, 115

Mackey, A. D. *et al.* 2007, *ApJ* (Letters), 655, 85

Makarov, D. I., Sharina, M. E., Karachentseva, V. E., & Karachentsev, I. D. 2015 *A&A*, 581, 82

Meylan, G., Sarajedini, A., Jablonka, P., Djorgovski, S., Bridges, T., & Rich, R. 2001, *AJ*, 122, 830

Sharina, M., Donzelli, C., Davoust, E., Shimansky, V., & Charbonnel, C. 2014, *A&A*, 570, 48

Sharina, M. E., Shimansky, V., & Davoust, E. 2013, *Astronomy Reports*, 57, 410

Sharina, M. E., & Davoust, E. 2009, *A&A*, 497, 65

Veljanoski, J. *et al.* 2013, *MNRAS*, 435, 3654

Formation, evolution, and survival of massive star clusters
Proceedings IAU Symposium No. 316, 2015
C. Charbonnel & A. Nota, eds.

doi:10.1017/S1743921316008693

Twins born in different environments? Nuclei of two dSphs: isolated galaxy KKS3 and E269-66, a close neighbor of NGC5128

Margarita Sharina[1], **Alexei Kniazev**[2,3,4] **and Igor Karachentsev**[1]

[1]Special Astrophysical Observatory, Russian Academy of Sciences, Nizhnii Arkhyz, 369167, Russia
[2]South African Astronomical Observatory, PO Box 9, Observatory 7935, Cape Town, South Africa
[3]Southern African Large Telescope Foundation, PO Box 9, Observatory 7935, Cape Town, South Africa
[4]Sternberg Astronomical Institute of Lomonosov Moscow 119992, Russia
email: sme@sao.ru, akniazev@saao.ac.za, ikar@sao.ru

Abstract. We present the results of age, metallicity and radial velocity determination for central massive globular clusters (GCs) in dwarf spheroidal galaxies: KKs3 and ESO269-66. KKS3 is a unique isolated galaxy. ESO269-66 is a close neighbor of the giant S0 Centaurus A. The results contribute to the knowledge about the origin of massive star clusters and their host dSphs. The structure and star formation histories of the two dwarf galaxies look rather similar. Both of them have experienced several star-forming events. The most recent ones occurred 1-2 Gyr ago, and most powerful bursts happened 12-14 Gyrs ago. Our analysis has shown that both GCs appear to be 1-2 Gyr younger and 0.1-0.3 dex more metal-rich than the most ancient metal-poor stars in the host dSphs. We examine signatures of multiple stellar population in the GCs using our data. Since central star-forming bursts were extended in time, the massive clusters might be considered as nuclei of the galaxies.

Keywords. galaxies: dwarf: individual (KKs3, ESO269-66) - galaxies: nuclei, formation

---

The close vicinity of giant neighbors determines the environmental mechanisms that have been considered responsible for the evolution of dwarf spheroidal galaxies (dSphs). In the recent years, Karachentsev and collaborators have reported on the discovery of a few truly isolated dSphs in the Local volume. This study focuses on one of these unusual objects, KKs3. It contains a massive globular cluster near its optical center. We have estimated its radial velocity using a medium-resolution spectrum obtained with the RSS spectrograph at the Southern African Large Telescope (SALT). The signal-to-noise ratio in the spectrum was sufficient to estimate the age and metallicity for the GC using simple stellar population models and the method of Lick index diagnostic diagrams (Worthey *et al.* /, 1994; Worthey & Ottaviani, 1997). In the same way we have analyzed another

Table 1. Main parameters of the studied galaxies and their globular clusters.

Dwarf galaxy	Distance [Mpc]	Diameter [kpc]	M_V [mag]	V-I [mag]	$[Fe/H]^{12-14\ Gyr}$ [dex]	**GC in**	M_V [mag]	V-I [mag]	[Fe/H] [dex]	Age [Gyr]	V_h [km/s]
KKs3	2.12[4]	1.5[3]	-12.5[3]	0.85[3]	-1.90[4]	KKs3	-8.5[4]	0.83[4]	-1.55[1]	12[1]	316±7[3]
E269-66	3.82[5]	1.5[5]	-14.4[5]	0.94[5]	-1.75[6]	E269-66	-9.9[2]	0.93[2]	-1.65[1]	12[1]	775±6[1]

Notes: 1: This work; 2: Georgiev *et al.* (2009); 3: Karachentsev *et al.* (2015a); 4:Karachentsev *et al.* (2015b)); 5: Karachentsev *et al.* (2007); 6: Makarova *et al.* (2007).

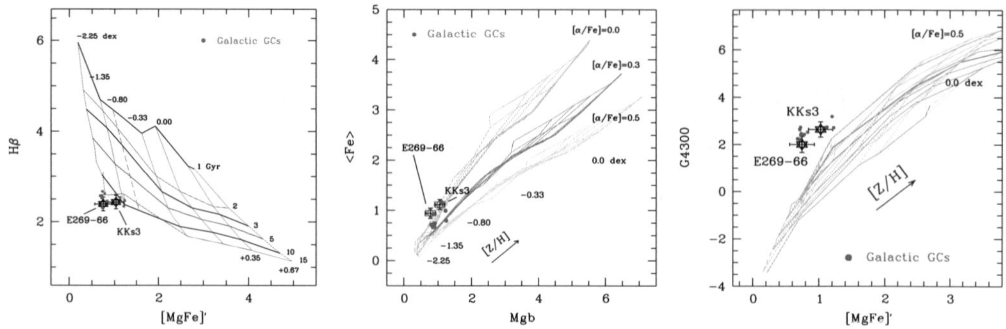

Figure 1. Lick-index diagnostic plots for GCs in KKs3 and E269-66: age vs. metallicity and [α/Fe] vs. metallicity. SSP models by Thomas et a. (2003, 2004) are overplotted.

luminous GC in the center of ESO269-66. The cluster was observed with SALT in the same instrumental configuration.

Medium-resolution spectroscopic analysis is a promising tool for breaking the age-metallicity degeneracy and studying light element abundance anomalies using integrated-light spectra of globular clusters (e.g. Schiavon *et al.* /, 2012, see also Sharina *et al.* (2010) for the description of the used methods). We selected Galactic GCs with the values of the Lick indices (Schiavon *et al.* /, 2012) close to the ones of our two GCs: NGC1904, NGC3201, NGC5946, NGC5986, NGC6254, NGC6752, NGC7089. Fig. 1 illustrates the similarity of the age, metallicity and light-element sensitive indices of Galactic GCs and the two nuclei of dSphs. Many of the selected Galactic GCs have blue horizontal branches, host multiple stellar populations and show light element abundance variations. The resulting ages, metallicities and radial velocities for GCs in KKs3 and E269-66, and photometric and spectroscopic data from the literature are summarized in Table 1. Star formation histories of KKs3 and E269-66 look similar. Determination of the detailed spectroscopic abundances of their nuclei is a subject of future studies.

Acknowledgements

This work is supported by the Russian Scientific Foundation grant 14-12-00965. A.Y.K. acknowledges the support from the National Research Foundation (NRF) of South Africa. Based on SALT observations under the program 2015-1-SCI-030.

References

Georgiev, I. Y., Hilker, M., Puzia, T. H., Goudfrooij, P., & Baumgardt, H. 2009, *MNRAS*, 396, 1075

Karachentsev, I. D., Kniazev, A. Yu., & Sharina, M. E. 2015a, *Astron.Nachr*, 336, 707

Karachentsev, I. D., Makarova, L. N., Makarov, D. I., Tully, R. B., & Rizzi, L. 2015b, *MNRAS* (Letters), 447, 85

Karachentsev, I. D., *et al.* 2007, *AJ*, 133, 504

Makarova, L. N., *et al.* 2007, in: F. Combes and J. Palous (eds.) *Galaxy Evolution Across the Hubble Time*, Proc. IAUS 235 (Cambridge: Cambridge University Press), p. 320

Schiavon, R. P., Caldwell, N., Morrison, H., Harding, P., Courteau, S., MacArthur, L. A., & Graves, G. J. 2012, *AJ*, 143, 14

Sharina, M. E., Chandar, R., Puzia, T. H., Goudfrooij, P., & Davoust, E. 2010 *MNRAS*, 405, 839

Thomas, D., Maraston, C., & Korn, A. 2004, *MNRAS* (Letters), 351, 19

Thomas, D., Maraston, C., & Bender, R. 2003, *MNRAS*, 343, 279

Worthey, G., & Ottaviani, D. L. 1997, *ApJS*, 111, 377

Worthey, G., Faber, S. M., Gonzalez, J. J., & Burstein, D. 1994, *ApJS*, 94, 687

Formation, evolution, and survival of massive star clusters
Proceedings IAU Symposium No. 316, 2015
C. Charbonnel & A. Nota, eds.

doi:10.1017/S1743921315009059

Variability Survey of ω Centauri in the Near-IR: Period-Luminosity Relations

C. Navarrete[1,2], M. Catelan[1,2], R. Contreras Ramos[1,2], F. Gran[1,2], J. Alonso-García[3,2] and I. Dékány[2,1]

[1]Instituto de Astrofísica, Pontificia Universidad Católica de Chile,
Av. Vicuña Mackenna 4860, 782-0436 Macul, Santiago, Chile
email: [cnavarre, mcatelan]@astro.puc.cl

[2]Millennium Institute of Astrophysics, Santiago, Chile

[3]Unidad de Astronomía, Fac. Cs. Básicas, Universidad de Antofagasta,
Avda. U. de Antofagasta 02800, Antofagasta, Chile

Abstract. ω Centauri is by far the most massive globular star cluster in the Milky Way, and possibly the remnant of a dwarf galaxy. As such, it contains a large number of variable stars of different classes. Here we report on an extensive, wide-field time-series study of ω Cen in the J and K_S bands, which has allowed us to study the near-IR period-luminosity relations for different variability classes, including the first such relations for the SX Phoenicis stars.

Keywords. stars: variables: RR Lyrae, stars: variables: SX Phoenicis, stars: variables: general, globular clusters: individual: ω Centauri, infrared: stars, surveys, stars: horizontal-branch

---

ω Centauri (NGC 5139) is by far the most massive globular star cluster in the Milky Way, and has even been suggested to be the remnant of a dwarf galaxy. As such, it contains a large number of pulsating variable stars. Among them, RR Lyrae (RRL) and Type II Cepheids (T2Cs) are well-known for following tight period-luminosity (PL) relations in the near-IR (Catelan & Smith 2015). SX Phoenicis stars (SX Phe) are also expected to follow PL relations, but no such relations have heretofore been reported in the near-IR. ω Cen is one of the best targets for this purpose, as it has a rich variable star content, hosting at least 270 pulsating stars, including T2Cs, RRL, and SX Phe stars.

A deep, wide-field, near-IR variability survey of ω Cen was carried out by our team using ESO's 4.1m VISTA telescope. Our time-series data comprise 42 and 100 epochs in J and K_S, respectively. Figure 1 shows the K_S, $(J-K_S)$ color-magnitude diagram for the cluster (based on data for the innermost 10 arcmin), with the detected variable stars overplotted (based on our entire dataset, covering 1.1×1.5 deg^2 around the cluster center). This unique dataset has allowed us to derive complete K_S-band light curves for more than 200 variable stars in the cluster's field.

Using this dataset, Navarrete *et al.* (2015) reported an updated census on the number, variability type and membership of all but 4 of the known RRL stars in ω Cen's field, and also discovered four new RRL. Moreover, reference Oosterhoff I and II loci in the Bailey (period-amplitude) diagram were derived for the J and K_S bands, in addition to expressions relating the light-curve amplitudes in the V, J, and K_S bandpasses.

This unprecedented variable star catalogue also allow us to perform a detailed analysis of the near-IR PL relations for different variability classes, including T2Cs, SX Phe, and RRL stars (Navarrete et al., in prep.). Based on our intensity-averaged magnitudes, and using calibrated PL relations from the literature (Matsunaga *et al.* 2006 for T2Cs, Alonso-García *et al.* 2015 for RRLs), an updated distance modulus for the cluster was also

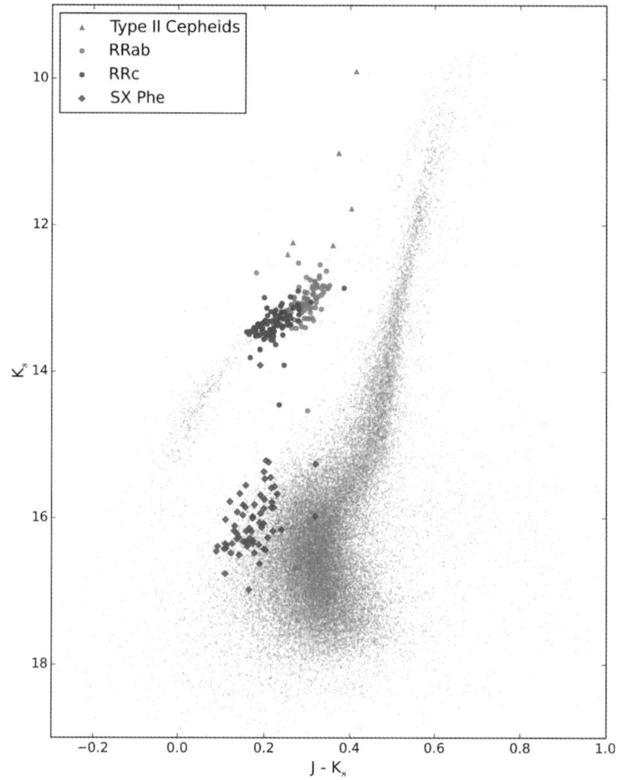

Figure 1. Near-IR CMD showing the non-variable ω Centauri stars in the cluster's innermost region (11 arcmin^2, *grey dots*). Variable stars over the entire cluster field (1.1×1.5 deg^2) are marked with *green triangles* (T2Cs), *red* and *blue circles* (RRab and RRc, respectively), and *purple diamonds* (SX Phe).

obtained. For SX Phe stars, fundamental-mode candidates (Olech *et al.* 2005; McNamara 2011) have been considered to derive, for the first time, calibrated near-IR PL relations.

Acknowledgements

This project is supported by FONDECYT grants #1130373 and #1141141, and by Basal-CONICYT grant PFB-06, awarded to the Centro de Astrofísica y Tecnologías Afines (CATA). Additional support is provided by the Ministry for the Economy, Development, and Tourism's Iniciativa Científica Milenio through grant IC 120009, awarded to the Millennium Institute of Astrophysics (MAS). C.N. acknowledges additional support from a SOCHIAS grant through Gemini/CONICYT Project #32140015, and CONICYT-PCHA/Doctorado Nacional 2015-21151643.

References

Alonso-García, J., Dékány, I., Catelan, M. *et al.* 2015, *AJ*, 149, 99
Catelan, M. & Smith, H. A. 2015, *Pulsating Stars* (Wiley-VCH)
Matsunaga, N., Fukushi, H., Nakada, Y., *et al.* 2006, *MNRAS*, 370, 1979
McNamara, D. H. 2011, *AJ*, 142, 110
Navarrete, C., Contreras Ramos, R., Catelan, M., *et al.* 2015 *A&A*, 577, 99
Olech, A., Dziembowski, W. A., Pamyatnykh, A. A., *et al.* 2005, *MNRAS*, 363, 40

Formation, evolution, and survival of massive star clusters
Proceedings IAU Symposium No. 316, 2015
C. Charbonnel & A. Nota, eds.

doi:10.1017/S1743921316007067

Searching for Possible Members of Star Moving Groups in the Kepler Field

Chang Yao Chen[1] W. P. Chen[1,2] and Ali Luo[3]

[1]Department of Physics, National Central University,
300 Zhongda Rd., Zhongli, Taoyuan 32001, Taiwan
email: a.e.johnny@hotmail.com

[2]Institute of Astronomy, National Central University, Taiwan

[3]National Astronomical Observatories, Chinese Academy of Sciences, Beijing, 100012, China

Abstract. We present identification of candidate members of the Beta Pictoris, AB Doradus, and TW Hydrae moving star groups. Our pilot study toward the *Kepler* field, where a multitude of stellar parameters are available, led to a list of 19 highly probable candidates that share the same range of distance and space (tangential and radial) motion as known members in a moving group. A few candidates lack radial velocity data, but are ROSAT X-ray sources, so are considered as likely young stars.

Keywords. star clusters and associations, kinematics

1. Moving Star Groups

It is known that stars are formed in groups out of molecular clouds, with dozens to thousands of members. These clusters dissociate with time (Zuckerman *et al.* 2001), but the then-members, when young, still share the same motion through space. There are eight known moving groups of stars, all young ($\sim$ 10–100 Myr) and nearby (within 120 pc) (Torres *et al.* 2008). Interestingly, most groups have known members scattered preferentially in the southern sky. It is unknown whether such a skew distribution is intrinsic, e.g., by Galactic tidal distortion, or a result of selection bias.

As a pilot program to identify candidate members of moving groups, we focused on the *Kepler* field in the northern sky to take advantage of the wealth of stellar data to search for possible members of the Beta Pictoris, AB Doradus, and TW Hydrae moving groups. The ϵ Chamaeleontis group is away from the line of sight to the *Kepler* field (see Fig. 1), so its candidate list serves to assess the false positive rate of our selection.

2. Possible Members

We searched in a 16×16 deg^2 region centering on the *Kepler* field, at R.A.=290.7 deg, and decl.=44.5 deg (J2000), in which there are some 1.2 million UCAC4 objects. Using the space location and motion of known members of a particular moving group, we computed the expected proper motion and the radial velocity of a member in the entire sky. Fig. 2 shows the example for the Beta Pic group. The stellar distance was estimated by the 2MASS K magnitude and $J-H$ color, assuming it is a dwarf (Cox 2000). No extinction was considered. A candidate is identified if its space tangential velocity, computed by its distance and proper motion, is within the range of those of known members.

While thousands of sources satisfied the tangential velocity plus distance criteria, about 1/5 of these can be further winnowed by the radial velocity data by LAMOST (Large Sky Area Multi-Object Fiber Spectroscopic Telescope). A total of 121 stars (34 of Beta

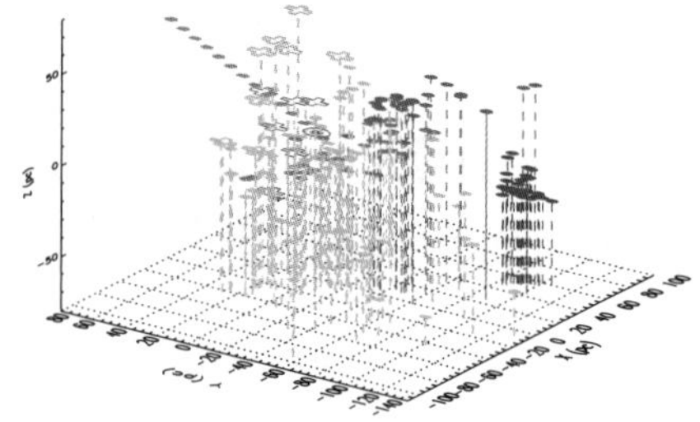

Figure 1. Spatial positions of known members of the Beta Pic (orange), AB Dor (green), TW Hyd (blue), and ϵ Cha (brown) moving groups (Torres *et al.* 2008). The cross symbols are members from Schlieder *et al.* 2012. Sun's location is at origin. Pink dots illustrates the line of sight toward the *Kepler* field.

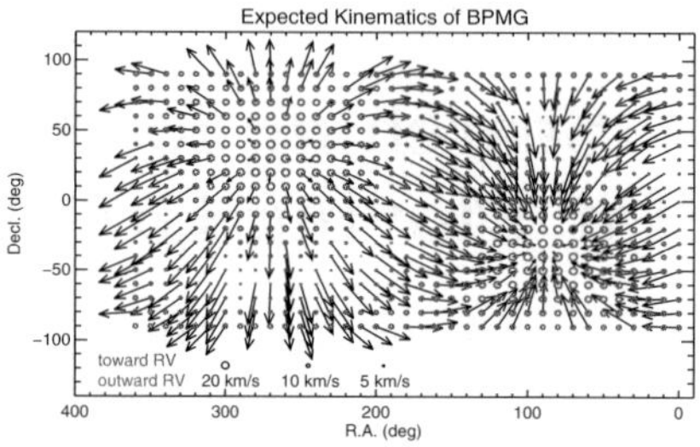

Figure 2. The expected space motion of the Beta Pic moving group on the whole sky in equatorial coordinates. The proper motions are shown as black arrows, whereas the radial velocities are depicted as color circles (blue or red shifted)

Pic, 76 of AB Dor, and 11 of TW Hyd groups) have distances, proper motions, and radial velocities consistent with being possible members. Among these, only 19 have been observed at multiple epochs and show no obvious variations caused by binary orbits; these are hence probable members. There are also a few ROSAT X-Ray sources (Voges *et al.* 1999), signifying stellar youth or activity, that lack radial velocity measurements. More spectroscopic radial velocity data are being collected.

The false positive rate was estimated based on the 121 candidates. For each group, the corresponding number density is, respectively, 1.6×10^{-3} pc^{-3} for the Beta Pic group, 3.5×10^{-3} pc^{-3} for AB Dor, and 1.5×10^{-3} pc^{-3} for TW Hyd. In comparison, the number density toward the ϵ Cha group is 1.5×10^{-3} pc^{-3}. This suggests that most candidates should be simply chance field stars. Additional diagnosis of membership includes, other than X-ray emission, stellar flares in the (*Kepler*) light curve, lithium absorption in the spectrum, or infrared excess arising from a circumstellar debris disk.

We acknowledge the grant by the Ministry of Science and Technology MOST103-2112-M-008-024-MY3. We also thank the LAMOST staff for help with the LAMOST data.

References

Cox, A. N. 2000, *Allen's Astrophysical Quantities*, 4th ed., Am. Inst. Phys., AIP Press, Springer-Verlag, New York

Schlieder, J. E., Lépine, S., & Simon, M. 2012, *AJ*, 144, 109

Torres, C. A. O., Quast, G. R., Melo, C. H. F., & Sterzik, M. F. 2008, *in Handbook of Star Forming Regions*, Volume II, Reipurth B., ed., p. 757

Voges, W., *et al.* 1999, *A&A*, 349, 389

Zuckerman, B., Song, I., Bessell, M. S., & Webb, R. A. 2001, *The ApJL*, 562, L87

Formation, evolution, and survival of massive star clusters
Proceedings IAU Symposium No. 316, 2015
C. Charbonnel & A. Nota, eds.

doi:10.1017/S1743921315008868

Synthetic clusters of massive stars to test stellar evolution models

Cyril Georgy[1,2] and Sylvia Ekström[1]

[1]Geneva Observatory, Geneva University
Chemin des Maillettes 51, 1290 Versoix, Switzerland
email: `cyril.georgy@unige.ch`
email: `sylvia.ekstrom@unige.ch`

[2]Astrophysics Group, Keele University
Keele, ST5 5BG, United Kingdom

Abstract. During the last few years, the Geneva stellar evolution group has released new grids of stellar models, including the effect of rotation and with updated physical inputs (Ekström *et al.* 2012; Georgy *et al.* 2013a,b). To ease the comparison between the outputs of the stellar evolution computations and the observations, a dedicated tool was developed: the SYCLIST toolbox (Georgy *et al.* 2014). It allows to compute interpolated stellar models, isochrones, synthetic clusters, and to simulate the time-evolution of stellar populations.

Keywords. Stars: evolution, Stars: rotation, Open clusters and associations: general

1. The SYCLIST toolbox

In cluster mode, SYCLIST† has (among others) the following capabilities: it accounts for a Salpeter IMF, includes several initial distributions for the angular velocity of the stars, as well as the effect of a random distribution of the angle of view on the gravity darkening. It can mimic in a simplified way the presence of a fraction of binary stars in the population.

The effect of fast rotation on the shape of the star is included in the Roche model approximation. The variation of the effective temperature and of the luminous flux over the surface (the so-called gravity darkening) is implemented using either the von Zeipel (1924) or Espinosa Lara & Rieutord (2011) relations. In this framework, the equator of a fast rotating star is cooler and dimmer than the average temperature and luminosity, and the poles are hotter and brighter.

As a consequence, an observer ignoring the orientation of the axis of rotation of a star will deduce an erroneous temperature and luminosity depending on its angle of view. For example, an observer looking at a fast rotating star pole-on will see principally the hot and bright polar regions and thus deduce that the star is bluer and more luminous than the average surface values.

The latitudinal variation of the gravity has also observational consequences. Assuming that the gravity deduced from a spectral fitting corresponds to the flux-averaged gravity of the visible hemisphere, a random distribution of the angle of view will produce a scatter in the $g_{\rm eff} - T_{\rm eff}$ plane (Fig. 1).

Extended main-sequence turn-off in the colour-magnitude diagram is a common feature of star clusters. One explanation is that these clusters, instead of having an instantaneous burst of star formation, have a spread (e.g. Goudfrooij *et al.* 2014; Correnti *et al.* 2014).

† `http://obswww.unige.ch/Recherche/evoldb/index/`

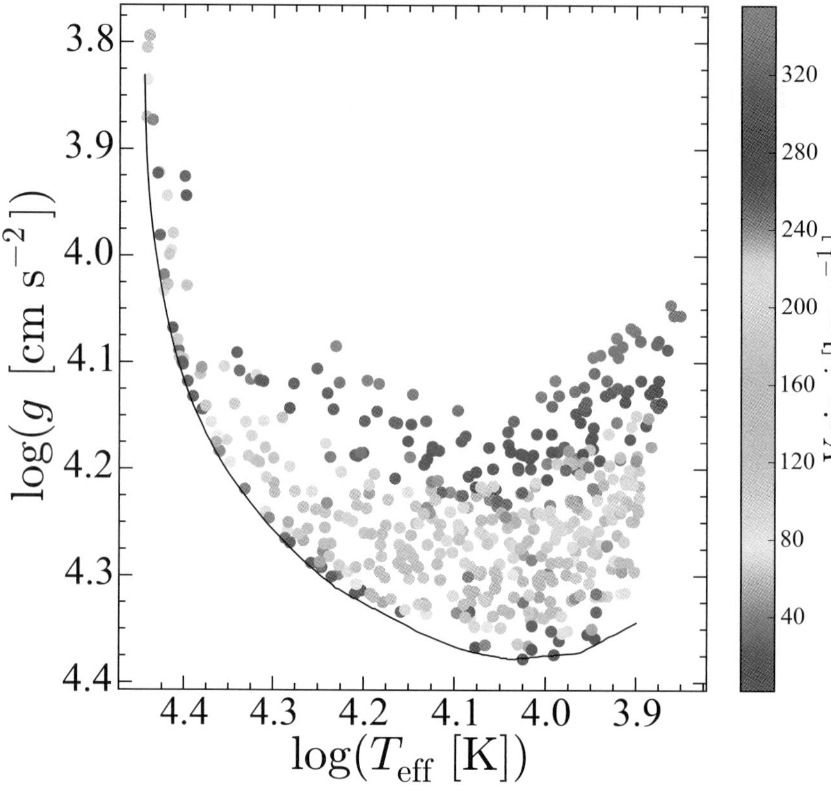

Figure 1. 'Observed' $g_{\rm eff}$ as a function of 'observed' $T_{\rm eff}$ for a cluster at 25 Myr at $Z_{\odot}$. The colour code shows the $V\sin(i)$ value. The black line is the corresponding isochrone for a non rotating population. No instrumental noise has been considered here.

Recently, we have shown that rotation could be a plausible alternative explanation, providing a natural framework to the observed relation between the duration of the spread in the star formation rate and the age of the cluster (Niederhofer *et al.* 2015). In this framework, the extension of the turn-off is no more produced by a spread, but by the distribution of initial rotation rates of the stars.

Acknowledgements

CG acknowledges support from the European Research Council under the European Union's Seventh Framework Programme (FP/2007-2013) / ERC Grant Agreement n. 306901.

References

Correnti, M., Goudfrooij, P., Kalirai, J. S., *et al.* 2014, *ApJ*, 793, 121
Ekström, S., Georgy, C., Eggenberger, P., *et al.* 2012, *A&A*, 537, A146
Espinosa Lara, F. & Rieutord, M. 2011, *A&A*, 533, A43
Georgy, C., Ekström, S., Eggenberger, P., *et al.* 2013a, *A&A*, 558, A103
Georgy, C., Ekström, S., Granada, A., *et al.* 2013b, *A&A*, 553, A24
Georgy, C., Granada, A., Ekström, S., *et al.* 2014, *A&A*, 566, A21
Goudfrooij, P., Girardi, L., Kozhurina-Platais, V., *et al.* 2014, *ApJ*, 797, 35
Niederhofer, F., Georgy, C., Bastian, N., & Ekström, S. 2015, *MNRAS*, 453, 2070
von Zeipel, H. 1924, *MNRAS*, 84, 665

Formation, evolution, and survival of massive star clusters
Proceedings IAU Symposium No. 316, 2015
C. Charbonnel & A. Nota, eds.

doi:10.1017/S1743921315010595

The Initial Mass Function of Young Open Clusters in the Galaxy: A Preliminary Result

Beomdu Lim[1], Hwankyung Sung[2], Hyeonoh Hur[2], and Byeong-Gon Park[1,3]

[1]Korea Astronomy and Space Science Institute, 776 Daedeokdae-ro, Yuseong-gu, Daejeon 305-348, Korea
email: bdlim1210@kasi.re.kr

[2]Dept. of Astronomy & Space Science, Sejong University, 209 Neungdong-ro, Gwangjin-gu, Seoul 143-747, Korea

[3]Korea University of Science and Technology, Gajeong-ro, Yuseong-gu, Daejeon 305-333, Korea

Abstract. The initial mass function (IMF) is an essential tool with which to study star formation processes. We have initiated the photometric survey of young open clusters in the Galaxy, from which the stellar IMFs are obtained in a homogeneous way. A total of 16 famous young open clusters have preferentially been studied up to now. These clusters have a wide range of surface densities ($\log \sigma = -1$ to 3 [stars pc^{-2}] for stars with mass larger than $5M_{\odot}$) and cluster masses ($M_{\rm cl} = 165$ to $50,000 M_{\odot}$), and also are distributed in five different spiral arms in the Galaxy. It is possible to test the dependence of star formation processes on the global properties of individual clusters or environmental conditions. We present a preliminary result on the variation of the IMF in this paper.

Keywords. stars: luminosity function, mass function, open clusters and associations: general

1. Introduction

It is well recognized that the evolution of stars mainly depends on their initial mass, partly on the chemical composition. However what determines the stellar mass is poorly understood. The origin of the stellar mass spectrum is thought to be closely related to star formation processes. And therefore the initial mass function (IMF), the mass distribution of stars from a single star formation event, is a crucial diagnostic tool with which to understand star formation processes.

Although the IMF seems to be universal according to a number of previous studies (Bastian, Covey, & Meyer 2010), many evidences for the variation of the IMF have been reported both from observational and theoretical studies (Lu et al. 2013; Marks *et al.* 2012; etc). One of main reasons for this debate may be caused by the use of inhomogeneous data. Thus, we focus on the results from a homogeneous imaging survey of young open clusters in the Galaxy ("Sejong Open cluster Survey" – Sung *et al.* 2013).

2. Preliminary Results and Conclusions

The young open clusters are useful objects in the study of the IMF because they are not only a coeval and co-spatial population but also an ecological indicator in the Galaxy. We investigated the IMF of 16 famous young open clusters by analyzing their Hertzsprung-Russell diagrams. The upper left-hand panel of Fig. 1 shows their IMF and its slope Γ (the data from Lim 2014, references therein, and ongoing work).

Variations of the slope Γ with respect to the surface density and total mass of the clusters are shown in the upper right and lower left-hand panels, respectively. The slope

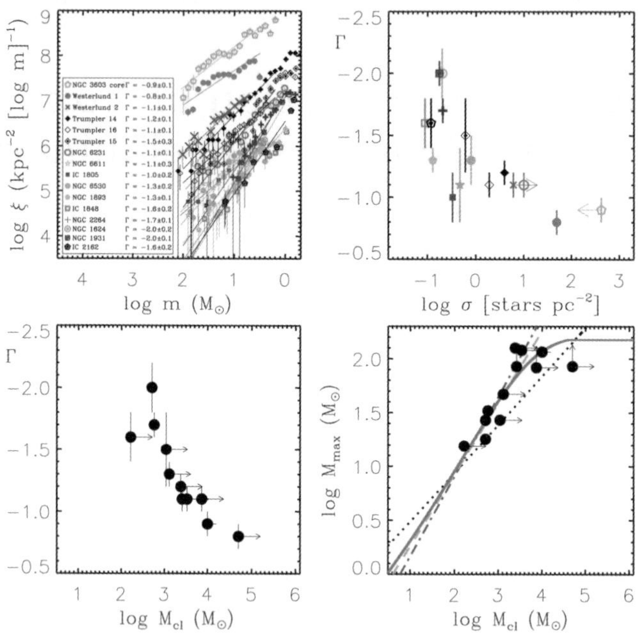

Figure 1. The initial mass functions (IMF) of 16 young open clusters in the Galaxy (upper left), variations of the IMF slope Γ with respect to the surface density of the clusters (upper right) and the cluster mass (lower left), and a dependency of maximum stellar masses on the mass of host clusters (lower right). In the lower right-hand panel, the dot-dashed, dotted, dashed, and solid lines show $M_{\rm max}$-$M_{\rm cl}$ relations from analytic approaches of Elmegreen (2000), Larson (2003), Bonnell, Bate, & Vine (2003), and Weidner & Kroupa (2004), respectively.

of the IMF in the high-mass regime ($> 1M_{\odot}$) appears to be shallow for massive dense clusters. The maximum stellar mass is also larger for more massive clusters (lower right-hand panel). This $M_{\rm max}$-$M_{\rm cl}$ relation is consistent with results from analytic approaches (Elmegreen 2000; Larson 2003; Bonnell, Bate, & Vine 2003; Weidner & Kroupa 2004). It may imply the deterministic origin of the stellar mass.

The Galactic variations of the IMF and the cluster properties were also investigated to simply test the environmental influences on the star formation processes. The massive dense clusters tend to be formed in the inner Galaxy, while the formation of less massive sparse clusters seem to be dominant in the outer Galaxy. We tentatively suggest that the star formation is likely to be controlled by local environmental conditions rather than a universal process.

References

Bastian, N., Covey, K. R., & Meyer, M. R. 2010, *ARAA*, 48, 339

Bonnell, A., Bate, M. R., & Vine, S. G. 2003, *MNRAS*, 343, 413

Elmegreen, B. G. 2000, *ApJ*, 539, 342

Larson, R. 2003, *ASP-CS* Galactic Star Formation Across the Stellar Mass Spectrum, 287, 65

Lim, B. 2014, *PhD thesis, Sejong Univ., Seoul, Korea*

Lu, J. R., Do, T., Ghez, A. M., Morris, M. R., Yelda, S., & Matthews, K. 2000, *ApJ*, 764, 155

Marks, M., Kroupa, P., Dabringhausen, J., & Pawlowski, M. S. 2012, *MNRAS*, 422, 2246

Sung, H., Lim, B., Bessell, M. S., Kim, J. S., Hur, H., Chun, M.-Y., & Park, B.-G. 2013, *J. Korean Astron. Soc.*, 46, 103

Weidner, C. & Kroupa, P. 2004, *MNRAS*, 348, 187

Formation, evolution, and survival of massive star clusters
Proceedings IAU Symposium No. 316, 2015
C. Charbonnel & A. Nota, eds.

doi:10.1017/S1743921315009230

Modeling Small Stellar Populations Using Starburst99

Gerardo A. Vázquez[1] **and Claus Leitherer**[2]

[1]Physics Department, Salisbury University,
1101 Camden Ave, Salisbury MD, 21801, USA
email: gavazquez@salisbury.edu

[2]Space Telescope Science Institute,
3700 San Martin Dr, Baltimore MD, 21218, USA
email: leitherer@stsci.edu

Abstract. In this work, we have developed a new approach to form stars from clusters first, where massive stars are formed from fractions of mass of small stellar clusters. This new approximation is based on the empirical power law found in recent years and the maximum stellar mass that can be formed in a cluster. To produce the new models we have used the most recent version of Starburst99 that incorporates the most recent stellar evolution models with rotation. At the verge of solving nearby stellar populations and observing small stellar populations across the universe, this new approach brings a new scope on trying to disentangle the nature of hyper and supermassive stars in small stellar populations. Models for NGC 3603 and NGC 604 are presented. Our most important result is a strong ionizing power from small clusters by forming enough supermassive stars in a cluster of $\sim 10^4$ M$_\odot$.

Keywords. supermassive star clusters, stellar population synthesis, galaxy evolution

---

1. Theory

We have used the empirical mass function of clusters ($MF_C \sim M_c^{-2}$) from 20 - 1100 M$_\odot$ found by Lada & Lada (2003). As the total mass of the system M$_T$ decreases, this mass function will form fractions of clusters of the more massive clusters. A condition to form at least one cluster of the upper mass limit for the MF_C is used to form the fractions of clusters. When only fractions of those massive clusters are formed, it means that smaller clusters, low mass stars, massive stars, and multiple stellar systems are formed. The maximum mass of a star formed in each cluster depends on the mass of the cluster as it was found by Pflamm-Altenburg, Weidner, & Kroupa (2007).

Clusters are formed in cells defined by the mass function of clusters. A change in scale occurs when the mass of the cell increases above or equal to the maximum mass of the mass function of clusters ($\sim$ 1100 M$_\odot$.) Then, the mass of the cluster becomes the mass of the cell.

2. Results

All models in Figure 1 used an IMF (0.1 to 100 M$_\odot$) from Kroupa (2001) using the new evolutionary tracks with rotation (V40, Maeder & Meynet 2012) with solar metallicity and the new version of Starburst99 (Leitherer *et al.* 2014.) Models shown in Figure 1 correspond to different M$_T$ for the system: 1.0 (1), 2.0 (2), 5.0 (3), 8.0 (4), 11.0 (5), and 14.0 (6) $\times 10^4$ M$_\odot$ as shown by the conventional models (black dashed lines) and models in colors using the new approach of cluster star formation. In the figure, it is possible to observe the increase of the ionizing photons when M$_T$ is small compared with the regular

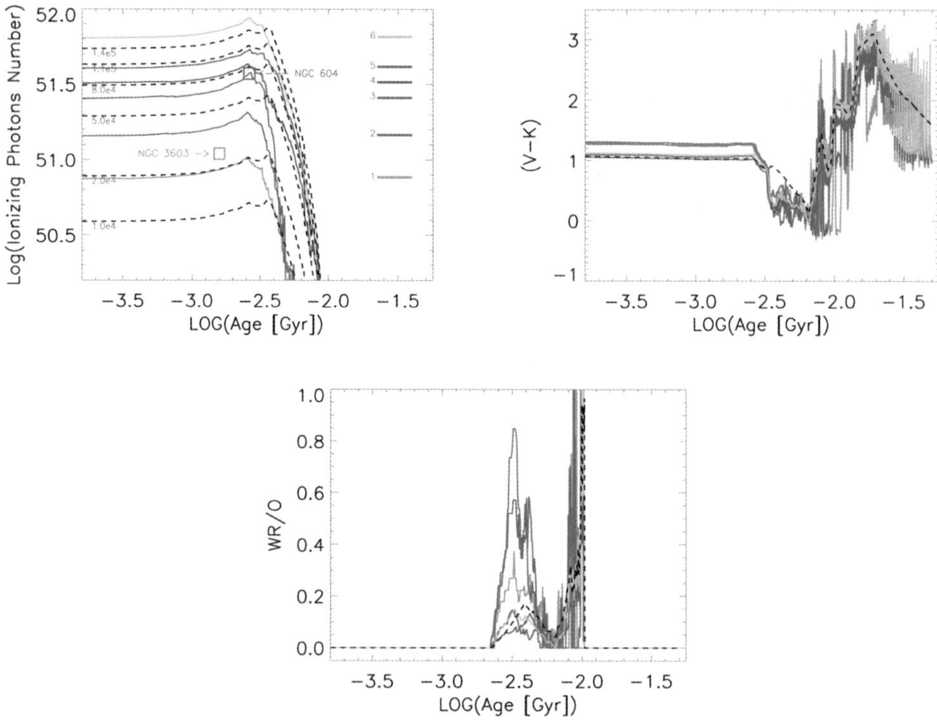

Figure 1. Color models were generated with this new approach. Dashed models are from the current version of Starburst99 with rotation.

models as in the case for a total mass of $\times 10^4$ $M_{\odot}$. As M_T increases the new models converge to the regular models. It is clear that there is now a dependence on M_T for the WR/O ratio and the color index $(V - K)$. The reason for this is that the number of massive stars does not scale with M_T.

3. Concluding Remarks

We have found an empirical relation between the M_T and the number of cells to produce the clusters. This relation is proportional to the number of supermassive stars generated. This new approach has allow us to generate a number of supermassive stars in small clusters that match the amount of UV energy seen in some of these small clusters such as NGC 3603 and NGC 604 (Kennicutt 1984) as seen plotted in Figure 1. Using these new models, we have found a new determination for the mass of NGC 604 between $5 - 8 \times 10^4$ $M_{\odot}$. Our most important result is a strong ionizing power from small clusters by forming enough supermassive stars in a cluster of $\sim 10^4$ $M_{\odot}$.

References

Kennicutt, R. C. 1984 *ApJ*, 287, 116

Kroupa, P 2001, *MNRAS*, 322, 231

Lada, C. J & Lada, E. A. 2003, *ARAA*, 41, 57

Leitherer, C., Ekstrom, S., Meynet, G., Schaerer, D., Agienko, K. B., & Levesque, E. M. 2014, *ApJS*, 212, 14L

Maeder, A. & Meynet, G. 2012, *RvMP*, 84,25

Pflamm-Altenburg, J., Weidner, C., & Kroupa, P. 2007, *ApJ*, 671, 1550

Formation, evolution, and survival of massive star clusters
Proceedings IAU Symposium No. 316, 2015
C. Charbonnel & A. Nota, eds.

doi:10.1017/S1743921316008681

Are Galactic globular cluster AGB stars rich or poor in Sodium?

Sodium abundance of AGB and RGB stars in NGC 2808

Yue Wang[1,2], Francesca Primas[1], Corinne Charbonnel[3,4], Mathieu Van der Swaelmen[5], William Chantereau[3] and Giuseppe Bono[6]

[1]European Southern Observatory, Garching, Germany

[2]Key Laboratory of Optical Astronomy, National Astronomical Observatories, Chinese Academy of Sciences, China (email:ywang@nao.cas.cn)

[3]Department of Astronomy, University of Geneva, Switzerland

[4]CNRS UMR 5277 IRAP & Université de Toulouse, France

[5]Observatoire de la Côte d'Azur, CNRS UMR 7293, BP4229, Nice, France

[6]Dipartimento di Fisica, Universit á di Roma Tor Vergata, Rome, Italy

Abstract. A spectroscopic study comparing the [Na/Fe] distributions of RGB and AGB stars in the Galactic globular cluster (GC) NGC 6752 found that there was no Na-rich, 2nd-generation star along the early-AGB of this cluster. This came as a surprise since in this GC, as well as other Galactic GCs studied so far, 1st- and 2nd-generation stars have usually been found from the main sequence turnoff up to the red giant branch. To investigate whether the failure of a significant fraction of stars to ascend the AGB also happens to other GCs, we studied a sample of AGB and RGB stars in NGC 2808 observed at the ESO/VLT with FLAMES. Contrary to NGC 6752, we find that the AGB and RGB stars we studied in NGC 2808 have comparable [Na/Fe] dispersions.

Keywords. globular cluster, abundances, AGB star

1. Introduction

Galactic globular clusters (GCs) are now known to have multiple stellar populations. These populations can be identified by their chemical properties, e.g. the O-Na anticorrelation. So far, stars with Na overabundances have been found at different evolutionary stages, from the main sequence turnoff up to the tip of the red giant branch (e.g. Carretta *et al.* 2006). Surprisingly however, in a spectroscopic study comparing the [Na/Fe] distributions of red giant branch (RGB) and asymptotic giant branch (AGB) stars in NGC 6752, Campbell *et al.* (2013) did not find Na-rich AGB stars, so they concluded that the Na-rich 2nd-generation stars cannot enter the AGB phase in this GC. To investigate whether this phenomenon also happens in other GCs, we observed a sample of AGB and RGB stars in the Galactic GC NGC 2808 whose multi-populations have been well studied (Carretta *et al.* 2006). Here we present the results of the chemical analysis of our sample.

2. Observation, data reduction and analysis

The high-resolution spectra of our sample of AGB and RGB stars in NGC 2808 were obtained with FLAMES at ESO/VLT adopting the combined mode of GIRAFFE + UVES. We achieved S/N ratios larger than 100, by combining multiple exposures. Seventy-four

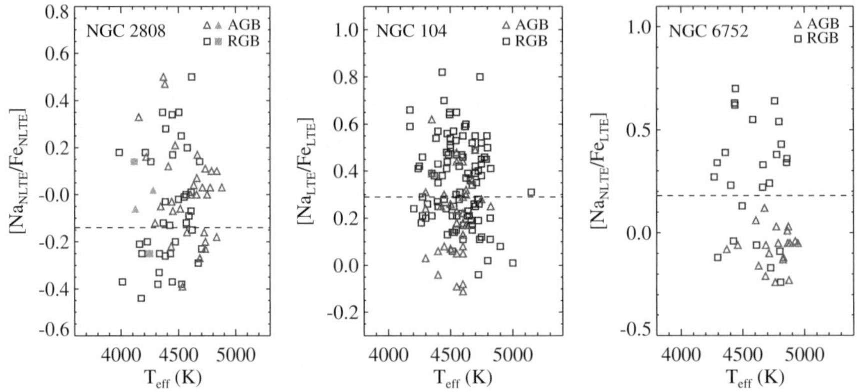

Figure 1. [Na/Fe] distributions of AGB and RGB stars in NGC 2808, NGC 104 and NGC 6752. Left: NGC 2808, our work where the Na and Fe abundances have been corrected for non-LTE effect. Middle: NGC 104, from Johnson *et al.* (2015) for AGB stars and Cordero *et al.* (2014) for RGB stars. Right: NGC 6752, from Campbell *et al.* (2013), with their Na abundances corrected for non-LTE effect. Open symbols represent results derived from GIRAFFE spectra, and solid ones from UVES spectra. The black horizontal dashed lines mark the critical [Na/Fe] ratio adopted to distinguish the 1st- and 2nd- generation stars in each cluster, following the definition in Carretta *et al.* (2009) that $[Na/Fe]_{cri} \sim [Na/Fe]_{min} + 0.3$ dex.

member stars (34 AGB stars + 40 RGB stars) of NGC 2808 are identified by their radial velocities and metallicities. The stellar parameters (effective temperature $T_{\rm eff}$, surface gravity log g, metallicity [Fe/H] and microturbulent velocity ξ_t) were determined with an iterative procedure. We synthesized the spectra in the region of the Na doublet at 6154-60Å to determine the Na abundance. NLTE corrections were applied to the Fe I and Na abundances (Bergemann *et al.* 2012; Lind *et al.* 2012; Lind *et al.* 2011). An average [Fe/H] = -1.17 ± 0.05 dex was derived for our sample.

3. Results

The distribution of [Na/Fe] of our sample of NGC 2808 stars is shown in the left panel of Fig. 1. We find that AGB and RGB stars exhibit comparable Na abundance dispersions. The ratios of Na-poor over Na-rich stars are respectively 24:76 for the AGB sample and 43:57 for the RGB sample. For comparison, we also show in Fig. 1 the [Na/Fe] distributions of AGB and RGB stars in NGC 104 (derived by Johnson *et al.* 2015 and Cordero *et al.* 2014, Na-poor:Na-rich = 63:37 for AGB, 45:55 for RGB) and NGC 6752 (derived by Campbell *et al.* 2013, Na-poor:Na-rich = 100:0 for AGB, 30:70 for RGB). Different from NGC 6752, NGC 2808 and NGC 104 both have Na-rich AGB stars. Because the three clusters span quite a range in metallicity (from [Fe/H]=-0.73 to -1.54), our next step will be to add more GCs at different metallicities to further constrain any possible metallicity dependence. Data on new GCs have already been secured.

References

Bergemann, M., Lind, K., Collet, R., Magic Z., & Asplund, M. 2012, *MNRAS*, 427, 27
Campbell, S. W., D'Orazi, V., Yong, D., Constantino, T. N., *et al.* 2013, *Nature*, 498, 198
Carretta, E., Bragaglia, A., Gratton, R. G., Leone, F., *et al.* 2006, *A&A*, 450, 523
Carretta, E., Bragaglia, A., Gratton, R. G., Lucatello, S., *et al.* 2009, *A&A*, 505, 117
Cordero, M. J., Pilachowski, C. A., Johnson, C. I., McDonald, I., *et al.* 2014, *ApJ*, 780, 94
Johnson, C. I., McDonald, I., Pilachowski, C. A., Mateo, M., *et al.* 2015, *AJ*, 149, 71
Lind, K., Asplund, M., Barklem, P. S., & Belyaev, A. K. 2011, *A&A*, 528, A103
Lind, K., Bergemann, M., & Asplund, M. 2012, *MNRAS*, 427, 50

Author index

Adamo, A. – 17, 44
Alexander, P. – 214
Aloisi, A. – 61
Alonso-García, J. – 351
Amram, P. – 44
Anderson, C. – 157
Apellániz, J. M. – 165

Bailey, III, J. I. – 341
Banerjee, S. – 228
Barnes, P. J. – 123
Barnes, P. – 127, 131
Bastian, N. – 302
Baumgardt, H. – 240, 257
Beaver, J. – 149
Beckman, J. – 135
Bernard, J.-P. – 133
Beuther, H. – 125
Bihr, S. – 125
Bik, A. – 44
Boesgaard, A. M. – 338
Boily, C. – 246
Bonanos, A. Z. – 55
Bonnell, I. A. – 184
Bono, G. – 36, 361
Boulanger, F. – 190
Braga, V. F. – 36
Bragaglia, A. – 287
Briley, M. – 149
Brodie, J. P. – 105
Bruzual, G. – 143

Caldwell, N. – 341
Camps-Faria, A. – 135
Carraro, G. – 345
Carter, D. – 253
Cassisi, S. – 320
Catelan, M. – 351
Cava, A. – 111
Chakrabarti, S. – 151
Chantereau, W. – 328, 361
Charbonnel, C. – 1, 328, 361
Chatterjee, S. – 234
Chen, C. Y. – 353
Chen, C.-H. R. – 167
Chen, L. – 163, 265
Chen, W. P. – 353
Chen, W.-P. – 163
Chiboucas, K. – 253
Chiosi, C. – 171
Christian, C. – 55
Cignoni, M. – 77
Clark, J. S. – 159, 165, 167
Colín, P. – 196
Cooper, J. – 151
Cox, E. – 153
Crane, J. D. – 341
Cropper, M. – 171

Dale, J. E. – 177
Dalessandro, E. – 345
Dallilar, Y. – 127
Davies, B. – 167
de Grijs, R. – 222
De Marchi, G. – 137
de Mink, S. E. – 55
de Zeeuw, P. T. – 50
de Zeeuw, T. – 240
Decressin, T. – 281
Dékány, I. – 351
Deliyannis, C. P. – 338
Dessauges-Zavadsky, M. – 111
Dorda, R. – 165
Dorval, J. – 246
Dottori, H. – 141

Ehlerová, S. – 251, 294
Ekström, S. – 355
Feldmeier-Krause, A. – 50, 240
Fenech, D. – 139, 169
Ferguson, P. – 253
Ferraro, I. – 36
Figer, D. F. – 167
Fiorentino, G. – 36
Fischer, C. – 153
Font, J. – 135
Frank, M. – 257
Frinchaboy, P. – 255
Fujii, M. S. – 25
Fujita, Y. – 171
Fukui, Y. – 173, 208
Garca-Lorenzo, B. – 135
Geballe, T. R. – 161
Gebhardt, K. – 240
Geller, A. M. – 222
Georgy, C. – 355
Gieles, M. – 214
Gilmozzi, R. – 36
Ginsburg, A. – 131
Golev, V. – 149
González-Fernández, C. – 165
González-Samaniego, A. – 196
Goodwin, S. P. – 159, 261
Gouliermis, D. A. – 117

Gran, F. – 351
Grebel, E. K. – 55, 171
Grosbøl, P. – 141

Haghi, H. – 257
Han, Z. – 316
Hasegawa, K. – 173
Hayes, M. – 44
Hénault-Brunet, V. – 310
Henning, T. – 125, 151
Hernández-Pérez, F. – 143
Herrera, C. N. – 161, 190
Hilker, M. – 50, 99, 263
Hofner, P. – 157
Hojaev, A. S. – 145
Hony, S. – 117
Hou, J. – 163, 265
Huang, Y.-F. – 129, 147
Hur, H. – 357

Iannicola, G. – 36
Iserlohe, C. – 153
Ivanov, V. D. – 167

Jiang, D. – 316
Johnson, C. I. – 341
Juvela, M. – 133

Kaltcheva, N. – 149
Kamiński, T. – 155
Karachentsev, I. – 349
Kim, S. S. – 261, 336
Kissler-Patig, M. – 50, 240
Klein, R. – 151, 153
Kniazev, A. – 349
Krabbe, A. – 153
Kramer, B. H. – 155
Kraus, A. – 155
Krause, O. – 125
Kroupa, P. – 257
Kruijssen, D. – 240
Kudritzki, R.-P. – 167
Kuntschner, H. – 50
Küpper, A. H. W. – 257

Lada, E. – 127
Larsen, S. S. – 91
Lee, J. – 336
Leitherer, C. – 359
Li, C. – 222
Li, J.-Z. – 129, 147
Li, L. – 316
Lim, B. – 357
Lin, C.-C. – 265, 163
Linz, H. – 125
Liu, D. – 343
Liu, H.-L. – 129, 147
Looney, L. W. – 153
Looney, L. – 151
Louvet, F. – 9
Lu'o'ng, Q. N. – 9
Lum, M. G. – 338
Lund, S. – 149
Luo, A. – 353
Lützgendorf, N. – 50, 240, 257

MacKenty, J. W. – 167
Magris C., G. – 143
Magurno, D. – 36
Marco, A. – 165
Marino, A. F. – 267
Marton, G. – 133
Mateo, M. – 341
Matsunaga, N. – 36
Mayer, L. – 111
Melinder, J. – 44
Menacho, V. – 44
Menten, K. M. – 155, 167
Messineo, M. – 167
Meynet, G. – 328
Mieske, S. – 257
Milone, A. P. – 275
Monelli, M. – 36
Monguió, M. – 165
Montes, V. A. – 157
Montier, L. – 133
Morford, J. – 139, 169
Morscher, M. – 234
Motte, F. – 9
Muñoz-Tuñón, C. – 294

Najarro, F. – 167
Navarrete, C. – 351
Negueruela, I. – 159, 165
Neumayer, N. – 50, 84, 240
Nota, A. – 55
Noyola, E. – 240

Ohama, A. – 173
Olczak, C. – 259
Ordonez, A. – 131
Östlin, G. – 44

Palouš, J. – 251, 294
Panagia, N. – 137
Pang, X. – 259
Park, B.-G. –357
Park, S.-M. – 261
Pasetto, S. – 171
Pasquali, A. – 55
Patel, N. A. – 155
Pattabiraman, B. – 234
Pelupessy, F. I. – 334

Pelupessy, I. – 240
Peng, E. – 253
Pérez González, P. G. – 111
Phillipps, S. – 253
Pilachowski, C. A. – 341
Pols, O. R. – 334
Primas, F. – 361
Prinja, R. – 139, 169

Ragan, S. E. – 125
Ramos, R. C. – 351
Randriamanakoto, Z. – 70
Rasio, F. A. – 234
Rastello, S. – 36
Rich, R. M. – 167, 341
Richard, J. – 111
Ristorcelli, I. – 133
Ritchie, B. W. – 159
Rodriguez, C. L. – 234
Romanowsky, A. J. – 105
Rosero, V. – 157
Ryder, S. – 127
Sabbi, E. – 55, 137
Schaerer, D. – 111
Schap, W. – 131
Schulz, C. – 263
Schödel, R. – 50
Serrano-Borlaff, A. – 135
Seth, A. – 50
Shao, Z. – 163, 265
Sharina, M. – 347, 349
Shenoy, S. – 151
Shimansky, V. – 347
Shin, J. – 336
Smilgys, R. – 184
Smith, R. J. – 125
Spurzem, R. – 259
Sung, H. – 357

Tachihara, K. – 173
Tackenberg, J. – 125
Tamburello, V. – 111
Tenorio-Tagle, G. – 251, 294
Thompson, B. – 255
Torii, K. – 173
Tosi, M. – 55
Toth, L. V. – 133
Trombley, C. – 167
Tsai, C.-W. – 161
Tully, R. B. – 253
Turner, J. L. – 31

Väisänen, P. – 70
Valenti, E. – 167
Vázquez, G. A. – 359
Vázquez-Semadeni, E. – 196
van der Helm, E. – 240
Van der Swaelmen, M. – 361

Walcher, C. J. – 50
Wang, B. – 343
Wang, Y. – 361
Whitmore, B. C. – 202
Wijnen, T. P. G. – 334
Wünsch, R. – 251, 294
Wu, Y. – 129

Xie, X. – 265

Yamamoto, H. – 173
Yeh, S. C. C. – 161
Yuan, J. – 129
Yuan, J.-h. – 147

Zahorecz, S. – 133
Zamora-Avilés, M. – 196
Zaragoza-Cardiel, J. – 135
Zeidler, P. – 55
Zhang, B. – 155
Zhang, C.-P. – 175
Zhong, J. – 265
Zonoozi, A. H. – 257
Zwart, S. P. – 25, 240, 334